Series Preface

It is a cliché to comment on the explosion of the scientific literature in recent decades. Most investigators focus on obtaining new data and communicating these data in specialized journals. The integration and interpretation of this new information has lagged behind the pursuit of experimental investigations. Novel studies are essential, but reviews are also important if a field is to remain healthy and advance in productive ways. This is an especially acute problem for inter-disciplinary fields. Investigators may have specialized knowledge about only one aspect of a field and fail to appreciate the significance of their work for the entire enterprise. The goal of this series is to address this problem for the field of behavioral neuroendocrinology.

Books in this series will provide reviews of important and timely topics concerning how hormones influence brain functions, especially behavior. It has been known for many years that hormones have profound effects on behavior. However, it is also the case that behavioral and physical experiences of various sorts, processed by the central nervous system, can influence endocrine activity. Thus hormone-behavior relationships are one component of a complex interrelation among hormones, brain and behavior. Scientists interested in this field are drawn from diverse disciplines in the biological and behavioral sciences including endocrinology, neuroscience, anatomy, psychology, and ethology. In recent years a new generation of truly interdisciplinary studies has emerged in which investigators simultaneously manipulate circulating hormone concentrations in the blood, assess the neural site of action of these hormones and measure the behavioral consequences of hormonally induced changes in neural activity. The maturity of this field is reflected by the founding of the Society for Behavioral Neuroendocrinology in 1996.

The aims of this series will be to highlight and integrate recent empirical advances in the field and discuss and analyze methodological challenges facing investigators working in this area. The clinical implications of this basic research for problems such as mood disorders, sexual dysfunction, violence and other behavioral problems will also be considered. Volumes will include books focussed on the neuroendocrine regulation of a particular behavior, a summary of the anatomical localization of hormone sensitive brain networks (e.g., localization of hormone receptors, metabolizing enzymes, etc.) and a consideration of the cellular and molecular mechanisms mediating hormone action in the brain. This series is planned to produce two to four volumes a year. The books will be aimed at advanced graduate students as well as established investigators working in the fields of neuroscience, physiology and psychology with a particular emphasis on behavioral neuroscience, neuroethology and neuroendocrinology. Some books in the series will also be useful for upper division neuroscience and neuroethology courses. It is also hoped that clinical investigators will find these books as useful summaries of basic research that is relevant to many health-related questions.

GREGORY F. BALL, Baltimore, Maryland
JACQUES BALTHAZART, Liège, Belgium
RANDY J. NELSON, Columbus, Ohio

The Neurobiology of
Parental Behavior

Springer
New York
Berlin
Heidelberg
Hong Kong
London
Milan
Paris
Tokyo

Michael Numan   Thomas R. Insel

The Neurobiology of Parental Behavior

With 80 Illustrations

Springer

Michael Numan
Department of Psychology
Boston College
Chestnut Hill, MA 02467
USA
numan@bc.edu

Thomas R. Insel
Department of Psychology
National Institute of Mental Health
Bethesda, MD 20892
USA
insel@mail.nih.gov

Series Editors
Gregory F. Ball
Department of Psychological and Brain
  Sciences
Johns Hopkins University
Baltimore, MD 21218
USA
gball@jhu.edu

Jacques Balthazart
University of Liège
Center for Cellular and Molecular
  Neurobiology
Behavioral Neuroendocrinology Research
  Group
B-4020 Liège, Belgium
jbalthazart@ulg.ac.be

Randy J. Nelson
Departments of Psychology &
  Neuroscience
Ohio State University
Columbus, OH 43210
USA
rnelson@osu.edu

Library of Congress Cataloging-in-Publication Data
Numan, Michael, 1946–
    The neurobiology of parental behavior / Michael Numan, Thomas R. Insel.
      p. cm.—(Hormones, brain, and behavior series)
    Includes bibliographical references and index.
    ISBN 0-387-00498-X (hbk. : alk. paper)
      1. Parental behavior in animals.   2. Animal behavior—Endocrine aspects.   I. Insel,
    Thomas R., 1951–   II. Title.   III. Series.
    QL762.N86 2003
    591.56'3—dc21   2003042473

ISBN 0-387-00498-X   Printed on acid-free paper.

Printed in the United States of America.

9 8 7 6 5 4 3 2 1   SPIN 10913116

www.springer-ny.com

Springer-Verlag   New York Berlin Heidelberg
A member of BertelsmannSpringer Science+Business Media GmbH

Preface

When we were asked by Greg Ball, Jacques Balthazart, and Randy Nelson if we could prepare a volume on parental behavior for Springer's Hormones, Brain, and Behavior book series, our initial response was one of trepidation because we knew the level of work and commitment such an endeavor would involve. However, we ultimately cleared our decks to make room for this project because we thought that the time was right for such a book, and we knew that if we developed it properly it would make an important scientific contribution.

Although there are several excellent review chapters and edited books dealing with various aspects of the biology of parental behavior, we are not aware of any recent monograph on this important topic. In *The Neurobiology of Parental Behavior* we take an integrative and multilayered approach to the study of the biology of parental behavior. The research that we review, which includes an analysis of the role of developmental factors, experiential factors, hormones, genes, neurotransmitters/neuromodulators, and neural circuits in the control of parental behavior, is all aimed at defining the neural underpinnings of this important behavior. Therefore, diverse sources of data are integrated in a unified way with a common goal in mind. In addition to filling a need within the field of parental behavior, this book also contributes importantly to the growing area of emotional and motivational neuroscience. A major part of neuroscience research at the whole organism level has been focused on cognitive neuroscience, with an emphasis on the neurobiology of learning and memory, but there has been a recent upsurge in research that is attempting to define the neural basis of basic motivational and emotional systems that regulate such behaviors as food intake, aggression, reproduction, reward-seeking behaviors, and anxiety-related behaviors. Our book nicely fits into this latter category.

As an undergraduate student at Brooklyn College, one of us (Michael Numan) read Lehrman's outstanding review on the role of hormones in the parental behavior of birds and mammals and was completely enthralled (Lehrman, 1961). That one chapter defined the field at that time, and it intelligently indicated the complexity and richness of the research issues that could be studied. Here was a behavior that was essential for the reproductive success of birds and mammals, that was influenced by genes and hormones, but that also had an ontogenetic course that allowed for the influence of experiential factors. After reading that chapter, one could become preoccupied with the question of how experience and/or hormones might modify the brain so that an organism's responsiveness to infant-related stimuli could change. We hope that the present book not only clearly reviews and integrates the large body of work that has appeared on the biology of parental behavior since Lehrman wrote his chapter, but that it is also a worthy descendant of Lehrman's work, shedding light on some of the important issues that he raised.

In this book we emphasize research findings obtained from rodents, sheep, and primates. Our goal, of course, is to provide a foundation that may help us understand the neurobiology of human parental behavior. Indeed, our last chapter attempts to integrate the nonhuman research data with some human data in order to make some inroads toward an understanding of postpartum depression, child abuse, and child neglect. Clearly, motivational and emotional neuroscience has close ties to psychiatry, and this will be very evident in the final chapter. By understanding the neurobiology of parental behavior we are also delving into neurobiological factors that may have an impact on core human characteristics involved in sociality, social attachment, nurturing behavior, and love. In this very violent world, it is hard to conceive of a group of characteristics that is more worthy of study.

In preparing this book, each of us owes a debt of gratitude to our former teachers and colleagues. Michael Numan was strongly influenced by David Raab and Barton Meyers while he was an undergraduate, by Howard Moltz, his doctoral dissertation advisor, and by Jay Rosenblatt, his postdoctoral research advisor. In fact, Jay Rosenblatt has remained an important influence and friend for almost 30 years. Don Pfaff, by inviting Michael Numan to write a major review chapter on maternal behavior (Numan, 1988, 1994), helped him hone those integrative skills needed to write this current contribution. Finally, Michael Numan deeply appreciates the advice, support, and encouragement that he has always received from his wife and colleague, Marilyn. In addition, Marilyn Numan has prepared most of the figures in this book, and she critically evaluated and commented upon many of the chapters.

Tom Insel has enjoyed the generosity of a number of colleagues, friends, and mentors including Paul MacLean, Myron Hofer, Michael Kuhar, and Dennis Murphy. In addition, he owes much to a generation of postdoctoral fellows who have worked on neuropeptides and maternal behavior, including Marianne Wamboldt, Allan Johnson, Zuoxin Wang, Larry Young, and Darlene Francis. Paul Plotsky and David Rubinow kindly critiqued parts of the text and Svetlana Gurvitch assisted with illustrations. Most of all, Deborah Insel has been a patient and supportive spouse who endured the prolonged labor and delivery of this volume.

During the course of writing this book, Michael Numan's research was supported in part by National Science Foundation Grant IBN-9728758 and Tom Insel was supported in part by the Center for Behavioral Neuroscience, an STC program of the National Science Foundation, under Agreement No. IBN-9876754, and by NIMH from RO1-MH-56538 and a NARSAD Distinguished Investigator Award.

Contents

1

Introduction

1. Introduction

In this book we will review existing knowledge pertinent to the neural basis of parental behavior in mammals. In doing so, we will also review the role of hormonal mechanisms and experiential factors with the aim of defining how hormones and experience operate on the brain to influence the occurrence and nature of parental responsiveness. Although parental behavior occurs in a variety of vertebrate and invertebrate species, it is most highly developed in birds and mammals (Clutton-Brock, 1991). Indeed, since all female mammals lactate, maternal behavior can be viewed as one of the defining characteristics of mammals. Our emphasis will be on mammalian parental behavior because most of the work on the neurobiology of this behavior has been done on mammals, primarily rodents and sheep. Our goal is to outline a core neural circuitry for parental behavior, which we conceive of as being highly conserved across species, and our analysis will show that the phylogenetically older parts of the brain (the limbic system and hypothalamus) are critical elements of this pathway. It is hoped, therefore, that the findings of the current literature, although largely limited to the study of only a few species, will have wide generality and be applicable to many other mammalian species, including primates. This is not to say that data from primates, including humans, do not exist. We will review the relevant studies of parental behavior in primates with the goal of ultimately understanding the biological basis of human maternal behavior.

Parental behavior can be defined as any behavior of a member of a species toward a reproductively immature conspecific that increases the probability that the recipient will survive to maturity. In mammals, since the female lactates, it is the mother who is the primary caregiver of her offspring, and we refer to such behavior as maternal. In fact, parental care systems can be either uniparental or biparental. In about 90% of mammals, a uniparental care system exists with sole maternal care of offspring (Kleiman & Malcolm, 1981). In the remaining mammalian species, usually in the context of a monogamous mating system, one finds both maternal and paternal behavior. Although parental behavior is usually shown by females and males toward their own offspring, in some species individuals who are not the biological parents of an infant provide care. Such behavior, referred to as alloparental behavior, may be critical for survival of the offspring and, when exhibited by siblings, may be important training for adult parental care. Because the dominant form of parental behavior in mammals is maternal, it should come as no surprise that most of our knowledge on the neurobiology of parental behavior concerns maternal behavior. However, some important work on the neural basis of paternal behavior does exist, and we will review these findings with an eye toward comparing paternal neural control mechanisms with their maternal counterparts.

In contrast to mammals, avian parental care is usually biparental. Birds lack both the specialization for internal gestation (pregnancy) and feeding (lactation), so males might be expected to have a more important role in parental care. Indeed, more than 90% of avian species are monogamous and biparental, providing a complementary picture to mammals. These differences in social organization in birds and mammals suggest an interesting opportunity to compare sex differences in neural organization that might be important for male parental care. This point is discussed more fully in Chapter 7.

The degree of development of the young at birth influences the patterns of maternal behavior among mammals (Numan, 1994; Rheingold, 1963; Rosenblatt, Mayer, & Siegel, 1985). Young may be altricial and immobile at birth, precocial and mobile at birth, or intermediate between these extremes (semiprecocial and semimobile). Rodents offer good examples of maternal care patterns in species with altricial young. Most rodent young are helpless, essentially immobile, and incapable of temperature regulation at birth. The young are kept in a secluded nest that the mother builds prior to parturition, and the nest serves to insulate the young in the mother's absence. When the female is in the nest area, she crouches over the young in order to warm them and nurse them. The female also licks her young, particularly in their anogenital region, and this stimulates urination and defecation. If the nest site is disturbed, or if pups become displaced from the nest, the rodent mother will engage in transport or retrieval behavior in which she carries pups (one at a time) in her mouth to a new nest site or back to the original nest. Another aspect of maternal care in rodents is the occurrence of maternal aggression toward intruders at the nest site, which presumably protects the young. In describing maternal care in rodents, one can therefore refer to pup-directed patterns (retrieving, licking, nursing) and non-pup-directed patterns (nest building, maternal aggression), all of which contribute to the successful rearing of the young.

Sheep, goats, and other ungulates provide examples of maternal behavior in mammalian species that give birth to precocial young. In such species, the young calf follows and stays with its mother soon after birth, and she allows it to suckle, licks/grooms it, and protects it from danger. Similar patterns are observed in pinnipeds (e.g., seals). In contrast to altricial species in which infants are nest-bound, parental care in precocial species may require rapid attachment of the mother to a specific infant as well as vigilant protection from predators. Although precocial infants generally do not require the mother for heat, they may be especially vulnerable to predation.

Finally, most primates show maternal care patterns adapted to young who are semiprecocial at birth. Initially, the mother is in constant contact with her infant, even when she moves around in her environment, and the infant depends on clinging to its mother for transport. As in nonprimate mammals, the mother nurses and grooms her infant and protects it from danger. During movement, for example when foraging for food, the primate mother carries the infant either ventrally or dorsally, depending on the species and the stage of infant development. As the infant develops, it begins to wander from its mother to explore its environment, but the mother will retrieve the infant to her ventral or dorsal surface when it is in danger. In many New World primates, fathers or siblings perform much of the carrying and grooming, yielding the infants to the mother for feeding.

Maternal care is essential for the growth and development of all mammalian young, which eventually become independent of their mothers. As offspring advance in age, their increasing independence and decreasing reliance on maternal care coincide with a decrease in maternal behavior, and these events permit weaning to occur (Rheingold, 1963; Rosenblatt et al., 1985).

As behavioral neuroscientists, our major interest in this volume is to address the "how", the "where", and the "when" questions to understand the proximate mechanisms of parental care. The reader should also note that behavioral ecologists have written extensively about the "why" questions, or the so-called ultimate mechanisms: Why has parental care evolved? Why is parental care observed in some species and not others? Why is parental care more evident in some individuals than others? While a thorough discussion of these questions is beyond the scope of this book, we feel that, at the outset, readers should know that from an evolutionary perspective the goal of parental care is to increase the inclusive fitness of the parent, that is, to increase the likelihood that the mother's (or father's) genes will survive into subsequent generations. Not all species require parental care for successful transmission of parental genes. Many vertebrate and invertebrate species reproduce with hundreds or thousands or even millions of offspring that ensure genetic survival stochastically. Generally, these species are ectothermic and thus do not require parental warmth for survival. Avian and mammalian species, which are typically endothermic and reproduce in smaller numbers, not only require parental care to maintain temperature but also represent a larger investment of parental resources. Both birds and mammals are anisogamous, meaning that males and females bring different kinds of investments to reproduction and face different selection pressures. For females, there are few opportunities for reproduction but there is great certainty of maternity, so that an investment in young in terms of parental care is essential to increase fitness. For males, the cost of producing gametes is less and the selection pressures to invest in young are balanced by the opportunity to mate with other females, the uncertainty of paternity, and the costs of protecting offspring. These issues are considered in much greater length in Clutton-Brock (1991) and Rosenblatt and Snowdon (1996).

Beyond the question of why parental care is more evident in birds and mammals than in other vertebrates is why, in those species that exhibit parental care, is parental care either highly selective for just the mother's offspring, or, instead, apparently "promiscuous"? What evolutionary factors determine whether a postpartum female will react maternally only toward her own (biological) offspring, while rejecting or ignoring alien young (young from another mother)? In an important review, Gubernick (1981) provides some guidelines, based on evolutionary considerations, which can be used to predict which species might recognize and respond preferentially to their own offspring and which might not. An evolutionary perspective would argue that mothers should generally care for their own young, and not unrelated young, because care offered to unrelated young would decrease the inclusive fitness of the altruistic mother. However, in cases where the ecology of a species makes it improbable that a mother would encounter alien young, we would not expect to see the evolution of mechanisms that would allow her to respond selectively to infant stimuli from her own young. Only under environmental conditions in which alien young and one's own young can intermingle, so that confusion could occur, would we expect to see the evolution of mechanisms that would permit selective maternal recognition. At least three factors should influence the evolution of maternal selectivity (Gubernick, 1981): (a) the maturity of the young at birth, with precocial young being capable of wandering from their mother; (b) the basic social structure within which the mother–young unit is embedded, which can vary along a continuum from a solitary mother–offspring group to a larger social group; and (c) the degree of genetic relatedness of the members of the social group. Given these parameters, an evolutionary perspective would predict that mothers who raise altricial young in a solitary group are not likely to recognize their own young because confusion between others' and their young is not likely to occur. This is the case for most rodents, which raise their altricial young in nest sites within isolated burrows, forming a solitary group (for a review, see Numan, 1994). Indeed, one can cross-foster rodent young from one mother to another, and the adoptive mother will care for both her own and the alien pups. It should be realized, of course, that although ecological conditions do not favor the evolution of maternal selectivity toward pup stimuli in most rodents, the rodent mother still needs to have a mechanism that will allow her to recognize and remember where her nest site is located. Interestingly, in some rodents (e.g., ground squirrels) the mother continues to care for her offspring after they become fully mobile and, therefore, during a time when they can both leave and return to the nest area. In these cases the mother does learn to recognize her own young, but this recognition develops only after the young become mobile (Holmes & Sherman, 1982).

In contrast to the above analysis, natural selection should favor the evolution of maternal selectivity in those species in which mothers raise their young in a large social group, particularly if the young are precocial and the group is composed of many unrelated individuals (kin selection mechanisms might counteract the evolution of maternal selectivity if the group were composed of related individuals). Many ungulate and pinniped mothers give birth to precocial young in social groups composed of many unrelated individuals and, therefore, it is not surprising that the maternal care system in these species involves the development of a selective maternal attachment to one's own offspring (Numan, 1994). For example, maternal sheep nurse, lick, and protect their own young from danger while rejecting the suckling attempts of alien young. A selective maternal attachment to one's own offspring also develops in many primate species, although such attachments may not be as absolute and rigid as that observed in ungulates, and this maternal selectivity is consistent with the semiprecocial nature of primate young and the fact that many primate species form large social groups that are composed of both related and unrelated individuals (see Numan, 1994).

An important section of this book will be devoted toward understanding the neural mechanisms that allow selective maternal attachments to occur, and we will see that learning mechanisms are involved. We will see that in certain species (many rodents, for example), a general infant stimulus can activate the neural mechanisms for maternal responsiveness in females, while for other species (sheep, for example), experiences that occur with a particular infant narrow the range of subsequent infant stimuli that can activate the neural mechanisms for maternal behavior.

2. Overview of the Book

A traditional view has been that the maternal behavior of nonprimate mammals is hormonally regulated by the endocrine events of late pregnancy, parturition, and lactation, while the maternal behavior of primates is emancipated from endocrine control and instead is influenced by a variety of social experiences (Keverne, 2001). Hormonal control implies that hormonal factors regulate the ability of infant stimuli to arouse maternal interest, and a lack of such control indicates that infant stimuli directly activate the neural circuitry regulating maternal responsiveness. In Chapters 2 and 3 we review the literature pertaining to the role of hormonal and experiential factors in the regulation of mammalian maternal behavior. The evidence indicates that such regulation in primate and nonprimate mammals may not be qualitatively different, but instead may differ only in the degree to which hormones and experience regulate maternal behavior. Importantly, we will show that hormones influence or modulate maternal motivation in primates and nonprimates, and that experiential factors, such as previous maternal experience or early adverse experiences (such as maternal deprivation during early postnatal development), influence subsequent maternal responsiveness in both nonprimates and primates. We also show that maternal behavior in nonprimates can have a nonhormonal basis. Therefore, one emphasis in these chapters is on the underlying similarities in the regulation of maternal behavior across mammalian species, which is suggestive of a commonality in the underlying neural regulatory mechanisms.

Chapter 4 is a central chapter in the book. It presents a motivational analysis of the occurrence of maternal behavior, and it shows for a variety of mammalian species, including primates, that the hormonal events of late pregnancy, parturition, and lactation have three major effects on the behavior of the mother: (a) they act on the nervous system to increase the attractive value of infant-related stimuli; (b) they decrease the fear-arousing properties of infant stimuli, particularly in first-time mothers; and (c) importantly, such physiological events also result in a general decrease in the fear/stress reactivity of the mother to a variety of stimuli. Such a general reduction in fearfulness is viewed as allowing the mother to effectively care for her offspring under challenging/demanding environmental conditions. Importantly, experiential factors may influence maternal behavior by also operating on these same motivational systems. This chapter sets the tone for the entire book because our approach is to uncover central neural motivational and emotional mechanisms that influence maternal responsiveness. We feel that such an approach will delineate neural regulatory mechanisms controlling maternal behavior, which will have wide generality across mammalian species.

Chapter 5 deals with the neuroanatomy of maternal behavior via a neural systems approach. We outline neural circuits controlling maternal behavior that map on to the motivational processes presented in Chapter 4. We show that there is a positive or excitatory neural system that promotes maternal motivation and attraction toward infant-related stimuli, and we also delineate a central fear/aversion system that has inhibitory effects on maternal responsiveness. In order for effective maternal behavior to occur, physiological, experiential, and genetic factors need to promote the positive system while down-regulating the central fear system. In contrast, factors that promote activity in the aversion system should disrupt maternal behavior.

Chapter 6 reviews the neurochemistry and molecular neurobiology of maternal behavior. This chapter outlines neurotransmitter/neuromodulator systems that influence maternal behavior, and it describes recent work on transgenic mouse strains that have begun to inform us about the genetic basis of maternal behavior. A key focus of this chapter is to relate these neurochemicals and molecular neurobiological factors to the neural circuits described in Chapter 5.

Up to this point, the book deals primarily with maternal behavior. Chapter 7 turns attention to an analysis of the regulation of paternal behavior. Since paternal males are not exposed to the hormonal events of late pregnancy, parturition, or lactation, we examine whether hormones influence paternal behavior and whether such hormonal mechanisms share any similarities with the hormonal events that stimulate maternal responsiveness. Since most mammalian males do not show paternal behavior under natural conditions, we will discuss processes that may act to suppress or prevent male interest in infant-related stimuli.

In Chapter 8, "Neural Basis of Parental Behavior Revisited", we integrate the evidence from hormonal, anatomic, and neurochemical studies to focus on the mechanisms by which experience might modify the neural circuits for maternal behavior. We describe the neural basis of three major experiential effects. First, for those species that form a selective attachment to their own young while rejecting the advances of alien young, we describe how experience with a particular infant near the time of parturition acts on the nervous system in order to narrow the range of infant stimuli that can activate maternal behavior. Second, for a variety of mammalian species, previous maternal experience decreases the dependence of future maternal behavior on hormonal mediation. It is as if previous maternal experience emancipates future maternal behavior from endocrine control. Using the neural circuits described in previous chapters, we outline a variety of neural modifications that might underlie such an experience-induced emancipation. Finally, we describe neural modifications that are caused by early adverse experiences and that appear to have an impact on the developing female so that her maternal behavior is disrupted in adulthood. Importantly, we show that one of the ways in which maternal deprivation might disrupt the adult maternal behavior of the affected offspring is by causing an up-regulation of central fear or aversion systems.

The final chapter discusses the human implications of the data we presented in the previous chapters. We make significant interconnections between the physiological and neural data that we described and pathological states related to the human maternal condition. In particular, we attempt to outline possible neural mechanisms that may underlie postpartum depression, child abuse and child neglect, and the intergenerational transmission of faulty parental behavior in humans. Although much of what we propose is preliminary in nature, the final chapter clearly shows that an understanding of the neurobiology of parental behavior has important psychiatric implications.

Within the field of biology, systems neuroscientists have been primarily interested in the neurobiology of sensory and motor systems, and this interest has helped foster the development of cognitive neuroscience, which has strong ties to neurology (see Kandel, Schwartz, & Jessell, 2000). In contrast, physiological psychologists have long been interested in the neurobiology of motivation and emotion (see Grossman, 1967), and the study of these processes has important implications for psychiatry. There has been a recent resurgence of such interests, and this has fostered the development of emotional and motivational neuroscience (see LeDoux, 1996). Our intent in writing this book is to present a worthy contribution to this latter field.

2

Hormonal and Nonhormonal Basis of Maternal Behavior

1. Introduction

This chapter will review research findings that show that hormonal factors associated with pregnancy termination stimulate increases in maternal responsiveness in females of many mammalian species. Such a process makes sense from an evolutionary perspective because it helps to ensure that the mammalian female will care for her own young at a point in time when she is also capable of lactating. We want to state from the outset, however, that stimulation by the specific hormonal events of late pregnancy is not the only mechanism through which parental behavior can be realized. Two natural examples will make this point clear. First, paternal behavior occurs in some mammalian species, and since males do not experience the particular endocrine events of pregnancy, some other process must arouse their parental responsiveness. Second, as we will review below, alloparental behavior occurs in certain mammalian species, including rodents and primates (Fairbanks, 1990; Mitani & Watts, 1997; Solomon & French, 1997). Alloparental (allomaternal or allopaternal) behavior is caregiving behavior directed toward a conspecific infant by an individual who is not the infant's genetic parent (Riedman, 1982), and in many cases of allomaternal behavior, the females are neither pregnant nor lactating.

As we will make clear, there are hormonal and nonhormonal routes of access to the neural mechanisms underlying parental responsiveness, and just which route is dominant in a particular individual depends on that individual's species, sex, age, and experiences during ontogeny. When comparing primates with nonprimate mammals, a dominant view (which will be critically evaluated in this book) has been that many nonprimate mammalian females are more dependent upon hormonal stimulation of maternal behavior than are their primate counterparts, while primate maternal behavior is more dependent upon social experiences during development (see Coe, 1990; Keverne, 1996; Pryce, 1996; and Ruppenthal, Arling, Harlow, Sackett, & Suomi, 1976, for reviews). Indeed, we know that nulliparous human females can adopt infants and show normal maternal behavior (Brodzinsky & Huffman, 1988). In a recent book that provides an excellent evolutionary perspective on maternal behavior, Hrdy (1999) has argued that human maternal behavior may be less dependent upon (but not independent of) hormonal stimulation because allomaternal behavior by non-pregnant/nonlactating females toward kin may have been an adaptive character-istic in human hunter-gatherer societies during a large part of hominid evolution. Therefore, if one's primary interest were in the neurobiology of primate, and particularly human, parental responsiveness, one might question the significance of examining the hormonal basis of maternal behavior in other mammals. Our view, however, is that hormones can serve as tools to help us explore and define the central neural, neurochemical, and molecular mechanisms underlying ma-ternal behavior in those species that depend upon hormonal stimulation, and that this information on the underlying neural basis of maternal behavior may then be applicable to all mammals (also see Bridges, 1990; Keverne, 1996; Numan, 1994, 1999; Stern, 1989). In subsequent chapters we will show that particular limbic and hypothalamic structures, and their neural connections, are critically involved in the regulation of parental behavior, and that such neural networks form a substrate upon which hormones can act to influence parental behavior. Since these regions are phylogenetically old, and since maternal behavior is a defining characteristic of mammals, it is highly probable that these regions rep-resent a common core neural circuitry that underlies parental behavior in all mammals.

Figure 2.1 portrays the ideas that we are developing (see Crews [1992] for similar ideas concerning reproductive behavior in reptiles). Based on the social and ecological characteristics of a particular species, natural selection favors the occurrence of certain genes in the gene pool of that species, and these genes influence the development and function of the neural circuitry underlying pa-rental behavior. The question is, how does a particular individual respond to infant-related stimuli? Although the basic neural network (the particular nuclei and their connections) that regulates parental behavior in mammals may be highly conserved across species, the degree to which this core neural circuitry is activated or affected by hormonal, sensory, and/or experiential factors may vary across species, and such variation is assumed to have adaptive significance. As a hypothetical example, suppose that in Species X infant-related stimuli gain immediate access to the underlying core neural circuitry without the necessity of hormonal mediation, while in Species Y the specific hormonal events of late pregnancy must first prime the critical neural regions before infant-related stim-uli can gain access to the parental circuitry. Therefore, under natural conditions, in Species X we might expect to see maternal, paternal, and alloparental behav-ior, while in Species Y we would only expect to see maternal behavior in post-partum females.

Some concrete examples will clarify these points. In many mammalian spe-cies, the biological mother is the sole caregiver to her infants (maternal behav-ior), but in some other mammalian species a cooperative or communal breeding system occurs (Solomon & French, 1997). Cooperative breeding systems are of two major types: singular and plural. In plural systems, many lactating fe-males may share in the care of each other's offspring (Lewis & Pusey, 1997;

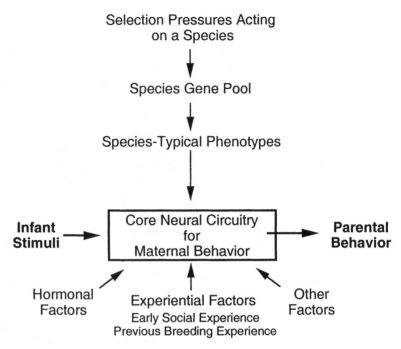

Figure 2.1. A diagrammatic representation of how species differences in mammalian maternal behavior may have evolved. We show a core neural circuitry, made up of the phylogenetically older parts of the brain (limbic system and hypothalamus), which regulates parental responsiveness to infant stimuli. The operation of this core substrate, in turn, can be influenced by genetic, hormonal, experiential, and other factors. We argue that the core neural circuitry that regulates maternal behavior may be highly conserved across mammalian species. However, as a result of different species being affected by differential natural selection pressures, what might vary among species is the degree to which this circuitry is dependent upon hormonal facilitation and whether, and to what degree, this neural network can be influenced by experiential factors during ontogeny.

Packer, Lewis, & Pusey, 1992). A good example of such a communal nursing species is the house mouse (*Mus musculus*) (Manning, Dewsbury, Wakeland, & Potts, 1995). Plural cooperative breeding is not a concern to us here because the nursing females would all be exposed to the endocrine events of late pregnancy. Singular systems are our concern at this juncture: in such systems, a single male and female do all the breeding, while other male and female nonbreeding members of the group help care for the young of the breeding pair (see Solomon & French, 1997, for many examples). In the case of singular cooperative breeding units we can observe maternal, paternal, and alloparental behavior. Clearly, some factors other than the exact endocrine events that occur at the end of pregnancy must be arousing parental interest in the fathers and the alloparents.

Descriptive/observational laboratory and field studies on various species of microtine rodents provide good examples of closely related species that vary in their social organization, with some species displaying sole maternal care of young and others showing singular cooperative breeding with maternal, paternal, and alloparental behavior (McGuire & Novak, 1984; Powell & Fried, 1992; Wang & Insel, 1996; Wang & Novak, 1992). A particularly instructive laboratory study was conducted by Wang and Novak (1992), who examined parental activities of meadow voles (*Microtus pennsylvanicus*) and prairie voles (*Microtus ochrogaster*) in a seminatural environment (also see McGuire & Novak, 1984). For meadow voles, a breeding pair mated and then separated, with each sex building separate nests. When the young were born, they were cared for exclusively by the mother, and when the young became juveniles and were weaned, they separated from the mother, stayed in a separate nest site, and played no role in the care of subsequent litters. If a weaned juvenile or the adult male approached the nesting area of a nursing female, she would chase the intruder away. The prairie vole system contrasts strongly with that of the meadow vole: mated prairie vole males and females shared the same nest site and the male showed paternal behavior. Juveniles also remained in the same nest site with their mother and father after the birth of a subsequent litter, and they helped care for their younger siblings. Although the father and the juveniles do not lactate, they were observed to help care for young pups in a variety of other ways: they retrieved displaced pups back to the nest area, they groomed pups and huddled over them to keep them warm, and they rebuilt nests, thus maintaining the nest site. These laboratory data fit with field study findings that suggest that prairie voles tend to be monogamous and usually form extended families while meadow voles usually have a promiscuous mating system, with sole maternal care of young and early dispersal of weaned young from the natal nest area (Getz & Carter, 1980; Madison, 1978).

Although the focus of this book is not on evolutionary processes, we want to stress that these variations in reproductive strategies across species are considered to have adaptive significance. Individuals in a population should behave in ways that maximize their inclusive fitness. What might be the adaptive advantage of alloparental behavior shown by juveniles or subadults? As others have suggested, such behavior might increase the inclusive fitness of alloparents if they are caring for their siblings (kin selection). Such alloparental behavior may be particularly adaptive if ecological conditions favor delayed breeding and delayed dispersal from the natal nest site on the part of the juvenile/subadult. Finally, caring for young may provide the alloparent with experiences that will be beneficial to its own reproductive efforts at some future point (we will revisit this particular issue in the next chapter). The reader is referred to the following to get more information on evolutionary approaches to alloparental behavior: Powell and Fried (1992), Snowdon (1996), and Solomon and French (1997).

Descriptive observational studies can take us just so far in understanding the proximate causes of parental behavior. From the Wang and Novak (1992) study one might want to conclude that the particular endocrine events associated with pregnancy termination are what is needed to trigger parental behavior in meadow voles, and this results in the occurrence of only maternal behavior, with no evidence of paternal and alloparental behavior. However, Wang and Novak reported that the meadow vole mother chased the father and the weaned juveniles away from the nest area containing young pups. Perhaps the juveniles and the father would have shown parental behavior if they were permitted access to the young. In contrast, for the prairie vole, although juveniles and the father showed parental behavior, does this exclude the involvement of endocrine factors? Although the adult male and juveniles were not pregnant or lactating, could they have shared some endocrine factor with the parturient female that promoted parental behavior? Finally, with respect to the juvenile prairie voles, do ontogenetic changes in parental behavior occur, particularly with respect to females? In other words, will an adult nulliparous female also show alloparental behavior? It is certainly possible that juvenile females show alloparental behavior but that some kind of ontogenetic change occurs so that parental responsiveness in adult females requires facilitation by the endocrine events of late pregnancy.

It is these kinds of issues that require experimental physiological investigations. For the remainder of this chapter we will explore the mechanisms—hormonal and nonhormonal—involved in the expression of parental behavior in female mammals. We will divide this literature into studies on rodents, rabbits, sheep, and primates. A discussion of paternal and alloparental behavior will be deferred until Chapter 7.

2. Hormonal and Nonhormonal Basis of Maternal Behavior in Rodents and Rabbits

2.1. Rats: Nonhormonal Basis of Maternal Behavior

Primiparous female rats (*Rattus norvegicus*), upon giving birth, show immediate maternal attention to their young (Rosenblatt & Lehrman, 1963). The question is: How does such maternal responsiveness come about? The simplest answer is that pup stimuli are the sole instigators of maternal responsiveness—when such stimuli are present, adult females will behave parentally. Such a view would suggest that any female, independent of physiological state, will show parental responsiveness upon exposure to pup stimuli. In their classic studies on maternal behavior, Weisner and Sheard (1933) showed that a pure pup stimuli activation hypothesis of *immediate* maternal responsiveness in adult rats is not accurate. They noted that although parturient females show immediate maternal attention to their own young or to *foster young obtained from another mother*, virgin (nonpregnant nulliparous) females and pregnant primigravid females (1–9 days prepartum) did not show immediate maternal behavior toward foster young (also see Stone, 1925). Weisner and Sheard (1933) also noted that neither the experience of parturition nor suckling stimulation was a necessary factor in stimulating maternal interest in young at parturition in primiparous females because neither near-term caesarean section nor prepartum thelectomies (removal of the nipples) disrupted subsequent maternal behavior (also see Moltz, Geller, & Levin, 1967; Moltz, Robbins, & Parks, 1966). They concluded, therefore, that "some internal changes occur in the highly pregnant or parturient organism which 'awaken the maternal drives' " (p. 177). In support of this interpretation, recent studies have shown that there is a prepartum onset of maternal behavior in the rat that occurs on the day of parturition (Mayer & Rosenblatt, 1984; Slotnick, Carpenter, & Fusco, 1973). While pregnant nulliparous females generally do not show immediate interest in young, Mayer and Rosenblatt (1984) found that 91% of such females retrieve pups during a 15-minute test if the pups are presented within 6.5 hours of parturition.

Weisner and Sheard (1933) also made another dramatic discovery. Although adult nulliparous females do not show immediate maternal responsiveness toward young, they can be induced to show such behavior as a result of cohabitating with pups over a series of days. That is, if freshly nourished pups are placed in a nulliparous female's cage on a daily basis, after a period of several days the virgin female begins to care for them: she retrieves them to a nest area, licks and grooms them, and hovers/huddles over them so as to expose her ventral nipple region (such females, of course, do not lactate). The process whereby continuous exposure to young pups can induce maternal responsiveness in nonlactating female rodents has come to be called sensitization (Leblond, 1938, 1940; Rosenblatt, 1967), and it is this process of pup-stimulated maternal behavior, which is suggestive of a nonhormonal route of access to the neural circuitry underlying maternal behavior, that we will now explore in more detail. (As we will see in Chapter 4, the term "sensitization" is not a fully accurate descriptor of pup-stimulated maternal behavior. Sensitization implies an increase in responsiveness to a stimulus that is repeatedly presented, but recent research indicates that pup-stimulated maternal behavior involves a dual process. First, a process similar to *desensitization* or tolerance develops to certain pup stimuli, and then this is followed by a period of sensitization to additional pup stimuli.)

The initial response of adult estrous-cycling nulliparous females from various laboratory strains of rat toward pups is either avoidance, infanticide, or parental care, with the typical response being avoidance (Fleming & Rosenblatt, 1974a; Jakubowski & Terkel, 1985; Rosenblatt, 1967; Weisner & Sheard, 1933), although the level of infanticide may increase under seminatural conditions (Mennella & Moltz, 1989). However, after approximately 4–7 days of continuous exposure to young, the nulliparous female will begin to care for them (Fleming & Rosenblatt, 1974a; Rosenblatt, 1967; Stern, 1983). Stern (1997a) has reported on her observations of continuous 24-hour videotape recordings of the sensitization process. Although most virgin females initially avoid young, this is followed by a period of tolerance toward pup proximity (also see Fleming & Luebke, 1981). In the 24-hour period prior to the onset of complete maternal behavior (retrieval of all young to a single nest site with licking and hovering over young), there was a dramatic increase in licking behavior, and partial retrievals (individual pups were carried short distances) also occurred. These results suggest that the onset of maternal behavior during sensitization in rats is a gradual and staged process, rather than an abrupt all-or-none event. First, a decrease in pup avoidance allows proximal contact to occur between the adult female and pups. Second, a certain amount of such proximal contact precedes the occurrence of full maternal behavior (also see Fleming & Rosenblatt, 1974a; Terkel & Rosenblatt, 1971).

Although maternal behaviors can be induced in nulliparous female rats through pup exposure, the question arises as to whether this behavior is similar in all respects to the maternal behaviors shown by postpartum females. The answer appears to be no. When the maternal behavior of sensitized virgin female rats is compared with that of primiparous lactating rats in home-cage tests, aspects of the behavior are remarkably similar between the two groups in terms of retrieval behavior, nest building, and time spent in the nest with pups (Fleming & Rosenblatt, 1974a; Reisbick, Rosenblatt, & Mayer, 1975). However, aspects of the particular nursing postures utilized by virgins and postpartum females differ, and this is most likely due to the fact that the nipples are not developed in virgin females and therefore do not receive the suckling input that postpartum lactating females receive (Lonstein, Wagner, & De Vries, 1999). Even more dramatic differences have been found between sensitized and lactating rodents when a test situation that is more demanding than home-cage testing is utilized. In particular, several investigators have examined whether a female would retrieve a test pup from a T-maze extension attached to the home cage (Bridges, Zarrow, Gandelman, & Denenberg, 1972; Stern & MacKinnon, 1976; also see Mayer & Rosenblatt, 1979). Virgin females that are retrieving young in home-cage tests are much less likely than lactating females to retrieve pups from the T-maze, although they do enter the maze. Perhaps this finding is related to the fact that postpartum lactating females show maternal aggression, while sensitized virgin females do not (see Chapter 4). Maternal aggression is a protective response in which lactating females attack intruders in order to keep them away from the nest site. Perhaps the increased likelihood of both maternal aggression and T-maze retrieval in lactating females, when compared to sensitized virgins, is indicative of the fact that the physiological events of late pregnancy and lactation not only enable a female to show immediate maternal responsiveness toward young, but also allow her to be more willing than her sensitized counterpart to take reasonable risks in caring for her young.

Although sensitized virgin females do not experience the particular endocrine events associated with pregnancy termination, one can still ask whether hormonal factors play any role in the maternal behavior that is induced in such females by pup stimulation. Three possibilities exist:

1. The sensitization process may represent the ability of persistent pup stimuli to ultimately have direct access to the neural mechanisms underlying maternal behavior independent of any hormonal mediation.
2. During sensitization, perhaps chronic pup stimulation induces endocrine

changes in the nulliparous female that share certain characteristics in common with the endocrine events associated with late pregnancy, and it is these hormonal changes that activate maternal responsiveness.

3. It is possible that the endocrine background that is naturally present in the adult estrous-cycling female, a background that is independent of pup stimulation, is essential for the maternal responsiveness that occurs in virgin females. Current evidence supports aspects of hypotheses 1 and 3.

Rosenblatt (1967) measured the latencies to onset of maternal behavior following pup presentation in adult nulliparous rats that were either intact and showing normal estrous cycles, ovariectomized, or hypophysectomized. The latencies to onset of retrieving, licking, and hovering behavior (5–7 days) were similar in these groups, which led Rosenblatt to conclude that there is a basic level of maternal responsiveness in rats that is directly activated by pup stimulation and is independent of hormonal mediation (hypothesis 1).

While Rosenblatt (1967) only measured onset latencies to various aspects of maternal behavior, other studies have investigated the level or intensity of pup-induced maternal behavior displayed by sensitized intact or ovariectomized nulliparae, and these studies have found that the intact female shows a higher level of maternal responsiveness than does her ovariectomized counterpart: Intact females retrieve more quickly, build better nests, and hover over pups at the nest site for longer durations than do ovariectomized females (LeRoy & Krehbiel, 1978; Mayer & Rosenblatt, 1979). These results support a modified hypothesis 3: The endocrine background of the adult estrous-cycling female, although not essential for the display of sensitized maternal behavior, increases the intensity of the behavior that is induced by pup stimulation.

The results reviewed indicate that the level of maternal responsiveness is highest in postpartum females, intermediate in intact sensitized females, and lowest in ovariectomized sensitized rats. As we will soon see, estradiol and prolactin are two hormones that play a stimulatory role in the onset of maternal behavior in rats at the time of pregnancy termination. Since estradiol and prolactin are also secreted during the estrous cycle, but are either absent or secreted at lower levels in ovariectomized females (Amenomori, Chen, & Meites, 1970), perhaps it is these hormones that allow the intact sensitized female to show more vigorous maternal responses than her ovariectomized counterpart.

One final topic on the sensitization process in rats remains to be discussed: ontogenetic mechanisms. It appears that an important ontogenetic shift occurs in the responsiveness of nulliparous female rats to young pups. Several studies (Bridges, Zarrow, Gandelman, & Denenberg, 1974; Brunelli, Shindledecker, & Hofer, 1985; Gray & Chesley, 1984; Mayer & Rosenblatt, 1979; Stern, 1987; Zaias, Okimoto, Trivedi, Mann, & Bridges, 1996) have shown that 24-day-old female rats have shorter sensitization latencies to the onset of maternal behavior than do older juvenile and adult females (30 days of age or older). For example, Zaias et al. (1996) report that 24-day-old females show parental responses toward young pups after about 2–3 days of cohabitation with the pups, while 30-

day-old females respond more like adult cycling females, requiring about 6–8 days of continuous association with young pups before they begin to act parentally. Although the parental behavior of juvenile females has sometimes been described as incomplete (only some of the pups may be retrieved) or inconsistent (parental responsiveness may be shown on one day, but not the next) (Brunelli & Hofer, 1990; Gray & Chesley, 1984), several studies have reported that the full pattern of parental responsiveness is induced in 24-day-old females as a result of continuous pup stimulation—the juvenile retrieves all pups to a single location, anogenital licking occurs, and the female lies over the pups (Stern, 1987; Zaias et al., 1996).

Rats are normally weaned between 21 and 28 days of age (Gilbert, Burgoon, Sullivan, & Adler, 1983; Rosenblatt & Lehrman, 1963), and puberty in females normally occurs at about 40 days of age (Mayer & Rosenblatt, 1979). Within this context we can ask about the nature of the ontogenetic processes that leads to shorter sensitization latencies in 24-day-old juveniles when compared to females that are 30 days of age or older. Clearly, going through the endocrinological changes associated with puberty is not essential for the decline in maternal responsiveness in 30-day-old females, since these females are prepubertal (also see Stern, 1987). Gray and Chesley (1984) exposed 22-day-old juvenile females to pups and made the interesting observation that even before these females began to show parental behavior they did not avoid the pups. That is, when pups were placed in the part of the cage where the female was resting or nesting, the female did not move to another part of the cage. This contrasts with adult females in the early phases of the sensitization process, during which avoidance responses, which include shifting the nest site to a part of the cage that does not contain pups, are a prominent feature (Fleming & Luebke, 1981; Gray & Chesley, 1984). Therefore, it is possible that maturational changes, which include the development of fear responses, play a role in the ontogenetic shift that causes a decrease in maternal responsiveness of female rats between 22–24 days of age and 30 days of age. Perhaps younger females receive more intense stimulation from the positive attributes of pups because they remain in close proximity to them, and this accounts for the shorter sensitization latencies of such females. Indeed, it has been reported that juvenile maternal responses toward pups, particularly retrieving, are often associated with playful responses directed toward the pups (for example, charging and pouncing behaviors: Brunelli & Hofer, 1990). It seems likely that for the 22–24-day-old female, young pups are initially viewed as positive social stimuli that are capable of eliciting affiliative play responses rather than avoidance responses. As a result of continued contact with pup stimuli, a more organized and complete set of maternal behaviors may subsequently be aroused. In line with this overall reasoning, Bronstein and Hirsch (1976) have reported that defensive responses mature to adult levels in juvenile rats around 30 days of age. That is, while threatening stimuli, such as a suddenly moving object, were unlikely to disrupt the open field activity of 20-day-old rats, such stimuli inhibited locomotion and caused freezing behavior in rats that were 30 days of age or older. Bronstein and Hirsch suggest that the development of defensive responses to potentially threatening objects at about 30 days of age is adaptive because it coincides with the approximate time at which feral rats begin to leave their natal burrow system and are, therefore, more likely to be exposed to predation. With respect to neural mechanisms, since the brain of the postnatal rat shows considerable developmental changes (Jacobson, 1991), it would be important to determine just which changes mediate or affect the ontogenetic depression in parental responsiveness during juvenile development. Such knowledge would help us understand the organization and function of neural systems involved in adult maternal behavior and would inform us about the involvement of central fear systems.

Therefore, to review, 20–24-day-old juvenile female rats show a high level of parental responsiveness to young pups, but by 30 days of age an inhibitory mechanism has been set up, and sensitization latencies increase and remain high as long as the female is nulliparous. If a female becomes pregnant, then near the time of parturition she becomes highly responsive to pup stimuli and shows immediate maternal behavior. It is now time for us to explore the physiological mechanisms associated with pregnancy termination that stimulate such immediate maternal responsiveness.

2.2. Rats: Hormonal Basis of Maternal Behavior

The classic studies of Terkel and Rosenblatt (1968, 1972) showed that humoral factors associated with the peripartum period stimulate maternal behavior in rats. They found that when blood was transfused from a parturient female (one that had given birth within 30 minutes of the onset of the transfusion) into a virgin female, the maternal behavior of the virgin toward test pups was facilitated when compared to the response of a virgin female that was transfused with virgin blood.

Figure 2.2 shows the peripheral plasma levels throughout pregnancy of the major hormonal factors that have been implicated in the control maternal behavior in rats. With respect to the steroids, note that plasma estradiol levels are low throughout the first part of pregnancy, then rise around day 15; this peak is subsequently maintained through the day of parturition. In contrast, progesterone levels are very high throughout the first part of pregnancy, with a peak on day 15, followed by a decline that becomes abrupt after day 20. Therefore, near the time of parturition (day 22) we see a reversal of the estrogen-to-progesterone ratio, moving from a long period of progesterone dominance to a short period of estradiol dominance. Lactogenic hormones, that is, protein hormones that promote mammary gland development and lactogenesis (Southard & Talamantes, 1991), are secreted from either the anterior pituitary (pituitary prolactin) or the placenta (placental lactogens) throughout most of pregnancy (Butcher, Fugo, & Collins, 1972; Freeman, 1994; Grattan, 2001; Klindt, Robertson, & Friesen, 1981; Morishige, Pepe, & Rothchild, 1973; Smith & Neill, 1976). Although these various lactogenic hormones are not structurally identical, they do share certain structural features (Nicoll, Tarpey, Mayer, & Russell,

1986; Southard & Talamantes, 1991), they appear to produce their biological actions through interaction with the prolactin receptor (Cohick et al., 1996), and, in addition to being lactogenic, pituitary prolactin and placental lactogens are luteotropic in rats and mice (Galosy & Talamantes, 1995).

As can be seen in Figure 2.2, during the first half of pregnancy, prolactin is released from the pituitary in two daily surges, while during most of the second half of pregnancy, pituitary prolactin secretion is low except for an increase in secretion that occurs on the final day of pregnancy. As prolactin levels decline during the second half of pregnancy, plasma levels of placental lactogens in-

A. Estradiol and Progesterone

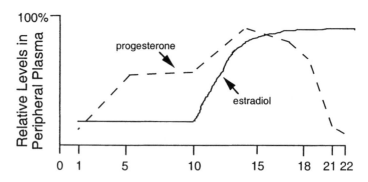

B. Prolactin and Placental Lactogens

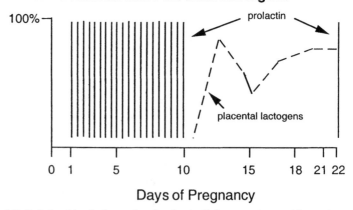

Days of Pregnancy

Figure 2.2. Relative blood plasma levels of estradiol, progesterone (shown in panel **A**), prolactin, and placental lactogens (shown in panel **B**) over the course of the rat's 22-day pregnancy. For each hormone, data are depicted as the percentage of the maximum value detected during pregnancy. Prolactin is secreted in daily surges over the first half of pregnancy and on day 22. We show these surges as vertical lines. Modified from Bridges (1996) and Grattan (2001).

crease. Although there are several different placental lactogens, the two major types are referred to as placental lactogen I and placental lactogen II (see Cohick et al., 1996).

In conclusion, during most of pregnancy rats are exposed to chronic lactogenic and progestin stimulation, and toward the end of pregnancy a steep withdrawal of progesterone levels is superimposed upon maintained lactogenic hormone levels and rising estradiol levels. What is the evidence that high lactogenic and estrogen levels superimposed on a background of progesterone withdrawal is stimulatory for maternal behavior in rats? Research that came out of the laboratories of Moltz (Moltz, Lubin, Leon, & Numan, 1970) and Zarrow (Zarrow, Gandelman, & Denenberg, 1971) were the first reports to show that this particular pattern of hormone secretion was indeed stimulatory for maternal behavior. Moltz's work will serve as an example. Ovariectomized nulliparous female rats were injected subcutaneously with estradiol benzoate (EB) over days 1–11, progesterone (P) over days 6–9, and prolactin (Prol) on days 9 and 10. Control females received only two of the three hormones (EB + P, or P + Prol, or EB + Prol) or were not treated with hormones. Sensitization tests were performed on all females, with the presentation of test pups beginning on day 10. The females that received the EB + P + Prol regimen had the shortest latencies to onset of maternal behavior (retrieving pups to a nest and hovering over them), showing such behavior after about 48 hours of pup exposure. Females in each of the remaining groups had significantly longer latencies to the onset of maternal behavior. The females in the P + Prol and the EB + Prol groups did not differ from untreated females, and the onset latencies in these groups averaged more than 4 days. Interestingly, the females in the EB + P group showed a moderate stimulation of maternal behavior, with an average onset latency of 3 days, which was significantly shorter than that of each of the remaining control groups.

This work clearly shows that exposing virgin females to high levels of estradiol and prolactin superimposed on a background of progesterone withdrawal can result in a dramatic stimulation of maternal behavior. The moderate facilitation of maternal behavior in the EB + P females most likely resulted from the fact that estradiol can stimulate endogenous prolactin release from the anterior pituitary (Ben-Jonathan, Arbogast, & Hyde, 1989; Leong, Frawley, & Neill, 1983). This latter point has been proved in an elegant series of studies by Bridges. By subcutaneously implanting ovariectomized nulliparous females with Silastic capsules that released *physiological amounts* of estradiol and progesterone into the circulation, Bridges (1984) confirmed that estradiol treatment superimposed on progesterone withdrawal could stimulate maternal behavior without the necessity of adding exogenous prolactin, and he observed onset latencies of 1–2 days (also see Stern & McDonald, 1989). Bridges demonstrated this through the use of two major hormone treatment schedules, which are shown in Figure 2.3A (see Bridges & Ronsheim, 1990). In the Type I schedule, also called the concurrent schedule, animals are concurrently treated with estradiol and progesterone over a given period of time, and then the progesterone implant

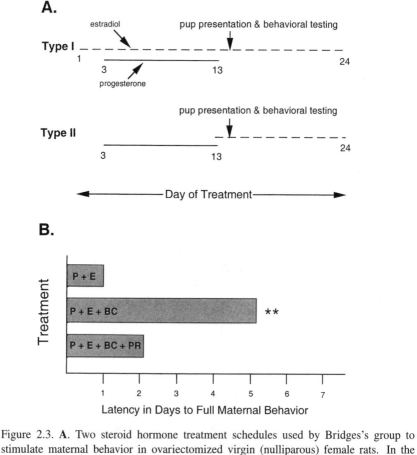

Figure 2.3. **A.** Two steroid hormone treatment schedules used by Bridges's group to stimulate maternal behavior in ovariectomized virgin (nulliparous) female rats. In the Type I concurrent schedule, females receive subcutaneous implants of Silastic capsules containing estradiol for 24 days. Over days 3–13 they are also implanted subcutaneously with progesterone. 24 hours following the removal of the progesterone capsules, on day 14 of treatment, females are presented with pups and daily observations of maternal behavior commence, with a freshly nourished litter being presented on each day. In the Type II sequential schedule, females are first exposed to progesterone for 10 days, the progesterone implant is removed and replaced with estradiol containing capsules, and 24 hours after the onset of estradiol treatment, the females are exposed to pups and the onset of maternal responsiveness is measured. **B.** Median latencies to the onset of maternal behavior in ovariectomized virgin female rats treated with the Type II hormone regimen. One group was treated with progesterone (P) and estradiol (E), and another was treated similarly, except that bromocriptine (BC, a dopamine agonist drug that inhibits the release of prolactin from the pituitary) was administered subcutaneously throughout the period of estradiol treatment. In addition to treatment with E, P, and BC, the third group also received subcutaneous prolactin injections during the period of estradiol and bromocriptine treatment. In the relevant groups, appropriate vehicle control injections were also administered. Rats treated with P + E + BC have significantly longer latencies to the onset of maternal behavior than each of the remaining two groups. Panels **A** and **B** are modified from Bridges and Ronsheim (1990).

is removed while the estradiol implant remained in place during the period of pup presentation and tests for sensitization. In the Type II schedule, also called the sequential schedule, the animals are first implanted with progesterone capsules, and then these capsules are removed before the female is exposed to estradiol and pups. Both of these schedules effectively stimulate maternal behavior without the need for exogenous prolactin administration (Bridges, 1984), and both of these schedules mimic a sequence of progesterone withdrawal followed by estradiol dominance. Importantly, however, these hormone schedules no longer stimulate maternal behavior if endogenous prolactin release from the anterior pituitary is inhibited either through hypophysectomy (Bridges, DiBiase, Loundes, & Doherty, 1985), or through administration of the drug bromocriptine (a dopamine agonist; also called CB-154) during the period of estradiol treatment in the Type II schedule (Bridges & Ronsheim, 1990). Significantly, the inhibitory effects of hypophysectomy or bromocriptine treatment can be reversed by exogenous prolactin administration (Bridges & Ronsheim, 1990; Loundes & Bridges, 1986). Some of these results are shown in Figure 2.3B. These studies show that the stimulation of maternal behavior in rats that occurs after exposure to estradiol treatment superimposed on progesterone withdrawal requires the induction of endogenous prolactin release by estradiol (see Bridges & Ronsheim, 1990). Therefore, the conclusion of Moltz et al. (1970) that maternal behavior is stimulated by high levels of estradiol and prolactin superimposed on a background of progesterone withdrawal is the most accurate.

In more recent studies, Bridges's lab has shown that placental lactogen I and placental lactogen II were each just as effective as pituitary prolactin in reversing the inhibitory effects of bromocriptine on the maternal behavior of rats that were also treated with the sequential Type II schedule (Bridges et al., 1996, 1997). Therefore, each of the lactogenic hormones is stimulatory for maternal behavior when associated with progesterone withdrawal and estradiol dominance.

Additional findings have indicated that the treatment of ovariectomized nulliparous females with each of the following is not capable of stimulating maternal responsiveness: (a) long-term progesterone treatment followed by progesterone withdrawal (Bridges, 1984; Moltz et al., 1970) and (b) long-term prolactin treatment (Baum, 1978; Beach & Wilson, 1963; Lott & Fuchs, 1962; Loundes & Bridges, 1986). Finally, it has been shown that if ovariectomized virgins are treated with estradiol and progesterone but progesterone is not withdrawn prior to behavioral testing, maternal behavior will not be stimulated (Bridges & Russell, 1981).

Although Moltz et al. (1970) reported that estradiol plus prolactin treatment was not capable of stimulating maternal behavior, Bridges (1984) has subsequently reported that treatment of ovariectomized virgin females with *supraphysiological* levels of estradiol (higher than the levels found during the end of pregnancy) was stimulatory for maternal behavior. Since these females retained their pituitaries, estradiol-induced prolactin release was probably also involved. What all these results suggest is that although supranormal levels of estradiol and prolactin may facilitate maternal behavior, the progesterone withdrawal that occurs at the end of pregnancy lowers the threshold for estradiol and lactogenic hormone stimulation of maternal behavior so that physiological levels of these hormones are effective. Additional research on pregnant rats supports this view.

Several investigators have shown that when the pregnancy of primigravid rats is terminated by hysterectomy (removal of the uterus that contains the fetuses and placentas) on about day 15 of a 22-day gestation period, maternal behavior is facilitated (Bridges, Rosenblatt, & Feder, 1978; Numan, 1978; Rosenblatt & Siegel, 1975). That is, when pups are presented to such females 48 hours after the hysterectomy, the females respond maternally to these young with median latencies of either 0 or 1 day. This short-latency sensitization period should be contrasted to the 4 days or greater latencies of pregnant females or hysterectomized virgin females (Rosenblatt & Siegel, 1975). These studies have also shown that hysterectomy of the pregnant rats was associated with a decline in serum progesterone levels (recall that placental lactogens are produced during the second half of pregnancy and that they are luteotropic in the rat) followed by the occurrence of ovulation and an estrogenized vaginal smear (Bridges et al., 1978; Rosenblatt & Siegel, 1975). What these results suggest is that hysterectomy of pregnant rats during the second half of pregnancy stimulates maternal behavior because it prematurely activates those endocrine events that would normally occur closer to term: progesterone withdrawal followed by rising estrogen and prolactin levels (because of estrogen-induced prolactin release). Evidence in support of this view is that: (a) ovariectomy at the time of the hysterectomy blocks the facilitatory effect of the hysterectomy, presumably because an estrogen and prolactin rise would be prevented (Siegel & Rosenblatt, 1975a); (b) the inhibitory effect of ovariectomy on hysterectomy-induced maternal behavior in pregnant rats is reversed if the animal is also treated with either 5 or 20 µg/kg of estradiol benzoate (EB) at the time of the hysterectomy/ovariectomy (Siegel & Rosenblatt, 1975a, 1978a); and (c) the facilitation of maternal behavior by hysterectomy, or by hysterectomy/ovariectomy and EB treatment, is blocked by appropriately timed progesterone injections that block progesterone withdrawal (Bridges et al., 1978; Numan, 1978; Siegel & Rosenblatt, 1978a).

In further confirmation of these findings, Siegel and Rosenblatt (1975b) and Siegel, Doerr, and Rosenblatt (1978) have shown that if *virgin* female rats are hysterectomized and ovariectomized and injected subcutaneously with 20 µg/kg or 50 µg/kg of EB at the time of surgery and then presented with pups 48 hours later, maternal behavior is *not* facilitated. However, a single injection of 100 or 200 µg/kg of EB, which is supraphysiological, does cause a modest facilitatory effect. Importantly, this facilitation of maternal behavior in hysterectomized and ovariectomized virgin rats treated with a single injection of a supranormal amount of estradiol can be blocked by *concurrent* progesterone administration (Siegel & Rosenblatt, 1975c). Most importantly, Doerr, Siegel, and Rosenblatt (1981) have shown that either 5 or 50 µg/kg of EB can stimulate maternal behavior in hysterectomized and ovariectomized virgin female rats if such treatment is preceded by a period of progesterone dominance followed by progesterone withdrawal.

These studies not only support the view that physiological levels of estradiol and prolactin can promote maternal responsiveness in rats when they are superimposed on a background of progesterone withdrawal, but they also illustrate the complex role played by progesterone. Progesterone appears to exert two effects: (a) the decline of progesterone from high levels (progesterone withdrawal), although having no stimulatory effect by itself, potentiates the facilitatory effects of the estradiol/lactogen complex and (b) high levels of progesterone can inhibit maternal behavior, and this effect can be dissociated from the first effect. More specifically, as shown by Siegel and Rosenblatt (1975c), *acute* progesterone administration can counteract the stimulatory effects of a supraphysiological dose of estradiol without the necessity of exposing the animal to prolonged progesterone treatment. That is, progesterone is not simply inhibitory because it is preventing the progesterone withdrawal potentiation effect.

As indicated in Figure 2.3, during mid-to-late pregnancy estradiol and lactogen levels are rising but progesterone is also high at this time. Perhaps these high levels of progesterone act to inhibit the potential facilitatory effects of estradiol and lactogens, in this way preventing a premature or early onset of maternal responsiveness. The subsequent decline in progesterone in the final stages of pregnancy not only removes this inhibitory effect, but also potentiates the positive effects of estradiol and lactogens, thus coordinating the onset of maternal behavior with parturition.

2.3. Rats: Conclusions

The results we have reviewed clearly indicate that *prolonged* exposure to the hormonal events of late pregnancy stimulates the immediate onset of maternal behavior at parturition. This overall evidence seems to conflict with the finding of Terkel and Rosenblatt (1972) that cross-transfusion of parturient blood into virgin females facilitates the maternal behavior of virgins. Since the transfusion occurred after the period of prolonged progesterone exposure, any direct facilitatory effects of progesterone withdrawal would have been eliminated. These conflicting results need to be reconciled, although it is possible that progesterone withdrawal causes some other chemical change that is present at parturition, and this humoral factor then synergizes with estrogen and lactogens to stimulate maternal behavior during the transfusion period.

Our review has emphasized the importance of estradiol, progesterone, and lactogens. One other hormone plays an important role in parturition and one might assume that it would also be involved in the onset of maternal behavior: oxytocin. This neuropeptide is synthesized by hypothalamic neurosecretory cells and released into the general circulation from the neural lobe of the pituitary (Swanson & Sawchenko, 1983). Oxytocin levels increase in the peripheral plasma of rats near the time of parturition (Higuchi, Honda, Fukuoka, Negoro, & Wakabayashi, 1985), and the hormone acts on uterine oxytocin receptors to stimulate the uterine contractions that facilitate the expulsion of the fetuses (Challis & Lye, 1994; Fuchs, 1983). In addition to its role in parturition, oxytocin is also released from the neurohypophysis in response to suckling stimulation, and the hormone then acts on the contractile tissue of the mammary gland to promote milk ejection (Wakerley, Clarke, & Summerlee, 1994).

Although hormonal oxytocin is closely associated with the events of parturition, there is no evidence linking systemic oxytocin with the stimulation of maternal behavior in rats. In addition, since circulating oxytocin has a low penetration across the blood-brain barrier (Jones, Robinson, & Harris, 1983; Mens, Witter, & Greidanus, 1983; Morris, Barnard, & Sain, 1984), it would probably not have effective access to neural regions regulating maternal behavior. However, as we will show in Chapter 6, there is a brain oxytocin system that does play an important role in maternal behavior. That is, oxytocin released at synapses within the brain near the time of parturition is a factor that stimulates maternal behavior in rats, but in this case oxytocin is acting as a neurotransmitter/neuromodulator rather than as a hormone.

Since peptides usually have difficulty crossing the blood-brain barrier (Banks & Kastin, 1985), one might question whether lactogenic hormones promote maternal behavior by acting on the brain. In this regard, it should be noted that there is a specialized, receptor-mediated transport system within the choroid plexus that transports circulating lactogens into the cerebrospinal fluid (CSF) (Bridges et al., 1996; Mangurian, Walsh, & Posner, 1992; Pihoker, Robertson, & Freemark, 1993; Rubin & Bridges, 1989). From the CSF, lactogens can presumably reach central neural sites.

Another point to note is that none of the specific hormone treatment regimens that have stimulated maternal behavior in virgin female rats has been able to induce immediate maternal responsiveness. Instead, the hormone-treated rats typically show maternal behavior after about 48 hours of pup exposure. Several possibilities may account for this: (a) the time of pup presentation may not have been synchronized properly with the timing of the hormonal effects, (b) the particular doses and sequences of hormone administration may not have faithfully mimicked the patterns that occur at the end of pregnancy, and (c) some factor(s) other than high levels of estradiol and lactogens superimposed on a background of progesterone withdrawal may be involved in stimulating maternal behavior in rats. One possibility, for example, is that the cervical stimulation (and its physiological effects) that accompanies a normal parturition may play a facilitatory role in maternal responsiveness (Yeo & Keverne, 1986).

Finally, since the evidence we have reviewed shows that the hormonal events of late pregnancy promote the *onset* of maternal behavior in puerperal female rats, one can ask the important question of whether the subsequent hormonal events associated with lactation are necessary for the continuance or maintenance of maternal behavior throughout the remainder of the postpartum period. As will be reviewed in Chapter 3, the answer to this question is no. The hormonal and other physiological events of late pregnancy appear to prime the neural circuits underlying maternal behavior so that the parturient female reacts appropriately to pup-related stimuli, but once the behavior is initiated, it continues without the need for hormonal regulation.

2.4. Hormonal and Nonhormonal Basis of Maternal Behavior in Other Rodents

Research on the physiological basis of maternal behavior in other rodents is scant in comparison to that performed on the rat. The work that exists for Mongolian gerbils (*Meriones unguiculatus*) and golden hamsters (*Mesocricetus auratus*) fits with our review of the research on rats. That is, adult virgin female gerbils and hamsters do not show maternal behavior toward alien young on their first encounter with them but instead usually attack them, while recently parturient primiparous females will respond maternally (Elwood, 1977, 1981; Marques & Valenstein, 1976; Richards, 1966; Rowell, 1960; Siegel & Greenwald, 1975; Swanson & Campbell, 1979a, 1981). These findings suggest that the physiological events of late pregnancy and lactation stimulate maternal behavior in these species. Indeed, during early pregnancy gerbils and hamsters are infanticidal toward pups, but during late pregnancy infanticide is inhibited and maternal behavior may be expressed (Buntin, Jaffe, & Lisk, 1984; Elwood, 1981; Siegel, Clark, & Rosenblatt, 1983a). Sensitization also occurs in gerbils and hamsters: Exposing nulliparous females to young over a period of several days induces maternal responsiveness (Clark, Spencer, & Galef, 1986; Swanson & Campbell, 1979a). Not much research has been done to elucidate the specific physiological events that stimulate the immediate onset of maternal responsiveness toward pups in the recently parturient gerbil and hamster (see Siegel & Rosenblatt, 1980, for research on hamsters), but some work suggests that prolactin may be one of the hormones that stimulates maternal behavior in hamsters (McCarthy, Curran, & Siegel, 1994a).

Research on the maternal behavior of several inbred and outbred strains of laboratory house mice contrasts strongly with the research on rodents that we have presented up to this point. From the earliest studies on mice (Leblond, 1940), it has been noted that nulliparous females from many laboratory strains show near-immediate maternal attention toward alien young (for a review, see Noirot, 1972). For example, approximately 80–90% of adult nulliparous female mice of the Rockland-Swiss outbred strain, whether they are intact or gonadectomized, retrieve alien pups to a nest site in a short 15-minute test (Gandelman, 1973a; Gandelman & vom Saal, 1975); approximately 50% of adult nulliparous females from two inbred mouse strains (DBA/2J and C57BL/6J) retrieve an alien pup to their nest sites in a 15-minute test (Mann, Kinsley, Broida, & Svare, 1983); and nearly all virgin females from another strain (hybrids of 129/Sv and C57BL/6J inbred strains) retrieve and group three alien pups to a nest site in a 30-minute test (Thomas & Palmiter, 1997; similar results for this hybrid strain have been reported by Lucas, Ormandy, Binart, Bridges, & Kelly, 1998). These results suggest that the specific endocrine events associated with pregnancy termination do not play a major role in the pup-directed maternal responsiveness of these strains of mice. It should be noted, however, that there are many more laboratory mouse strains than those indicated above, and differences in the maternal responsiveness of *postpartum* females, in terms of speed of retrieval of displaced pups back to the nest area, have been found in a comparison of 11 inbred strains (Carlier, Roubertoux, & Cohen-Salmon, 1982). These results suggest that a more careful and complete comparison of the maternal responsiveness of nulliparous females from a larger variety of mouse strains might yield some interesting differences (cf. Wang, Crombie, Hayes, & Heap, 1995). In this context, the findings of Jakubowski and Terkel (1982) are important. They found that while all adult virgin females of the C57BL/6J inbred strain were spontaneously maternal toward unfamiliar pups, showing maternal behaviors within 30 minutes of pup exposure, wild (feral) virgin mice consistently killed young, even when tested over a period of 7 days (also see McCarthy, 1990). In subsequent studies, it was shown that feral females even killed alien pups during pregnancy, but did care for such pups during the postpartum period (Soroker & Terkel, 1988). These results clearly suggest that genetic factors influence maternal responsiveness in mice, and they are further supported by the findings of Perrigo, Belvin, Quindry, Kadir, Becker, van Look, Niewoehner, and vom Saal (1993), who compared the responses of adult nulliparous female house mice to unfamiliar pups. As expected, wild females showed infanticide and CF-1 inbred females did not, but instead cared for the pups during a 30-minute test. Importantly, CF-1 X wild hybrid females exhibited the behavior typical of the CF-1 genotype, irrespective of their mothers' genotype. Overall, these results indicate that the endocrine events associated with pregnancy termination and lactation undoubtedly do stimulate maternal responsiveness and inhibit infanticide in certain stocks of wild house mice but that the genetic selection that has occurred in the production of many inbred and outbred strains of laboratory house mice has substantially decreased the necessity of a strong involvement of such hormonal changes in the maternal behavior of these strains. Just which genetic changes, and their corresponding neural/physiological effects, have occurred in particular laboratory strains that allow for the more or less spontaneous maternal responsiveness in virgin females should be an important area of future study. Additionally, researchers will be obliged to increase their understanding of the biology of mouse maternal behavior if they want to take advantage of the modern advances in molecular biology that are revolutionizing our understanding of brain function through the use of transgenic mouse models (see Chapter 6).

As indicated in Chapter 1, maternal behavior can be characterized as infant-directed and non-infant-directed. Our approach in this section on rodents has been to concentrate on hormonal involvement in pup-directed maternal activities. However, one aspect of non-pup-directed maternal behavior in mice (nest building) is clearly influenced by pregnancy hormones. Pregnant mice begin to build a maternal nest during early to mid-pregnancy, well before their young are born, and it has been shown that such nest building is under the influence of high progesterone levels and low estradiol levels—both of these hormones act together to induce the behavior (Lisk, 1971). Interestingly, while intact or gonadectomized virgin females do not build maternal nests, they can be induced to show such behavior by presenting them with pups that they care for (Gandelman, 1973b). These results are important because they show that pup stimuli can affect aspects of the same neural circuitry that hormones act upon to induce a particular behavior, in this case nest building.

Although the hormones of pregnancy and lactation do not appear to play a major role in pup-directed maternal responsiveness during home-cage tests in virgin females from many laboratory strains of mice, some studies have shown that when the testing situation is modified, hormones can exert an influence. For example, in a manner similar to that observed in rats, intact virgin laboratory mice that are showing maternal behavior in home-cage tests are much less likely than lactating females to retrieve pups that are placed in a T-maze extension attached to the home cage (Gandelman, Zarrow, & Denenberg, 1970; also see Ehret & Koch, 1989; Koch & Ehret, 1989a). These results suggest that the willingness of the lactating mother to take risks in caring for her young is greater than that of the maternally behaving virgin female. Additionally, a study by Hauser and Gandelman (1985) found that although virgin and postpartum Rockland-Swiss laboratory mice both cared for young in home-cage tests, postpartum females pressed a lever at a much higher rate than did virgin females in an operant task where each lever press was reinforced by the presentation of a 1-day-old pup. Subsequent experiments implicated the endocrine changes associated with pregnancy termination as the cause of this increased responding. These results suggest that although virgin laboratory mice may show near-immediate maternal responsiveness to pups in standard home-cage tests, the hormonal events associated with the end of pregnancy may act to increase the level of maternal motivation so that pup stimuli become more salient incentive stimuli, exerting a strong attractive influence over the behavior of the mother.

We will conclude this section by discussing what is known about the involvement of pregnancy hormones in the maternal responsiveness of female microtine rodents. Recall that species in this genus show a variety of social organizations. Some species are monogamous cooperative breeders that show evidence of maternal, paternal, and alloparental behavior (prairie vole), while other species tend to be promiscuous breeders with only the postpartum female showing parental behavior (sole maternal behavior: meadow voles, montane voles [*Microtus montanus*]) (Insel, 1992; Wang & Novak, 1992). One might predict, therefore, that pregnancy hormones might play a facilitatory role in the maternal behavior of meadow voles and montane voles, while female prairie voles, which show alloparental behavior, may not be dependent on pregnancy hormones for their parental responsiveness. The scant evidence that exists for the promiscuous breeders fits this prediction. Naive virgin montane voles do not respond parentally to an unfamiliar pup, while postpartum females do (Insel, 1992; Insel & Shapiro, 1992). The evidence for prairie voles is more complicated. Roberts, Zullo, Gustafson, and Carter (1996) have reported that virtually all naive virgin females that were either 21 or 42 days of age acted parentally toward unfamiliar pups in a short 10-minute test (sniffed, licked, and hovered over pups). These data fit with the findings of Wang and Novak (1992) that *juvenile* female prairie voles show high levels of alloparental behavior in a seminatural setting. In contrast to these findings, Lonstein and De Vries (1999a) have reported that virtually all 90–100-day-old naive virgin female prairie voles show infanticide when presented with unfamiliar pups. These results indicate that, as in the rat, there appears to be an ontogenetic shift in the parental responsiveness of nulliparous prairie voles, and recent work by Lonstein and De Vries (2001) has clearly substantiated this point. Therefore, while younger naive females do not appear to require a facilitation from pregnancy hormones, older naive females may require such a facilitation. In this regard, it should be noted that under natural conditions, if female prairie voles disperse from their natal nest area they do so at about 45 days of age (McGuire, Getz, Hoffmann, Pizzuto, & Frase, 1993). Therefore, pregnancy hormone–independent parental responsiveness may be restricted to the ontogenetic period when the young female is likely to remain in her natal family group. A complicating factor is that a recent study has indicated that there are differences in the parental responsiveness of prairie voles that have been derived from different geographical populations (Roberts, Williams, Wang, & Carter, 1998b). Therefore, more detailed studies need to be conducted on the maternal responsiveness of adult naive prairie voles derived from a variety of geographical stocks before any general conclusions can be reached about the involvement of pregnancy hormones.

In both the Roberts et al. (1996) and Lonstein and De Vries (1999a, 2001) studies, the females were weaned at about 21 days of age. That is, they were not allowed to remain in a family group, and therefore were not exposed to any of the subsequent litters of their parents. In other words, they were truly naive, except for exposure to their own littermates prior to weaning. An interesting question is whether alloparental experience as a juvenile would have any effect on the subsequent parental responsiveness of an adult nulliparous female. We will explore this issue in the next chapter.

2.5. Hormonal Basis of Maternal Behavior in the Rabbit

Some of the earliest laboratory investigations on maternal behavior were conducted by Zarrow, Denenberg, and colleagues on rabbits (*Oryctolagus cunuculus*) (Ross, Sawin, Zarrow, & Denenberg, 1963), and these investigators raised many of the seminal questions concerning the influence of hormonal, genetic, and experiential factors on maternal behavior. A recent resurgence of experimentation on maternal behavior in rabbits has been conducted by González-Mariscal and Rosenblatt (1996). Concerning the hormonal regulation of maternal behavior in rabbits, both of these groups of investigators have concentrated on non-pup-directed maternal activities, primarily maternal nest building. More recent research, however, has begun to focus on those hormonal factors that might incite the mother rabbit to care for pups (González-Mariscal & Poindron, 2002).

Gestation lasts approximately 30 days in rabbits, and during the latter part of pregnancy dramatic changes occur in nest-building behavior (González-Mariscal, Díaz-Sánchez, Melo, Beyer, & Rosenblatt, 1994). Beginning at about

8 days prepartum females show increases in digging behavior, and such behavior allows the doe to dig a burrow that will contain the future nest site. Digging behavior subsequently declines about 3 days prior to parturition and is followed by the carrying of straw into the nest site, the construction of a nest with the straw, and the lining of the maternal nest with hair that the doe pulls from her body (González-Mariscal et al., 1994; Ross et al., 1963). The increased digging behavior is temporally correlated with high serum levels of estradiol and progesterone, while the termination of digging and the onset of maternal nest building is correlated with progesterone withdrawal and high serum levels of estradiol and prolactin (González-Mariscal et al., 1994). Experimental studies have shown that it is just these endocrine changes that influence these particular behavioral changes over the course of pregnancy (Anderson, Zarrow, Fuller, & Denenberg, 1971; González-Mariscal, Melo, Jiménez, Beyer, & Rosenblatt, 1996; Zarrow, Farooq, Denenberg, Sawin, & Ross, 1963; Zarrow et al., 1971).

Pup-directed maternal activity in postpartum rabbits is interesting because the duration of such activity is so short. After giving birth in the maternal nest, the postpartum doe basically nurses her young for about only 5 minutes each day, and for the remainder of the time she is absent from the nest area (González-Mariscal et al., 1994; Lincoln, 1974; Zarrow, Denenberg, & Anderson, 1965). In addition, the postpartum doe will not retrieve scattered young (Ross, Denenberg, Frommer, & Sawin, 1959). Therefore, pup-directed maternal activities in rabbits basically consist of a short nursing episode that occurs each day of the 30-day postpartum period, and the question is whether hormones play a role in allowing such behavior to occur. In comparison to rodents, only a few studies are available that will allow us to address this question. One indication that the hormonal events that occur at the end of pregnancy allow the female rabbit to respond maternally to her young comes from a recent study (González-Mariscal et al., 1998). Two groups of rabbits gave birth and remained with their young for approximately 4–8 hours postpartum. Females in one group (early separation) were separated from their young for the next 7 days, and then tested for maternal responsiveness on days 8, 9, and 10 postpartum. Females in the second group (late separation) were allowed to remain with their young through day 10 postpartum, were then separated from their young over days 11–17, and were subsequently tested for maternal behavior over days 18–20. (Note that pups in each group were allowed to suckle from their anesthetized [unconscious] mothers on each day during the "separation" interval.) The maternal behavior tests indicated that females in the late separation group were much more likely to enter the nest box and nurse their young than were females in the early separation group. When the latter females entered the nest box, they would sniff the young, show a startle response, and leave the nest box without caring for their young. One interpretation of these findings is that the hormonal events of late pregnancy prime the female doe to act maternally toward young at parturition, but that this hormonal facilitation effect is time-limited and if the female does not interact with her young for a sufficient amount of time after parturition, then her maternal motivation will wane so that at a later time she will no longer be maternally responsive (also see Findlay & Roth, 1970). In contrast, if the female interacts with her young for a sufficient amount of time under the influence of hormone priming, then her maternal behavior is maintained at some later point, even after a period of separation from pups (see Chapter 3). It would certainly be interesting to know whether the hormonal treatments that have been found to stimulate maternal nest building in ovariectomized female rabbits (González-Mariscal et al., 1996) are also capable of stimulating pup-directed maternal activity in such females, but these experiments have not been done. There is other evidence, however, that indicates that prolactin is one of the hormones involved in regulating the onset of pup-directed maternal responsiveness of rabbits during the early postpartum period (González-Mariscal, Melo, Parlow, Beyer, & Rosenblatt, 2000): The administration of daily injections of bromocriptine to rabbits over the last 5 days of pregnancy prevented the normal increase in plasma prolactin levels observed at this time and also disrupted the onset of nursing behavior. These results are significant because rabbits do not appear to produce placental lactogens (Grattan, 2001; Numan, 1994; Tucker, 1994; but see Grundker, Hrabe de Angelis, & Kirchner, 1993), and therefore bromocriptine administration presumably was able to disrupt the only endogenous source of lactogenic hormone.

3. Hormonal Basis of Maternal Behavior in Sheep

The context within which maternal behavior occurs in sheep (*Ovis aries*) differs from that in rats and rabbits in two important respects (Carter & Keverne, 2002; González-Mariscal & Poindron, 2002; Kendrick et al., 1997a; Levy, Porter, Kendrick, Keverne, & Romeyer, 1996; Numan, 1994). First, sheep and other ungulates give birth to precocial young that are mature and mobile at birth, which contrasts with the altricial young born to most rodents and rabbits. Second, sheep are grazing/herding mammals with a synchronized breeding season, which results in the temporally restricted birth of numerous mobile and genetically unrelated lambs within the herd. Since it would not be advantageous, in terms of reproductive success, for a ewe to care for a lamb that is not her own, and since the particular context within which sheep maternal behavior occurs increases the likelihood that lamb confusions might occur, mechanisms have evolved in sheep that result in a heightened maternal responsiveness near the time of parturition, followed by the formation of a highly selective maternal bond between a mother and the particular offspring she interacted with at birth. The focus of the present chapter is on the hormonal basis of the heightened maternal responsiveness that occurs in sheep around the time of parturition, and subsequent chapters (Chapters 5 and 8) will deal with the mechanisms that underlie the formation of the selective maternal attachment of a ewe to its lamb. However, we mention the selectivity of maternal behavior in sheep here because it is probably related to the fact that for sheep, unlike rodents and rabbits, there is no evidence that sensitization processes occur (Levy et al., 1996). In other words, the evolution of maternal behavior neural mechanisms in sheep appears to have prevented the occurrence of sensitization processes, and this is probably related to ensuring that a female will only care for her own young even though she is surrounded by many other lambs for long periods of time.

Maternal behavior in postpartum sheep can be categorized as containing high levels of lamb acceptance behaviors and low levels of lamb rejection behaviors (Kendrick et al., 1997a). In contrast, estrous-cycling ewes are not maternally responsive, and they show the reverse behavioral tendencies when exposed to lambs (Levy et al., 1996; Numan, 1994). Postpartum maternal ewes also show high rejection and low acceptance to lambs that are not their own, as would be required once the development of maternal selectivity has occurred (Levy et al., 1996). Acceptance behavior includes the occurrence of low-pitched bleats, approaching the lamb, sniffing and licking the lamb, and allowing the lamb access to the udders. Rejection behavior includes avoidance of the lamb and aggression toward an approaching lamb (head butts), along with the occurrence of high-pitched bleats.

In sheep, pregnancy lasts about 150 days, and as parturition approaches plasma levels of progesterone decline while estradiol levels rise (see Numan, 1994, for a review). As with rat studies, some sheep studies have emphasized that progesterone withdrawal followed by an estradiol rise might be a hormonal trigger for maternal behavior in sheep (Dwyer, Dingwall, & Lawrence, 1999; Numan, 1994). However, when such a hormonal regimen is administered to nonpregnant/nonlactating multiparous ewes, only about 50% of such females show some maternal behavior toward a lamb, and supraphysiological levels of steroids are needed to produce this partial effect, which has suggested that under natural conditions other factors must be operative to produce full maternal behavior in parturient sheep (Kendrick et al., 1997a; Levy et al., 1996). Additional research has shown that steroid priming (sequential treatment with progesterone and estradiol) followed by vaginocervical stimulation (produced with a probe) results in a much higher level of maternal responsiveness toward a lamb in ovariectomized multiparous ewes (Kendrick & Keverne, 1991; Keverne, Levy, Poindron, & Lindsay, 1983). Vaginocervical stimulation in the absence of steroid priming has no such stimulatory effect, suggesting that the steroid treatment results in a central neural modification that allows vaginocervical stimulation to be effective. The importance of naturally occurring vaginocervical stimulation for maternal behavior in ewes that have been primed with pregnancy hormones is further supported by the finding that peridural spinal anesthesia near the time of parturition, which would block the transmission of labor-induced vaginocervical sensory stimulation to the brain, disrupts the onset of maternal behavior in parturient ewes (Krehbiel, Poindron, Levy, & Prudhomme, 1987). Overall, these results suggest that increases in the estradiol-to-progesterone ratio at the end of pregnancy prime the brain to be responsive to vaginocervical stimulation, which then produces effects that trigger optimal maternal responsiveness.

Although vaginocervical stimulation may play a minor role in the maternal behavior of rodents (Yeo & Keverne, 1986), it appears to play a preeminent role in that of sheep. Just what effects such stimulation is having on the brain of sheep will be a subject for future chapters. However, the fact that vaginocervical stimulation appears to play such an important role in sheep may be related to the strict requirement of synchronizing maternal responsiveness in sheep with the birth of the young, which would allow a ewe to form a selective maternal bond with her own young rather than with an unrelated lamb in the herd. Another difference between the physiological regulation of maternal behavior in rats and sheep may relate to the involvement of lactogenic hormones. Although estradiol causes the release of prolactin from the anterior pituitary in sheep, as it does in the rat, preliminary evidence suggests that such release is not important for maternal behavior in sheep (Poindron, Orgeur, LeNeindre, Kann, & Raksanyi, 1980). We need to be cautious about interpreting these data, however, since the results were obtained on parturient ewes who were treated with bromocriptine during late pregnancy. Although bromocriptine disrupts pituitary prolactin release, it would not interfere with the effects of placental lactogens (sheep produce placental lactogens during pregnancy: Grattan, 2001; Tucker, 1994). Also, Bridges and Ronsheim (1990) have shown that bromocriptine is effective in disrupting steroid-induced maternal behavior in rats only under certain specific conditions. In particular, bromocriptine is only disruptive if it is administered concurrently with estradiol, so that all estradiol-induced prolactin release is blocked. It is unlikely that this was the case in the study by Poindron et al. (1980).

4. Hormonal and Nonhormonal Basis of Maternal Behavior in Primates

As previously mentioned, traditional views contrast the regulation of maternal behavior in primates with that in nonprimate mammals by emphasizing that maternal behavior in adult primates is primarily influenced by social experiences acquired during development and is relatively free from endocrine control in adulthood. However, recent evidence, although still preliminary, is suggesting a modification in this perspective (Fleming, Ruble, Krieger, & Wong, 1997a; Maestripieri, 1999; Pryce, 1996). This new evidence suggests that the hormonal events accompanying pregnancy termination are one of probably many factors (including early social experience) that influence the intensity of maternal responsiveness in primates, including humans.

The information that fostered the traditional view that primate maternal behavior is emancipated from endocrine control included the fact that adoption occurs frequently in human populations and that alloparental behavior occurs to varying degrees in diverse primate societies. Because of such observations, one might argue that selective forces acting on primates may have resulted in the evolution of a nervous system that allows infant stimuli direct access to the neural circuitry underlying parental responsiveness without the necessity of hormonal intervention. Indeed, we have already seen that in laboratory mice, inbreeding and selective breeding (which is experimental rather than natural selection) can create strains whose basic (home-cage) maternal responsiveness has been largely emancipated from control by the endocrine events associated with pregnancy termination. The question remains as to whether the forces of natural selection could have resulted in the complete emancipation of primate maternal behavior from hormonal control. Pryce (1996) has been a strong advocate of the view that such a general evolutionary development in primates is unlikely since it is hard to conceive how such a phenotype could be adaptive. Most primates live in social groups that contain many females. In spite of this, most primate mothers develop a specific attachment to their particular infant (for reviews, see Numan, 1994; Pryce, 1996). It is difficult to imagine the normal and adaptive functioning of a primate group in which all females, irrespective of their physiological condition, were equally attracted to infants, including those to whom they were not genetically related. As emphasized by Pryce (1996), it appears more likely that the hormonal events associated with pregnancy termination should help synchronize the onset of a high level of maternal responsiveness with parturition so that such responsiveness would be directed toward a female's own young.

Concerning adoption in human societies, although such behavior suggests that maternal motivation may be relatively high even in nulliparous women, this does not rule out a role for hormonal factors and processes akin to sensitization in human maternal responsiveness. Interestingly, Hrdy (1999) describes some historical evidence that is relevant to these issues. In Europe in the 1800s, the level of abandonment of infants by their natural mothers was high in destitute populations. In a particular hospital in Paris, a subset of such indigent women were asked to remain with their infants for 8 days after birth prior to making a decision on whether they would give up their young. This requirement substantially reduced the incidence of abandonment. Since these females gave birth naturally and also breastfed their infants, it is conceivable that infant stimulation coupled with hormonal events played a role in reversing the initial intent of these mothers to abandon their young.

With respect to the occurrence of alloparental behavior in nonhuman primates, these data have been reviewed by several investigators (Hrdy, 1977; Maestripieri, 1994a; Mitani & Watts, 1997; Numan, 1994; Pryce, 1996; Silk, 1999; Snowdon, 1996). These studies show that nonlactating females of several primate species, observed either under natural free-ranging conditions or in captive social groups, may show maternal responses toward the infants of other females. The type of responsiveness observed can vary from simply touching the infant to actual infant carrying, and some have preferred to refer to such behavior as infant handling rather than as alloparental behavior (Maestripieri, 1994a). Maestripieri (1994a) has listed two important general characteristics of infant handling in primates: (a) female primates of all ages interact more frequently with infants that are not their own than do their male counterparts and (b) juvenile and subadult females are typically the most frequent infant handlers. Another important point is that the frequency and duration of infant handling varies widely across species. Within Old World monkeys, for example, relatively high levels of allomaternal behavior are observed in langur monkeys (*Presbytis entellus*) (Hrdy, 1977; Jay, 1963) and vervet monkeys (*Ceropithecus aethiops*) (Fairbanks, 1990; Lancaster, 1971), while relatively low levels of such behavior occur in baboon (*Papio*) and macaque (*Macaca*) species (Breuggeman, 1973; Maestripieri, 1994a).

A special case of alloparental behavior occurs in the New World marmoset (*Callithrix*) and tamarin (*Saguinus*) species (Pryce, 1993; Snowdon, 1996). In these species the group size can vary between 2 and 10 (mother, father, infants, juveniles, and subadults), and there is typically only one adult breeding female per group, who usually gives birth to twins. Importantly, all group members of both sexes show substantial amounts of infant carrying, and the intensity and duration of such behavior clearly warrants the descriptive term of alloparental behavior rather than simply infant handling.

The occurrence of alloparental behavior and infant handling in primates clearly indicates that there is interest in and attraction toward infants by non-pregnant/nonlactating females under natural or seminatural conditions. This finding, of course, does not mean that the intensity of maternal responsiveness is equivalent between natural mothers and alloparents. Also, as in rodents, ontogenetic mechanisms may be operative in primates, as juvenile and subadult female primates usually show the highest levels of alloparental behavior/infant handling. Perhaps, as in rats, maternal interest in primates declines once the typical female primate becomes reproductively mature and such motivation is then subsequently boosted by the endocrine events associated with pregnancy termination when the adult female gives birth to her own young. It also needs to be emphasized that most primates grow up in social groups and are constantly surrounded by and observing natural mothers caring for their offspring. It is highly probable that observational learning and processes similar to sensitization are acting on nonpregnant/nonlactating females to affect their maternal responsiveness. Such processes do not rule out the possibility that the endocrine factors associated with pregnancy termination further enhance maternal responsiveness in the female primate. Therefore, we will now turn our attention to those studies that specifically support the view that endocrine factors contribute to primate maternal behavior.

Laboratory-type studies of nonhuman primates have presented evidence that interest and attraction toward infants increases in the latter part of a female primate's pregnancy, suggesting that the endocrine events associated with pregnancy termination act to enhance maternal responsiveness. Rosenblum (1972) compared the responses of adult nonpregnant and late pregnant female squirrel monkeys (*Saimiri sciureus*) to an infant introduced into their cages. Squirrel monkeys carry young by lowering their dorsal regions, in this way allowing the young to climb on their backs and cling. Significantly, 80% of the late pregnant females showed such a response while less than 20% of the nonpregnant females did so. Somewhat similar results have been obtained by Maestripieri and Zehr

(1998), who studied a captive social group of pigtail macaque monkeys (*Macaca nemestrina*) under seminatural conditions. The responsiveness of multiparous females in various stages of pregnancy to the infants of other females was recorded, and the frequency of various types of infant handling was noted. Infant-handling responses included touching, grooming, and carrying, and the most common response was brief touching. An important aspect of this study was that blood samples were drawn at weekly intervals from the pregnant females and the serum was assayed for estradiol and progesterone. The 24-week pregnancy was divided into three 8-week periods, and as shown in Figure 2.4A, the rate of infant handling increased significantly during the last 8-week period. Importantly, the mean rate of infant handling was positively correlated with mean serum estradiol levels and with the estradiol-to-progesterone ratio. Figure 2.4B shows the plasma estradiol and progesterone levels over the course of pregnancy in the pigtail macaque. It is worth noting that in contrast to most nonprimate mammals, there is not a major prepartum drop in serum progesterone levels; the late pregnancy rise in the estradiol-to-progesterone ratio is primarily due to rising estradiol levels. This hormone profile not only occurs in pigtail macaques, but also in many other primate species, including humans (Bahr, Martin, & Pryce, 2001).

Pryce, Dobeli, and Martin (1993) have examined the maternal responsiveness of primigravid common marmoset monkeys (*Callithrix jacchus*) over the course of pregnancy (these females had alloparental experience in their family groups during their prior development) while also measuring their serum levels of estradiol and progesterone. Maternal responsiveness was measured with an operant conditioning procedure, and the females learned to press a bar in order to trigger the *onset* of a 15-second visual presentation of a model of an infant marmoset coupled with the *termination* of an auditory stimulus that presented infant distress vocalizations. The rationale here was that as maternal responsiveness increases, females should be more likely to press the bar to see a replica of a baby marmoset while also terminating distress calls. Indeed, this view was supported by the fact that postpartum females pressed the bar at a very high rate. The results of this experiment are shown in Figure 2.5. Bar-pressing increased dramatically over the last 25 days of pregnancy, and this increase coincided with increases in serum estradiol levels.

Pryce, Abbott, Hodges, and Martin (1988) have presented additional evidence that late-term estradiol levels may be important for the development of maternal responsiveness in another New World monkey, the red-bellied tamarin (*Saguinas labiatus*). In a captive social group, *postpartum* maternal behavior directed toward one's own infants was rated as either good or poor, based on infant survival over the first postpartum week (poor mothers tended to reject their infants and would not allow them to suckle). Urinary estradiol levels were also determined *prepartum* during the last 5 weeks of pregnancy. This study found that estradiol levels remained stable at the end of pregnancy for the good mothers, but declined for the poor mothers, and by 1 week prepartum urinary estradiol was significantly higher in the good mothers. (See Pryce, Mutschler, Dobeli, Nievergelt,

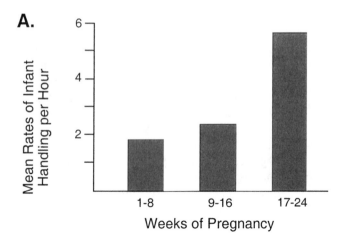

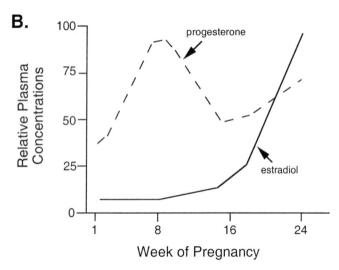

Figure 2.4. **A**. Mean rates of infant handling (touching, grooming, or carrying an infant) directed toward the offspring of other females in the social group over the course of pregnancy in pigtail macaques. Infant handling was significantly greater during the last 8 weeks of pregnancy when compared to the first 8 weeks. **B**. Relative blood plasma levels of estradiol and progesterone over the course of pregnancy in pigtail macaques. For each hormone, data are depicted as the percentage of the maximum value detected during pregnancy. Panels **A** and **B** are modified from Maestripieri and Zehr (1998).

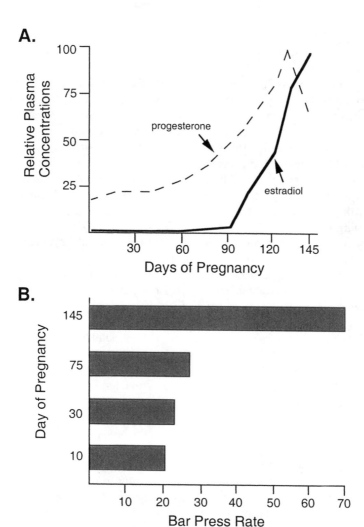

Figure 2.5. **A**. Relative blood plasma levels of estradiol and progesterone over the course of pregnancy in common marmosets. For each hormone, data are depicted as the percentage of the maximum value detected during pregnancy. **B**. Mean rate of bar presses over the course of pregnancy when marmoset infant stimuli are used as the reinforcing event. The highest bar-pressing rate, which corresponded to a high estradiol-to-progesterone ratio, occurred over the last days of pregnancy. Panels **A** and **B** are modified from Pryce et al. (1993).

and Martin [1995] for similar work on primiparous marmoset mothers, but also see Fite and French [2000] for conflicting findings on another marmoset species.)

Early studies by Holman and Goy (1980; also see Holman & Goy, 1995) suggested that hormonal factors may also influence maternal behavior in first-time primiparous rhesus macaque monkeys (*Macaca mulatta*). Infant monkeys were singly presented to individual nonpregnant/nonlactating adult nulliparous or multiparous females, and it was found that the nulliparous females did not interact with the infant and appeared indifferent, while the multiparous females retrieved the infant and cradled it in a ventroventral position. One interpretation of these results is that without parturient hormones or previous maternal experience, the nulliparous female will not show infant-handling responses under this testing situation.

Only two studies have directly manipulated hormones and examined the effects of such manipulation on maternal responsiveness. In a captive *social group* of rhesus macaques, Maestripieri and Zehr (1998) examined the infant-handling responses of adult intact nonpregnant/nonlactating multiparous females, ovariectomized nulliparous females, and ovariectomized nulliparous females treated with subcutaneous Silastic implants of estradiol toward the infants of other females. The ovariectomized females treated with estradiol had the highest serum estradiol levels and also showed the highest number of infant-handling episodes. Serum progesterone levels were low and did not differ between the three groups. These results suggest that when tested under seminatural conditions in the presence of infants and their natural mothers, ovariectomized nulliparous females treated with estradiol are more motivated to interact with infants than are either their ovariectomized counterparts or intact multiparous females with previous breeding experience. One might not have predicted this result from the previous findings of Holman and Goy (1980), but it needs to be emphasized that plasma estradiol levels were highest in the nulliparous estrogen-treated females and that it may take a high level of maternal responsiveness for rhesus macaque females to interact with infants in the presence of the infant's natural mother. Another point worth mentioning is that the actual level of infant handling displayed by the ovariectomized, estrogen-treated females was actually quite low, with a mean number of only about 2 infant-handling episodes per hour. This should be contrasted to the mean number of about 6–9 such episodes observed in pigtail macaques at the end of pregnancy (see Figure 2.4).

Finally, Pryce et al. (1993) have reported that if common marmoset adult nulliparous females with previous alloparental experience are treated with steroids that simulate the endocrine profile occurring over the last 20 days of pregnancy (an increase in the estradiol-to-progesterone ratio resulting from declining progesterone and rising estradiol), they will bar-press for infant-related stimuli at a significantly higher rate than untreated nulliparae.

These studies, taken as a whole, suggest that for most of the nonhuman primate species that have been examined, the increases in estrogen and in the estrogen-to-progesterone ratio that occur toward the end of pregnancy are two factors that stimulate maternal responsiveness and interest in infants (for possible exceptions see Bahr et al., 2001; Fite & French, 2000). The New World marmosets and tamarins show an endocrine profile at the end of pregnancy similar to that of the rat, with declining progesterone and rising estradiol (Pryce, 1993, 1996). Although macaque species do not show a major prepartum decline in progesterone, estradiol levels do rise prepartum, in this way increasing the estradiol-to-progesterone ratio (Maestripieri & Zehr, 1998; see Figure 2.4). As far as we are aware, no one has yet directly explored the role of lactogenic hormones in the maternal behavior of parturient primates (but see the next section in reference to alloparental behavior), although high levels of placental lactogens and prolactin are present both prepartum and postpartum (Numan, 1994).

Fleming and her colleagues have done some of the most important work on the hormonal correlates of maternal behavior in humans (*Homo sapiens*) (Corter & Fleming, 1990; Fleming et al., 1997a; Fleming, Steiner, & Corter, 1997b). The finding we want to emphasize here is the relation observed by Fleming, Ruble, Krieger, and Wong (1997a) between prepartum steroid levels and postpartum levels of maternal attachment. All females in this study were primiparous at the time of birth. Blood samples were taken throughout the course of pregnancy and a questionnaire was administered to each subject on day 1 postpartum in order to measure the degree of attachment of the mother to her newborn infant. Importantly, females showing high attachment displayed increases in their plasma estradiol-to-progesterone ratio over months 5–9 of pregnancy, while low-attachment mothers showed decreases in this ratio over the same time period. It is interesting to note that the actual estradiol-to-progesterone ratio at parturition did not distinguish the two groups. Instead, it was the pattern of change over pregnancy that was different, with declining ratios occurring for the low-attachment females, and rising ratios occurring for the high-attachment mothers. These results are shown in Figure 2.6.

5. General Conclusions

Although there are clear species differences, a significant aspect of parental behavior in mammals is that it is not rigidly tied to the specific physiological events associated with late pregnancy and lactation. The occurrence of alloparental behavior, sensitization processes, and paternal behavior can all be used as evidence favoring this view. In the next chapter we will explore the data that show that in a variety of species, experiential factors, in particular previous maternal experience, can also play a role in decreasing the involvement of the specific hormonal events associated with pregnancy and lactation in maternal behavior control. The physiological events of late pregnancy and parturition are best viewed as increasing the intensity of maternal behavior, increasing the commitment of the mother to her offspring so that she will take risks to protect them, and decreasing the latency needed before a female begins to show parental

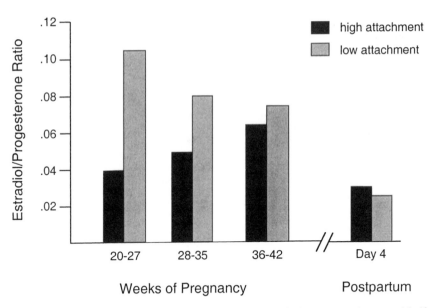

Figure 2.6. Mean estradiol-to-progesterone ratios in blood plasma over the second half of pregnancy in women who subsequently expressed either high or low levels of attachment to their infants during the first few postpartum days. Modified from Fleming et al. (1997a).

responses to infants. A large portion of this book will be devoted to exploring just where hormones and other physiological factors act within the brain to produce these effects on maternal behavior, with an emphasis on uncovering the multiple functional systems that are modified so that maximal maternal responsiveness occurs at parturition.

What appears to vary between species is the baseline level of parental responsiveness that occurs in the absence of being exposed to the physiological events associated with pregnancy and lactation. In some species, like sheep, such responsiveness is low, while in other species, such as laboratory mice, some nonhuman primates, and humans, it is relatively high. Further exploration of the neural basis of this high level of parental responsiveness in naive nulliparous laboratory mice, with the aid of transgenic mouse models, should be very instructive and possibly relevant to other species, including humans. However, even in laboratory mice and in primates, the hormonal events of late pregnancy can be seen as further enhancing maternal behavior by increasing maternal motivation and by increasing the risks that a female will expose herself to in order to gain access to young.

The occurrence of alloparental behavior in several mammalian species clearly shows that parental behavior can occur quickly following infant exposure outside the physiological boundaries of pregnancy and lactation. However, such a temporal occurrence of parental responding does not rule out the involvement of some hormonal mediation. Indeed, evidence exists that shows that pituitary prolactin may be involved in the short-latency parental behavior shown by juvenile rats (Kinsley & Bridges, 1988a), and in the alloparental behavior shown by parentally inexperienced marmosets, since bromocriptine administration disrupts these behaviors (Roberts, Jenkins, Lawler, Wegner, & Newman, 2001a; Roberts et al., 2001b). This is a good place to indicate that prolactin not only plays an important role in regulating maternal behavior in mammals, but it also plays a central role in regulating parental responsiveness in birds (Ball & Balthazart, 2002; Buntin, 1996). Additionally, prolactin appears to be involved in the alloparental or helping behavior that is observed in several species of birds (Khan, McNabb, Walters, & Sharp, 2001; Schoech, 1998). Although prolactin may be involved in certain instances of heightened alloparental responsiveness, it is not involved in the short-latency maternal behavior shown by virgin laboratory female mice toward pups in home-cage tests, since neither hypophysectomy nor a null mutation of the prolactin gene disrupts such behavior (Horseman et al., 1997; Leblond & Nelson, 1937). However, prolactin receptors exist within the brain (Grattan, 2001), and as we will show in Chapter 6, such receptors appear critical for the spontaneous maternal behavior shown by laboratory mice, suggesting that there may be an endogenous central nervous system ligand, in addition to prolactin, that is capable of binding to central prolactin receptors to stimulate maternal behavior in such mice. These sorts of analyses are important because they can suggest how evolutionary forces might act on a basic hormonally regulated maternal response system to alter it so that parental responsiveness might occur outside the postpartum period.

For nonprimate mammals, with the strongest evidence coming from rats, we have emphasized the involvement of prepartum progesterone withdrawal as being one component in the hormonal complex that promotes the onset of maternal behavior. For many primates, including humans, however, this exact mechanism could not be operative because progesterone remains relatively high until after parturition (for a review, see Bahr et al., 2001; also see Figure 2.4). Therefore, although the estradiol-to-progesterone ratio may increase as parturition approaches, primarily as a result of increases in estradiol, for most primates a progesterone withdrawal effect, which in the rat has been proposed to prime neural circuits for the actions of estradiol and lactogens, is not likely to occur.

3

Experiential Factors Influencing Maternal Behavior

1. Introduction

In the previous chapter we investigated the common assumption that primate maternal care is not dependent on hormonal regulation, being primarily determined by experience. We presented evidence arguing against this assumption, evidence that suggests a role for hormones in specific aspects of maternal care in primates. A converse assumption is that maternal care in nonprimate species is determined completely by hormones or genes and largely independent of experience. In this chapter, we will examine this assumption by reviewing the evidence that experiential factors influence both rodent and primate maternal behavior.

Experiential factors that have been found to influence maternal behavior can be broken down into four broad categories: (a) the effects of an initial adult maternal experience on future episodes of maternal behavior; (b) the effects of juvenile parental experience on subsequent adult maternal behavior; (c) the effects of maternal treatment of her female offspring on the subsequent maternal behavior of those offspring; and (d) the effects of experience with one's offspring on the development of maternal selectivity, resulting in a mother who will only care for her own young and will reject alien young. This latter effect, which only occurs in certain species (see Chapter 1), has been investigated primarily in sheep. In the present chapter, we will discuss the first three experiential influences on maternal behavior while deferring a discussion of the formation of selective maternal attachments until Chapter 5, where we will analyze the development of maternal selectivity in sheep in the context of the sensory mechanisms involved.

2. The Effects of Previous Adult Maternal Experience on Future Maternal Behavior

The first issue that concerns us is whether the experience of rearing an infant or litter during an initial reproductive cycle has an impact on future maternal behavior. One aspect of this issue relates to the question of whether postpartum primiparous females (first birth cycle) differ in their maternal behavior from postpartum multiparous females (more than one birth cycle)? It makes sense to predict that an initial rearing experience should have an impact on future maternal competence and efficiency during subsequent breeding episodes because the experienced mother would be expected to have learned how to best handle her infant under a variety of stressful situations. In seeming support of this view, Hrdy (1999) notes that, under natural conditions, firstborn primate infants die at higher rates than do future offspring. Indeed, mortality rates of firstborn infants can be higher than 60% in some populations of monkeys and apes. It needs to be emphasized, however, that this kind of correlational field data does not allow us to definitely conclude that the maternal experience gained by first-time mothers makes them more proficient in their future maternal behavior. First-time mothers are generally younger than multiparous females, and therefore maternal age rather than maternal parity might be the important variable related to infant mortality. Perhaps the lower mortality rate of offspring of multiparous females is related to maturational changes in lactational proficiency, or perhaps older and larger females can more easily carry their young for long distances without tiring or becoming less vigilant. The point to note here is that careful analysis is needed before we can conclude that maternal experience per se has an impact on the adequacy of future maternal behavior.

Several investigators, under laboratory test conditions, have examined parity effects on the actual performance of maternal behavior in postpartum rodents, and the results are equivocal. Beach and Jaynes (1956a) compared the time it took for primiparous and multiparous female laboratory rats (*Rattus norvegicus*) to retrieve their scattered pups back to a nest site, and they found no differences in retrieval speed. Moltz and Robbins (1965) expanded on these findings by showing in laboratory tests that primiparous and multiparous female rats were not only similar in retrieving behavior, but they also did not differ from one another in nest-building behavior, nursing duration, and litter growth over the course of the postpartum period. Somewhat different results have been obtained by Carlier and Noirot (1965) for rats and Swanson and Campbell (1979b) for hamsters (*Mesocricetus auratus*). These investigators found that the retrieval behavior of multiparous females is better than that of primiparous females: Multiparous females retrieved more quickly and dropped fewer pups. The problem with each of these latter two studies is that the same group of subjects was compared over successive parturitions and therefore the multiparous females were older than the primiparous females. Perhaps the older females were also physically stronger and thus were more able to successfully carry pups back to the nest site. The results would have been more convincing if the primiparous and multiparous females were similar in all respects except for maternal experience. A more recent study by Wang and Novak (1994) also supports the view that multiparity is associated with improved maternal behavior, but the underlying mechanism of the parity effect again cannot be convincingly determined. When compared to primiparous prairie voles (*Microtus ochrogaster*), an independent group of multiparous voles (these females were studied after they gave birth to their *third* litter) spent more time caring for their young, and the young of the multiparous females developed more rapidly and had a higher survival rate. These results could be the result of one or more of the following: maternal experience, nonmaternal social experiences, maternal age and physical maturation, and changes in lactational proficiency.

Overall, for rodents the evidence suggests that if maternal experience acquired during an initial reproductive episode actually influences the performance, competence, and efficiency of maternal behavior during a subsequent reproductive cycle, then this experience effect is mild. It needs to be emphasized, however, that laboratory tests of maternal behavior are not terribly demanding, and are certainly not analogous to the stresses that the lactating female must deal with under natural conditions. We hypothesize that a more demanding, ecologically valid test situation would probably demonstrate clear differences in the performance of the maternal behavior of primiparous and multiparous females who are matched on all variables except for maternal experience.

A similar picture emerges from a detailed examination of the possible effects of parity on maternal performance in primates. These parity-related effects are of two general types, one of which varies on a dimension of maternal adequacy or competence and the other on a dimension of maternal style. With respect to the dimension of maternal adequacy, research on captive groups of marmosets (*Callithrix* species), tamarins (*Saguinus* species), and pigtail macaques (*Macaca nemestrina*) suggests a higher incidence of infant rejection, neglect, abandonment, and infant mortality in primiparous than in multiparous females during the early postpartum period (Epple, 1978; Johnson, Petto, & Sehgal, 1991; Maestripieri, Wallen, & Carroll, 1997; Snowdon, 1996; Tardif, Richter, & Carson, 1984). In these studies many factors either varied or may have varied with parity and therefore it is difficult to relate the parity effect on maternal adequacy to the effects of maternal experience per se. Again, the possible influences of maturational factors, lactational competence, and social factors need to be considered. The involvement of social factors in the maternal adequacy of marmosets and tamarins may be particularly important: In captive family groups, primiparous females do not receive alloparental aid from juveniles while multiparous females would receive such aid from the juveniles of the previous birth cycle (see Johnson et al., 1991).

Primate maternal behavior also varies on a dimension of maternal style. That is, although all individuals in a group of females may show adequate maternal behavior (their young survive and mature), they may exhibit differences in maternal style (see Fairbanks, 1996, for a review). One aspect of maternal style that has received much attention is that of protectiveness (see Section 4.2 of current chapter). Mothers that are high in protectiveness tend to remain in contact with their young and restrict or restrain movement of their young to a higher degree than do mothers that are low in protectiveness. Studies on captive social groups of macaque (*Macaca* species) and vervet monkeys (*Ceropithecus aethiops*) have found that protectiveness varies with parity: Multiparous females are less protective than are primiparous females (Fairbanks, 1996; Kemps, Timmermans, & Vossen, 1989). However, an interesting study performed on a captive social group of Japanese macaques (*Macaca fuscata*), which employed a multivariate statistical analysis, provided evidence that the most important factor related to the degree of maternal protectiveness was maternal age (Schino, D'Amato, & Troisi, 1995). When maternal experience or maternal parity is held constant, protectiveness decreases with maternal age. These results suggest that the previous reports of a decrease in protectiveness as one moves from primiparity to multiparity were possibly due in part to the fact that the multiparous females were older than the primiparous females. Schino et al. (1995) speculate that age may directly influence maternal protectiveness because maturational changes in physiological and psychological characteristics may make a female macaque less anxious and more competent in coping with stressful situations related to infant care. In addition, in a stable social group, older females may be less protective than younger ones because they would be more accurate in predicting the behavior of other group members. This latter possibility would be an example of general social experience, rather than maternal experience, exerting an influence on subsequent maternal behavior.

In summary, as in rodents, a specific effect of adult maternal experience on subsequent maternal competence and performance in primates, although likely, is difficult to prove with the available evidence. Correlational studies, by their very nature, make cause-effect relationships difficult to discern. We will have to await future experimental studies to accurately settle this important issue.

Although it is still not proven that adult maternal experience during an initial birth cycle can influence future maternal *performance* in terms of competence, efficiency, or stylistic criteria, there is strong and definitive evidence that adult maternal experience can influence subsequent maternal behavior in other ways. More specifically, research has indicated, for certain species, that adult maternal experience can render the onset of future maternal behavior less dependent upon hormonal stimulation. The first evidence along these lines came out of Moltz's laboratory (Moltz, Levin, & Leon, 1969; Moltz & Weiner, 1966). Moltz et al. (1969), for example, found that systemic administration of progesterone to late-term primigravid (first pregnancy) female rats disrupted the onset of maternal behavior. This finding is expected since such treatment would have prevented the near-term decline in progesterone levels that, as shown in the previous chapter, is critical for maternal behavior onset in rats. Importantly, however, similar progesterone treatment to multiparous females, who were age-matched to the primiparous females, had no such inhibitory effect on the onset of their maternal behavior, suggesting that progesterone withdrawal is no longer essential for the onset of maternal behavior in multiparous females.

Subsequent research by others has elaborated on this important finding. In a series of studies, Bridges (1975, 1977, 1978) allowed primiparous mother rats to care for their young for the first 1 or 2 days postpartum, and then removed the pups. A second group of primiparous females had each of their pups removed immediately upon its delivery. Twenty-five days later he exposed both types of females to foster test pups and measured their sensitization latencies to the onset of maternal behavior. The latencies of the females who were allowed to interact with their young during the postpartum period averaged 1–2 days, and these latencies were significantly shorter than those of the primiparous females who were not allowed postpartum maternal behavior experience. The latencies of this latter group, which averaged 5 days, did not differ from the sensitization latencies of naive nulliparous females. These findings indicate that if a primiparous female rat is allowed a minimal amount of postpartum maternal experience, her future maternal responsiveness to pup-related stimuli is facilitated at a point in time that is far removed from the specific endocrine events associated with pregnancy termination.

Fleming and her colleagues have continued this line of research and have clarified some of the issues (Orpen & Fleming, 1987). In an initial study, the aim was to determine the duration of time, from the termination of pregnancy to a point into the postpartum period of primiparous females, that the hormonal stimulation of maternal behavior lasts. To this end, primigravid female rats were caesarian-sectioned on day 22 of pregnancy and presented with test pups after various time intervals postsurgery. The latencies to onset of maternal behavior were compared among these groups, and these latencies were also compared to those of a control group of nulliparous nonpregnant females. Some of the results are shown in Figure 3.1A, where it can be seen that females presented with pups between 1 and 3 days post–caesarian section showed a facilitated onset to maternal behavior in comparison to females first exposed to pups at either 8 or 10 days postsurgery. The latencies to onset of maternal behavior in these latter two groups did not differ from the onset latencies of virgin female rats. These results indicate that the hormonal events associated with pregnancy termination result in a heightened maternal responsiveness for a few days postpartum, but that in the absence of pup exposure this hormonal facilitation subsequently declines so that by 8 days post–caesarian section the maternal responsiveness of the postpartum females is no different from that of virgins (also see Rosenblatt & Lehrman, 1963). In a subsequent experiment, Orpen and Fleming (1987) explored the effect of a maternal experience 24 hours post–caesarian section on future maternal responsiveness. Primigravid female rats were caesarian-sectioned on day 22 of pregnancy and were either left untreated 24 hours later, or were exposed to pups at this time and allowed to show maternal behavior for 15 minutes, 30 minutes, or 24 hours, after which the pups were removed. Following this, at 10 days post–caesarian section, all females were exposed to pups and their sensitization latencies to the onset of full maternal behavior were observed. The results are shown in Figure 3.1B, where it can be seen that if females had as little as 30 minutes of maternal experience on the day following the caesarian section, then their maternal responsiveness was facilitated 9 days later. In a final experiment it was shown that full interaction with pups on the day following caesarian section was necessary for the development of heightened maternal responsiveness: Females that received exteroceptive or distal stimuli from pups (they could see, hear, and smell the pups, but could not interact

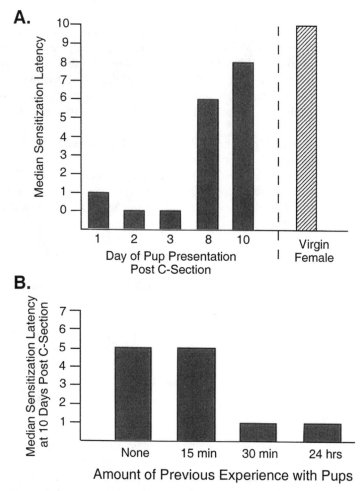

Figure 3.1. **A**. Median sensitization latencies to onset of maternal behavior in primigravid female rats who were caesarian-sectioned on day 22 of pregnancy and then presented with pups for the first time at either 1, 2, 3, 8, or 10 days post–caesarian section. For comparison, the latency to onset of maternal behavior in virgin females is also included. **B**. Median latencies to onset of maternal behavior at 10 days postsurgery for primigravid female rats who were caesarian-sectioned on day 22 of pregnancy and allowed either no exposure to pups or 15 minutes', 30 minutes', or 24 hours' exposure to pups at 24 hours postsurgery. Maternal behavior was shown toward pups at 24 hours post–caesarian section. Panels **A** and **B** are modified from Orpen and Fleming (1987).

with them nor show maternal behavior because the pups were enclosed in a perforated plastic box) for a full 24 hours beginning on the day following the caesarian section did not show shortened latencies to the onset of maternal behavior when tested on day 10 post–caesarian section. Similar results have been reported by Jakubowski and Terkel (1986), although their data indicate that a longer period of full interaction with pups after parturition (48 hours) is necessary for the facilitation of future maternal responsiveness in their particular strain of rat.

The combined results from the work of Moltz, Bridges, and Fleming suggest the following for rats: In first-time mothers the hormonal events associated with pregnancy termination are essential for the immediate onset of maternal behavior, but this hormonal stimulatory effect wanes over the first few postpartum days. However, if such females are exposed to pups and show maternal behavior for as little as 30 minutes during the early postpartum period, then the future maternal responsiveness of these females is facilitated so that a short-latency onset to maternal behavior occurs in the absence of the specific hormonal events associated with pregnancy termination. This effect has been referred to as the maternal experience effect, the long-term retention of maternal responsiveness, or maternal memory. It is as if maternal experience in primiparous rats partly emancipates future maternal behavior from control by the endocrine events associated with pregnancy termination. We use the term "partial emancipation" because the maternally experienced females rarely show immediate maternal behavior (0-day latencies) when tested with pups between 9 and 25 days after their initial maternal experience (see Figure 3.1B).

An issue that arises here is whether an initial maternal experience must occur under the influence of hormonal stimulation (e.g., at the time of pregnancy termination) in order for the development of a long-term retention of maternal responsiveness. In other words, if virgin female rats were induced to show maternal behavior through normal sensitization procedures (without hormonal priming) and then reinduced after a period of separation from pups, would their reinduction latencies be shorter than those of virgin females without prior sensitization experience? The jury is still out on this important question. Some studies have observed an effect of adult sensitization experience on future sensitization latencies (Bridges, 1977; Cohen & Bridges, 1981), while others have not (Bridges, 1975). The best conclusion to reach at this point is that the effects of an initial maternal experience on the level of future non-hormone-primed maternal responsiveness are greater if the initial experience occurs under conditions of hormone priming (Bridges, 1977; Fleming & Sarker, 1990).

The effect of prior maternal experience on subsequent maternal behavior in rats is probably related to another well-known fact. There is an important dichotomy in the regulation of maternal behavior in rats in that the immediate onset of the behavior at parturition in *primiparous* females is hormone-dependent, but its maintenance or continuance during the postpartum period is not (for reviews, see Numan, 1994; Rosenblatt et al., 1985). More specifically, a variety of endocrinological manipulations, when applied to the postpartum dam, do not disrupt maternal behavior once it has become established. These manipulations include hypophysectomy, ovariectomy, adrenalectomy, treatment with prolactin release–inhibiting drugs that block lactation, and the administration of high doses of progesterone. The general model that has emerged (Rosenblatt et al., 1985) is that the hormonal events of late pregnancy alter the brain so that at parturition pup-related stimuli elicit immediate maternal responsiveness. This hormonal modulation of reactivity to pup stimuli then wanes: There is a short time window after parturition within which the primiparous female must be exposed to pups in order to show a relatively immediate onset of maternal behavior (see Figure 3.1A). If pups are present within this time window then maternal behavior occurs, becomes established, and continues at high levels, and this continuance does not require any known hormonal mediation. One way of viewing established postpartum maternal behavior is that it is regulated by processes that are similar to those that regulate maternal behavior in sensitized virgin rats: Once the behavior is initiated, pup stimuli have direct access to the underlying neural mechanisms without the necessity of continued hormonal mediation. However, this sensitization view of the maintenance of maternal behavior would not necessitate that such established maternal behavior should also survive a long period of mother-young separation. An expanded view of maintenance mechanisms would be based on the experiential influences described above: Once a female interacts with pups while under the influence of the hormonal events associated with pregnancy termination, her brain undergoes relatively long-lasting modifications so that pup stimuli can more easily access the neural mechanism underlying maternal behavior independent of hormonal mediation, and, importantly, such a direct access process is maintained, along with the concomitant brain changes, even after a relatively long period of mother-pup separation, lasting up to 25 days. To put the distinction between these two views of the maintenance of postpartum maternal behavior more succinctly, the first view (sensitization view) would argue that the brain is temporarily changed by hormonal mechanisms (or by *prolonged* pup stimulation during sensitization), and then this change, which allows pup stimuli direct access to central maternal mechanisms, is maintained only for as long as the female is exposed to young pups. The second view (experience-based view) argues that the brain is temporarily changed by hormonal mechanisms, but more permanent changes occur if the female engages in maternal behavior during the period of hormonal stimulation, and these changes persist even if the female is separated from pups for a significant period of time. These two views are schematically presented in Figure 3.2.

The experience-based view of the continuance of maternal behavior after an initial maternal experience in the early postpartum period seems most appropriate for the data collected on rats (see Figure 3.2B). An analysis of other rodent species, however, clearly indicates that species differences exist. For example, wild female house mice (*Mus musculus*) (Soroker & Terkel, 1988),

A. Sensitization View

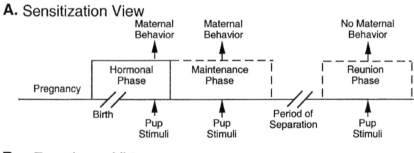

B. Experience View
(Long-Term Retention)

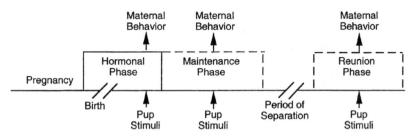

Figure 3.2. Two views of the mechanisms underlying the maintenance of maternal behavior in rats after its hormonally induced initiation. In the sensitization view (**A**), the hormonal events that occur at the end of pregnancy act on the brain so that pup stimuli have access to maternal neural circuitry and so that maternal behavior occurs. Once maternal behavior occurs, then hormonal facilitation of maternal behavior is no longer required, and maternal behavior is maintained for as long as the female remains with pups. However, if a female is subsequently separated from pups for an extended period, then the brain reverts to the naive state so that during a later reunion pup stimuli no longer have immediate access to the maternal neural circuitry. In the experience view (long-term retention view) (**B**), experience with pups and the display of maternal behavior during the hormonal phase permanently modify the brain so that even after a prolonged period of mother-infant separation, the brain does not revert back to a naive state, but instead pup stimuli retain the ability to have relatively immediate access to the neural circuitry regulating maternal behavior. Research on rats supports the long-term retention view of the maintenance of maternal behavior.

hamsters (Rowell, 1961; also see Swanson & Campbell, 1981), and gerbils (*Meriones unguiculatus*) (Elwood, 1977, 1981) do not appear to be affected by a previous maternal experience in the same way as is the rat: Nonlactating/nonpregnant females in these species, irrespective of their breeding histories, attack and eat young pups. That is, although the females in these species will care for alien pups while they are lactating and caring for their own pups, once they are separated from their pups for a prolonged period of time they will no longer care for pups, but instead will attack them. Therefore, perhaps a sensitization view of the continuance of maternal behavior is most appropriate for these species (see Figure 3.2A). With respect to rabbits (*Oryctolagus cuniculus*), more research needs to be done in order to define the mechanisms that regulate the continuance of maternal behavior after an initial postpartum experience (González-Mariscal et al., 1998, 2000).

Research on sheep (*Ovis aries*) further complicates the picture we are developing. First, estrous-cycling multiparous ewes, like nulliparous ewes, will not care for lambs (Le Neindre, Poindron, & Delouis, 1979). Therefore, there are no obvious long-term effects of previous breeding experience on future maternal responsiveness in nonlactating ewes. However, for all of the treatments that have been found effective in stimulating the onset of positive maternal responses in nonpregnant/nonlactating ewes (reviewed in Chapter 2), which include treatment with progesterone withdrawal followed by estradiol and vaginocervical stimulation, it should be noted that these treatments are only effective in multiparous ewes that have had previous maternal experience. These treatments do not work in nulliparous ewes (Kendrick & Keverne, 1991). What these results suggest is that although progesterone withdrawal followed by estradiol treatment and vaginocervical stimulation are involved in maternal behavior in sheep, additional factors are absolutely necessary in first-time mothers.

The little research that has been done on primates suggests that a primiparous maternal experience influences future maternal responsiveness in a manner similar to that in rats. Recall from Chapter 2 that when Holman and Goy (1980) presented an infant rhesus monkey (*Macaca mulatta*) to nonlactating multiparous females, these females immediately retrieved the infant and held it in a ventroventral position. Multiparous females behaved in this manner whether they were intact, ovariectomized, or postmenopausal. In contrast, under similar testing conditions, nulliparous females did not care for the infant, but instead were indifferent (also see Holman & Goy, 1995). One interpretation of these data is that in rhesus monkeys, previous adult maternal experience emancipates future maternal responsiveness from hormonal control. It is probably more accurate to argue, however, that previous maternal experience simply decreased the dependence of maternal behavior on the hormonal changes associated with pregnancy termination. Holman and Goy (1980) tested the maternal responsiveness of their adult females in cages where only one infant and one adult female were present. As noted in Chapter 2, the data obtained from this situation should be contrasted with the data obtained by Maestripieri and Zehr (1998), who studied infant handling in a captive *social group* of rhesus monkeys. These latter results indicated that when an infant's natural mother is present, only low levels of maternal responses are directed toward the infant by other females. These responses basically involve touching the infant, and are referred to as infant handling. Significantly, nulliparous females who were treated with high doses of estradiol showed greater amounts of infant handling than did intact nonlactating *multiparous* females. These results indicate that the social context is important, that the maternal responsiveness of multiparous nonpregnant/nonlactating rhesus females is not as high as the Holman and Goy (1980) study would lead one to predict, and that hormonal stimulation would likely increase the level of maternal responsiveness in these multiparous females. It may require a high level of maternal responsiveness to approach and interact with an infant in the presence of the infant's biological mother, and estradiol seems to play a role in increasing such responsiveness.

Clearly, in both rats and rhesus monkeys previous postpartum maternal experience decreases, but does not eliminate, the involvement of the endocrine changes associated with pregnancy in the onset of maternal responsiveness. We are arguing, therefore, that it is incorrect to view the maternal behavior of experienced rats and rhesus monkeys as being completely emancipated from endocrine control. Although the sensitization latencies of experienced rats are shorter than those of naive nulliparous females, such rats do not show immediate maternal responsiveness toward test pups—they need to be exposed to pups for 1, 2, or even 3 days before they begin to show maternal behavior. The comparison of the Holman and Goy (1980) study with that of Maestripieri and Zehr (1998) also suggests that although previous maternal experience increases maternal responsiveness to infant-related stimuli in rhesus monkeys, hormonal factors can still operate to further increase such responsiveness. In fact, it would not make evolutionary sense for previous maternal experience to completely liberate maternal behavior from endocrine control. We conceive of the endocrine events associated with pregnancy termination as synchronizing the onset of maternal behavior with parturition and lactation. Such a process would allow a physiologically prepared female to care for her own offspring. It would be maladaptive for a female to show *a high level of immediate maternal responsiveness* to any young she encounters irrespective of her physiological condition and her genetic relatedness to that young. Therefore, it is probably best to view the research on the effect of previous maternal experience on the subsequent involvement of hormones in the maternal behavior of rats and rhesus monkeys as informing us about the operation of the underlying neural mechanisms that control maternal behavior in these species, and we will explore this issue in future chapters.

Another issue should be considered at this point. Although previous maternal experience in adult rats and primates decreases the dependence of future maternal behavior on the specific hormonal changes associated with pregnancy termination, does this mean that hormonal factors play absolutely no role in the maternal behavior shown by these experienced nonpregnant/nonlactating females? In Chapter 2 we reviewed preliminary evidence that prolactin might be involved in the parental behavior of juvenile rats and in the alloparental behavior of inexperienced marmosets. Perhaps infant-induced prolactin release is also involved in the facilitated maternal behavior that is shown by adult rats and rhesus monkeys who have had previous maternal experience. More research needs to be done on this and similar issues (see Holman & Goy, 1980).

3. The Effects of Juvenile Parental Experience on Adult Maternal Behavior

A variation on the themes discussed above is related to the question of the degree to which juvenile parental experience influences adult maternal behavior. Recall that juveniles in several species tend to show a heightened parental responsiveness. In laboratory rats this has been shown through sensitization experiments, while in certain species of voles and primates, juvenile and subadult parental responses have been observed to occur in social groups under natural or captive conditions. The question is whether such juvenile parental experience has effects on future maternal behavior, causing the performance of adult maternal behavior to be more competent and efficient, and also partially emancipating adult maternal behavior from endocrine control. In the context of such concepts as critical or sensitive periods in behavioral development, one might predict that juvenile parental experience would have a larger impact on future maternal behavior than does adult maternal experience. The evidence that exists, however, does not allow us to reach this conclusion, but instead will only permit us to say that the occurrence of juvenile parental experience probably influences adult maternal behavior in a manner similar to the effect of an initial adult maternal experience on subsequent maternal behavior.

Both Gray and Chesley (1984) and Stern and Rogers (1988) have reported that when juvenile laboratory rats are sensitized to show maternal behavior and then are retested in adulthood after a prolonged period of separation from pups, the adult sensitization latencies of these females to the onset of full maternal behavior are shorter than those of adult females without prior juvenile parental experience. Similar effects appear to operate in prairie voles (Lonstein & De Vries, 2001). It is not clear, however, whether these results are any different from the effects of an initial adult sensitization experience on subsequent adult sensitization latencies.

It is worth noting that juvenile parental experience probably occurs naturally in wild rats. Rats usually mate during a postpartum estrus (Gilbert, 1984), and therefore it is highly likely that an older litter that has not yet been weaned will be present for a short time after the birth of a second litter. The heightened parental responsiveness of juvenile rats might simply be a mechanism that evolved via kin selection to ensure that an older litter would not attack its younger siblings. However, it is certainly also possible that such naturally occurring juvenile parental responsiveness is part of the rat's developmental program, and would have long-term positive consequences for the parental competence of primiparous postpartum females under demanding/stressful environmental conditions.

In certain vole species, and in marmoset and tamarin New World monkeys, high levels of juvenile alloparental behavior are observed as a normal part of behavioral development within the social groups of the relevant species (Lonstein & De Vries, 2001; Pryce, 1996; Snowdon, 1996; Wang & Insel, 1996).

In primates other than marmosets and tamarins, infant handling by juvenile females occurs in a variety of species, and one such species that has received much attention in this regard is the vervet monkey. With vervet monkeys, what is the impact of such juvenile parental behavior on the adequacy of maternal behavior in adult primiparous females? Lancaster (1971), while studying free-ranging vervet monkeys, was one of the first to suggest that juvenile maternal responses are a form of play mothering that provides the young primate with critical experiences necessary for the development of its ability to show adequate maternal behavior toward its own young. Subsequent evidence has indicated that juvenile parental experience in certain vole and primate species may influence adult maternal behavior, but such experience is probably not essential for the occurrence of maternal behavior in primiparous postpartum females of these species:

1. Primiparous female prairie voles that have had juvenile alloparental experience showed higher levels of maternal behavior toward their own litter (spent more time in the nest with their young) than did primiparous females that did not have such juvenile experience (Wang & Insel, 1996).
2. Pryce (1993, 1996) and Snowdon (1996) have reviewed the evidence for tamarins and marmosets, which suggests that juvenile alloparental experience increases the percentage of infants that are successfully reared by primiparous adult females. It should be noted, however, that these results were simply calculated from colony breeding records; actual behavioral observations were not taken. For example, a female was defined as having juvenile alloparental experience if she was present in a group at a time when younger siblings were born (see Tardif, Richter, & Carson, 1984). It should also be realized that such an approach does not take into account other factors that may have influenced the infant mortality of the primiparous females, such as maternal age.
3. Fairbanks (1990), while studying a captive social group of vervet monkeys, established the following correlation: Those females that spent more time carrying infants as juveniles were more likely to raise a surviving infant as a primiparous adult. The point to note here is that this is a correlation and therefore the nature of any cause-effect relationship is unknown. It is possible that the primiparous females that were more likely to successfully rear young were simply better mothers and would have been this successful even without the juvenile parental experience. Indeed, the possibly higher native maternal behavior of such females may have also accounted for their higher levels of attraction toward infants while they were juveniles.

Although these results suggest that juvenile parental experience can influence the adequacy of adult maternal behavior in primiparous postpartum females of a variety of species, the results are certainly not definitive in all cases. In addition, the following points should be noted: (a) The nature of the effects of juvenile maternal experience on subsequent adult maternal behavior may be no different from the effects of an initial period of adult maternal experience on subsequent maternal behavior, and (b) preadult parental experience is clearly not a general requirement for the development of adequate maternal behavior in postpartum mammals. Primiparous female rats typically show normal maternal behavior upon their first exposure to baby pups (Moltz & Robbins, 1965; Thoman & Arnold, 1968). Similar results have been observed for primiparous rhesus monkeys that were not exposed to infants during their juvenile development (Gibber, 1986). Finally, in several vole species, such as meadow voles (*Microtus pennsylvanicus*), adequate maternal behavior develops in the absence of juvenile alloparental experience (Wang & Novak, 1992).

Based on the above considerations, we would like to offer the following hypothesis. Preadult parental experience probably has the same effects as adult maternal experience on future maternal behavior, *but for certain species in which a naturally high level of juvenile alloparental behavior occurs (prairie voles, pine voles* [Microtus pinetorum], *marmosets, tamarins), the positive effects of such preadult maternal experience on future maternal behavior probably substitute for the corresponding positive effects of a primiparous maternal experience, and this substitution process may have adaptive significance.* This hypothesis is based on the still tentative conclusion that postpartum females who have had previous maternal experience may be more successful or competent in rearing young under stressful and demanding conditions than are females without such experience. We are suggesting that for certain species in which adult breeding is *delayed* because of ecological conditions, resulting in nonbreeding females who remain in their natal group for a long period of time prior to dispersal (see Powell & Fried, 1992), it may be adaptive for such females to acquire maternal experience through alloparental behavior so that when they do reproduce on their own after dispersal as *delayed* primiparous females, they will be more successful in rearing their young than if they lacked such experience. In other words, in terms of lifetime reproductive success, such delayed breeders cannot afford to use a primiparous maternal experience to make future breeding attempts more successful.

Given the available evidence, the last point we want to make in this section is that, just like the case for adult maternal experience, juvenile parental experience does not completely emancipate adult maternal behavior from hormonal influences. For example, rats with juvenile parental experience do not show immediate maternal responsiveness to pups in adulthood, although their sensitization latencies are shortened (Gray & Chesley, 1984; Stern & Rogers, 1988). Also, as indicated in the previous chapter, adult nulliparous marmoset females with previous alloparental experience who are treated with hormones that simulate the endocrine changes occurring at the end of pregnancy will perform an operant bar-pressing task to gain access to infant stimuli at a higher rate than will females with similar alloparental experience but without steroid treatment (also see Holman & Goy, 1995, for evidence that preadult maternal experience does not liberate the adult maternal behavior of rhesus monkeys from endocrine control).

4. The Effects of Maternal Treatment of Her Female Offspring on the Subsequent Maternal Behavior of Those Offspring

Another experiential factor that influences the adult maternal behavior of a female mammal is the nature of her relationship with her own mother during her early development. In the most extreme case, we can ask what the adult maternal behavior of a female mammal would be like if she were raised without being mothered by a female of her own species (sometimes referred to as maternal deprivation). This was the question that was examined in the classic studies of rhesus monkey maternal behavior emanating from Harry Harlow's laboratory in the 1960s and 1970s (Ruppenthal et al., 1976). Harlow's work on motherless mother monkeys gave rise to a rich line of additional primate research that has focussed not so much on the effects of the absence of mothering as on the impact of variations in the manner in which a mother treats her offspring on the subsequent maternal behavior of those offspring. After we review the primate research in these areas, we will turn our attention to recent research that has uncovered similar maternal treatment effects on the development of maternal behavior in rodents. The basic theme of all this work is that the manner in which a mother interacts with her offspring can influence the development of the offspring's emotionality or temperament, and this in turn may have an impact on the offspring's subsequent maternal behavior or maternal style toward its own young.

4.1. Motherless Mother Monkeys

"Motherless mothers" was the term used by Harlow's group to refer to laboratory-born adult rhesus monkey mothers who had been raised without their own mothers. Such females were separated from their natural mothers at birth and hand-reared (bottle-fed) by humans in a neonatal nursery until they could eat on their own. The subsequent developmental histories of these motherless monkeys were varied in the following ways (Ruppenthal et al., 1976; Suomi, 1978; Suomi & Ripp, 1983): *Social isolates* were completely socially deprived of interaction with conspecifics until at least 6 months of age, while *peer-reared* females were placed with similarly aged conspecifics shortly after birth but were not allowed to interact with mothers or other adult females. In adulthood, such motherless monkeys were either mated or artificially inseminated, and their subsequent maternal behavior toward their own offspring was studied and categorized as follows:

Adequate—the offspring of such females were able to survive without the need for intervention by the laboratory staff.

Indifferent—these females did not allow their young to nurse and staff intervention was needed to hand- or bottle-feed the infants, these females were disinterested in their young and failed to initiate physical contact with them, and their indifferent maternal behavior might also be categorized as maternal neglect.

Abusive—such females not only failed to nurse, but also physically abused their young, which included violent rejection of infant contact and biting, jumping, and stepping on the infant. Staff intervention was necessary in order to both feed and protect the infant.

A summary of the results of the two major types of rearing conditions on the maternal behavior of primiparous females is shown in Figure 3.3. The most important finding is that although 75% of peer-reared females showed adequate maternal behavior, only 25% of the social isolates did so. In this regard, it should be noted that the incidence of adequate maternal behavior toward firstborn young exceeds 95% for laboratory-born mother-reared rhesus females (Ruppenthal et al., 1976). It should also be noted that the socially isolated females not only showed a severe disruption of maternal behavior, but also showed disruptions in other aspects of social behavior, which included aggressive and sexual behaviors (Ruppenthal et al., 1976). The following conclusions can be reached from these findings: (a) Even in the presence of appropriate hormone priming, which is assumed (but not proved) on the basis of the occurrence of normal pregnancies and parturitions, early social deprivation severely disrupts maternal behavior in the primiparous rhesus monkey; (b) peer-peer interactions early in development can ameliorate the se-

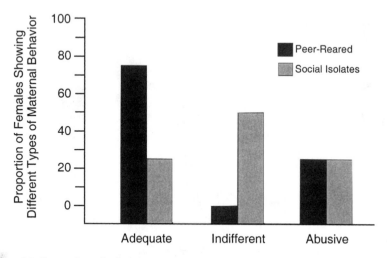

Figure 3.3. Proportion of primiparous rhesus monkeys that showed adequate, indifferent, or abusive behavior toward their own infants. All of these mothers were separated from their mothers at birth ("motherless mothers"), but some were raised in total social isolation (social isolates), while others were raised in social peer groups (peer-reared). Modified from Suomi (1978).

vere effects of maternal deprivation on the development of maternal responsiveness, and therefore severe disruptions in the development of adequate maternal behavior may be generally related to a lack of social stimulation, rather than specifically related to a lack of maternal care; (c) the disruption of maternal behavior caused by social deprivation is not specific to maternal behavior, but impacts other social behaviors as well, suggesting a disruption of the socialization process; and (d) although primiparous peer-reared monkeys show much better maternal behavior than their socially isolated counterparts, the maternal behavior of such peer-reared monkeys is still inferior to mother-reared monkeys. This finding suggests that there is a developmental effect of being reared by a normal mother that cannot be substituted for by peer-rearing.

Another interesting finding coming out of this work on motherless mother monkeys was that the maternal behavior of the social isolates improved over subsequent parturitions. Multiparous females showed a higher incidence of adequate maternal behavior than did primiparous females (Ruppenthal et al., 1976). These results are shown in Figure 3.4, where it can be seen that the incidence of adequate maternal behavior increased from 25% in primiparous females to about 65% in multiparous females. Interestingly, the level of abusive maternal behavior does not vary significantly with parity, remaining at about 25%; what decreases is the incidence of indifferent maternal behavior. Additional findings showed that if a primiparous socially isolated female remained with its young for at least 48 hours, its maternal behavior improved dramatically on subsequent

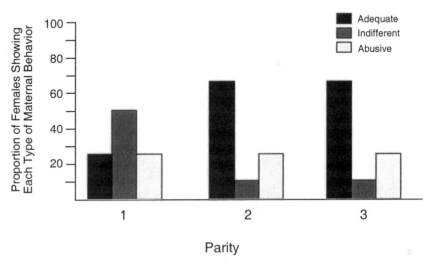

Figure 3.4. Proportion of rhesus monkeys that showed adequate, indifferent, or abusive behavior toward their own infants as a function of parity. All of these mothers were separated from their mothers at birth ("motherless mothers") and were raised in total social isolation (social isolates). Modified from Ruppenthal et al. (1976).

parturitions, while if the infant was removed earlier, maternal behavior did not improve.

Although the exact causes of the abnormal maternal behavior of socially isolated and peer-reared monkeys are not clear, Suomi (1997, 1999) has interpreted the results within the framework of attachment theory: Under normal conditions, the infant rhesus monkey becomes attached to its mother and then uses her as a secure base from which to explore its novel social and nonsocial environment. If the infant were to be frightened by something in its environment, it could return to its mother to decrease its anxiety. Activation of the sympathetic nervous system and the hypothalamic-pituitary-adrenal axis (HPA axis) have long been used as physiological indices of stress, fear, anxiety, and arousal (Sapolsky, 1992). Suomi (1999) cites evidence that shows that when an infant monkey initiates ventral contact with its mother, there are decreases in the infant's sympathetic activation as indicated by a lowered heartrate, and decreases in HPA activity as measured by decreases in serum cortisol levels, both of which are suggestive of a calming effect. It is assumed that this kind of mother-infant interaction allows the infant to gradually increase its orbit of exploratory activity and also helps the infant adapt to novel situations. Following this line of reasoning, it could be argued that socially isolated monkeys never have a secure base from which to explore their environment, while peer-reared monkeys have a less than optimal secure base (perhaps peers are less effective anxiety reducers than are mothers). Such a lack of a secure attachment base may result in the development of a more anxious/timid monkey, and this, in turn, may have an impact on the offspring's future maternal behavior. What is the evidence that socially deprived and peer-reared females are more anxious than mother-reared rhesus monkeys? Champoux, Coe, Schanberg, Kuhn, and Suomi (1989) measured the serum cortisol levels of socially isolated females and mother-reared females at one month of age and found that the basal serum cortisol levels were higher in the isolated rhesus females. In addition, in comparison to mother-reared monkeys, peer-reared monkeys are less inclined to explore novel objects and they are more hesitant in initial interactions with unfamiliar peers (Suomi, 1999). When peer-reared monkeys are grouped with mother-reared monkeys of similar age, the peer-reared monkeys fall to the bottom of a dominance hierarchy (Suomi, 1999). Furthermore, peer-reared monkeys exhibit a greater acoustic startle response when compared to their mother-reared counterparts (Parr, Winslow, & Davis, 2002). Finally, peer-reared monkeys show greater HPA activation when they are separated from their social group than do mother-reared monkeys (Higley, Suomi, & Linnoila, 1992; Suomi, 1999; for disparate results, see Clarke, 1993). What all this suggests, not surprisingly, is that a female's temperament and response to novelty may influence her maternal behavior. The socially deprived rhesus female may be indifferent, avoidant, and abusive to its *first* offspring because an infant that is trying to maintain contact with her and is also attempting to suckle may be an extremely stressful, fear-eliciting stimulus, even in the face of appropriate hormonal priming. If such a primiparous female is able to remain with its infant for at least 48 hours, then on a subsequent parturition perhaps the novelty of the second infant would be reduced, increasing the likelihood that an adequate mother-infant relationship would be established. Peer-reared females, although better suited for social interaction than socially isolated females, still appear to be more anxious/timid than mother-reared females, and perhaps this accounts for the relatively high rates of infant abuse observed in these females (see Figure 3.3). Future research should focus on more conclusively substantiating that this relationship between emotionality and maternal behavior is indeed a causal one.

Offering some support for the idea that the increased abusive behavior of peer-reared postpartum monkeys may be related to increased levels of maternal anxiety is the finding of Maestripieri (1998) who studied mother-infant interactions in captive social groups of rhesus monkeys. Although all the mothers that were sampled were themselves raised by mothers, Maestripieri noted that a small percentage of the sampled rhesus mothers showed abusive behavior toward their infants. This abuse was not so severe as to warrant any intervention, and the abusive responses primarily consisted of infant dragging, crushing, and throwing. Importantly, when abusive mothers were compared to a group of mothers that did not abuse their young, it was found that the abusive females spent more time in contact with their young and were also more likely to restrain their young. In other words, such females were more protective, and others have shown that high maternal protectiveness is associated with maternal anxiety (Maestripieri, 1993; also see Section 4.2 of this chapter). Finally, there is a preliminary report that treatment of an abusive Japanese macaque mother with diazepam, an antianxiety agent, decreased the abusive behavior shown by this female (Troisi & D'Amato, 1991, 1994).

4.2. The Influence of a Primate Mother's Maternal Style on the Subsequent Maternal Behavior of Her Offspring

In an excellent review, Fairbanks (1996) has analyzed the factors that influence spontaneously occurring individual differences in the maternal style of postpartum females within various ceropithecine primate species (e.g., macaques, baboons, vervet monkeys). Maternal style varies along two independent dimensions: protectiveness and rejection. A highly protective mother tends to remain in contact with her infant and restricts or restrains the movement of her infant. Infants of protective females are not allowed to explore their environments or wander from their mothers to the same degree as are the infants of less protective mothers. When averaged across all mothers, protectiveness, of course, is higher for mothers with younger infants than for those with older infants. "Rejection" refers to the mother's tendency to prevent infant contact, particularly with respect to suckling, and such behavior rises dramatically around the time of weaning, which usually occurs around 5–6 months of age. Mothers high in rejection tend to wean their young at earlier ages. Fairbanks (1996) notes that although the particular maternal style of a female monkey can be influenced by social and ecological variables, individual differences in maternal style across females remain relatively constant when females are examined under the same conditions. For example, as described previously, although the maternal protectiveness of a primiparous female may be higher than that of a multiparous female, if female A displayed higher protectiveness than female B when primiparous, she would also do so when both females were compared under similar multiparous conditions. As another example, females tend to show lower protectiveness when they are in a troop that contains long-term resident males (RM), but their protectiveness increases when new males (NM) are introduced into the troop, since such males are likely to harass the infant. However, as schematized in Figure 3.5, those females that showed the highest protectiveness under the RM condition would also do so under the NM condition.

Important for our present purposes, Fairbanks (1989) has shown for vervet monkeys that there is an intergenerational consistency in maternal style (also see Berman, 1990, for data on rhesus monkeys): The protectiveness of adult daughters toward their own offspring correlates strongly with the protectiveness level of the daughters' mothers. Although a genetic hypothesis would predict a similarity in maternal style between parents and their offspring, Fairbanks offers strong evidence, based on a statistical analysis using partial correlations, that the type of mother-infant contact received by a particular infant was the best predictor of that infant's maternal protectiveness as an adult. In other words, the maternal style of an individual female may be partly learned as a result of the experience of that female with its own mother. Fairbanks (1989) offers the following hypothesis to explain the mechanisms underlying such an experience-based intergenerational consistency in maternal style:

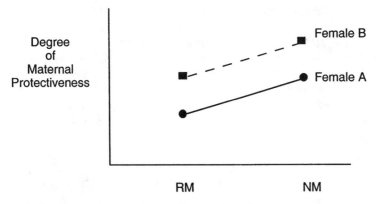

Figure 3.5. The degree of maternal protectiveness shown by two postpartum vervet monkeys toward their infants under two conditions: when a long-term familiar resident male (RM) was in the group, and when a new male (NM) entered the group. Although protectiveness increases under the NM condition, individual differences are still evident, with female A showing lower protectiveness under both conditions. Modified from Fairbanks (1996).

Highly protective mothers have lower thresholds than do less restrictive mothers for perceiving a particular environmental situation as threatening. In other words, they are more fearful or timid. Because they perceive greater threat in their infant's environment, they tend to maintain greater contact with the infant, restraining its movement. This behavior, in turn, may teach the infant to be more fearful of its environment, particularly in novel and unpredictable situations, and this increased fearfulness will then similarly influence the adult maternal behavior of the infant.

There is some evidence that supports Fairbanks's (1989) overall hypothesis. First, there is evidence that the level of maternal protectiveness correlates strongly with objective measures of maternal anxiety in rhesus monkeys (Maestripieri, 1993). Second, juvenile vervet monkeys that had highly protective mothers as infants are more timid and less likely to explore a novel environment than are juveniles that had less protective mothers (Fairbanks, 1996). Third, the experimental induction of higher levels of maternal protectiveness through the introduction of new males into a troop of vervet monkeys results in the infants of the more protective mothers being more fearful and cautious as juveniles (Fairbanks & McGuire, 1993). Finally, for rhesus monkeys, Suomi (1997) has reported the preliminary data of a cross-fostering study where he finds that the maternal style of cross-fostered females is more similar to the style of their foster mothers than to the style of their biological mothers.

4.3. Conclusions on Primates

The work on motherless mother monkeys and on experience-based intergenerational consistency in the maternal behavior of nonhuman primates all point in the same direction: The early experiences of the infant monkey with its mother influence the temperamental/emotional characteristics of the developing organism, and these characteristics, in turn, influence the adult social (and nonsocial) behavior of the individual, including its maternal behavior. We are not arguing that genes play no role in this developmental process. Clearly, primate temperament has a genetic basis, as shown by the work of Suomi (1991, 1997). However, given a genetic predisposition, it appears that early mother-infant interactions have a strong influence on the development of the infant's future emotional reactivity and social behavior. Future research is likely to uncover a variety of contexts within which maternal treatment effects on offspring development manifest themselves, while at the same time discovering the relationships between specific experiences and particular developmental outcomes (see Rosenblum & Andrews, 1994). In the next section we will review the rodent research on this important topic.

4.4. The Influence of a Rodent Mother's Maternal Behavior on the Subsequent Maternal Behavior of Her Offspring

Thoman and Arnold (1968) reared female rat pups in incubators without mothers or peers, and such social isolation was then continued into maturity except for mating. When these socially isolated female rats gave birth, they did not show any gross deficits in maternal behavior, as indicated by normal levels of retrieving and nest building, and these findings seemed to contrast sharply with the data on the effects of social deprivation on the maternal behavior of rhesus monkeys. Thoman and Arnold (1968) did find, however, that the growth rate of the pups raised by the socially isolated mothers was subnormal and that the mortality rate of those pups was higher than that of pups raised by normal mothers. These latter effects could have been caused by deficient nursing behavior on the part of the incubator-reared moms. A recent study by Gonzalez, Lovic, Ward, Wainright, and Fleming (2001) supports this contention. They found that incubator-reared rats who did not have conspecific social experience until they were 21 days of age showed deficits in their adult maternal behavior when compared to maternally reared controls. In particular, they showed decreases in the amount of time they engaged in nursing behavior, and in the time they spent licking and grooming their pups. In other words, although such maternally deprived mother rats were responsive to their young, their maternal behavior was reduced. Starting from this platform, additional work has further supported the view that the nature of the interactions of a rodent mother with her pups (which, in the case of incubator-reared pups, would include a lack of such interaction) influences the subsequent maternal behavior of the offspring.

The initial work of Levine (1967) and Denenberg (1964) found that neonatal experiences influenced the subsequent emotional reactivity of the adult rodent (also see Levine, 2002). This initial work led to the development of two major paradigms used to study the developmental effects of postnatal experiences on the physiology and behavior of rodents: The early handling (infantile stimulation) paradigm and the prolonged maternal separation paradigm (for reviews see Francis, Caldji, Champagne, Plotsky, & Meaney, 1999a; Ladd et al., 2000; Meaney et al., 1996; Plotsky & Meaney, 1993; Walker, Welberg, & Plotsky, 2002). In the early handling procedure, rat pups are removed from their mother and placed in a small container and then returned to their mother after a 10–15-minute separation period. This brief separation period occurs daily over the first 2 or 3 weeks postpartum. In the prolonged maternal separation paradigm, rat pups are treated similarly except that they are separated from their mothers for 3 hours (or more) per day over the first 2 weeks postpartum. Importantly, *as adults*, rats that were exposed to early handling display reduced emotional reactivity and those that were exposed to prolonged maternal separation exhibited increased emotional reactivity when compared to control rats that were not separated from their mother during infancy (nonhandled controls) or were normally reared (which involved brief separations of dam and pups during cage cleaning). For example, early handling decreases and prolonged maternal separation increases hypothalamic-pituitary-adrenal activation in response to restraint stress or mild foot shock stress (Ladd, Owens, & Nemeroff, 1996; Liu, Caldji, Sharma, Plotsky, & Meaney, 2000; Plotsky & Meaney, 1993). In addition, behavioral measures of emotionality/anxiety are decreased as a result of early handling and increased as a result of prolonged maternal separation (Boccia & Pedersen, 2001; Caldji, Francis, Sharma, Plotsky, & Meaney, 2000; Ladd et al., 2000; Wigger &

Neumann, 1999): As adults, rats exposed to neonatal early handling are more active in a novel open-field arena, exhibit reduced startle responsivity to a loud sound, and show less novelty-induced suppression of feeding than do nonhandled controls, while rats exposed to prolonged maternal separations as neonates show increased startle responsivity and greater novelty-induced inhibition of feeding than do the nonhandled controls. When compared to normally reared controls, rats exposed to prolonged maternal separations as neonates show increases in an even broader range of anxiety-related behaviors in adulthood (see Ladd et al., 2000).

How are the effects of these early neonatal separation experiences mediated? Levine (1975) was one of the first to hypothesize that the effects of early handling on development were mediated by alterations in mother-litter interactions that follow each daily reunion. Early work by Lee and Williams (1974) supported this view by finding that neonatal rats exposed to early handling were licked more by their mothers than were nonhandled controls. In a more recent study, Liu, Diorio, Tannenbaum, Caldji, Francis, Freedman, Sharma, Pearson, Plotsky, and Meaney (1997) examined the behavior of the mothers of early handled and nonhandled litters over the first 10 postpartum days and found that the mothers of handled rat pups showed increased amounts of pup licking and grooming and arched-back nursing than did the mothers of nonhandled pups. (During arched-back nursing, the female nurses the pups by hovering over them with an arched back; Liu et al. [1997] contrasted such nursing with a more passive blanket-nursing posture in which the female simply lies over the suckling pups. Arched-back nursing is probably composed of the following nursing postures as described by Stern [1996a]: hover, low crouch, and high crouch.) This effect of early handling on the nursing behavior of the mothers of the handled pups has been confimed by Pryce, Bettschen, and Feldon (2001). Additional research has indicated that there are natural (rather than handling-induced) individual differences in the amount of licking, grooming, and arched-back nursing (LG-ABN) that a mother rat spontaneously exhibits toward her pups, that these individual differences in "maternal style" are consistent across birth cycles, and that such differences are related to the subsequent emotional reactivity of the adult offspring (Caldji et al., 1998; Francis, Diorio, Liu, & Meaney, 1999b). As adults, the offspring of mothers that show spontaneously high levels of LG-ABN exhibit reduced novelty-induced suppression of feeding and a blunted HPA response to restraint stress when compared to the offspring of mothers that show spontaneously low levels of LG-ABN (Caldji et al., 1998; Liu et al., 1997). Furthermore, high LG-ABN dams are themselves less emotionally reactive (as measured in an open-field apparatus) than low LG-ABN mothers, even when emotionality is measured during lactation (Francis, Champagne, & Meaney, 2000). Finally, the offspring of high LG-ABN mothers also showed high levels of LG-ABN to their own offspring in adulthood, while the offspring of low LG-ABN mothers responded with low levels of these behaviors to their own offspring in adulthood (Francis et al., 1999b). Overall, these results support the following hypothesis: Maternal rats that show spontaneously high levels of LG-

ABN are less emotionally reactive than are mothers who show low levels of LG-ABN, and the differential maternal stimulation received by the young from these different mothers plays a role in the development of the offsprings' adult emotional reactivity and maternal behavior. In support of this hypothesis, Francis et al. (1999b) report that if pups from low LG-ABN mothers are cross-fostered to high LG-ABN mothers, then in adulthood these young show reduced emotional reactivity while also displaying high levels of LG-ABN to their own offspring. The converse is true when one examines the offspring of high LG-ABN mothers that are cross-fostered to low LG-ABN mothers. Also supportive of the hypothesis was the finding that early handling of the offspring of naturally low LG-ABN mothers increases the level of LG-ABN shown by these mothers with the result that in adulthood their offspring are less emotionally reactive and show higher levels of LG-ABN (Francis et al., 1999b).

These results strongly suggest that early handling reduces the emotional reactivity and enhances LG-ABN of the affected offspring because such treatment modifies LG-ABN shown by their mothers (or modifies some other aspect of the maternal environment that is closely related to the level of LG-ABN). But why should prolonged maternal separations increase the emotional reactivity of the affected offspring, and does such increased emotionality influence the adult maternal behavior of the offspring? Much less research has been done utilizing the prolonged maternal separation paradigm, but it has been suggested that such offspring show higher levels of emotional reactivity in adulthood because they received suboptimal levels of licking, grooming, and nursing from their mothers during development (Francis et al., 1999a; Ladd et al., 2000; also see Pryce et al., 2001, for a conflicting point of view). Significantly, Lovic, Gonzalez, and Fleming (2001) have shown that rats that have been exposed to prolonged maternal separations as neonates show deficits in maternal behavior in adulthood. Consistent with the argument that is being developed here, such adult rats showed reduced levels of maternal licking and crouching over their pups (also see Boccia & Pedersen, 2001). These results conform with the view that the effects of prolonged maternal separations on the development of maternal behavior in the affected offspring may be mediated by effects on the adult emotional reactivity of these offspring. Figure 3.6 presents a summary of the data on the effects of early handling and prolonged maternal separations on rodent development.

It needs to be emphasized that although the level of a rat's maternal LG-ABN appears to influence both the emotional reactivity and level of LG-ABN of her offspring, it has yet to be proved that it is indeed emotional reacitivity that influences LG-ABN. Evidence supporting such a view would be forthcoming if it could be shown, for example, that an antianxiety agent could convert a low LG-ABN female into a high LG-ABN female. If increased emotional reactivity does cause low levels of LG-ABN, what might be the underlying mechanism? Although we can only conjecture at this point, perhaps increased fearfulness results in a mother that is constantly distracted by even minor disturbances in her environment, and this, in turn, may then cause decreases in LG-ABN. In

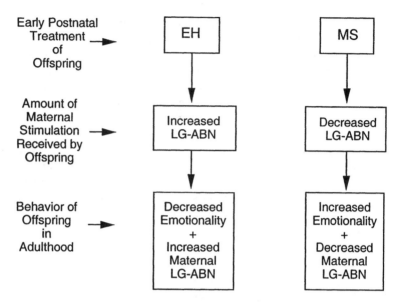

Figure 3.6. Summary diagram of the effects of early postnatal experiences on adult behavior in rats. Neonatal female rats who are exposed to early handling (EH; brief separations from their mother during the first two postpartum weeks) receive high levels of licking, grooming, and arched-back nursing (LG-ABN) from their mothers on their return to their nests, and in adulthood these females show increased levels of LG-ABN to their own offspring and decreased levels of fearfulness. In contrast, neonatal female rats who are exposed to prolonged maternal separations (MS) over the first few postpartum weeks receive less maternal attention (LG-ABN) from their mothers, and in adulthood these females are more fearful and show low levels of LG-ABN to their own offspring.

support of this idea, Gonzalez et al. (2001) present data that suggest that maternal rats that had been incubator-reared as neonates (complete maternal deprivation) are easily distracted by external stimuli. It is interesting to speculate on the possible adaptive significance of the relationships we have been describing. Perhaps the more emotionally reactive mothers would also be subordinate females under natural social conditions. On the assumption that such females would have inferior nest sites, it would be advantageous for them to be wary of their surrounding environment. Finally, although perhaps not immediately apparent, it is worth considering that the relationship between high emotionality and low levels of LG-ABN in rats may be analogous to the relationship between high emotionality and high levels of protectiveness in nonhuman primates.

That maternal treatment of offspring influences the development of the pups suggests the primacy of early infantile stimulation in affecting the development of emotional reactivity and aspects of maternal behavior. However, we should not exclude the importance of genetic factors with which the effects of early life experiences must interact, as indicated by a recent report that studied two mouse strains (Anisman, Zaharia, Meaney, & Merali, 1998). The C57BL/6ByJ strain shows low emotional reactivity, a reduced HPA stress response, and high levels of LG-ABN in comparison to the BALB/cByJ strain. Interestingly, when BALB strain offspring were cross-fostered to C57BL mothers, the cross-fostered BALB offspring showed a lowered level of emotional reactivity in adulthood. However, cross-fostering the C57BL pups to BALB mothers did not affect the adult emotional reactivity of the cross-fostered C57BL offspring.

One final study highlights the important role of the parent-litter interaction in the development of maternal behavior in rodents. Recall that various vole species show important differences in their social behaviors. Prairie voles tend to be monogamous and both maternal and paternal behavior occurs, while meadow voles are promiscuous and the maternal female usually does not let the father have access to the nest site. It should also be noted that female meadow voles spend less time in the nest nursing their young than do prairie vole females, and that meadow vole mothers wean their young at an earlier age than do prairie vole mothers (McGuire, 1988). In an interesting study, McGuire (1988) fostered meadow vole pups to either meadow vole parents (in-fostered young) or to prairie vole parents (cross-fostered young). In adulthood, the maternal behavior of the two groups of females was compared (each female was mated with a meadow vole male with a similar rearing history). Cross-fostered meadow voles behaved more like prairie vole mothers while in-fostered meadow vole mothers displayed the typical meadow vole maternal pattern. In particular, cross-fostered meadow vole mothers spent more time in the nest nursing and grooming their young than did the in-fostered females, and the cross-fostered females also allowed the father of the offspring into the nest site, while this was not the case for in-fostered females (see Figure 7.2).

In conclusion, these studies show that the early maternal environment experienced by the neonatal female rodent influences the development of her maternal behavior, and that aspects of this developmental influence may be mediated through effects on emotional reactivity. In addition, the similarities between the effects of the early maternal environment on the development of maternal behavior and emotional reactivity in rats and primates are greater than the differences, suggesting that there may be some commonality in the underlying control mechanisms.

5. General Conclusions

In this chapter we explored three major ways in which experiential factors operate to influence maternal behavior. First, previous maternal experience, either as an adult or as a juvenile, may result in subsequent episodes of adult maternal behavior that are more competent and efficient. It is as if a female learns to cope with the diversity of infant-rearing problems, and this results in more adequate maternal behavior. Second, we have clearly shown that for certain species an initial maternal experience decreases the involvement of the hormones associated with pregnancy termination in the occurrence of future episodes of maternal behavior. This finding is significant because it suggests that maternal experience can, in part, substitute for the effects of hormones. As we will see in future chapters, this effect of maternal experience will inform us about some of the functions performed by endocrine mechanisms in the activation of maternal behavior in first-time mothers. Finally, evidence indicates that the interactions a female infant rodent or primate has with her mother influence both the emotional development of the infant and the development of her maternal behavior, and that the relationship between emotional and maternal development is probably a causal one rather than simply a parallel one.

The theme that the emotional reactivity of a mother influences her maternal behavior is an important one. A highly emotional mother may be more protective and restrictive with respect to her infant, and she may also be constantly vigilant with respect to the other aspects of her environment. Under stressful environmental conditions, it may be difficult for such a female to cope well with the demands of infant rearing. As we have seen for primates, there appears to be a continuum of maternal behavior styles ranging from permissive to protective to abusive, and this continuum appears to be mirrored by corresponding increases in emotional reactivity. Perhaps when emotional reactivity becomes extremely high, as would be expected for socially isolated motherless mother monkeys, maternal behavior becomes wholly inadequate.

It is worth considering that the three experiential influences we have discussed in this chapter are interrelated in that they all may influence maternal behavior, in part, by affecting emotional mechanisms. A female who has had a previous infant may be calmer under challenging environmental conditions, leading to more competent and effective maternal behavior, because she can anticipate childcare problems that may arise; a female who has had a previous infant may be less dependent upon hormones for the activation of her maternal behavior because one of the functions of parturient hormones might be to decrease the possible fear-arousing properties of novel infant stimuli in first-time mothers; lastly, the evidence suggests that the early rearing experiences of infants affect their subsequent maternal behavior through the mediation of influences on emotional reactivity.

4

Motivational Models of the Onset and Maintenance of Maternal Behavior and Maternal Aggression

1. Introduction

Through which processes do the hormonal events of late pregnancy promote maternal responsiveness, and what are the processes that allow for its continuance during the postpartum period? Natural or species-typical behaviors, such as maternal behavior, have often been divided into appetitive and consummatory components (Everitt, 1990; Pryce et al., 1993; Timberlake & Silva, 1995). The appetitive component is made up of those behaviors that bring the organism into contact with an attractive or desired stimulus object or goal, while the consummatory component is composed of those behaviors that are performed once the goal is achieved. Wallen (1990) has made an analogous distinction by referring to these categories as desire versus ability: Appetitive behavior reflects an underlying motivation, appetite, drive, or desire to engage in a behavioral interaction with a specific goal object, while consummatory behavior reflects the ability to perform specific behavioral responses once the goal object is attained. With respect to maternal behavior in rats (*Rattus norvegicus*), for example, one would usually classify those behaviors that the female performs in order to gain access to her pups (typically, various types of approach responses) as appetitive, while retrieving and nursing behavior would be viewed as consummatory responses. However, an alternative, less traditional view might classify the various rodent maternal responses differently: If nursing were conceived of as the primary goal of maternal behavior, then it would clearly be the major consummatory component, while approaching a displaced pup *and* retrieving it to the nest site might both be considered as reflecting an underlying appetitive state. Stern's (1996a) formulation, in which she refers to pronurturant and nurturant aspects of maternal behavior, approximates this latter perspective. In other words, retrieving can be conceived of as both a consummatory response toward pup stimuli and as reflecting an underlying appetitive state with respect to nursing behavior.

The physiological events of late pregnancy and parturition may promote the onset of maternal behavior by influencing neural mechanisms that regulate both the motivation to behave maternally and the ability to perform particular maternal responses. Such changes may then be maintained during the postpartum period as a result of postpartum mother-infant interactions (see Chapter 3). Clearly, if a mother is disinterested in her infants (low motivation or desire) she will show poor or neglectful maternal behavior, even if she is capable of performing the relevant responses. Alternatively, even if maternal motivation is high, maternal behavior will not occur if the appropriate responses cannot be performed. Although hormones may influence both appetitive and consummatory components of maternal behavior in nonprimates, in primates the primary mode of hormonal influence may be via affecting motivational processes. Such a taxonomic distinction appears to be the case when exploring the influence of hormones on rodent and primate female sexual behavior (Wallen, 1990), and an analogous situation may exist for parental behavior.

Although hormones may act to promote maternal responsiveness solely by affecting appetitive and consummatory aspects of maternal behavior, such a view

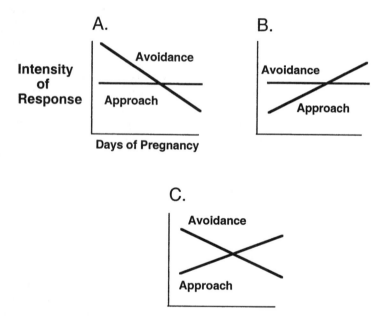

Figure 4.1. Three alternative approach-avoidance models of the onset of maternal behavior at parturition. Maternal behavior occurs when central neural approach systems are more active than central neural avoidance systems with respect to infant-related stimuli. In **A** the physiological events of pregnancy primarily decrease avoidance, in **B** these physiological factors promote approach responses, and in **C** avoidance systems are depressed and approach systems are activated toward the end of pregnancy. Research on the rat is most consistent with model **C**. (Reproduced with permission from Numan M, Sheehan TP [1997] Neuroanatomical circuitry for mammalian maternal behavior. Ann NY Acad Sci 807:101–125. Copyright 1997 by New York Academy of Sciences.)

looks at maternal behavior in isolation from other behaviors and motivations. In order for optimal maternal behavior to occur, it may be necessary for hormones not only to specifically increase maternal behavior but also to counteract competing or inhibitory behaviors and motivations. Within this context, several investigators have recently proposed motivational models that have argued that maternal behavior is facilitated when the tendency to approach infant stimuli and engage in maternal behavior is greater than the tendency to avoid or withdraw from such stimuli (Fleming & Orpen, 1986; Pryce, 1992; Rosenblatt & Mayer, 1995). Such a view, which has been influenced by the ideas of Schneirla (1959), may be particularly applicable to first-time mothers for whom novel infant stimuli might elicit fear and withdrawal. Therefore, in terms of motivational processes relevant to the onset of maternal responsiveness, the hormonal and other physiological events of late pregnancy and parturition may stimulate the occurrence of maternal behavior by acting on the brain to decrease fear/ aversion of infant stimuli, increase attraction/approach toward infant stimuli, or both. These possibilities are illustrated in Figure 4.1. It is the purpose of this chapter to critically evaluate the evidence behind such motivational models of maternal behavior. Therefore, in this chapter we are interested in two types of motivational changes that may promote the occurrence of maternal behavior: increases in maternal motivation or attraction toward young and decreases in fearfulness. In other words, fearfulness is being viewed as a motivational state that can oppose maternal motivation.

2. Approach-Avoidance Models of the Regulation of Maternal Responsiveness in Rats

2.1. Sensitization in Female Rats

Recall that an adult virgin female rat does not show maternal behavior upon initial exposure to young pups. However, as a result of continuous association with such pups a sensitization process occurs that results in the initiation of maternal behavior after several days of pup exposure. As outlined by Rosenblatt and Mayer (1995), there are two ways one can look at the sensitization process. The *single system activation threshold view* argues that the induction of maternal behavior by pup stimulation occurs because continuous exposure to young slowly builds up maternal motivation until a threshold is reached that allows maternal behavior to occur (it is from this historically older view that the term "sensitization" was probably derived). In contrast, the *approach/withdrawal biphasic processes view* proposes that when a naive female rat is first exposed to pups, both avoidance and approach processes are activated but avoidance tendencies are greater than approach tendencies. Maternal behavior subsequently occurs for two reasons. First, continuous association with pups results in a habituation of neophobic avoidance responses so that approach responses come to dominate. Once approach tendencies are dominant, then proximal pup stimulation further builds up appetitive responses (maternal motivation) until complete maternal behavior eventually occurs.

Several pieces of evidence support the biphasic processes view of sensitization. This understanding is relevant to an analysis of how hormones stimulate the immediate onset of maternal behavior at parturition because it indicates what the baseline naive virgin condition is, and this, in turn, informs us that pregnancy hormones may stimulate maternal behavior by one of the processes outlined in Figure 4.1. As indicated in Chapter 2, the descriptive course of sensitization (Stern, 1997a) conforms with the biphasic processes view in that during the early stages of pup exposure the female avoids the young. This is then followed by a period of tolerance toward pup proximity, and after this period, pup licking and partial retrievals occur that herald the onset of full maternal behavior. Also, during the early stages of sensitization females may show what has been called the stretched attention posture, or they may alternate between approaching and withdrawing from pups (Rosenblatt & Mayer, 1995). Furthermore, several laboratories, including our own (Sheehan, Cirrito, Numan, & Numan, 2000), have indicated that nonmaternal rats tend to cover (or bury) the pups they are exposed to with the bedding material that is usually in the cage. The occurrence of burying behavior has been used by others as a measure of anxiety (Picazo & Fernández-Guasti, 1993), and the stretched attention posture has been used as an ethological measure of an approach-avoidance conflict (Molewijk, van der Poel, & Olivier, 1995). More direct evidence in favor of the biphasic processes view and the proposal that pup stimuli are initially predominantly aversive to the adult virgin female rat comes from the work of Fleming and Luebke (1981). Virgin females were observed in their home cages over a series of days, and it was noted that such females typically slept or rested in the same corner of the cage on each day, this corner being referred to as the preferred quadrant. However, if pups were presented to these females in their preferred quadrant, such females would switch their sleeping/resting area to another quadrant. If pups were presented to virgins in this manner over several days, then the switching behavior persisted for the first 3–4 days of pup exposure. This switching behavior was then followed by a period of tolerance in which the female did not switch her preferred quadrant, but also did not show maternal behavior. After a day or two of tolerance, maternal behavior began to occur.

Probably the most convincing evidence that during the sensitization process the female rat must first overcome some inhibitory influence of pup stimulation before she shows maternal behavior comes from research that indicates that interference with the ability of the virgin female to detect olfactory-related chemosensory stimuli actually *facilitates* the onset of sensitized maternal behavior. Figure 4.2 shows a sagittal section through the rodent cranium indicating the nasal cavity and the olfactory chemosensory apparatus. Within the nasal cavity lies the primary olfactory epithelium, which contains the sensory receptors of the primary olfactory nerve, which projects to the main olfactory bulb. Also

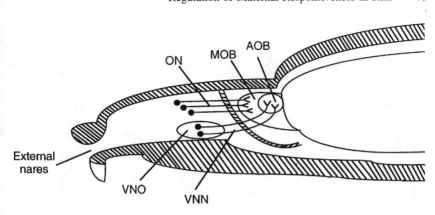

Figure 4.2. Sagittal section through the rodent cranium showing the neural connections between the olfactory and vomeronasal systems with the olfactory bulb. Abbreviations: AOB = accessory olfactory bulb; MOB = main olfactory bulb; ON = olfactory nerve; VNN = vomeronasal nerve; VNO = vomeronasal organ.

contained within the nasal cavity is the vomeronasal organ, whose axons travel centrally to terminate in the accessory olfactory bulb. A traditional view has been that the vomeronasal organ is involved in detecting pheromones while the primary olfactory system is involved in detecting volatile nonpheromonal odorants, but recent evidence suggests that both of these systems may be involved in pheromone detection (Buck, 2000; Johnston, 1998; Keverne, 1999; Wysocki, 1979). Several findings show that damage to the primary olfactory system and the vomeronasal system can facilitate the sensitization process (Fleming & Rosenblatt, 1974b, 1974c; Fleming, Vaccarino, Tambosso, & Chee, 1979): Olfactory bulbectomy, vomeronasal nerve destruction, or intranasal application of zinc sulfate (which destroys the primary olfactory epithelium) all shorten sensitization latencies. Figure 4.3 shows some of the results from the Fleming et al. (1979) study. Prior to pup exposure, virgin females received (a) *partial* lesions of the main olfactory bulb, (b) knife cuts severing the vomeronasal nerves, (c) combined damage to both the main olfactory bulb and the vomeronasal nerves, or (d) sham lesions. The females with disruption of both chemosensory systems showed the shortest sensitization latencies, becoming maternal after only about one day of pup exposure, which contrasted strongly with the standard sensitization latency of about one week that was shown by the sham females. These results suggest that inputs from both chemosensory systems are involved in inhibiting maternal responsiveness in virgin rats.

Fleming and Rosenblatt (1974c) have interpreted these results in terms of the biphasic processes view of sensitization. They suggest that under standard testing conditions, the virgin female finds the novel odors/pheromones of test pups aversive, and before maternal behavior can occur this neophobia must be overcome through a habituation or familiarization process. By disrupting chemosensory detection systems the aversive qualities of the pups are reduced, and

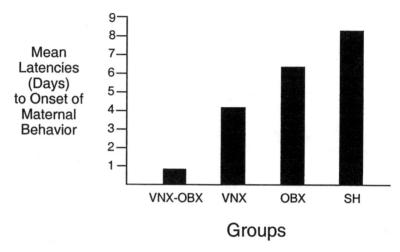

Figure 4.3. Mean latency in days to become maternal in virgin female rats that received cuts to the vomeronasal nerves (VNX), main olfactory bulbs (OBX), both chemosensory systems (VNX-OBX), or control sham cuts (SH). VNX-OBX differs significantly from each of the remaining groups. Modified from Fleming et al. (1979).

this, in turn, allows the female to respond more quickly to the positive characteristics of pups that favor maternal responsiveness.

Although a detailed analysis of the neural mechanisms of maternal behavior will be presented in subsequent chapters, there is one finding that is very relevant to the present discussion. Figure 4.4 shows in schematic form some of the projections of the main and accessory olfactory bulbs to the corticomedial amygdala. The point to note is that both olfactory systems project to the corticomedial amygdala, each to a different part, but because of interconnections between the various corticomedial amygdalar nuclei, both primary olfactory and vomeronasal input can reach the medial amygdaloid nucleus (Canteras, Simerly,

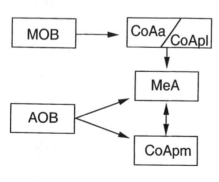

Figure 4.4. Schematic representation of the neural connections between the main olfactory bulb (MOB), accessory olfactory bulb (AOB), and corticomedial amygdala to show how vomeronasal and primary olfactory input can converge onto the medial amygdaloid nucleus (MeA). Additional abbreviations: CoAa = anterior part of the cortical amygdaloid nucleus; CoApl = posterolateral part of the cortical amygdaloid nucleus; CoApm = posteromedial part of the cortical amygdaloid nucleus.

& Swanson, 1995; Licht & Meredith, 1987; Scalia & Winans, 1975; Swanson
& Petrovich, 1998). Also relevant is the long-standing knowledge of the involvement of the amygdala in emotion, fear, and anxiety-related processes (Davis, 1992; LeDoux, 2000; Pitkänen, Savander, & LeDoux, 1997). These facts
have led to the proposal that novel olfactory stimuli from pups delay the onset
of maternal behavior in naive virgin females because such input activates fear-
inducing mechanisms mediated by the amygdala. In support of this hypothesis,
it has been shown that lesions of the corticomedial amygdala, produced with
either electric current (Del Cerro, 1998; Fleming, Vaccarino, & Luebke, 1980)
or with an excitotoxic amino acid that destroys neuronal cell bodies while sparing fibers of passage (Numan, Numan, & English, 1993), shorten sensitization
latencies in virgin female rats (see Figure 4.5). Such females show maternal
behavior after about 3 days of pup exposure in comparison to the typical 7–8-
day exposure periods required by control females. Importantly, Fleming et al.
(1980) report that virgin rats with corticomedial amygdala lesions do not show

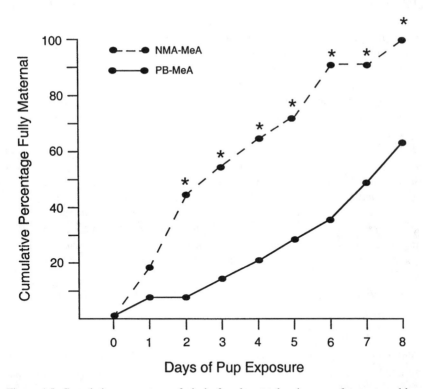

Figure 4.5. Cumulative percentage of virgin female rats showing complete maternal behavior on each test day after receiving injections of either the excitotoxic amino acid N-
methyl-D,L-aspartic acid (NMA) or the phosphate buffer (PB) vehicle solution into the
medial amygdaloid nucleus (MeA). *Significantly different from PB-MeA group (Fisher
exact probability test). Modified from Numan et al. (1993).

the typical switching of their sleeping/resting corner when pups are placed in their preferred quadrant. Instead, and in contrast to control females, amygdala-lesioned females tolerated proximity to pups and also allowed pups to crawl under them, even though they had not yet displayed maternal behavior. This result suggested to Fleming et al. (1980), in accordance with the biphasic processes view, that amygdala lesions eliminated the avoidance stage, but that proximal pup stimulation was still needed over a series of days to increase maternal motivation to a level that would allow maternal behavior to occur.

As outlined in Chapter 2, the shorter sensitization latencies of juvenile female rats when compared to those of adults also support a biphasic processes view of the *adult* sensitization process: The shorter juvenile sensitization latencies are associated with an absence of an avoidance phase during sensitization (Gray & Chesley, 1984). Additional evidence shows that general defensive and fear responses are not yet completely developed in juvenile rats, and therefore such females do not show avoidance and defensive responses to a variety of novel or threatening stimuli (Bronstein & Hirsch, 1976; Moretto, Paclik, & Fleming, 1986). It should also be considered that the olfactory systems may not be completely mature in juvenile rats (Herrada & Dulac, 1997). In the absence of an avoidance phase the young juvenile tolerates pup proximity, and this presumably builds up maternal motivation so that full maternal behavior is displayed after only 2–3 days of pup exposure.

As reviewed above, juvenile females, and adults with either olfactory damage or amygdala lesions, do not show immediate maternal behavior. What this suggests is that the mere absence of avoidance responses to pup stimuli is not sufficient to mimic the immediate onset of maternal behavior that occurs at parturition. Such females must still be exposed to positive proximal stimuli from pups for a few days before full maternal behavior occurs, suggesting that such stimulation is needed to boost maternal motivation. Therefore, in reference to the models depicted in Figure 4.1, the findings support the hypothesized processes indicated in Figures 4.1B and 4.1C. That is, in the absence of aversion, basal maternal motivation is not high enough to allow maternal behavior to occur immediately, suggesting that the hormonal events of late pregnancy should include effects on approach/attraction tendencies.

2.2. Do Hormones that Stimulate the Onset of Maternal Behavior in Rats Affect Approach and Avoidance Systems?

Within the context of the biphasic processes model, how do hormones act to stimulate maternal behavior in rats? Just because experimental procedures that eliminate or depress avoidance tendencies (amygdala lesions, for example) facilitate sensitization does not mean that hormonal action includes such a mechanism. In examining Figure 4.1, one can see that it is theoretically possible for hormones to stimulate maternal behavior simply by increasing the appetitive aspects of maternal behavior. In this section we will review the evidence that suggests that hormones probably do stimulate the onset of maternal behavior in rats by affecting *both* approach and avoidance tendencies (Figure 4.1C).

Fleming and Luebke (1981) compared the responses of nulliparous female rats and parturient primiparous females to environmental novelty and threatening situations. The nulliparous females were never exposed to pups and the parturient females were separated from their pups soon after birth and tested 2 days later. The parturient females were found to be less fearful in that they emerged more quickly into a novel open-field arena; they moved around more in the open field, particularly in its central portions that are viewed as being less protected; and they were less likely to run away from an intruder female rat that was placed in their home cages. These results are important because they suggest that a *general* reduction in fearfulness is associated with the onset of maternal behavior in parturient rats. If hormones do act to promote maternal behavior by decreasing avoidance of novel pup stimuli, they could act in either of two ways: by specifically decreasing avoidance tendencies just to pup stimuli or by generally decreasing defensive and withdrawal responses to all novel and threatening stimuli. The results of the Fleming and Luebke (1981) study suggest that the latter is a likely mechanism. A problem with the interpretation of this study, however, is that the parturient females showed maternal behavior for about 12 hours prior to being tested for emotional reactivity. Therefore, it is possible that it was only the occurrence of maternal behavior that caused the fear reduction, or that hormonal stimulation plus the occurrence of maternal behavior caused the fear reduction. If we want to argue that pregnancy hormones promote the onset of maternal behavior in part by decreasing avoidance responses, we need to show that such an effect occurs prior to pup exposure. This information was provided by Fleming, Cheung, Myhal, and Kessler (1989). Ovariectomized virgin female rats were treated with a hormone regimen known to facilitate maternal behavior (Bridges, 1984), but they were not exposed to pups. Such females were found to emerge more rapidly into a novel field than were ovariectomized females who were not treated with hormones (also see Picazo & Fernández-Guasti, 1993).

Neumann et al. (1998a) compared the responses of pregnant and virgin female rats in an elevated plus maze apparatus, and they present evidence that day-21 pregnant rats show a higher level of anxiety-related behavior than do virgin females in that they show fewer entries into the open (less protected) arms of the maze. Although these results appear to be in conflict with the view that the hormonal stimulation of maternal behavior is associated with a general reduction in fearfulness, it needs to be realized that parturition in rats usually occurs on day 22 or 23 of pregnancy, so that although anxiety may have been increased on day 21, closer to term it may have actually been decreased. A prepartum onset of maternal behavior occurs in rats on the day of parturition (see Chapter 2) so it would be important to determine whether a general reduction in fear-related responses also occurs at that time. In this context, it should be noted that when rats are tested in an elevated plus maze, if such rats are exposed to high levels of progesterone followed by progesterone withdrawal, then an *anx-*

iogenic effect is observed—such rats spend less time in the open arms of the maze than do females not exposed to progesterone withdrawal (Smith et al., 1998). Importantly, this anxiogenic effect is observed in rats that are not exposed to an estrogen rise following the progesterone withdrawal. Recall that progesterone withdrawal alone does not facilitate maternal behavior, while progesterone withdrawal superimposed on high estradiol levels does. Perhaps this latter condition, which would occur near the time of parturition, would be associated with an *anxiolytic* effect, as suggested by the data of Fleming et al. (1989).

In addition to the evidence that maternal hormones may cause a general reduction in fear responsiveness, there is also evidence that hormonal events associated with pregnancy termination may increase the attraction value of pup-related olfactory stimuli. Bauer (1983) was the first to report that late pregnant female rats, when given a 2-choice test, preferred (spent more time in) a compartment that contained bedding from a lactating female than one that contained clean bedding. Virgin females showed no such preference. A similar finding has been reported by Kinsley and Bridges (1990), except that in this case the bedding material was either clean or soiled by pups only (a lactating female did not contact the bedding). The results of this latter study are shown in Figure 4.6, where it can be seen that females on day 22 of pregnancy show a substantial preference for pup-related odors. It should be noted that in both the Bauer (1983) and Kinsley and Bridges (1990) studies, although virgin females did not show a preference for pup-related odors, they also did not show an aversion. Finally, Fleming et al. (1989) have reported that when ovariectomized female rats are treated with a hormone regimen that stimulates maternal behavior, such females are more attracted to pup-related odors than are untreated females.

Since these findings show that virgins do not show an aversion to pup-related odors, how can this be related to the fact that virgin females tend to avoid live pups (Fleming & Luebke, 1981), and that olfactory deficits promote sensitization (Fleming et al., 1979)? One possibility, as suggested by Rosenblatt and Mayer (1995), is that olfactory input from pups is aversive only when presented in the context of other pup stimuli (auditory, for example).

The fact that the physiological events of late pregnancy cause the female to have a preference for pup-related odors clearly shows that hormones do not stimulate maternal behavior by causing anosmia. Instead, the hormonal events of late pregnancy appear to shift the valence of pup-related odors from neutral (or negative, when in the context of other pup stimuli) to positive, and the increased attraction value of these odors is probably involved in the facilitation of the onset of maternal behavior. However, such a proposed facilitation would occur not because pup odors are attractive per se, but because the valence shift results in pup stimuli no longer exerting an inhibitory effect. Anosmic female rats show maternal behavior (Fleming & Rosenblatt, 1974b, 1974c), and therefore attraction to pup odors is not a necessary prerequisite for maternal

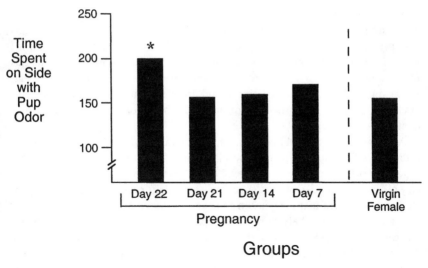

Figure 4.6. Olfactory preferences for pup odors in virgin female rats, and in rats during different stages of pregnancy. In a 5-minute test, rats are placed in a 2-compartment cage, with 1 side containing bedding that was soiled by pups, and the other containing clean bedding. The amount of time spent in each compartment is recorded. 150 seconds is a chance level of responding. *Significantly greater than the remaining groups and significantly different from chance. Modified from Kinsley and Bridges (1990).

behavior to occur. Interestingly, if the hormonal events of late pregnancy result in pup odors becoming attractive, there may be no essential need, with respect to the onset of maternal responsiveness, for hormones to also cause a *general* reduction in fearfulness. Perhaps the general reduction in fearfulness shown by the parturient female is related to other aspects of the maternal condition (see section 2.5 of this chapter and section 5.3 of Chapter 4). However, it is also possible that a general reduction in fear responsiveness is part of the process that results in pup-related odors becoming attractive. As one example, perhaps chemosensory input from pups in the naive virgin female reaches both positive (appetitive processes) and negative (aversion processes) brain centers, with the aversion areas being more dominant. If hormones at the end of pregnancy generally depress the aversion system, then withdrawal responses would not occur and olfactory input from pups would be attractive.

2.3. Evidence for a Biphasic Processes Motivational Model Obtained from Postpartum Female Rats

Several studies have indicated that postpartum female rats are attracted to pups and find interactions with them rewarding, while naive virgins do not, and that postpartum females also show a *general* reduction in fear-related behaviors when compared to their naive virgin counterparts. Although both of these types of motivational changes could have been brought about solely as a result of interacting with pups after birth, in the context of the work already reviewed, it is more likely that the hormonal events associated with late pregnancy set up these motivational changes, which are then maintained as a result of mother-young interaction. This latter perspective would argue that such motivational changes not only support the initiation of maternal behavior, but also support its continuance.

A relevant motivational question, which was explored in Chapter 2, is whether a female will retrieve pups that are placed in a T-maze extension to the home cage. By placing the pups in a novel environment, one tests whether a female can overcome her fear of a strange environment in order to retrieve the pups. Research indicates that postpartum females and *hormone-induced* maternal virgins are much more likely to retrieve pups under such stressful conditions than are *sensitized* virgins (Mackinnon & Stern, 1977; Stern & Mackinnon, 1976). The cause of this increase in T-maze retrieval in hormone-primed females is ambiguous, however. Although it is possible that this effect is solely the result of an increased maternal motivation in the hormone-stimulated rats (the pups become highly attractive stimuli as a result of hormonal stimulation), the results also fit the view that the hormonal events of late pregnancy result in females being less fearful. It appears likely that both processes are affecting the T-maze retrieval results, as suggested by the evidence below.

With respect to a general reduction in fear-related responsiveness in postpartum females, in addition to the work of Fleming and Luebke (1981), other important work exists. Bitran, Hilvers, and Kellogg (1991) found that lactating female rats showed increases in exploration of the open arms of an elevated plus maze when compared to diestrus, proestrus, ovariectomized, and pregnant (day-19) females (see Figure 4.7). Since pups were not present in the maze, these results suggest a general decrease in fearfulness. Hard and Hansen (1985) examined the acoustic freezing response as a measure of fear: Rats were presented with a loud sound, and the amount of time spent freezing (inhibition of movement) after the sound was measured. Lactating females showed a marked reduction in the duration of the freezing reaction when compared to virgin females who were not exposed to pups. In a related finding, Hansen and Ferreira (1986a) have reported that the induction of maternal behavior in virgin female rats *through hormone treatment* was also associated with a decrease in the acoustic freezing response. Interestingly, these reductions in freezing responses were dependent upon the presence of pups in the cage at the time of the presentation of the acoustic stimulus. In contrast, Toufexis, Rochford, and Walker (1999) report that in comparison to cycling females, lactating rats show a reduced acoustic startle response, which measures the jumping response to a loud acoustic stimulus, and this reduction was not dependent upon pup presence in the test cage. These authors also obtained results consistent with Fleming and Luebke's and Bitran et al.'s findings: In open-field tests, lactating females showed a shorter latency to emerge, less freezing behavior, and more entries into the field than

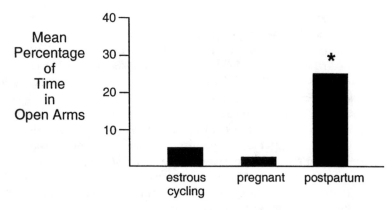

Figure 4.7. Mean percentage of total test time spent in the open arms of an elevated plus maze for female rats in different endocrine states. *Significantly different from remaining groups. Modified from Bitran et al. (1991).

did cycling females. In another measure of anxiety, Ferreira, Hansen, Nielsen, Archer, and Minor (1989) examined the punished drinking response in virgin and postpartum females. In this test, licking from a water spout is punished by electric shocks, and it was found that lactating females drank more than virgins. Evidence was provided that this difference was not due to differences in thirst or pain perception, and the authors suggested it resulted from decreased fear or anxiety responses resulting from the shock.

All of these results suggest that the hormonal events that induce maternal behavior at parturition are associated with a reduction in general fearfulness. Except for the T-maze retrieval data, which are difficult to interpret, no one has examined whether sensitized (non–hormone treated) virgins show a reduction in general fear responsiveness: Would such females emerge more quickly into an open field, show increased exploration of the open arms of an elevated plus maze, or show reduced acoustic freezing and startle responses in comparison to their nonmaternal virgin counterparts? If a low level of T-maze retrieval is at least in part a reflection of fearfulness, one would predict that sensitized virgins would not show a general reduction in fearfulness when tested with these other procedures. Additional support for the idea that sensitized virgins do not show a general decrease in fearfulness comes from findings on maternal aggression, which will be reviewed in Section 5.3 of this chapter. The occurrence of maternal aggression is associated with decreased fearfulness, and postpartum female rats show much higher levels of such aggression than do sensitized virgins. This perspective on sensitized maternal behavior suggests that a reduction in general fearfulness is not an essential requirement for the occurrence of maternal behavior. It should be understood, however, that sensitized maternal behavior may in part be mediated by a *specific* reduction in the fear of novel pup-related stimuli through a familiarization or habituation process. The idea that postpartum females show a general reduction in fearfulness, while sensitized virgins may not, suggests that the physiological events associated with late pregnancy, parturition, and lactation have influences on the maternal condition that go beyond the occurrence of maternal responsiveness, and that such influences probably allow the postpartum female to take reasonable risks when caring for her young under challenging environmental conditions.

With respect to the appetitive aspects of maternal behavior, research utilizing the conditioned place preference (CPP) paradigm has clearly shown that pups can serve as strong reinforcers to postpartum females (Fleming, Korsmit, & Deller, 1994a). In the exposure phase of the CPP paradigm, female rats are exposed to two distinctively different cage compartments, and one of these compartments (termed the "positive compartment") contains pups while the other does not. In the test phase, the female is exposed to the two different compartments in the absence of pups, and it is determined whether the female has developed a preference for the compartment previously paired with pups, as measured by the amount of time the female spends in each compartment. The basic idea here is that if pups are strong positive primary reinforcers, then the compartment that was paired with pups should acquire secondary reinforcing characteristics through classical conditioning and should therefore attract the female even in the absence of the pups. Fleming et al. (1994a) report that postpartum females do develop a preference for a compartment that was previously paired with pups while naive virgin females do not. Not surprisingly, during the exposure phase the postpartum females showed maternal behavior toward the pups that were present in the positive compartment while the virgin females did not. That some aspect related to the actual performance of maternal behavior is important for the formation of the place preference is shown by the fact that if postpartum females are exposed to pups in a perforated box that allows the female access to distal exteroceptive cues from the pups but prevents proximal interactions with the pups, then a place preference does not develop. Although proximal interactions with pups appear essential for the development of a conditioned place preference, such interactions may not be sufficient: It is also necessary to deprive the maternal female of her pups for a period of time prior to the 1-hour exposure phase in order for a CPP to develop (Fleming et al., 1994a). These results indicate that proximal pup stimuli, which appear to be both olfactory and tactile in nature (Magnusson & Fleming, 1995), are potent positive reinforcers for postpartum females, particularly if such females have been deprived of their pups for several hours. In an important extension of this work, Mattson, Williams, Rosenblatt, and Morrell (2001) have shown that when one compartment of a CPP apparatus is associated with pups while the other is associated with cocaine administration (which is highly reinforcing), postpartum rats prefer the compartment associated with pups if tested on day 8 postpartum, but prefer the cocaine-associated compartment when tested later in the postpartum period (day 10 or 16). These results show that during the early postpartum period, pups are highly reinforcing (see Figure 4.8A).

In related research, Lee, Clancy, and Fleming (2000) have shown that post-

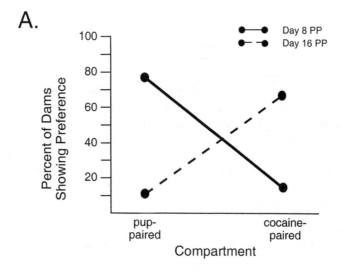

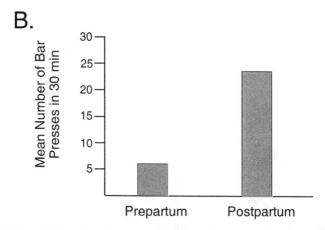

Figure 4.8. Two findings that show that pups are highly attractive and rewarding to post-partum female rats. Panel **A** shows data from rats who were tested in a conditioned place preference apparatus on either day 8 or day 16 postpartum. There were three compartments in the apparatus: one that had previously been paired with pups, one that had been previously paired with cocaine, and a neutral compartment. The data show the percentage of rats that preferred the pup-paired or the cocaine-paired compartment. On day 8 postpartum, when maternal motivation is high, most females preferred the compartment that had previously been paired with pups. On day 16 postpartum, when maternal motivation is waning, most females preferred the compartment that had previously been paired with cocaine. Panel **B** shows that pups can reinforce a bar-pressing response when postpartum rats are tested in an operant conditioning apparatus. Rats were first trained to bar-press for a highly palatable food, and then pups were used as the potential reinforcing stimulus. Prepartum rats, who were not showing maternal behavior toward pups placed in their cages, did not bar-press at a high rate for pups. In contrast, postpartum rats, who were highly maternal in home-cage tests, bar-pressed at a high rate for pups. Data in **A** modified from Mattson et al. (2001). Data in **B** modified from Lee et al. (2000).

partum rats will press a bar in an operant chamber at a high rate in order to obtain rat pups as a reward. However, such pup presentation was only rewarding if the female could subsequently interact with the pups. If such interaction was prevented by a partition, then the bar-pressing response did not develop. Importantly, when full interaction with pups was allowed, late pregnant females, who were not showing maternal behavior toward pups in home-cage tests, did not develop the strong bar-press response seen in postpartum females when pups were used as a reward, although such females were capable of pressing at a high rate when a highly palatable food was used as a reward. These results also support the view that the hormonal events of parturition are associated with an increase in the attractiveness and rewarding capability of rat pups for the postpartum dam (see Figure 4.8B).

2.4. The Effects of Maternal Experience on the Hormonal Dependency of Maternal Behavior in Rats May Be Mediated by Motivational Influences

Recall that in rats and other species, previous maternal experience results in the onset of future maternal behavior being less dependent upon hormonal stimulation. As indicated in Figure 4.1C, if during an initial pregnancy (primigravid females) hormones act to decrease aversion and increase attraction with respect to infant stimuli, then how might an initial postpartum maternal experience decrease the need for these hormonal effects during subsequent maternal episodes? More specifically with respect to the rat data, why should sensitization latencies in females with previous maternal experience be shorter than those in naive females? Clearly, one can hypothesize that previous maternal experience facilitates future maternal behavior because maternal interaction with pups permanently increases the attractive value of pups and/or permanently decreases the aversive qualities of pup stimuli. The facts that experiential influences on maternal behavior are dependent upon full interaction with pups (mere exposure to exteroceptive pup stimuli are ineffective) (Orpen & Fleming, 1987) and that the rewarding value of pup stimuli is also dependent upon full interaction between the mother and her pups (Fleming et al., 1994a) suggest that one mechanism through which experience facilitates future maternal behavior is by affecting the reward value or attractive characteristics of pups. It also seems likely that prior exposure to pups would result in a familiarization process, which, if maintained over time, would decrease the future fear-eliciting properties of infant stimuli. Figure 4.9, adapted from Rosenblatt and Mayer (1995), shows various models of how experiential effects might operate. As shown in Figure 4.9A, for the naive virgin female, exposure to pups first decreases the fear-eliciting effects of pups (through a familiarization process), and this allows the female to obtain proximal contact. Proximal contact increases the attraction value of pups until a threshold is reached that activates full maternal behavior, and this occurs at approximately 7 days from the onset of pup exposure. The other modules in this figure show how previous maternal experience might decrease future sensitization latencies by decreasing pup avoidance (Figure 4.9B), increasing attraction to pups (Figure 4.9C), or both (Figure 4.9D).

2.5. Critical Evaluation of the Role of Fear Reduction

The results we have just reviewed for rats are certainly consistent with the view that the hormonal events that occur at the end of pregnancy stimulate the onset of maternal behavior in part by causing a general reduction in fearfulness coupled with an increase in the attraction value and rewarding capability of pup-related stimuli. (The hormonal events of late pregnancy may also operate to promote specific maternal responses by affecting consummatory mechanisms in addition to appetitive processes, but such an action is not a focus of this chapter.) Such motivational changes may then be maintained during the postpartum period as a result of continued mother-young interactions. However, as reviewed above, it is conceivable that the hormonal events of late pregnancy could stimulate the immediate onset of maternal behavior at parturition in primiparous females by simply shifting the valence of pup-related stimuli from negative to positive. That is, in the naive cycling virgin female, pup stimuli may primarily access brain centers that mediate fear and withdrawal responses, while in the parturient female the access of pup-related stimuli to such hypothesized central aversion centers might be prevented and, at the same time, the access of such stimuli to appetitive brain centers related to maternal behavior might be promoted. In support of this argument, the sensitized virgin female rat ultimately shows adequate maternal behavior in home-cage tests, and such behavior appears to occur in the absence of a reduction in general fearfulness. With respect to fearfulness, if a hormone-induced specific reduction in the ability of pup stimuli to arouse fear and withdrawal is all that would be needed for maternal behavior to occur at parturition, what is the function of the more general reduction in fear and anxiety that is observed in postpartum females? One possibility is that a general decrease in fear to a large variety of stimuli, although not absolutely essential for the immediate onset of maternal behavior, may serve ancillary or supportive functions during the postpartum period. Such a process may allow the maternal female to take acceptable risks in caring for her young. Therefore, we propose that the primary function of hormones in inducing the onset of maternal behavior in primiparae is to promote a decrease in the fear-arousing capability of pup stimuli coupled with an increase in the attractive value of such stimuli, and that a secondary function of pregnancy hormones is to induce a more general reduction in fearfulness. We hypothesize that pregnancy and parturient hormones are doing all of these things, but that under relatively undisturbed conditions only the primary function would be necessary for maternal behavior to occur at parturition. The secondary function is hypothesized to be essential under demanding or stressful conditions. Indeed, under natural conditions both functions may be essential for effective caretaking activities to occur, which would ultimately lead to the survival of the maximum number of young. At the present

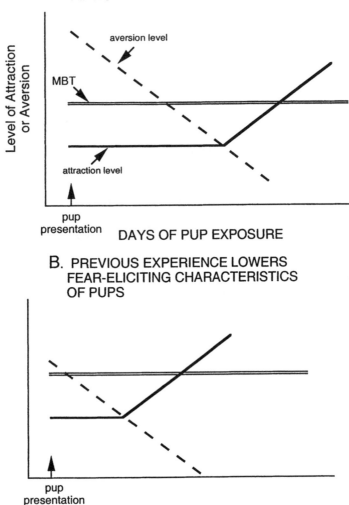

A. NORMAL SENSITIZATION

aversion level

MBT

attraction level

pup
presentation DAYS OF PUP EXPOSURE

B. PREVIOUS EXPERIENCE LOWERS FEAR-ELICITING CHARACTERISTICS OF PUPS

pup
presentation

Figure 4.9. Various models to show how previous maternal experience might operate on motivational and emotional mechanisms to shorten subsequent sensitization latencies in rats. Panel **A** shows the course of normal sensitization in naive virgin rats. Exposure to pups initially reduces aversion or avoidance levels (dashed line) via a process of familiarization. Once avoidance reaches a certain low level, the female tolerates proximal contact with pups, and this proximal contact increases attraction to pup-related stimuli

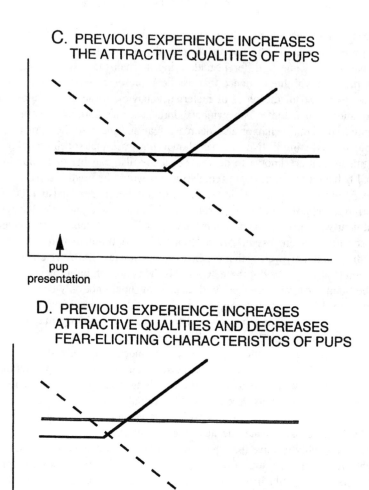

C. PREVIOUS EXPERIENCE INCREASES THE ATTRACTIVE QUALITIES OF PUPS

pup
presentation

D. PREVIOUS EXPERIENCE INCREASES ATTRACTIVE QUALITIES AND DECREASES FEAR-ELICITING CHARACTERISTICS OF PUPS

pup
presentation

Figure 4.9 *Continued*. (solid line). Once attraction reaches a certain point (MBT = maternal behavior threshold, shown as a double line), then full maternal behavior occurs. The other panels show how previous maternal experience might decrease future sensitization latencies by either decreasing pup avoidance (**B**), increasing attraction to pups (**C**), or both (**D**). Panel **A** is adapted from Rosenblatt and Mayer (1995).

time it is not clear whether the postpartum reduction in general fearfulness and the postpartum elimination of the fear-arousing qualities of novel pup stimuli are part of the same or different underlying mechanisms.

A problem with the evidence that has been presented in favor of the biphasic processes model of the onset of maternal behavior in rats is that most of it is correlational in nature. The evidence indicates that hormone treatments that stimulate maternal behavior also decrease fear and increase the attraction value of pup-related stimuli. No one has shown, however, that specific nonhormonal treatments that are known to decrease fear are also capable of facilitating maternal behavior in naive virgin female rats, as would be predicted by the model. Indeed, Ferreira, Picazo, Uriarte, Pereira, and Fernandez-Guasti (2000) have referred to unpublished findings from their laboratory that indicate that systemic treatment with an antianxiety agent, diazepam, was not capable of accelerating the sensitization process in virgin female rats. A problem with using a drug like diazepam, however, is that it generally potentiates GABAergic neural transmission (Ballenger, 1995). Since GABA is the major inhibitory neurotransmitter in the brain, systemic treatment with diazepam should not only depress anxiety-related neural regions, but might also depress those neural regions playing a positive role in maternal behavior. Therefore, this particular finding is not a serious blow to the model we are developing.

If fearful responses to novel pup stimuli are indeed a factor that delays the onset of maternal behavior in naive cycling female rats, one might predict that rats who are generally more fearful should have longer sensitization latencies. One might expect virgins who were exposed to prolonged maternal separations as neonates to display longer adult sensitization latencies than virgins who were briefly handled as neonates because the former rats would be expected to be more emotionally reactive than the latter females (see Chapter 3). Similarly, one might predict that the adult offspring of high LG-ABN mothers (low emotionality offspring) would have shorter sensitization latencies than would the offspring of low LG-ABN mothers (high emotionality offspring). Three findings support these predictions. First, Champagne, Diorio, Sharma, and Meaney (2001) found that the offspring of high LG-ABN mother rats do indeed have shorter sensitization latencies (averaging about 4 days) than do the offspring of low LG-AGN mothers (averaging about 8 days). Second, data from Rees and Fleming (2001) indicate that brief handling of neonatal rats reduces the adult sensitization latencies of these rats. Finally, Kinsley and Bridges (1988b) have reported that *prenatally* stressed female rats (their mothers were exposed to heat and restraint stress during the last week of pregnancy) had longer adult sensitization latencies than did their control counterparts. Importantly, other studies have shown that prenatally stressed rats show high anxiety-like behavior on a battery of behavioral tests in adulthood (Lehmann, Stohr, & Feldon, 2000; Valee et al., 1997). Interestingly, mothers who are stressed during late pregnancy show lower levels of LG-ABN to their offspring once they are born, suggesting that the effects of prenatal stress may be partly mediated by postnatal alterations in the behavior of the mother toward her pups (Champagne & Meaney, 2000).

Although emotionality levels influence sensitization latencies in virgins, *lactating* rats who are higher on the emotionality spectrum still show maternal behavior, and their young survive. For example, mothers who show low levels of LG-ABN and high levels of emotional reactivity during lactation are still capable of raising their young (see Chapter 3). Similarly, Neumann, Wigger, Leibsch, Holsboer, and Landgraf (1998b), who used selective breeding to create two lines of rats that showed either high or low levels of anxiety-related behavior (as measured in the elevated plus maze), have cited unpublished work that these two strains do not show major differences in maternal behavior. These results support the view that a reduction in general fearfulness is not absolutely essential for the onset and maintenance of maternal behavior under standard laboratory conditions. These results also appear to imply that the underlying mechanism that shifts the valence of pup-related stimuli from negative to positive differs from that which regulates the level of general fearfulness. It should be noted, however, that although one group of postpartum rats may be more emotional than another group, the hormones of pregnancy and lactation may still lower the absolute level of general emotionality in *both* groups, and some evidence for this supposition exists (Boccia & Pedersen, 2001).

Although the above findings tend to support a view that argues that high levels of general emotionality/fearfulness are consistent with adequate maternal behavior, one can still ask what the maternal behavior of high-emotionality mothers would be like if they were challenged or exposed to a stressful situation. In this case, maternal behavior might be disrupted, and if it were, evidence would be provided for the view that lower levels of general fearfulness probably play a supportive role in the maternal care system. Importantly, Fride, Dan, Gavish, and Weinstock (1985) have reported on the adult maternal behavior of females that were stressed prenatally. Prenatally stressed rats did not differ from their control counterparts when administered standard retrieving tests. However, when a more challenging test was administered, the prenatally stressed rats showed major deficits. In the standard test the lactating female was required to enter an alley attached to the home cage in order to retrieve displaced pups, while in the more challenging test the mother had to pass through an airstream in the alley before reaching the pups. Although all animals retrieved their pups in the standard test, and 96% of the control females did so in the challenging test, only 52% of the prenatally stressed mothers retrieved in the latter test. The implication is that the maternal performance of the prenatally stressed rats breaks down under anxiety-provoking situations, and this is presumably related to the higher level of anxiety-like behavior of these females, which may have prevented them from dealing with a difficult situation. Also relevant to this issue is the occurrence of low levels of licking, grooming, and arched-back nursing (LG-ABN) of pups by postpartum rats who are more emotionally reactive. Perhaps such rats are so hypervigilant and wary of their environment that they cannot focus their full attention on their pups.

2.6. Conclusions

Overall, we feel that the best conclusion to reach from the evidence reviewed up to this point is as follows: Virgin female rats do not show immediate maternal responsiveness toward test pups because pup stimuli are primarily aversive to them. The hormonal and other physiological events of late pregnancy and parturition act, in part, to promote immediate maternal behavior at birth by doing at least three things: (a) increasing the attractive value and rewarding potential of pups, (b) decreasing the access of pup-related stimuli to central aversion or fear-related brain regions, and (c) depressing the function of central aversion systems with the result that there is a general reduction in fear to a wide variety of stimuli. These changes, which are induced by the physiological changes that accompany pregnancy termination, persist into the postpartum period, possibly as a result of continued mother-young interaction. Factors (a) and (b) are probably critical for the onset and maintenance of normal maternal behavior under standard nonstressful conditions, while factor (c) probably allows the maternal female to effectively care for her young under more demanding and stressful situations. An important implication of these views is that maternal behavior is likely to be effective within a fairly broad range of emotional reactivity, but if a female were to have a degree of fearfulness or anxiety above the upper limit of the hypothesized critical range, then maternal responsiveness might completely break down under stressful environmental conditions. As an interesting test of this hypothesis, future research could examine the emotional and maternal behavior of lactating females who were both selectively bred for high anxiety-like behavior *and* exposed to prenatal or postnatal environmental conditions that increase emotional reactivity. What kind of maternal behavior would be shown by such females under normal and stressful environmental conditions?

3. The Relevance of the Reduced Hypothalamic-Pituitary-Adrenal (HPA) Stress Response during Lactation to Motivational Models of the Regulation of Maternal Responsiveness

A variety of physical and psychological stressors activate the release of adrenocorticotropic hormone (ACTH) from the anterior pituitary, which, in turn, stimulates glucocorticoid (corticosterone in rats, cortisol in primates) release from the adrenal cortex (Van de Kar & Blair, 1999). Indeed, as reviewed in the previous chapter, HPA activation has often been used as a physiological index of stress, fear, or anxiety, and this neuroendocrine response is highly correlated with behavioral measures of emotional reactivity and fearfulness (also see Rodgers et al., 1999). Importantly, when one compares lactating rats with nonlactating rats in their response to stress-induced HPA activation (the stressors employed have included ether exposure, restraint stress, forced swimming, noise stress, and exposure to a novel environment such as an elevated plus maze), one finds that the magnitude of the neuroendocrine stress response is reduced in lactating rats (Neumann et al., 1998a; Stern, Goldman, & Levine, 1973; Walker, Lightman, Steele, & Dallman, 1992; Windle et al., 1997c). Figure 4.10 shows the results of an early report on this effect in rats exposed to ether stress (Stern et al., 1973). Note that the HPA hyporesponsiveness to stress in lactating rats is dependent upon suckling stimulation: The magnitude of the depression increases as litter size increases (Myers, Denenberg, Thoman, Holloway, & Bowerman, 1975), pup removal reverses the HPA hyporesponsiveness within 48 hours following the dam-litter separation (Lightman & Young, 1989), and the stress response is not depressed in thelectomized mothers who continue to care for their young when they are provided with healthy foster pups on a daily basis (Stern & Levine, 1972).

In the context of the focus of this chapter, an interesting question is whether the reduced HPA responsiveness to stress observed during lactation is a necessary prerequisite for the reduction in general fearfulness that is observed in lactating females. In other words, does the same underlying physiological and neural mechanism regulate the decreased behavioral fearfulness and the decreased HPA neuroendocrine responsiveness observed in lactating females? The answer to this question appears to be no. First, although thelectomized lactating females do not show HPA hyporesponsiveness, they will retrieve pups from a T-maze extension attached to their cages (Stern & MacKinnon, 1976). To the extent that such T-maze retrieval behavior is an indication of decreased fearfulness, it suggests that postpartum decreases in anxiety can occur in the absence of HPA hyporesponsiveness. Second, recent work has shown that HPA hypo-

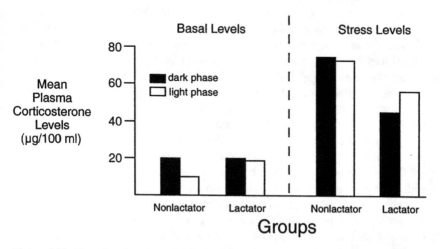

Figure 4.10. Mean basal and stress levels of plasma corticosterone in lactating and non-lactating female rats. Stress levels of corticosterone are significantly lower in lactating females. Modified from Stern et al. (1973).

responsiveness to stress can first be detected during the last week of pregnancy, and this neuroendocrine effect during pregnancy has actually been associated with increases in anxiety-related behavior, as measured in the elevated plus maze (Neumann et al., 1998a). Neumann et al. (1998a) conclude that there is an independent regulation of HPA axis activity and fearfulness (also see Neumann, Torner, & Wigger, 2000).

A possible explanation of the adaptive significance of HPA hyporesponsiveness during lactation is that it is a mechanism that conserves the adrenocortical system for uses related to lactation (Numan, 1994). Suckling is capable of stimulating ACTH release (Walker et al., 1992), normal lactation is dependent upon adrenocortical secretions (Tucker, 1994), and *basal* corticosterone levels are higher in lactating than in nonlactating females during the light phase of the light-dark cycle, presumably as a result of suckling stimulation (Stern et al., 1973; Walker et al., 1992; see Figure 4.10). This explanation conforms with the finding that thelectomized females, who do not receive a suckling stimulus, do not develop HPA hyporesponsiveness.

Although the HPA hyporesponsiveness and the decreased fearfulness observed during lactation do not share the exact same underlying regulatory mechanism, there may still be similarities and overlap between how these two processes are regulated. More specifically, decreases in general fearfulness and HPA hypo-responsiveness may be regulated by similar neurochemical changes acting in *different* parts of the brain, perhaps with different temporal courses. Also, although suckling stimulation may regulate changes in the HPA system, it may not be essential for neurochemical changes elsewhere. Figure 4.11 shows a simplified diagram of HPA regulation. Cells in the parvocellular division of the paraventricular hypothalamic nucleus (PVNp) manufacture the peptide corticotropin releasing factor (CRF), which is released into the hypothalamic-pituitary portal system upon appropriate stimulation of the PVNp (by neural processes activated by stress or suckling, for example). Through the portal system, CRF reaches the anterior pituitary corticotrophs, where it stimulates the release of ACTH into the blood, and ACTH, in turn, stimulates glucocorticoid release from the adrenal cortex (Sapolsky, 1992; Van de Kar & Blair, 1999). One of the functions of glucocorticoid secretion is to increase blood levels of glucose, which would be an appropriate physiological response to a stressful situation (Sapolsky, 1992). Although the underlying processes that cause HPA hyporesponsiveness are probably complex and varied, there are some recent data on five possible mechanisms that contribute to this effect. First, Neumann et al. (1998a) have reported that HPA hyporesponsiveness to stress during pregnancy and lactation is related to decreased responsiveness of anterior pituitary corticotrophs to CRF, which, in turn, may be mediated by decreases in CRF binding to its receptor at the level of the corticotrophs. Important for the argument that we are developing here, CRF is not only a neurohormone serving neuroendocrine functions, but it is also a neurotransmitter utilized by other neural systems within the brain, and one of the processes that it stimulates is anxiety-related or fearful behavior (Dunn & Berridge, 1990). Therefore, perhaps decreased

Figure 4.11. Schematic representation of the hypothalamic-pituitary-adrenal neuroendocrine system. Parvocellular neurons in the hypothalamic paraventricular nucleus (PVNp) release corticotropin releasing factor (CRF) into the hypophyseal portal system, which carries the CRF to the anterior pituitary where CRF stimulates the release of adrenocorticotropic hormone (ACTH). ACTH enters the general blood supply and reaches the adrenal cortex, where it stimulates the release of corticosterone or cortisol.

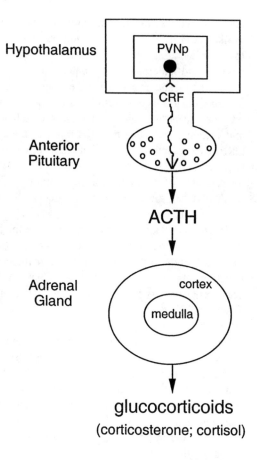

pituitary responsiveness to CRF is one of the factors that regulates HPA hyporesponsiveness, while decreased responsiveness to CRF released as a neurotransmitter at synapses in particular parts of the brain is one of the mechanisms that mediates the decreased fearfulness observed in lactating females. In support of such a view, it has been reported that the intracerebroventricular injection of CRF into virgin and lactating female rats has differential neural effects, with many more brain regions being activated by this neuropeptide in virgin females than in lactating females (Da Costa, Kampa, Windle, Ingram, & Lightman, 1997b; also see Da Costa, Wood, Ingram, & Lightman, 1996b). Interestingly, some of these differentially activated neural regions are known to be involved in mediating fearful behavior, such as the corticomedial amygdala, which is also involved in inhibiting maternal behavior.

Second, HPA hyporesponsiveness to stress during lactation may in part be regulated by decreased synthesis of CRF by parvocellular PVN neurons (Walker, Toufexis, & Burlet, 2001). Likewise, the reduction in fearfulness observed in postpartum females might also be affected by a reduction in the synthesis of

CRF by neurons in brain circuits that regulate anxiety-related behavior, since Walker et al. (2001) have reported that CRF mRNA levels in the amygdala are lower in lactating rats than in virgins. We will have more to say about CRF, anxiety, and maternal behavior in subsequent chapters.

A third mechanism that has been implicated in the HPA hyporesponsiveness to stress during lactation involves norepinephrine (NE). Brainstem NE neurons provide ascending input to a variety of telencephalic and diencephalic nuclei, and evidence indicates that NE input to the PVNp plays a stimulatory role in CRF release, and that this stimulatory effect is depressed during lactation (Toufexis et al., 1998; Toufexis & Walker, 1996; Windle et al., 1997a). Importantly, a depression of noradrenergic transmission in other parts of the brain may also be involved in the decreased fearfulness observed in postpartum rats (Toufexis et al., 1999).

Fourth, a prolactin system may promote the HPA hyporesponsiveness to stress during the postpartum period, since intracerebral administration of an antisense oligonucleotide to the prolactin receptor increases the HPA response to stress in lactating rats (Torner, Toschi, Nava, Clapp, & Neumann, 2002). Importantly, a brain prolactin system may also function to produce anxiolytic effects (Torner, Toschi, Pohlinger, Landgraf, & Neumann, 2001; also see Section 6 of the current chapter).

Finally, it has been proposed that HPA hyporesponsiveness to stress and decreased emotional reactivity during the postpartum period are both influenced by the release of oxytocin from neurons in the brain, with each effect presumably being mediated by oxytocin's action at different central neural sites (Lightman et al., 2001). Although other work indicates that a role for centrally released oxytocin in lactational HPA hyporesponsiveness is unlikely (Neumann et al., 2000), its role in postpartum fear reduction is likely, and we will have more to say about this later.

4. The Application of Motivational Models to the Maternal Behavior of Other Species

Compared to rats, not much work has been done on a motivational analysis of maternal behavior in other species. For hamsters (*Mesocricetus auratus*), rabbits (*Oryctolagus cuniculus*), and sheep (*Ovis aries*), there is evidence that the physiological events associated with late pregnancy and parturition modify the valence of olfactory input from infants so that such input becomes attractive rather than aversive:

1. Virgin female hamsters, which usually cannibalize test pups, tend to retrieve them after depression of the virgin female's primary olfactory and vomeronasal systems (Marques, 1979).

2. González-Mariscal (2001) has reported that lesions of the accessory olfactory bulb, which would interfere with vomeronasal input, promoted maternal re-

sponsiveness in virgin female rabbits. After several days of pup exposure, such females began to crouch over young pups while controls did not. To our knowledge, this is the first report of the experimental activation of maternal responsiveness in the nulliparous rabbit.

3. Nonpregnant ewes are repelled by the smell of amniotic fluids, but this repulsion is converted to a strong attraction at the time of parturition (Levy, Locatelli, Piketty, Tillet, & Poindron, 1995b; Levy, Poindron, & LeNeindre, 1983). This aversion and attraction to amniotic fluid has been shown to be dependent upon the main olfactory system, and is not dependent upon the vomeronasal system (Levy et al., 1995b). These findings are pertinent to the fact that the presence of amniotic fluid on the newborn lamb plays a facilitatory role in the establishment of maternal behavior in *primiparous* ewes (Levy & Poindron, 1987).

Therefore, as in the rat, for each of these species it is possible that the physiological events of late pregnancy and parturition modify the influence of olfactory input from infants so that such input no longer affects brain centers mediating aversion, but instead engages brain centers that mediate approach tendencies.

Related research on sheep is also consistent with a biphasic processes view of the onset of maternal behavior (Kendrick & Keverne, 1991). When nonpregnant multiparous ewes are treated with progesterone followed by its withdrawal, and then with estradiol and vaginocervical stimulation, they show increases in positive maternal responses (approach toward the lamb, emission of low-pitch bleats, sniffing and licking the lamb) and decreases in negative responses toward lambs (withdrawal from the lamb, aggression [head butts] directed at the lamb) when compared to untreated females. In contrast, when nonpregnant nulliparous females receive the same facilitatory treatment, they only show decreases in the negative responses without the concomitant increases in the positive maternal responses. One interpretation of these results is that the physiological events associated with natural pregnancy termination in primiparous ewes increase the attractive value and decrease the aversive properties of lamb-related stimuli, and that such physiological events are not completely replicated by progesterone withdrawal followed by estradiol and vaginocervical stimulation unless the treated females also have previous breeding experience. In other words, additional factors may be necessary to cause the positive motivational changes in the inexperienced ewe. Another interpretation is based on the fact that the females in the Kendrick and Keverne (1991) study were tested with dry lambs (not covered in amniotic fluid). Perhaps this resulted in the lack of positive responses in the nulliparous females, while for the multiparous females, other lamb characteristics had acquired attractive value as a result of experience.

A study on gerbils (*Meriones unguiculatus*) also supports elements of the motivational model we have presented. Clark et al. (1986) examined the responses of nulliparous gerbils to novel odors, such as the scent of lemon. They found that females who spent more time investigating a novel odor (low neo-phobia) also exhibited shorter sensitization latencies to the onset of maternal behavior when subsequently tested with pups. Presumably, the novel odor of the pups was more or less aversive to different females, and this influenced the onset of maternal responsiveness.

The case of laboratory strains of mice (*Mus musculus*) is a special one. As you will recall, nulliparous females from many strains of laboratory mice show spontaneous maternal behavior, so at first glance it would appear that the phys-iological events of pregnancy termination play no motivational role in such strains. However, as reviewed in Chapter 2, there is evidence that suggests otherwise. First, postpartum females are more likely to retrieve pups from a T-maze extension attached to the home-cage than are virgin females who are showing maternal behavior in home cage tests (Gandelman et al., 1970). Sec-ond, in an operant conditioning paradigm, postpartum females press a lever at a much higher rate than do maternal virgins when live pups are used as a reinforcing stimulus (Hauser & Gandelman, 1985). These results can be inter-preted as follows: As a result of selective breeding and inbreeding, the neural mechanisms underlying maternal behavior in many strains of laboratory mice have been modified such that pup stimuli can gain direct access to the neural circuitry mediating maternal behavior (recall that *wild* mice are not spontane-ously maternal). Pup-related stimuli, therefore, do not elicit aversive and with-drawal responses in virgin females of these strains, and these stimuli are also attractive to such females (such females do bar-press for pups, but at a lower rate than do postpartum females). However, the physiological events associated with the natural maternal condition (parturient and postpartum) appear to further increase the appetitive aspects of maternal behavior, as measured in the operant test, and may also cause a general reduction in fearfulness, as indicated by the T-maze retrieval data. (Also see Maestripieri & D'Amato [1991] for additional evidence that postpartum laboratory mice are generally less fearful than virgins.) Indeed, we would predict that under stressful environmental conditions, virgin mice would be poor mothers in comparison to postpartum females. The labo-ratory mouse seems to be an excellent model system for an animal population in which the primary effect of hormones and other physiological factors asso-ciated with pregnancy termination appears to be on appetitive and emotional, rather than consummatory, mechanisms.

Is there any evidence for nonhuman primates that the physiological events associated with pregnancy termination decrease the adult female's neophobic responses to infants, increase her attraction toward infants, and cause a general reduction in the postpartum female's general fearfulness? Some evidence does exist. First, it needs to be emphasized that most primates live in social groups and therefore are surrounded by young infants. Something related to sensiti-zation may occur under such conditions, particularly for those species in which the natural mother is permissive and allows other individuals to interact with her infant. Therefore, in many primate societies, a nulliparous female's fear of young and attraction toward them may be influenced by her exposure to infants during her ontogeny. However, at a female's first parturition, hormones may act on this "sensitized base" to further increase maternal responsiveness. Holman and Goy (1995) compared the response of nulliparous rhesus monkey (*Macaca mulatta*) females to infants that were introduced into the adult's home cage. Some of these nulliparous females were raised in feral groups and were therefore exposed to young during their development, while the remaining females were laboratory reared in association with their mother and peers but were not exposed to young infants during their development. Although neither the feral-reared nor the laboratory-reared females showed species-typical maternal behavior toward the test infant, the feral-reared females were clearly more attracted to and less fearful of the infant. The feral-reared females were more likely to approach and touch the infant and remain near it, while the laboratory-reared females appeared apprehensive in the presence of the infant and showed lip-smacking behavior, which Holman and Goy (1995) note is indicative of anxiety. These results should be contrasted with those of Gibber (1986), where laboratory-reared rhesus females, who were not exposed to young during their development, showed normal species-typical maternal behavior after their first parturition. These results suggest that for rhesus monkeys the physiological events associated with pregnancy termination can decrease neophobic responses to infant stimuli in naive females, while also increasing the appetitive (and perhaps consummatory) aspects of maternal behavior.

In related work, Pryce (1993) has compared the response of adolescent common marmoset (*Callithrix jacchus*) females to young infants. Some of these adolescents were raised in family groups in the presence of other infants, while the remaining experimental females did not have experience with younger infants. Both groups of females showed carrying behavior because the infants would climb on the adolescent and cling to her. However, the adolescents with previous infant experience carried the infants for longer durations, while the inexperienced females appeared to view the clinging infant as an aversive stimulus and pushed the infant off. These results should be compared to those of Pryce et al. (1993), who found that adult nulliparous marmoset females that had preadult alloparental experience and were also treated with progesterone and estradiol to mimic late-pregnancy hormone patterns showed higher operant bar-pressing rates for infant stimuli than did control females that also had preadult alloparental experience but that were not treated with steroids. Clearly, in the common marmoset, although alloparental experience increases maternal responsiveness, hormonal factors can promote even greater responsiveness.

With respect to a general reduction in fearfulness in nonhuman primates during the postpartum period, we can only provide very indirect evidence (but also see Section 5 in this chapter on maternal aggression). First, recall that adult *cycling multiparous* rhesus females will care for infants when such infants are introduced singly in the adult's home cage (Holman & Goy, 1980), but that such females show very low levels of infant handling when they are in a social group that contains mothers and their infants (Maestripieri & Zehr, 1998). In contrast, within a social group of macaques, infant handling by *pregnant* multiparous females of other females' infants does increase toward the end of pregnancy (Maestripieri & Zehr, 1998). Perhaps in multiparous macaques, hormonal changes associated with pregnancy termination are necessary to decrease fearfulness, in this way permitting a female to approach and contact an infant that is being guarded by its natural mother (it is also possible, of course, that such endocrine changes increase infant attractiveness, in this way causing approach behavior to be higher than avoidance). Second, as reviewed in Chapter 3, rhesus monkeys that are raised in social isolation and those that are peer-reared show deficits in maternal behavior as primiparous females. These deficits have been related to an increase in general fearfulness among these females. One interpretation of these results is that there is a normal spectrum of general fearfulness (temperament) that is compatible with normal maternal behavior in rhesus females, but if anxiety to novel stimuli becomes too great, maternal behavior (and other social behaviors) breaks down or deteriorates. Although the physiological events associated with pregnancy termination may act to decrease anxiety, in the case of social isolates and peer-reared monkeys, this effect may not occur or may not be sufficient to allow for normal maternal responsiveness. In Chapter 3 we also reviewed the evidence that in normally reared monkeys there is a natural variation in temperament, with high-anxiety females showing increased protectiveness toward their infants. Furthermore, we reviewed the evidence that normally reared high-anxiety female macaques tend to show higher levels of abusive behavior toward their infants. In the context of the other work reviewed in this chapter, we tentatively conclude that a decrease in general fearfulness allows for better maternal behavior in nonhuman primates, that the physiological events associated with the postpartum condition probably act to decrease general fearfulness in primates, and that maternal behavior is likely to break down under highly stressful environmental conditions in those female primates that are on the high end of the anxiety/fearfulness spectrum.

With respect to humans even less data are available, and therefore our conclusions with respect to the influence of physiological variables on motivational events related to the maternal condition are even more tentative. First, we have the correlational data of Fleming et al. (1997a), which indicated that primiparous women felt more attached to their infants on day 1 postpartum as their estradiol-to-progesterone ratio increased toward the end of pregnancy. Second, in related work, Fleming, Corter, Franks, Surbey, Schneider, and Steiner (1993) have found that primiparous women on days 2–4 postpartum rate infant-related odors as more attractive than do nulliparous women. Third, we note the reviews of Carter and Altemus (1997) and Uvnas-Moberg (1997), who have been strong proponents of the view that lactation in women is associated with decreases in general fearfulness and stress reactivity, and that these effects may help mothers care for and protect their offspring under stressful and threatening situations (also see Altemus, Deuster, Galliven, Carter, & Gold, 1995). For example, Sjogren, Widstrom, Edman, and Uvnas-Moberg (2000) have found that in primiparous women, somatic anxiety measures decrease between pregnancy and the postpartum period, and anxiety levels are lower in those females who breastfeed postpartum in comparison to those females who choose to bottle-feed their infants (also see Nissen, Gustavsson, Widstrom, and Uvnas-Moberg, 1998). Importantly, nursing and nonnursing mothers did not differ in their anxiety levels during pregnancy. In another study, Wiesenfeld, Malatesta, Whitman, Granose, and Uli (1985) found that nursing women showed a lower level of sympathetic activation in response to infant stimuli than did woman who were bottle-feeding their babies, suggesting that the former females were calmer and more relaxed. A problem with this study, however, is that prepartum measures of autonomic arousal were not taken. Finally, in a preliminary report, Klein, Skrobala, and Garfinkel (1994/1995) note that for patients with a history of panic disorder, the incidence of the disorder decreases during pregnancy (also see Cowley & Roy-Byrne, 1989) and then remains low during the postpartum period if the mother is lactating, but increases postpartum in mothers who bottle-feed their infants. Note that for all of these studies on postpartum anxiety in women, the suggestion is that the physiological events associated with nursing, and not other apsects of postpartum maternal behavior, are related to anxiety reduction (also see Uvnas-Moberg, Widstom, Nissen, & Bjorvell, 1990). Finally, in Chapter 9 we will review additional relevant data that indicate that abnormally high levels of anxiety (and depression) during the postpartum period are related to deficits in human maternal behavior, suggesting that postpartum anxiety levels need to be maintained within a normal range for adequate maternal behavior to occur.

5. Maternal Aggression

5.1. Introduction

We have presented evidence that the physiological events associated with pregnancy termination and the postpartum period act to cause a reduction in the maternal female's general fearfulness and that this motivational change may play an important supportive or ancillary function for the postpartum female by allowing her to function effectively under stressful environmental conditions. In this context, we want to discuss the phenomenon referred to as maternal aggression, which is the heightened aggressiveness shown by lactating females in a variety of species (Ostermeyer, 1983; Rosenblatt, Factor, & Mayer, 1994; Svare, 1981, 1990). After characterizing maternal aggression and describing aspects of its regulation, we want to explore whether the decrease in general fearfulness shown by the postpartum female is a necessary prerequisite for the occurrence of such aggression.

Although the increased aggressiveness of postpartum females has been documented in a variety of species, most of the important experimental work has been done on rodents. In laboratory-housed rhesus monkeys, Harlow, Harlow, and Hansen (1963) noted the high level of experimenter-directed threats displayed by maternal rhesus monkeys. For free-ranging langurs (*Presbytis entellus*), among whom high levels of allomothering occur, Jay (1963) pointed out that the natural mother can take her infant back from any female, irrespective of the allomother's rank. Also, for free-ranging langurs, Hrdy (1977) reported that mothers with infants are particularly aggressive toward strange (non–troop member) males that approach the troop. Weisbard and Goy (1976) have reviewed evidence that indicates that for a variety of primate species (baboons, macaques, langurs, patas monkeys) the rank of the female within the troop increases postpartum, and that such females may act aggressively toward group members that threaten their infants even if the mother was subordinate to such individuals prior to parturition (also see Maestripieri, 1994b). Finally, Troisi, D'Amato, Carnera, and Trinca (1988) have presented quantitative data, along with a statistical analysis, which indicate a clear increase in aggression shown by lactating females when compared to their nonlactating counterparts in a troop of Japanese macaque (*Macaca fuscata*) monkeys.

In standard laboratory tests for aggression in rodents, an adult intruder male or female is placed in the home cage of a resident female, and the number of resident females fighting, the latency to attack, the number and duration of attacks, and the display of submissive postures by the intruder are usually recorded. In a variety of rodent species, lactating resident females have been found to be much more aggressive than nonlactating resident females: house mice (Svare, 1981, 1990), rats (Erskine, Barfield, & Goldman; 1980b; Mayer, Reisbick, Siegel, & Rosenblatt, 1987b), hamsters (Siegel, Giordano, Mallafre, & Rosenblatt, 1983b; Wise, 1974), and various *Peromyscus* mouse species (Wolff, 1985).

With respect to the functions served by maternal aggression, a dominant view is that such behavior protects the young from infanticide by conspecific males and females (Hausfater & Hrdy, 1984; Parmigiani, Palanza, Mainardi, & Brain, 1994). Male rodents and primates may increase their fitness by killing alien young because the mother of the young would then cease to lactate and she would therefore resume mating and ovulating, in this way allowing the infanticidal male to sire his own offspring. Infanticide by nonlactating female rodents toward alien young may be a type of female-female competition for safe nesting sites. Importantly, there is some observational and experimental work which indicates that maternal aggression can operate to prevent the occurrence of infanticide (for review, see Numan, 1994).

5.2. Factors That Regulate the Occurrence of Maternal Aggression

In rats and mice, maternal aggression is particularly high during the first two postpartum weeks (Mayer et al., 1987b; Svare, 1990), suggesting that the presence of young pups might be a factor promoting its occurrence. Research has shown that although short-term removal of young pups from the postpartum female's cage does not depress aggression, long-term removal (5 to 24 hours) does so in mice (Gandelman, 1972; Svare & Gandelman, 1973), rats (Erskine,

Barfield, & Goldman, 1978; Ferriera & Hansen, 1986; Stern & Kolunie, 1993), and hamsters (Giordano, Siegel, & Rosenblatt, 1984). Importantly, this depression of aggression is reversed if pups are returned to the home cage several minutes prior to the aggression test.

Given that the presence of young promotes the occurrence of maternal aggression, a hypothesis is that the young stimulate endocrine secretions in the dam and that postpartum aggression is based on this pup-induced hormonal state. For example, perhaps suckling-induced prolactin release is involved in stimulating maternal aggression. Such a hypothesis is not likely given the findings that postpartum hypophysectomy does not disrupt maternal aggression in mice (Svare, Mann, Broida, & Michael, 1982; also see Broida, Michael, & Svare, 1981; Mann, Michael, & Svare, 1980) and rats (Erskine, Barfield, & Goldman, 1980a). The dominant view, therefore, is that pup stimulation has direct effects on the neural circuits underlying maternal aggression.

With respect to the nature of pup stimuli that promote maternal aggression, research has emphasized the importance of suckling stimulation during the early postpartum period in mice and the importance of ventral somatic sensory input from nuzzling pups in rats. For example, thelectomy performed on day 1 postpartum disrupts maternal aggression in mice (Svare & Gandelman, 1976b), and although thelectomy does not disrupt maternal aggression in rats, cutaneous anesthesia of the mother's ventral trunk region does (Mayer et al., 1987a; Stern & Kolunie, 1993). Importantly, these treatments that disrupt maternal aggression in mice and rats do not disrupt retrieval behavior in home-cage tests. These results suggest that in rats and mice, ventral trunk somatic sensory stimulation activates neural pathways in the central nervous system that ultimately promote the occurrence of maternal aggression. Additional research indicates that olfactory input from pups may play a subsidiary role in maintaining postpartum maternal aggression (Ferreira & Hansen, 1986; Kolunie & Stern, 1995; Mayer & Rosenblatt, 1993), while visual and auditory inputs from pups do not play a significant role (Ferreira & Hansen, 1986; Kolunie, Stern, & Barfield, 1994).

Although hormonal factors during the postpartum period do not appear to be involved in the display of maternal aggression, this does not rule out the possibility that the hormonal events associated with pregnancy and pregnancy termination act to set up or prime the relevant neural circuits so that they become responsive to ventral somatic sensory and olfactory inputs from pups, in this way allowing pup stimulation to maintain maternal aggression during the postpartum period. Indeed, the evidence supports this possibility. First, as shown in Figure 4.12, the maternal behavior that is shown by *virgin* rats and mice (sensitized maternal behavior) is not associated with the display of high levels of aggression toward intruders (Erskine et al., 1980b; LeRoy & Krehbiel, 1978; Mayer & Rosenblatt, 1987; Svare & Gandelman, 1976a, 1976b). Such maternal virgins lack exposure to the endocrine events of pregnancy. Second, in both rats and mice, levels of aggression increase during the last third of pregnancy, which is prior to pup stimulation, suggesting that the endocrine events of late pregnancy may be setting up the neural circuitry that promotes aggressive ten-

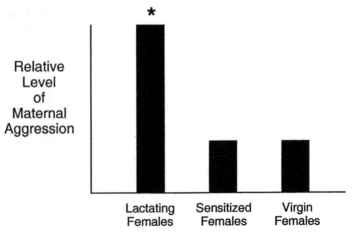

Figure 4.12. Relative level of maternal aggression in maternal lactating female rats, maternal (sensitized) virgin female rats, and nonmaternal virgin rats. In a 20-minute test, maternal aggression toward a male intruder in the female's home cage is highest for lactating females, who showed a mean number of about 17 attacks, compared to a mean number of about 5 attacks for each of the remaining groups. *Significantly different from remaining groups. Modified from Erskine et al. (1980b).

dencies toward intruders (Mann, Konen, & Svare, 1984; Mayer et al., 1987b; Noirot, Goyens, & Buhot, 1975). The aggression shown by rodents during pregnancy is sometimes referred to as pregnancy-induced aggression to distinguish it from postpartum maternal aggression, and the former behavior is usually less intense than the latter. Third, for both rats and mice, hormonal treatments that mimic pregnancy hormone profiles are capable of stimulating aggression in virgins (Mann et al., 1984; Mayer, Ahdieh, & Rosenblatt, 1990a; Mayer, Monroy, & Rosenblatt, 1990b). The essential hormonal requirements that induce this increased aggression in virgin rats and mice in the absence of pup stimulation appear to be high (pregnancy-like) levels of progesterone on a background of estradiol. Importantly, in rats this endocrine facilitation of aggression declines quickly following the termination of the hormone treatment if the females are not exposed to pups, but is maintained (and increased) if the females are exposed to pups and show maternal behavior (Mayer et al., 1990b).

In rats, many of the hormone treatments that have been found to activate the onset of aggression in nulliparous females (simulating pregnancy-induced aggression) are similar to those that activate maternal behavior in such females. This raises the possibility that pregnancy-induced aggression in rats and the onset of maternal behavior may share a common hormonal basis. Recent work suggests that this is not the case, as progesterone withdrawal and estradiol-induced prolactin release are necessary requirements for the hormonal induction of maternal behavior in virgin rats (see Chapter 2), but they are not necessary for activating aggressive behavior (Mayer et al., 1990a, 1990b).

In summary, the available evidence suggests that the hormonal events of mid-to-late pregnancy increase aggression in both rats and mice and also allow pup stimulation to further increase and maintain the aggression during the postpartum period in the absence of further direct hormonal involvement. The following should be noted: Maternal behavior can occur in rats and mice in the absence of maternal aggression (sensitized maternal behavior), and hormone treatments that activate aggression in virgin rats do not necessarily induce short-latency maternal behavior. These results suggest that the mechanisms underlying the increased aggression in periparturitional rodents are different from those underlying maternal behavior, and that such mechanisms are neither necessary nor sufficient for the occurrence of maternal behavior.

5.3. The Relationship between Maternal Aggression and Fearfulness in Postpartum Rodents

In this section we want to explore the possibility that one role for a reduction in general fearfulness during the postpartum period is to allow the rodent dam to protect her litter from intruders at the nest site. Hansen, Ferreira, and Selart (1985) were the first to draw attention to the possible relationship between postpartum reductions in fear and the occurrence of maternal aggression. They noted that virgin rats that are treated with antianxiety agents such as diazepam show both decreased fearfulness and increased aggression. Since diazepam and other benzodiazepines act at the benzodiazepine binding site on GABA$_A$ receptors to facilitate GABAergic transmission, in this way presumably increasing inhibition on neural areas regulating anxiety (Ballenger, 1995), they suggested that certain aspects of motherhood may be associated with increased GABAergic transmission in certain central neural sites. In support of this view, they found that when postpartum females were treated systemically with benzodiazepine antagonists, such treatment decreased maternal aggression and increased fearfulness (as measured by the duration of the freezing response to a loud acoustic stimulus). Mos and Olivier (1986) were not able to replicate the depressing effect of a benzodiazepine antagonist on maternal aggression in postpartum rats, but since they did not have an independent measure of fearfulness we are not able to determine whether the inverse relationship between fear and aggression was disrupted in their study. Perhaps their drug treatment was not effective in increasing anxiety.

Since the report of Hansen et al. (1985), other reports have appeared that support the idea that decreased postpartum fearfulness may be a factor involved in promoting maternal aggression:

1. Several studies have shown that benzodiazepine administration has biphasic effects on maternal aggression in rodents, with low doses increasing aggres-

sion, and high doses decreasing aggression (Olivier, Mos, & van Oorschot, 1985, 1986; Parmigiani, Ferrari, & Palanza, 1998; Yoshimura & Ogawa, 1991). It has been suggested that low doses increase aggression because of antianxiety effects, while high doses decrease aggression because of the sedative properties of global inhibition.

2. Maestripieri and D'Amato (1991) measured both maternal aggression and fearfulness in postpartum mice. Fearfulness was measured by a light/dark choice test, in which the amount of time spent in a lighted compartment was taken as a measure of anxiety (increased time in light indicative of decreased anxiety). They found that the mice that were the least anxious were also the most aggressive.

3. Parmigiani, Palanza, Rodgers, and Ferrari (1999) measured anxiety and maternal aggression in lactating females from several inbred and outbred mouse strains. They found that those strains that were least fearful (as measured by the amount of exploration in a novel open field) were also those that showed the highest level of maternal aggression.

4. Maestripieri, Badiani, and Puglisi-Allegra (1991) found that prepartum stress in mice (restraint stress over days 4–15 of pregnancy), while not interfering with postpartum maternal behavior, does decrease maternal aggression while at the same time increasing postpartum anxiety, as measured in the light/dark choice test (also see Pardon, Gerardin, Joubert, Perez-Diaz, & Cohen-Salmon, 2000).

5. Lonstein, Simmons, and Stern (1998a) have reported that brain lesions that decreased fearfulness in postpartum rats (as measured by increased exploration in the open arms of an elevated plus maze) caused an increase in the already high levels of maternal aggression shown by these females.

6. Boccia and Pedersen (2001) found that postpartum females who were exposed to prolonged maternal separations as neonates were more anxious (as measured on the elevated plus maze), licked and groomed their pups less, and showed lower levels of maternal aggression than did control females.

There is one piece of evidence that is not congruent with the view that a decrease in general fearfulness may be important for the occurrence of maternal aggression. Recall that Neumann et al. (1998a) reported that day-21 pregnant rats are more anxious than virgin females when tested on the elevated plus maze. In contrast, Mayer et al. (1987b) have reported that pregnancy-induced aggression is apparent in rats on day 21 of pregnancy. A more careful analysis needs to be done on this issue, particularly with respect to measuring both anxiety and pregnancy-induced aggression in the same animals. It is possible, however, that the low-intensity pregnancy-induced aggression does not require a reduction in fearfulness, while the high-intensity maternal aggression that is shown by the postpartum female does.

Most of these results are consistent with the view that the decreased fearfulness of the postpartum female may be a factor in enabling her to defend her young and nest site against both male and female intruders. It should be noted that we are not saying that the decreased fearfulness of the postpartum female is a sufficient condition for maternal aggression. Indeed, various neural interventions have been found to decrease maternal aggression in rats without affecting their fearfulness (Ferreira, Dahlof, & Hansen, 1987; Hansen & Ferreira, 1986a, 1986b). It should be realized, however, that maternal aggression probably has its own neural underpinnings that are separate from neural structures regulating the animal's fear state. We do think that the evidence presented here does support the view that a certain reduction in general fearfulness may be necessary for the occurrence of effective levels of maternal aggression.

Progesterone has been found to have anxiolytic effects in many animal models of anxiety (Bitran, Shiekh, & McLeod, 1995; Picazo & Fernandez-Guasti, 1995). Importantly, the anxiolytic effects of progesterone have been found to be the result of its conversion to reduced metabolites such as allopregnanolone (Bitran et al., 1995; Smith, 2002; Smith et al., 1998; Steimer, Driscoll, & Schulz, 1997). In addition, the anxiolytic effects of these reduced metabolites of progesterone have been found to be mediated by an action at the GABA receptor: Like benzodiazepines, allopregnanolone potentiates the effects of GABA at the GABAA receptor (Bitran et al., 1995; Brot, Akwa, Purdy, Koob, & Britton, 1997). Since progesterone levels are high in plasma in the postpartum rat (Kellogg & Barrett, 1999), one can ask whether progesterone's metabolites are important for both the reduced anxiety and increased aggression of postpartum rats. Since hypophysectomy does not disrupt maternal aggression in rats (Erskine et al., 1980a), it is unlikely that progesterone or its reduced metabolites are necessary for the occurrence of aggression. Consistent with the view that decreased fearfulness may be necessary for maternal aggression, Kellogg and Barrett (1999) have reported that reduced metabolites of progesterone are also not involved in the decreased fearfulness (measured in the elevated plus maze) that is displayed by postpartum rats. It would be interesting to explore the role of the reduced metabolites of progesterone in pregnancy-induced aggression, during which progesterone (along with estrogen) has been shown to play a positive role.

6. General Conclusions

In this chapter we have analyzed the motivational changes associated with the onset and maintenance of maternal behavior. In doing so, we have been very vague about the specific peripheral physiological events associated with pregnancy, parturition, and lactation that appear to underlie the motivational effects we have described. This vagueness results from the fact that research has not focussed on these issues. The best that can be said is that the hormonal and other physiological events associated with late pregnancy and parturition (which, for most species, would include rising estradiol and lactogen levels, declining progesterone levels, and vaginocervical stimulation) probably act on the brain so that infant stimuli become attractive rather than aversive. Aspects of these same physiological events may also cause a reduction in general fearfulness, although differences in the underlying basis of these two effects are likely to exist. All of these motivational changes may then be maintained during the postpartum period as a result of the interaction of the mother with her young.

A recent finding has suggested that prolactin action at the level of the brain might contribute to the reduction in general fearfulness that is shown by postpartum females. Torner et al. (2001) found that intracerebral infusion of prolactin into the lateral ventricle of virgin female rats caused an anxiolytic effect on the elevated plus maze. However, it is unlikely that pituitary prolactin, which is released by suckling stimulation and maintains lactogenesis (Numan, 1994), is necessary for the reduction in general fearfulness shown by postpartum rats since hypophysectomized female rats show both postpartum maternal aggression and hormone (estradiol plus progesterone)-induced aggression (Erskine et al., 1980a; Mayer et al., 1990a). Additionally, postpartum rats who are treated with a drug that blocks the release of prolactin from the anterior pituitary are still capable of retrieving pups from a T-maze extension attached to their home cage (Stern, 1977). However, as we will see in Chapter 6, there is also a brain prolactin system in the sense that central neurons exist that synthesize and release a prolactin-like compound as a neurotransmitter/neuromodulator. Perhaps the physiological events of late pregnancy and parturition promote the development and activity of central neural prolactin circuits, that activity in this system is then subsequently maintained by postpartum infant stimulation, and that this system contributes to fear reduction.

A similar possibility might also apply to a brain oxytocin system. Recent work has shown that the action of oxytocin at central neural sites produces an anxiolytic effect (Bale, Davis, Auger, Dorsa, & McCarthy, 2001; McCarthy, McDonald, Brooks, & Goldman, 1996; Neumann et al., 2000; Uvnas-Moberg, Eklund, Hillergaart, & Ahlenius, 2000; Windle, Shanks, Lightman, & Ingram, 1997b). Although oxytocin released from the neurohypophysis is required for the milk-ejection reflex, it is unlikely that significant amounts of systemic oxytocin can cross the blood-brain barrier. However, during the postpartum period, activation of some of the neural circuits that comprise the brain oxytocin system may contribute to fear reduction (see Chapters 6 and 8).

Brain oxytocin and prolactin systems might not only be involved in the reduction in general fearfulness that occurs in postpartum females, but they may also participate in the motivational changes that result in shifting the valence of infant-related stimuli from negative to positive. But we are getting ahead of ourselves at this point. In the next two chapters, these issues will become more resolved as we define the central neural circuits that are involved in maternal responsiveness and define the loci within these circuits where hormones act to produce their effects (Chapter 5), and as we also delineate the contribution of central oxytocin, prolactin, and other neurochemical systems to the control of maternal behavior (Chapter 6).

5

Neuroanatomy of Maternal Behavior

1. Introduction

In a review of the neural structures and circuits that regulate the occurrence of maternal behavior in mammals, this chapter will begin with an analysis of the sensory basis of maternal behavior, and will then move to a discussion of central neural mechanisms. This chapter's emphasis will be on presenting a functional neuroanatomy of maternal behavior. We will be interested in determining the neural circuits that are influenced by hormones and by sensory stimuli from infants, and in determining the functions served by these circuits so that maternal responsiveness is altered. This functional analysis will take place within the context of the motivational models that were presented in the previous chapter. In short, we will present evidence that there are dual neural mechanisms regulating maternal behavior: an inhibitory one that promotes avoidance and withdrawal from infant stimuli, and an excitatory system that promotes approach and interaction with infants. Maternal behavior occurs when the excitatory system is dominant over the inhibitory system. Once this analysis is complete, the next chapter will present the data on the neurochemistry and molecular neurobiology of maternal behavior, integrating that data into the neural circuits developed here.

2. Sensory Basis of Maternal Behavior

2.1. Introduction

When a primiparous female mammal gives birth and shows immediate maternal responsiveness toward her young, what sensory stimulus or stimuli from her young is she responding to that allows for the performance of maternal behavior? And in those species, such as sheep, that form a specific attachment to their own young so that they will not care for alien young (young from another mother), what are the sensory cues that allow the mother to discriminate her own from alien young? Since most of the research on these issues has been performed on rodents (rats [*Rattus norvegicus*] and mice [*Mus musculus*]) and sheep (*Ovis aries*), we will focus our attention on these species.

An understanding of the sensory cues from young that are utilized by maternal females in the performance of their infant-directed responses is important because such cues serve as the afferent limb directed toward central neural mechanisms. Therefore, knowledge of such sensory mechanisms is not only important for its own sake, but is also important because it can help us define central neural structures involved in the regulation of various aspects of maternal behavior.

2.2. Sensory Basis of Maternal Behavior in Rodents

Laboratory strains of rats and mice typically do not form a specific attachment to their own young. Although they are capable of discriminating their own young from the young of another mother, this discrimination does not prevent them from caring for alien young (Beach & Jaynes, 1956b; Ostermeyer & Elwood, 1983; Rosenblatt & Lehrman, 1963). Therefore, in an analysis of the sensory basis of maternal behavior in rodents we are primarily interested in those sensory cues used by mothers to generally recognize pups as such so that they can then perform the appropriate maternal responses.

An influential position on the sensory basis of maternal behavior in rats was taken by Beach and Jaynes (1956c) when they developed their concept of multisensory control. According to this view, although many infant-related stimuli may influence maternal responses in rats, no single sensory modality is essential for the performance of adequate maternal behavior. The main study that supported this position investigated the sensory cues from pups that were necessary for the performance of retrieving behavior in lactating rats. Beach and Jaynes (1956c) found that the surgical elimination of vision, olfaction, or tactile sensitivity of the snout and perioral region did not prevent the retrieval response. Additionally, elimination of any two or all three of these sensory modalities, although causing deficits in retrieving, did not abolish the behavior. The multisensory view was further supported in a study by Herrenkohl and Rosenberg (1972). Different groups of primigravid female rats were blinded, deafened, or rendered anosmic during pregnancy and it was found that their subsequent maternal behavior (retrieving and nursing) remained intact during the postpartum period.

Although behavioral studies have clearly shown that both auditory and olfactory stimuli from pups are utilized by maternal rats in the performance of retrieval responses (for example, such stimuli aid the maternal female in locating pups that are displaced from the nest) and other aspects of maternal behavior (Allin & Banks, 1972; Smotherman, Bell, Starzec, Elias, & Zachman, 1974; White, Adox, Reddy, & Barfield, 1992), the surgical elimination studies suggest that neither of these sensory cues is essential for the occurrence of maternal behavior. The multisensory view argues, therefore, that multiple cues are used by a maternal rat to respond to her young, and as long as some of these cues can gain access to the maternal brain, the appropriate behavior will occur.

The early studies of Beach and Jaynes (1956b) and Herrenkohl and Rosenberg (1972) need to be reevaluated in the context of more recent data. Subsequent research has basically confirmed the data on vision and hearing, neither of which is essential for maternal behavior in rats (Ihnat, White, & Barfield, 1995; Kolunie et al., 1994). However, we will need to update the roles of olfaction and tactile sensitivity in the regulation of maternal behavior in rodents.

2.2.1. Olfaction and Maternal Behavior in Rodents

Recall that there are two major chemosensory structures in the nasal cavity of rodents: the primary olfactory sensory neurons that project centrally to the main olfactory bulb, and the vomeronasal sensory neurons that project to the accessory olfactory bulb (see Figure 4.2). Two findings have confirmed the earlier work that indicated that neither of these sensory systems is essential for maternal behavior in rats: (a) Peripheral damage to the primary olfactory sensory neurons with intranasal infusions of zinc sulfate does not disrupt the maternal behavior of postpartum rats, except perhaps for minor retrieval deficits (increased latencies to retrieve) because females take slightly longer to locate pups (Benuck & Rowe, 1975; Jirik-Babb, Manaker, Tucker, & Hofer, 1984; Kolunie & Stern, 1995); and (b) cutting the vomeronasal nerves or removal of the vomeronasal organ does not interfere with postpartum maternal behavior in rats (Fleming, Gavarth, & Sarker, 1992; Jirik-Babb et al., 1984; Kolunie & Stern, 1995), except perhaps for modifications in the organization of anogenital licking behavior (Brouette-Lahlou, Godinot, & Vernet-Maury, 1999). In each of these studies, the primary olfactory system or the vomeronasal system was disrupted prior to parturition in primigravid females.

Some controversy has arisen when the effects of bilateral olfactory bulbectomy on the maternal behavior of rats have been explored. In confirmation of the work of Beach and Jaynes (1956b) and Herrenkohl and Rosenberg (1972), Fleming and Rosenblatt (1974b) have reported that primiparous female rats that had been bulbectomized during pregnancy showed normal maternal behavior during the early postpartum period. In contrast, Benuck and Rowe (1975) and Kolunie and Stern (1995) have reported moderate deficits in the maternal behavior of primiparous rats that were bulbectomized prior to parturition. A major deficit appeared to be in parturition behavior: Not all pups were cleaned of their membranes and placental attachments, and within a few hours after parturition there was a higher level of pup mortality in bulbectomized females when compared to controls. It is worth noting, however, that in both of these studies most of the pups were eventually retrieved to a nest site, nursed, and raised to weaning. Therefore, bulbectomy did not eliminate maternal behavior.

What can account for the possibly larger deficits in postpartum maternal behavior observed after bilateral olfactory bulbectomy when compared to the effects of peripheral deafferentation of either the primary olfactory sensory neurons or the vomeronasal system? One possibility, of course, is that bilateral bulbectomy destroys both systems at the same time since both the main and accessory olfactory bulbs are destroyed. Another possibility is that in addition to causing anosmia, bulbectomy may interfere with nonsensory functions. The olfactory bulbs are part of the central nervous system and their neurons receive afferent input from other central nervous system nuclei as far away as the brainstem (Shipley & Ennis, 1996). Destruction of axon terminals from afferent neurons outside the bulb could disrupt neural functions in ways unrelated to olfaction. Finally, Hatton's group has shown that the supraoptic nucleus of the hypothalamus receives glutamatergic inputs from both accessory and main olfactory bulb efferents (Smithson, Weiss, & Hatton, 1989; Yang, Smithson, & Hatton, 1995). The supraoptic nucleus contains magnocellular oxytocin neurons that project to the neural lobe of the pituitary, and one function of oxytocin release from the neurohypophysis is to stimulate the uterine contractions associated with parturition (Swanson & Sawchenko, 1983). Therefore, a likely possibility is that damage to both the main and accessory olfactory bulb interferes with oxytocin release during parturition. As females sniff and lick partially expelled pups during parturition, the olfactory input may promote oxytocin release, which in turn aids in the expulsion of the pup (cf. Hatton & Yang, 1990). Without such stimulation, parturition may be difficult and prolonged, leading to increases in pup mortality and pup debilitation. In support of such a view, we have found that when primiparous bulbectomized females are given a healthy foster litter at the time of parturition, their behavior toward this healthy litter is normal (Numan & Numan, 1995). We tentatively conclude, therefore, that the overall results support the original findings of Beach and Jaynes (1956c) that olfactory input, whether primary, vomeronasal, or both, is not essential for the occurrence of adequate maternal behavior in rats. The increased pup mortality observed in primiparous parturient females with bilateral olfactory bulbectomies is likely due to an interference with sensory regulation of a neuroendocrine event rather than sensory regulation of neurobehavioral phenomena.

The conclusion that chemosensory input from the nasal cavity is not essential for maternal behavior in rats is also strongly supported by the data presented in the previous chapter: Damage to both the primary olfactory system and the vomeronasal system actually *facilitates* the sensitization process in virgin female rats (see Figure 4.3). In Chapter 4 we also reviewed the evidence for rats that the physiological events of late pregnancy shift the valence of pup-related odors from negative (within the context of other pup stimuli) to positive, and this shift seems important for the onset of maternal behavior at parturition. However, as noted in Chapter 4, the increased attraction and preference for pup-related odors per se are not essential for normal postpartum maternal behavior since anosmic females show maternal behavior. In rats, this valence shift is important because it prevents, decreases, or overcomes the ability of pup-related odors to activate brain areas that arouse fear and withdrawal from pups in naive females. In other words, the primary and essential function of the valence shift appears to be to prevent the inhibition of maternal behavior by novel pup odors rather than to directly promote the occurrence of maternal behavior.

Can Beach and Jaynes's (1956c) multisensory view of maternal behavior, which was proposed from data on the rat, be extended to other rodent species? The answer is no, at least when we consider the involvement of olfaction in the maternal behavior of mice. Olfactory bulbectomy, whether performed during pregnancy or the postpartum period, eliminates pup-care behaviors in mice (Gandelman, Zarrow, & Denenberg, 1971a; Gandelman, Zarrow, Denenberg, & Myers, 1971b). Such females either cannibalize their pups or ignore them. Similar disruptions in the maternal behavior of primiparous mice are observed after intranasal application of zinc sulfate, which damages the primary olfactory neurons (Seegal & Denenberg, 1974; Vandenbergh, 1973). Since removal of the vomeronasal organ does not interfere with maternal behavior in mice (Lepri, Wysocki, & Vandenbergh, 1985), these results suggest that olfactory detection of pups via the primary olfactory system is essential for the occurrence of maternal behavior in mice. The mouse results clearly show that the sensory basis of maternal behavior needs to be examined separately in each species and that broad generalizations from a single species should not be made. With respect to research on the role of olfaction in the maternal behavior of other rodent species, it should be noted that the hamster (*Mesocricetus auratus*) seems to react to olfactory deafferentation in a manner similar to the rat (Marques, 1979).

2.2.2. The Role of the Somatosensory System and Tactile Input in the Control of Maternal Behavior in Rodents

Stern (1996a) has argued that since mammalian mothers spend so much time in physical contact with their young, one would expect somatic sensory inputs to play an important role in the regulation of maternal responsiveness. Although Beach and Jaynes (1956c) reported that surgical elimination of tactile sensitivity of the snout and perioral region did not prevent the retrieval response in postpartum rats, they did not indicate the time interval between their trigeminal nerve transections and their tests for retrieval, nor did they verify that their deafferentation procedure actually caused a lack of perioral somatosensation. Therefore, it is certainly possible that peripheral regeneration of trigeminal connections to perioral regions, with a concomitant recovery of function, resulted in their negative findings. Indeed, subsequent research has indicated that perioral tactile sensitivity is much more important for the occurrence of retrieval behavior and other aspects of maternal responsiveness than was originally suggested in the classic work of Beach and Jaynes.

Kenyon, Cronin, and Keeble (1981, 1983) reported that injection of lidocaine (a local anesthetic) into the mystacial pads (which desensitize the snout region) or section of the infraorbital branch of the trigeminal nerve (which blocks sensory input from the snout and upper lip) disrupts the retrieval response in lactating rats. The dams approach displaced pups and nose them but, presumably because of a deficit in tactile sensory feedback, do not retrieve them. These findings have been confirmed by work from Stern's laboratory (Stern, 1996b; Stern & Kolunie, 1989, 1991). Some of these results are shown in Figure 5.1. Figure 5.1A shows that injection of lidocaine into the mystacial pads interferes with retrieval in primiparous females (3–5 days postpartum) for at least 60 minutes following the injection, with recovery by 120 minutes (Kenyon et al., 1983). Recovery of the retrieval response coincided with recovery of perioral tactile sensitivity, as indicated by various neurological tests. However, with repeated daily lidocaine injections followed by retrieval tests, retrieval behavior eventually recovers during the time period when the perioral region is still anesthetized (Kenyon et al., 1981). Likewise, one can see in Figure 5.1B that the retrieval behavior of postpartum primiparous dams recovers within 24 hours after infraorbital trigeminal transection even though neurological tests indicate a lack of tactile sensitivity of the perioral region (Kenyon et al., 1983). The recovery from the effects of infraorbital trigeminal transections has been confirmed by Stern and Kolunie (1991), and Stern and Kolunie (1989) have also reported that pretreatment retrieval test experience prevents the drastic effects of mystacial pad lidocaine injections on the retrieval response (also see Stern, 1997b). These results have led Kenyon et al. (1981, 1983) to conclude that although perioral tactile sensation is important for retrieval, it is not indispensable. This conclusion would seem to conform to Beach and Jaynes's (1956c) multisensory view of the sensory control of maternal behavior in rats. However, in the face of continued perioral desensitization, recovery could be mediated by the use of tactile sensory inputs from the mandibular region (lower lip and chin) served by the trigeminal nerve, or through the use of other nontactile sensory modalities (vision, for example). If the latter occurred, this would be consistent with the multisensory view, but if the former were the case, it would suggest the primary importance of trigeminal somatosensation for the regulation of retrieval behavior in rats. In partial support of the former position, Stern and Kolunie (1991) found that trigeminal deafferentation of somatic sensory inputs from the snout, upper lip, lower lip, and chin resulted in more severe retrieval disturbances than did infraorbital deafferentation alone, which would only desensitize the snout and upper lip. However, retrieving was still not permanently abolished in the postpartum females with the more extensive deafferentations, but instead recovered by the third postoperative day. Stern (1990) has suggested that even more extensive deafferentations, which would include intraoral afferents, would probably be needed to permanently interfere with retrieving in rats. However, such extensive deafferentations would also interfere with other oral motor responses. For example, extensive trigeminal deafferentations have been found to interfere with ingestive behavior (Zeigler, Jacquin, & Miller, 1985). If an animal cannot open its mouth properly to eat food, it would not be surprising to find that it would also be unable to perform the more delicate response of retrieving pups. Clearly, many oral motor responses are not simply centrally programmed, but require reflexive sensory-motor integration at the level of the trigeminal complex in the brainstem. If one severely interferes with the afferent

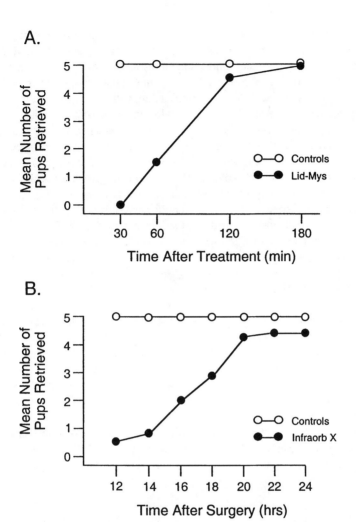

Figure 5.1. Mean number of pups retrieved by postpartum female rats after control treatments, lidocaine injections into the mystacial pads (Panel **A**: Lid-Mys), or section of the infraorbital branch of the trigeminal nerve (Panel **B**: Infraorb X). In both experiments, each retrieval test trial was 5 minutes in duration, after which pups were returned to the nest area until the next test. Modified from Kenyon et al. (1983).

limb of the trigeminal system, a variety of oral motor responses will be disrupted. Therefore, we can tentatively conclude that the total perioral and oral trigeminal somatic sensory input, operating at the level of the brainstem, is indispensable for the occurrence of retrieving behavior and other oral motor responses in the rat, and that other sensory modalities, like vision or olfaction, would not be able to substitute for a total elimination of such input.

Does trigeminal somatic sensory input only influence the rat retrieval response, and does such input only operate at the level of the brainstem as part of a sensory-motor reflex? Some exciting work from Stern's laboratory (Stern & Johnson, 1989; Stern & Kolunie, 1989, 1991) suggests that tactile input from the perioral region may have more general effects on maternal responsiveness by operating at higher levels of the central nervous system. In particular, they have reported that postpartum dams that are not retrieving their pups because of either infraorbital nerve transection or lidocaine injections into the mystacial pads are also not likely to nurse their young, even when the pups are placed back into the female's nest. Some of these results are shown in Figure 5.2. After initial investigations of the pups, the females tend to move away from the pups, resting in a part of the cage outside the nest area. Nursing behavior recovers with the recovery of the retrieval response, which is presumably mediated either by metabolic degradation of the lidocaine or by the substitution of additional trigeminal afferents in the infraorbital deafferented females. One interpretation of these results is that trigeminal cutaneous inputs reach higher levels of the brain (above the trigeminal complex) to influence the appetitive aspects of maternal behavior, fostering interest in pups and the maintenance of proximity-seeking behavior. Unfortunately, other investigators have not been able to replicate this interesting effect (that perioral desensitization disrupts nursing: Lonstein & Stern, 1997b; Magnusson & Fleming, 1995; Morgan, Fleming, & Stern, 1992), so a definitive conclusion cannot be reached at this time.

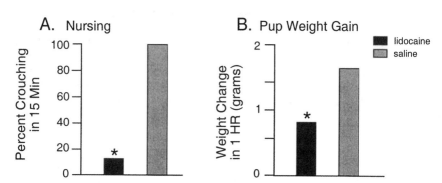

Figure 5.2. Percent of postpartum female rats crouching over their pups in a 15-minute test after either lidocaine or saline injections into the mystacial pads (**A**), and the mean weight gain of the pups of these females after a 1-hour reunion with the dams (**B**). *Significantly different from saline group. Modified from Stern and Johnson (1989).

However, some of these latter investigators did supply information that is generally consistent with the idea that tactile inputs from pups have motivational influences on the dam's maternal responsiveness. Magnusson and Fleming (1995) applied an anesthetic cream or a control cream to the perioral region of primiparous rats during the early postpartum period. Although both groups hovered over their pups and cared for them when placed in a novel cage setting, only the control females learned a conditioned place preference based on interacting with the pups in the distinct cage area (see Chapter 4 for a discussion of the conditioned place preference paradigm). This data suggest that perioral cutaneous sensations have rewarding properties for postpartum dams, reinforcing their ability to learn a conditioned place preference. In the Morgan et al. (1992) study, although postpartum dams with lidocaine injections into their mystacial region hovered over their young while their perioral region was anesthetized (the pups were placed directly in the female's nest), females that received similar lidocaine injections and also wore nylon jackets that prevented the female from detecting ventral somatic sensory inputs showed very little maternal behavior. Importantly, females that just wore the nylon jackets were observed to hover over their young. These findings suggest that tactile inputs from both the perioral region and the ventrum may play positive roles in maintaining overall maternal responsiveness, and that only when *both* are eliminated are all aspects of maternal responsiveness severely disrupted. However, since these findings were obtained from a single 1-hour test, more extensive work needs to be done before any firm conclusions can be reached (cf. Walsh, Fleming, Lee, & Magnusson, 1996).

Further research from Stern's laboratory has explored the additional involvement of ventral trunk somatic sensory inputs in the control of specific aspects of nursing behavior in female rats. Once a female has retrieved all of her pups to her nest site, she will begin to nurse them. As described by Stern and Johnson (1990), initially the female *hovers* over her litter, resting on her hindlimbs while engaging in other activities, such as licking, nuzzling, or mouthing the pups. The female subsequently becomes quiescent and engages in a *low crouch*, which is eventually replaced by a *high crouch*. Stern (1996a) has referred to crouching behavior (either low or high) as kyphosis. During kyphosis the female uses all four limbs to support a relatively immobile upright posture with her back either flat or slightly arched during the low crouch or intensely arched with legs rigid and splayed during the high crouch. In addition to these postures, a female can also nurse her young from a supine position, while laying prone on top of the litter, or while laying on her side. That kyphosis is an important aspect of maternal behavior in rats is indicated by the fact that brain lesions that interfere with kyphosis, but not with the hover posture or with supine/prone nursing, depress litter growth, particularly during the early postpartum period (Lonstein & Stern, 1997a). Therefore, the upright crouch that occurs during kyphosis seems important for allowing all members of a young litter access to the nipple region.

Research has shown that suckling stimulation and other aspects of ventral stimulation caused by nuzzling pups are important in regulating the occurrence of kyphosis: (a) Although female rats will retrieve dead pups and hover over them, they will not engage in crouching behavior (Stern & Johnson, 1990); (b) as shown in Figure 5.3, if pups have their mouths sutured closed, or if pups are injected with lidocaine into their mystacial pads (both procedures prevent suckling), their dams will retrieve, lick, and hover over the affected pups, but crouching does not occur (Stern & Johnson, 1990); and (c) thelectomy, although not

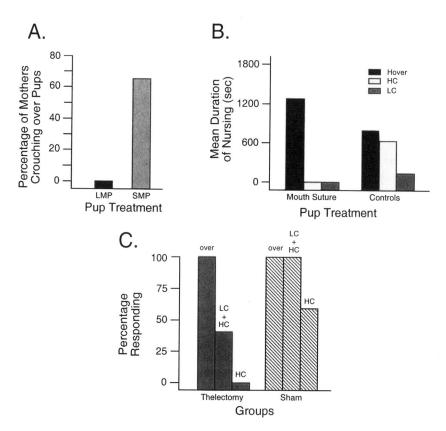

Figure 5.3. Various experiments that show that disruption of suckling stimulation interferes with crouching behavior (low-crouch [LC] and high-crouch [HC] kyphosis type of nursing behavior) in postpartum rats. In **A**, lidocaine injections into the mystacial pads of the pups (LMP), which blocks the pups' ability to suckle, interferes with the mother's crouching (SMP = saline injections into the mystacial pads of pups). In **B**, maternal display of HC and LC is suppressed in postpartum females if their pups have their mouths sutured closed. In **C**, postpartum females who have been thelectomized (nipple removal) show lower levels of crouching behavior than do sham controls. The terms "over" and "hover" refer to nonkyphosis nursing behavior. **A** and **B** are modified from Stern and Johnson (1990). **C** is modified from Stern et al. (1992).

interfering with hovering, depresses the low crouch and eliminates the high crouch, as shown in Figure 5.3 (Stern, Dix, Bellomo, & Thramann, 1992).

In analyzing the importance of suckling stimulation for the occurrence of kyphosis, it is worth emphasizing that in each of the above findings the criterion used for the occurrence of kyphosis was that the upright crouch had to be maintained for at least 2 minutes. Therefore, it is certainly possible that non-suckling tactile stimulation from nuzzling, rooting pups can activate the initiation of the low and high crouch, but that such behavior is dependent upon suckling stimulation for its continuance. Indeed, when kyphosis of any duration is recorded, such behavior is observed in lactating females whose pups have received lidocaine injections into their mystacial pads, and kyphosis is also displayed by virgin female rats who have been sensitized to show maternal behavior, but whose nipples are undeveloped so that suckling does not take place (Lonstein et al., 1999). However, the *duration* of kyphosis is longer in lactating females that receive suckling stimulation (Lonstein et al., 1999; also see Lonstein & De Vries, 1999b). In further support of the view that suckling stimulation is not essential for the initiation of kyphosis, Stern (1991) has reported that when virgin female rats and male rats (who do not have nipples) are injected with a catalepsy-inducing drug and then placed over nuzzling 1-week-old pups, both sexes are capable of showing kyphosis. These results further suggest that kyphosis can be viewed as a reflex that is induced by ventral somatic sensory stimulation even in the absence of any particular hormonal priming. Therefore, suckling or nonsuckling ventral tactile stimulation is capable of inducing kyphosis, but the subsequent maintenance of the upright crouch for relatively long durations is highly dependent upon suckling stimulation.

2.2.3. Conclusions on the Sensory Regulation of Maternal Behavior in Rodents

Beach and Jaynes's (1956c) view of the multisensory basis of maternal behavior, which was based on an analysis of retrieval in rats, clearly has to be modified in the light of more recent evidence. A summarizing resynthesis (also see Stern, 1990, 1997a) can be outlined as follows:

1. Distal stimuli from pups (visual, auditory, olfactory) are probably involved in attracting maternal female rats to their pups and, as suggested by Beach and Jaynes, neither of these inputs, acting alone, appears essential for the occurrence of maternal behavior in rats. As long as one of these stimulus inputs remain intact, the female can probably locate her pups from a distance and approach them.

2. Although the presence of olfaction is not essential for the occurrence of maternal behavior in postpartum rats, a modification in the effects of olfactory input appears essential for the initiation of maternal behavior. As reviewed in this chapter and the previous one, maternal behavior in naive virgin females is under olfactory inhibition, and this inhibition must be overcome,

either through hormonal priming or through sensitization, in order for maternal behavior to occur. There appears to be a shift in the central effects of olfactory input from pups as one moves from the virgin, naive condition to the parturient condition. In virgins, the novel odors from pups appear to activate brain regions that inhibit maternal behavior, while in late-pregnancy and postpartum females such inputs are actually attractive.

3. Although the detection of olfactory input from pups is not essential for the occurrence of maternal behavior in postpartum rats, such input is essential for the occurrence of maternal behavior in mice. It is difficult, therefore, to generalize about the sensory basis of maternal behavior from one species to another.

4. Once a postpartum female rat approaches her pups from a distance, tactile stimuli from pups appear to be primary in influencing both appetitive and consummatory aspects of maternal behavior. At a reflexive level, trigeminal somatic sensory inputs are probably essential for retrieval responses to occur, and ventral somatic sensory inputs trigger and maintain kyphosis.

A significant modification of the multisensory view, which is in need of further validation, is that somatic sensory input (*either* perioral or ventral trunk) appears essential for the maintenance of the appetitive aspects of maternal behavior. Although a postpartum female may be attracted to pups on the basis of distal stimuli, if she does not receive either perioral or ventral trunk tactile feedback from the pups, her interest in them appears to wane. This last statement, if verified by future research, indicates that tactile inputs from pups have motivational effects on maternal responsiveness that appear to maintain interest in pups and proximity-seeking behavior. In this sense, then, somatic sensory inputs from pups are essential for the display of normal maternal behavior in rats. Table 5.1 summarizes the sensory events that appear to underlie maternal behavior in rats.

One sensory modality that has received little attention with respect to its role in maternal behavior in rodents is taste. Since rodents and other mammals spend so much time licking their young, one would think that taste might influence several aspects of maternal behavior. The evidence that exists suggests that this

---

Table 5.1. Sensory basis of maternal behavior in rats.

A. Virgin Females
1. Olfactory inputs (primary and vomeronasal) inhibit maternal behavior.

B. Postpartum Females
1. Olfactory inhibition over maternal behavior is removed.
2. Exteroceptive distal stimuli (olfactory, visual, or auditory) from pups attract the mother and aid her in locating them.
3. Tactile perioral and oral stimuli are necessary for reflexive retrieval behavior.
4. Tactile ventral trunk stimuli are necessary for the reflexive kyphosis response.
5. Tactile ventral trunk or tactile perioral stimuli may maintain maternal motivation (interest in pups).

is not the case. Stern and Johnson (1990) found that injections of lidocaine into the tongue, which presumably interfered with taste sensations, did not prevent retrieval and nursing behavior in rats, although pup licking was decreased. They also found that when a postpartum dam's mouth was sutured closed, nursing behavior was not disrupted when the pups were placed in the female's nest.

2.3. Sensory Basis of Maternal Behavior in Sheep

Maternal behavior in sheep is different from that in most rodents because sheep form a selective bond with their lamb at birth and will eventually only care for that lamb and will reject the advances of other lambs (for reviews, see Carter & Keverne, 2002; González-Mariscal & Poindron, 2002; Poindron & Le Neindre, 1980). At parturition a ewe will accept *any* lamb and care for it: She will emit low-pitched bleats, she will lick the lamb, and she will allow it access to her udders. However, after several hours of contact with a particular lamb, the ewe will no longer accept any lamb, but will devote her exclusive attention to the lamb she was exposed to at parturition. After this exclusive bond is formed, the ewe will reject alien lambs: She will emit high-pitched bleats, will head-butt such lambs, and will not lick alien lambs nor allow them access to her udders. Importantly, if a ewe is separated from her lamb at birth before an exclusive bond forms, then her maternal responsiveness wanes so that by 12–24 hours postpartum she will reject the advances of all lambs. In other words, she will revert to the typical nonmaternal responsiveness of nonpregnant estrous-cycling ewes. Therefore, there is a sensitive period in the immediate postpartum period when the nervous system of the ewe is changed in such a way so that maternal behavior is aroused, and if a female is exposed to a lamb during this period she not only cares for it, but she also learns its characteristics so that her future maternal responsiveness is directed exclusively to that particular lamb.

With respect to the sensory basis of maternal behavior in sheep, we have two important questions: What are the sensory stimuli that the ewe responds to at parturition that allow her to indiscriminately accept all lambs? What are the sensory stimuli that the ewe responds to after a selective bond forms so that she only shows maternal responsiveness to a particular lamb? One way of viewing the process that occurs in sheep is as follows: As a result of the hormonal changes associated with parturition, coupled with the occurrence of the vagino-cervical stimulation that also occurs at this time (see Chapter 2), a ewe is responsive to a general stimulus or group of stimuli from lambs, and this sensory input is capable of gaining access to the central mechanisms regulating maternal acceptance so that the female is maternally responsive to any lamb. In the absence of lamb stimulation at parturition, this physiologically induced generalized maternal responsiveness wanes so that by 12–24 hours postpartum lamb stimuli can no longer gain access to positive maternal mechanisms, but instead activate central mechanisms that lead to the rejection of all lambs. Importantly, if a ewe is exposed to a particular lamb at parturition, then learning mechanisms come into play. Experience with a particular lamb during the period of generalized maternal responsiveness modifies the nervous system in such a way as to *narrow* the range of lamb stimuli that can activate central mechanisms controlling positive maternal responses. Such a process results in a specific maternal response to a unique lamb stimulus, with stimuli from alien lambs evoking rejection responses.

Most of the research on sheep has emphasized the importance of olfaction. The first point to note is that at parturition there is an important olfactory-mediated change in the preference of a ewe for amniotic fluid (Levy et al., 1983). Prior to parturition a ewe will not eat food that is contaminated with amniotic fluid, but at around the time of parturition she prefers to eat such food (the amniotic fluid is obtained from alien newborn lambs). Both the aversion and the preference for amniotic fluid are mediated by the primary olfactory system, as Levy et al. (1983) have shown that ewes that receive intranasal zinc sulfate treatment show neither an attraction nor an aversion to amniotic fluid. These results suggest that the presence of amniotic fluid on the newborn lamb may be a general positive olfactory stimulus that allows the parturient ewe to respond to all lambs. There is some evidence favoring this view, at least for primiparous ewes (Levy & Poindron, 1987): The lambs of primiparous ewes were removed at birth and 10 minutes later the ewes were presented with their lambs that were either still covered with amniotic fluid (controls) or that had been washed to remove the amniotic fluid. In a 20-minute test it was found that although 7 out of 10 control ewes let their lambs suckle, only 1 out of 11 females that had been exposed to washed lambs did so. The only criticism we have of this study is the short duration of the acceptance test. It is difficult to conclude that olfactory detection of amniotic fluid is absolutely essential for the display of maternal behavior in the recently parturient primiparous ewe. Instead, it is more likely that the presence of highly attractive amniotic fluid on the newborn facilitates acceptance and that the absence of such fluid merely delays maternal acceptance. This view is supported by the work of Levy, Locatelli, Piketty, Tillet, and Poindron (1995b). Primigravid ewes received either vomeronasal nerve cuts (to destroy the accessory olfactory system), intranasal treatment with zinc sulfate (to destroy the primary olfactory system), or sham procedures. Within the first hour postpartum sham females and females with vomeronasal nerve cuts were showing maternal behavior, but the zinc sulfate–treated females were not. However, by 6 hours postpartum the anosmic zinc sulfate–treated females were showing normal levels of maternal behavior. These results show that the vomeronasal system is not necessary for maternal responsiveness in the ewe, and neither is the primary olfactory system, although the latter system does play a supportive role in that it is involved in detecting a positive olfactory stimulus—amniotic fluid—which facilitates the onset of maternal acceptance. This general conclusion that olfaction is not essential for the display maternal behavior in the ewe is supported by other studies that show that olfactory bulbectomy, when performed prior to parturition in multiparous ewes, does not prevent the occurrence of maternal behavior (Baldwin & Shillito, 1974; Levy et al., 1995b; Poindron, 1976).

What about the role of olfaction in the formation of the selective bond that a ewe forms with her lamb? The general procedure that investigators have used to study the formation of selective bonds is as follows: A ewe gives birth and is allowed to interact and show maternal behavior toward her lamb for a few hours postpartum. The lamb is then removed from the ewe and a few hours later a selectivity test is given. At this time, the female is exposed to her own lamb and an alien, and acceptance and rejection responses are recorded. Under normal conditions the ewe will accept the lamb she was exposed to at parturition while rejecting the alien lamb. However, if a ewe receives an olfactory bulbectomy or intranasal application of zinc sulfate prior to parturition she will show nonselective maternal responsiveness, acting in a positive manner to both the lamb she was exposed to at parturition and toward the alien lamb (Baldwin & Shillito, 1974; Levy et al., 1995b; Poindron, 1976). Importantly, if the vomeronasal nerves are cut prior to parturition the ewe will form a selective attachment (Levy et al., 1995b). Therefore, under these conditions, although olfaction is not necessary for the occurrence of maternal behavior per se, the primary olfactory system is essential for the formation of an exclusive mother-infant bond in sheep.

In all of the above studies on the role of olfaction in the maternal behavior of sheep, interference with olfactory function occurred prior to parturition. An interesting question concerns the response of ewes to lambs when the ewe is deprived of proper olfactory input *after* she has formed a selective bond with a particular lamb. Alexander and Stevens (1981) separated ewes from their 2–8-day-old lambs for a few hours and then presented females with a choice between their own and an alien lamb. In some cases the stimulus properties of the young were altered. Most importantly, in order to eliminate specific odors, the lambs were washed with a detergent. When lambs were not manipulated, a mother would approach her lamb from a distance and allow it to suckle. However, if the stimulus lambs were washed, then the mother would approach her lamb from a distance, but upon making contact would not allow it to suckle. These results indicate that after a selective bond is formed during the early postpartum period, although auditory and visual cues can be used to recognize lambs from a distance, once contact is made, learned olfactory cues are primary in determining whether a mother will care for a lamb (also see Ferreira et al., 2000; Poindron & Le Neindre, 1980). It should be noted that the specific odors associated with the lamb's wool and skin are the likely candidates used for this discrimination, and such discrimination certainly cannot be based on amniotic fluid since this stimulus is not present in 2–8-day-old lambs (also see Levy, Kendrick, Keverne, Porter, & Romeyer, 1996a). Also worth mentioning is that this evidence suggests that it is the presence of a familiar odor, rather than the absence of a foreign odor, which determines maternal acceptance (Levy et al., 1996a).

Similar recognition mechanisms probably operate in goats (*Capra hircus*) in the regulation of their selective mother-infant bonds (Klopfer & Gamble, 1966; Romeyer, Poindron, & Orgeur, 1994). In particular, although destruction of the primary olfactory system prior to parturition does not disrupt maternal behavior at parturition, it does disrupt the formation of selective maternal bonds (Romeyer et al., 1994); however, interference with olfaction after an exclusive bond has already been formed results in a female who will not care for any young (Klopfer & Gamble, 1966).

One other group of findings must be presented before we can draw some conclusions (Poindron & Le Neindre, 1980). Recall that if a ewe is separated from her lamb for about 12 hours after parturition, her maternal responsiveness wanes. Importantly, if she receives distal (primary olfactory, visual, auditory) stimuli from her lamb over this 12-hour interval, her maternal behavior is maintained upon full reunion with her lamb. However, if she only receives visual and auditory stimuli she shows a lack of maternal responsiveness 12 hours later. These results suggest that if a female can smell her lamb during the early postpartum period, this is sufficient for her to form a selective bond with that lamb.

In an attempt to summarize these results on the sensory basis of maternal behavior in sheep, we offer the following tentative proposal.

1. As a result of the physiological events associated with parturition, the puerperal ewe responds to a variety of general stimuli from lambs, probably from more than one sensory modality. These stimuli (the full nature of which remain to be determined) can gain access to the central neural mechanisms regulating positive maternal responses, and this allows the female to show maternal behavior toward any lamb. Although a shift in olfactory preference in favor of amniotic fluid plays a facilitatory role with respect to maternal responsiveness, such an attraction is not essential for the occurrence of maternal behavior. In a manner analogous to the rat, however, the removal of the aversive qualities of amniotic fluid may be critical for the occurrence of maternal behavior at parturition.

2. If a parturient female interacts with her lamb for a few hours under conditions in which lamb odors can be detected by the primary olfactory system, she learns the specific odor of that lamb. This odor now has exclusive access to the central mechanisms underlying maternal responsiveness, and if this odor is not detected, maternal behavior (allowing the lamb to suckle) will not occur, but instead rejection and avoidance behavior will be displayed. In this sense we can argue that *after* a female forms an exclusive bond with her young, primary olfactory input becomes an essential stimulus for the occurrence of maternal behavior. This is an extremely important conclusion because we have a tight linkage between a specifc sensory input and the maternal response. If we could uncover the central neural networks that are activated in a maternal ewe in response to the scent of her familiar lamb, we would be taking an important step toward outlining the neural ciruitry of maternal responsiveness in sheep.

3. If a parturient ewe does not interact with a lamb over the first 12–24 hours postpartum, then the heightened generalized maternal responsiveness caused by the physiological events of parturition shuts down. Once the window of

this sensitive period closes, lamb stimuli can no longer gain access to the neural circuitry regulating positive maternal responses.

4. If a female is anosmic at parturition, then maternal behavior can be maintained as long as the female receives both distal *and proximal* stimulation from the lamb during the early postpartum sensitive period. As a result of such experience, these nonolfactory stimuli can subsequently gain access to the central mechanisms regulating positive maternal responses. The nature of these stimuli remains to be determined, but it is important to note that they are general in nature in that they do not allow the ewe to form a selective bond, at least when measured over the first several postpartum weeks (cf. Ferreira G. et al., 2000).

In addition to olfaction, one other source of afferent sensory stimulation has been shown to play a critical role in the maternal behavior of sheep: vaginocervical stimulation (VCS). The evidence that VCS is one of the factors involved in facilitating the onset of maternal behavior at parturition in sheep was reviewed in Chapter 2. Indeed, peridural anesthesia, which blocks genitosensory information from reaching the brain, both disrupts the onset of maternal behavior in sheep and prevents the normal attraction to amniotic fluid that occurs at parturition (Krehbiel et al., 1987; Levy, Kendrick, Keverne, Piketty, & Poindron, 1992). Importantly, not only does VCS promote maternal behavior at parturition in sheep, but it also appears to play a role in promoting the olfactory learning that underlies the formation of selective mother-infant bonds. Keverne et al. (1983) found that ewes who remained with their lambs for 2 hours postpartum would reject alien lambs presented to them after this time. However, if such ewes received experimenter-applied VCS when the alien young were presented, they would accept these young and allow them to nurse. Additional work has shown that VCS can induce the formation of a new selective bond even when it is administered to the ewe as late as 27.5 hours postpartum (Kendrick, Levy & Keverne, 1991). Although VCS can establish a new bond between a ewe and a lamb, it does not disrupt the original bond that was formed between the mother and her own lamb.

In an interesting study, Kendrick, Levy, and Keverne (1992b) recorded from single neurons in the main olfactory bulb of sheep prior to parturition and between 3 days and 4 weeks postpartum (after the formation of a selective mother-infant bond) while presenting the sheep with various odors such as food, amyl acetate, their own lambs' wool odor, and alien lambs' wool odor. During pregnancy, out of 105 recorded cells, none of these showed a preferential response to either amniotic fluid or lamb odors. Interestingly, Kendrick et al. report that in only 11 of these cells were these odors capable of causing any change in firing rate. This finding is surprising given the fact that pregnant ewes show a strong aversive response to amniotic fluid prepartum, and given that such behavioral responsiveness depends on olfactory input to the main olfactory bulb. Perhaps neurons in regions of the olfactory bulb that Kendrick et al. did not record from play a role in mediating the aversive responses toward amniotic fluid. Importantly, during the postpartum period there was a large increase in the proportion of cells that responded preferentially to lamb odors (of the 83 cells recorded from during the postpartum period, 50 responded preferentially to lamb odors). Although the majority of these cells responded to any lamb odor, some showed a selective response in that they only responded preferentially to the odor of the ewe's own lamb. Some of these results are shown in Figure 5.4. Microdialysis measures also indicated selective neurochemical changes in the main olfactory bulb during the postpartum period. Own lamb odor, but not alien lamb odor, increased the release of both glutamate (an excitatory amino acid neurotransmitter) and GABA (an inhibitory amino acid neurotransmitter).

Although the full implications of these findings will be discussed in another chapter, it is clear that changes in the function of the main olfactory bulb coincide with the occurrence of maternal behavior in sheep and with the formation of a selective mother-infant bond. It is tempting to speculate that the olfactory bulb neural population that responds in a general way to any lamb odor (in a maternally selective ewe) may ultimately influence those parts of the brain that promote rejection behavior. It is certainly possible, however, that nonolfactory inputs mediate rejection of an alien lamb. In contrast, the main olfactory bulb neural population which responds selectively to own lamb odors might actually interface with those neural regions that promote maternal acceptance.

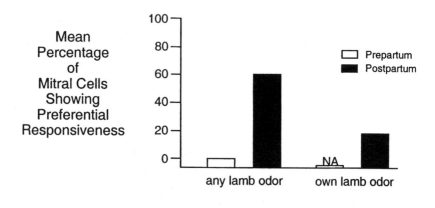

Figure 5.4. Mean percentage of mitral cell neurons in the main olfactory bulb that show preferential responsiveness to lamb odors in prepartum pregnant and postpartum sheep. NA signifies not applicable, since prepartum females, by definition, cannot be tested with the odor of their own lamb. Preferential responsiveness to lamb odors refers to a significantly greater number of action potentials elicited by such odors in comparison to the neuron's response to other types of odors (food, for example). During pregnancy, mitral cells do not show a preferential response to lamb odors, while during the postpartum period a substantial proportion do. Modified from Kendrick et al. (1992b).

2.4. *Comparisons of the Role of Sensory Factors in the Maternal Behavior of Rats and Sheep*

As a summarizing statement, along with a view of what is yet to come, we present Figure 5.5. This figure shows some of the known anatomical connections of both the main and accessory olfactory bulb (Brennan & Keverne, 1997; Canteras, Simerly, & Swanson, 1992, 1995; Kevetter & Winans, 1981a, 1981b; Licht & Meredith, 1987; Scalia & Winans, 1975; Shipley & Ennis, 1996), and is an elaboration of Figure 4.4. The following points are worth noting: (a) Input from both the main olfactory bulb (MOB) and the accessory olfactory bulb (AOB) converge on the medial amygdala (MeA); and (b) two important projection sites for MeA efferents are the medial preoptic area/bed nucleus of the stria terminalis region (MPOA/BST) and the anterior hypothalamic/ventromedial nucleus region (AHA/VMN). As will be shown, mainly from research derived from the rat, the former region plays an excitatory or positive role in the regulation of maternal responsiveness, while the latter region plays a negative or inhibitory role. In rats, although olfaction is not essential for maternal behavior, we will argue that removal of main and accessory olfactory bulb input to the AHA/VMN may be critical for the onset of maternal behavior at parturition. This view fits with the facts that damage to the vomeronasal system and the

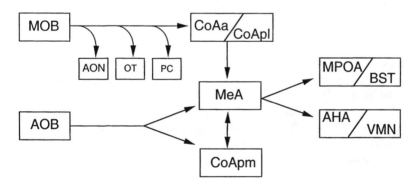

Figure 5.5. An elaboration of Figure 4.4 to show neural pathways through which olfactory input from the main and accessory olfactory bulbs (MOB and AOB, respectively) can reach the hypothalamus. The medial amygdaloid nucleus (MeA), which receives AOB and MOB input, is shown as projecting to the medial preoptic area (MPOA) and the bed nucleus of the stria terminalis (BST, which adjoins the MPOA, but is part of the telencephalon), and to the anterior hypothalamic area (AHA) and ventromedial nucleus (VMN) of the hypothalamus. We propose that MeA projections to MPOA may play a positive role in maternal behavior, while MeA projections to AHA/VMN may be inhibitory. Other abbreviations: AON = anterior olfactory nucleus; CoAa = anterior part of the cortical amygdaloid nucleus; CoApl = posterolateral part of the cortical amygdaloid nucleus; CoApm = posteromedial part of the cortical amygdaloid nucleus; OT = olfactory tubercle; PC = piriform cortex. Adapted from Brennan and Keverne (1997).

primary olfactory system facilitates the maternal responsiveness of naive virgin rats. Indeed, a parturient shift in olfactory mediated neural activity away from MeA-to-AHA/VMN dominance and toward MeA-to-MPOA/BST dominance may underlie the strong attraction to pup-related odors that occurs in rats at parturition, and this shift may promote maternal responsiveness primarily by removing inhibition, and secondarily by promoting attraction. Similar phenomena within the primary olfactory system may occur in sheep with respect to the valence of amniotic fluid odors. In sheep, we also speculate that after a selective maternal bond has formed, only a specific lamb odor is capable of affecting MPOA/BST activity, while perhaps unfamiliar primary olfactory and other sensory stimuli activate the inhibitory AHA/VMN region. (This model is tentative, and it should be noted that MOB output is capable of influencing hypothalamic mechanisms independent of an influence on the corticomedial amygdala [Price, Slotnick, & Revial, 1991].)

Since olfaction is not essential for maternal behavior in the rat, and since it is also not essential in ewes that have been made anosmic prior to parturition, we assume that other sensory inputs must be capable of interfacing in some way with MPOA/BST circuits in the regulation of maternal responsiveness under certain conditions. For the rat, as will be shown, it is possible that tactile input may provide important positive afferent input to the MPOA/BST region.

3. MeA and MPOA/BST Exert Opposing Influences on the Maternal Behavior of Rats

Most of the research on the neuroanatomy of maternal behavior using classical lesioning, electrical stimulation, and hormone-implant procedures has been performed on rats. In this section the evidence will be reviewed that indicates that the MeA exerts an inhibitory influence on maternal responsiveness in naive virgin rats while the MPOA/BST exerts an essential positive influence on maternal responsiveness. Where appropriate, data from other species will be integrated into the discussion.

3.1 The Inhibitory Role of MeA

Figure 5.6 shows a frontal section through the rat amygdala, indicating the location of some of its nuclei. Please refer to this figure in conjunction with Figure 5.5. As indicated in Chapter 4, Fleming et al. (1980) found that electrical lesions of the corticomedial amygdala, which damaged MeA, the anterior part of the cortical amygdala (CoAa), the posterolateral part of the cortical amygdala (CoApl), and the posteromedial part of the cortical amygdala (CoApm), shortened the sensitization latencies of estrous-cycling virgin female rats from 9 days to about 3 days. These findings have been replicated by Numan et al. (1993) with the important differences that the lesions were restricted to MeA and that the lesions were produced with an excitotoxic amino acid that destroyed cell

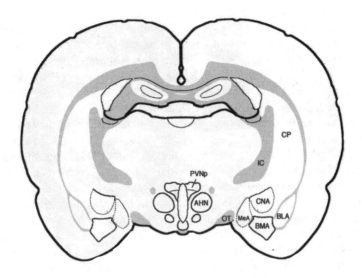

Figure 5.6. Frontal section through the rat brain showing the location of some of the amygdaloid nuclei. Abbreviations: AHN = anterior hypothalamic nucleus; BLA = basolateral amygdaloid nucleus; BMA = basomedial amygdaloid nucleus; CNA = central nucleus of the amygdala; CP = caudate-putamen; IC = internal capsule; MeA = medial amygdaloid nucleus; OT = optic tract; PVNp = parvocellular division of the paraventricular hypothalamic nucleus. Adapted from Swanson's rat brain atlas (1992).

bodies while sparing fibers of passage (see Figure 4.5). The MeA sends projections to the hypothalamus and other brain regions through two major pathways, the stria terminalis (ST) and the ventral amygdalofugal pathway (ansa peduncularis) (Canteras et al., 1995; De Olmos & Ingram, 1972). In this regard, it should be noted that Fleming et al. (1980) found that ST lesions were just as effective as corticomedial amygdala lesions in facilitating the onset of maternal behavior in virgin rats. Based on these results, it should not be surprising that ST lesions do not interfere with the established maternal behavior of postpartum lactating females (Numan, 1974).

Given that MeA lesions shorten sensitization latencies, one might expect that electrical stimulation of MeA would lengthen sensitization latencies in estrous-cycling female rats. This is exactly what was found by Morgan, Watchus, Milgram, and Fleming (1999), who employed a kindling procedure: The amygdala was first electrically stimulated in order to evoke an epileptic seizure. Such a procedure is expected to result in long-lasting increases in the excitability of the kindled nucleus in the absence of any further kindling stimulation, and Morgan et al. (1999) presented evidence that this was the case after kindling of MeA. Importantly, when MeA-kindled rats and nonkindled controls were exposed to pups 9 days after the conclusion of the kindling procedure, the kindled rats were found to have longer sensitization latencies than the controls.

There is also strong evidence that the corticomedial amygdala is involved in fear and anxiety-related processes: (a) Kindling of MeA in rats is associated with increased anxiety or fearfulness as measured by open-field activity or elevated plus maze performance (Adamec & McKay, 1993; Morgan et al., 1999); (b) lesions of the corticomedial amygdala appear to tame wild rats, decreasing their flight and defensive responses to threatening stimuli (Kemble, Blanchard, & Blanchard, 1990); and (c) MeA lesions decrease avoidance and freezing by a subordinate rat in response to exposure to a dominant individual who recently defeated it (Luiten, Koolhaas, de Boer, & Koopmans, 1985). These findings, along with the anatomy presented in Figure 5.5 and the knowledge that anosmia facilitates maternal behavior in naive virgin rats, are certainly consistent with the motivational model presented in Chapter 4: Novel olfactory input from pups inhibits maternal behavior in virgins because such stimuli activate fear-arousing mechanisms in the corticomedial amygdala. It should be noted that research on rodents has shown that MeA lesions do not cause a general anosmia, and individuals with such lesions can detect pheromones (Petrulis & Johnston, 1999). What seems to be disrupted by such lesions is the translation of such pheromonal detection into specific responses that are mediated by hypothalamic motivational and emotional mechanisms.

Dr. Teige Sheehan, while a graduate student in Michael Numan's laboratory, developed an important preparation to explore the role of the corticomedial amygdala and its efferent projections in the inhibition of maternal behavior in rats. First, he developed what we will refer to as a suboptimally hormone-primed rat model to explore aspects of the neural regulation of maternal behavior. Primigravid female rats were hysterectomized and ovariectomized on day 12 of pregnancy and injected with either 1.25 or 20 μg/kg of estradiol benzoate (EB). Forty-eight hours later, these females were exposed to pups and their latencies to the onset of maternal behavior were measured. Females injected with the higher dose of EB became maternal significantly faster than the females injected with the lower dose (median latencies of 0 days and 5 days, respectively) (Sheehan, 2000). Next, he wanted to determine whether excitotoxic amino acid lesions of the corticomedial amygdala could potentiate the maternal responsiveness of rats treated with the lower EB dose. Primigravid female rats were hysterectomized and ovariectomized on day 12 of pregnancy and injected with 1.25 μg/kg of EB. At this time, half the rats received bilateral injections of N-methyl-D-aspartic acid (NMDA), an excitotoxic amino acid, into the corticomedial amygdala, while the remaining females received injections of N-methyl-L-aspartic acid (NMLA, an inactive stereoisomer). Forty-eight hours later the females were presented with pups and latencies to onset of maternal behavior were measured. NMDA lesions of the corticomedial amygdala clearly facilitated maternal behavior, with NMDA lesioned females showing a median onset latency of 0 days, compared to the median latency of 6 days for the NMLA group (Sheehan, Paul, Amaral, Numan, & Numan, 2001). Additionally, although over 60% of the NMDA-injected females were behaving maternally on their first day of pup exposure, none of the NMLA-injected females were maternal. The lesions in this study were not confined to MeA, but also damaged parts of the cortical amygdalar nuclei.

The short onset latencies observed in this study should be contrasted to the 3-day onset latencies observed after corticomedial amygdala lesions in estrous-cycling virgins (Fleming et al., 1980; Numan et al., 1993). What this suggests is that the disinhibitory effect of corticomedial amygdala lesions on maternal responsiveness is potentiated if the lesioned females are exposed to some of the hormonal events associated with pregnancy and pregnancy termination, even though such events are suboptimal by themselves in terms of stimulating maternal behavior.

3.2. The Facilitatory Role of MPOA/BST

Starting with the work of Fisher (1956), research from several laboratories has shown that the MPOA and the adjoining ventral part of the BST (vBST) play an essential positive role in the regulation of maternal behavior in rats. Figure 5.7 shows a frontal and sagittal section of the rat brain, indicating the location of the MPOA and BST. The MPOA lies just rostral to the anterior hypothalamus and just caudal to the diagonal band-septal region. Lateral to the MPOA is the lateral preoptic area (LPOA), dorsolateral to the MPOA is the vBST (that part of the BST below the anterior commissure), and dorsal to the MPOA is the anterior commissure, dorsal BST, and septal region. Several studies have shown that electrical lesions of the MPOA disrupt maternal behavior under a variety of conditions: (a) When performed during the postpartum period, such lesions disrupt established maternal behavior (Jacobson, Terkel, Gorski, & Sawyer, 1980; Numan, 1974); (b) when performed during pregnancy, such lesions disrupt the onset of maternal behavior at parturiton (Lee et al., 2000); (c) such lesions also disrupt the onset of maternal behavior in pregnancy-terminated (hysterectomized and ovariectomized) females treated with estradiol (Numan & Callahan, 1980); and (d) finally, electrical lesions of the MPOA disrupt sensitized maternal behavior, and the lesions can be performed either prior to the beginning of the sensitization procedure, or after sensitization has occurred (Fleming, Miceli, & Moretto, 1983; Gray & Brooks, 1984; Lee et al., 2000; Numan, Rosenblatt, & Komisaruk, 1977; Oxley & Fleming, 2000).

Electrical lesions of the MPOA destroy MPOA neurons and fibers of passage through the MPOA. Importantly, research has shown that excitotoxic amino acid lesions of the MPOA, which produce neuronal cell body specific lesions, are just as effective as electrical lesions in disrupting maternal behavior in rats, indicating that it is MPOA neurons and their efferents that are essential for maternal behavior (Kalinichev, Rosenblatt, & Morrell, 2000a; Numan, Corodimas, Numan, Factor, & Piers, 1988). Studies utilizing specific knife cuts have attempted to localize the trajectory of these critical MPOA efferents, and the results have indicated that it is the lateral efferent projections from the MPOA that are most important for maternal behavior in rats (and hamsters) (Franz, Leo, Steuer, & Kristal, 1986; Miceli, Fleming, & Malsbury, 1983; Miceli & Malsbury,

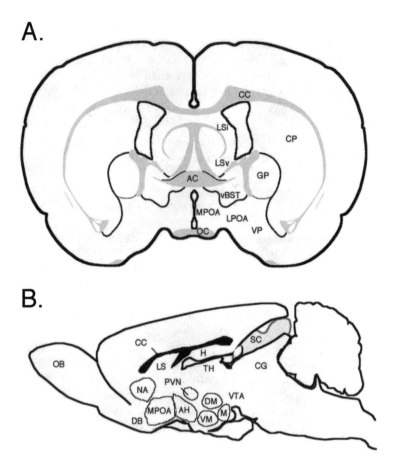

Figure 5.7. Frontal (**A**) and sagittal (**B**) sections of the rat brain at the level of the medial peroptic area. Abbreviations: AC = anterior commissure; AH = anterior hypothalamic nucleus; CC = corpus callosum; CG = central gray (periaqueductal gray); CP = caudate-putamen; DB = nucleus of the diagonal band of Broca; DM = dorsomedial hypothalamic nucleus; GP = globus pallidus; H = hippocampus; LPOA = lateral preoptic area; LS = lateral septum; LSi = intermediate nucleus of lateral septum; LSv = ventral nucleus of lateral septum; M = mammillary bodies; MPOA = medial preoptic area; NA = nucleus accumbens; OB = olfactory bulb; OC = optic chiasm; PVN = paraventricular hypothalamic nucleus; SC = superior colliculus; TH = thalamus; vBST = ventral bed nucleus of stria terminalis; VM = ventromedial hypothalamic nucleus; VP = ventral pallidum; VTA = ventral tegmental area. Adapted from Swanson's rat brain atlas (1992).

1982; Numan, 1974; Numan & Callahan, 1980; Numan & Corodimas, 1985; Numan, McSparren, & Numan, 1990; Terkel, Bridges, & Sawyer, 1979). For example, Numan and Callahan (1980) examined the maternal behavior of female rats that received knife cuts that severed either the lateral, dorsal, anterior, or posterior connections of the MPOA, or sham surgeries (see Figure 5.8), and they found that only the lateral cuts *selectively* disrupted maternal behavior (the anterior cuts also disrupted maternal behavior, but these animals were physically debilitated by the surgery). More recent work has found that it is a particular subset of the lateral MPOA efferents that appear to be most critical for maternal behavior—the dorsolateral projections: Knife cuts that sever the dorsolateral connections of the MPOA disrupt maternal behavior while those that sever the ventrolateral projections do not (see Figure 5.8) (Numan et al., 1990; Terkel et al., 1979). This finding, emphasizing the importance of the dorsolateral MPOA projections, fits with the results of Jacobson et al. (1980), who found that electrical lesions aimed at the dorsal MPOA were more effective in disrupting maternal behavior than were lesions of the ventral MPOA.

In examining Figure 5.8, note how disruptive knife cuts that sever the lateral connections of the MPOA would also interfere with the lateral connections of the vBST. This finding, in conjunction with the fact that excitotoxic lesions of the MPOA also frequently damage neurons in the adjoining vBST (Numan et al., 1988), suggests the possible importance of this latter region. Indeed, Numan and Numan (1996) have reported that excitotoxic amino acid lesions that specifically destroy the vBST without damaging the MPOA also disrupt maternal behavior in lactating rats, although the deficits are not as severe as those that occur after damage that is restricted to the MPOA. Numan and Numan (1996) suggested that neurons in the dorsolateral MPOA and vBST may form a common functional system important for maternal behavior in rats.

Since the MPOA is involved in regulating anterior pituitary function (Freeman, 1994; Gunnet & Freeman, 1983; Silverman, Livne, & Witkin, 1994), the question arises as to whether the disruptive effects of MPOA damage on maternal behavior are the result of a neuroendocrine, rather than a neurobehavioral, dysfunction. The available evidence gives a negative response to this question, arguing that the primary dysfunction is neurobehavioral in nature: (a) Preoptic damage disrupts maternal behavior in hormonally primed females (Numan & Callahan, 1980); (b) preoptic damage disrupts already established maternal behavior in postpartum lactating females (Numan, 1974; Numan et al., 1988), and the occurrence of maternal behavior during this maintenance phase is not dependent upon hormones (see Chapter 2); and (c) preoptic damage disrupts sensitized maternal behavior in intact and ovariectomized female rats (Gray & Brooks, 1984; Kalinichev et al., 2000a; Miceli et al., 1983; Numan et al., 1977; Oxley & Fleming, 2000). Such pup-stimulated maternal behavior in virgin females is relatively free of hormonal control (see Chapter 2).

Whenever lesions disrupt a particular behavior, one must ask whether the lesion selectively and directly interfered with the behavior, or whether the particular behavior was indirectly affected by some other effect of the lesion. In

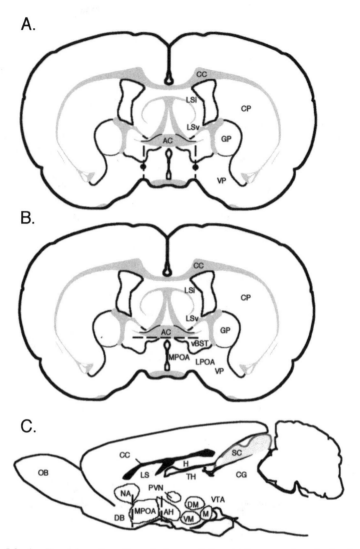

Figure 5.8. **A**. Frontal section through the medial preoptic area of rat brain showing knife cuts (dashed lines) severing the lateral connections of the medial preoptic area and adjoining ventral bed nucleus of the stria terminalis. Full cuts, which travel from the level of the anterior commissure to the base of the brain, selectively disrupt maternal behavior. The black dots separate the lateral cuts into dorsal and ventral parts. Dorso-lateral cuts also disrupt maternal behavior, while ventrolateral cuts do not. **B**. Frontal section through the medial preoptic area showing a knife cut (dashed lines) severing the dorsal connections of the medial preoptic area. **C**. Sagittal section through the level of the medial preoptic area showing the placement of a knife cut severing the anterior connections of the medial preoptic area (dashed line between DB and MPOA) and a knife cut severing the posterior connections of the medial preoptic area (dashed line posterior to MPOA). For abbreviations, see Figure 5.7. Data modified from Numan and Callahan (1980) and from Numan et al. (1990). Brain sections were adapted from Swanson's rat brain atlas (1992).

its simplest case, one wants to make sure that the maternal behavior disruption caused by preoptic damage is not the result of a general physical debilitation. In this regard, we note that rats that show disrupted maternal behavior after preoptic damage show normal: (a) sexual behavior (Numan, 1974), (b) locomotor activity in an open field (Kalinichev et al., 2000a; Numan & Callahan, 1980; Numan et al., 1988), (c) body temperature regulation (Numan et al., 1988; also see Miceli et al., 1983; Numan & Callahan, 1980), and (d) body weight regulation (Numan & Callahan, 1980). Also worth noting is that the disruption of retrieval behavior after preoptic damage cannot be attributed to a general oral motor deficit since females that are not retrieving their young are capable of hoarding (carrying in their mouth) pieces of candy that approximate the size and weight of baby pups (Kalinichev et al., 2000a; Numan & Corodimas, 1985).

What is the nature of the maternal behavior deficit that occurs after preoptic damage? All studies agree that retrieving behavior and nest-building behavior are abolished or severely disrupted after electrical or excitotoxic MPOA lesions or after knife cuts that sever the dorsolateral connections of the MPOA. Studies differ with respect to their findings on nursing behavior. Many studies report a severe disruption of all components of maternal behavior, including nursing, after preoptic damage (Kalinichev et al., 2000a; Lee et al., 2000; Numan, 1974; Numan & Corodimas, 1985; Numan et al., 1988), while other studies report that nursing behavior does occur in preoptic-damaged females that are not retrieving their young or building nests, although when the duration of such nursing is measured it is found to be significantly lower than that shown by control females (Franz et al., 1986; Jacobson et al., 1980; Miceli et al., 1983; Numan & Callahan, 1980; Terkel et al., 1979). Importantly, all studies that have measured the litter weight gains of females who have received preoptic damage during the postpartum lactational period have reported poor litter growth (freshly nourished pups are usually provided to all females on each test day to prevent pup debilitation). This finding supports the view that even when nursing behavior does occur in preoptic-lesioned females, its duration and quality are inferior to that of controls.

In most of the studies cited above, maternal behavior was studied within a few days of producing preoptic damage. These studies, therefore, examined the acute effects of preoptic damage. Numan (1990) examined the long-term or chronic effects of severing the lateral connections of the MPOA/vBST on maternal behavior. Females rats received knife cuts or control cuts 2 weeks prior to mating, and after mating they underwent a 3-week pregnancy period. After giving birth, their maternal behavior was studied for 2 weeks. Therefore, maternal behavior testing commenced approximately 5 weeks after the knife cuts were produced. In these females, retrieving was abolished, inferior nests were built, but nursing behavior was normal. Importantly, the females were capable of lactating and their pups gained weight daily, albeit at a reduced rate in comparison to controls. These results indicate that the most permanent effect of lateral MPOA cuts on maternal responsiveness is that they disrupt retrieval behavior. Perhaps the recovery of normal nursing behavior was the result of some type of neural reorganization within MPOA/vBST circuits that remained intact after the cuts. It would be interesting to determine whether nursing behavior would recover after long-term excitotoxic lesions that damage most of the neurons in the MPOA/vBST region.

Jacobson et al. (1980) and Terkel et al. (1979) initially proposed that the dorsolateral connections of the MPOA may be particularly important for the active components of maternal behavior, that is, those components (retrieving and nest building) actively initiated by the female. According to these researchers, nursing occurs because it is primarily initiated by the pups and is therefore viewed as a passive maternal response. Before we develop these ideas, we might similarly ask whether preoptic damage interferes with the appetitive or consummatory aspects of maternal behavior. In other words, are females with preoptic damage not interested in pups, or are they motivated to interact with the pups but are not capable of performing certain maternal responses? First, in those cases when preoptic damage interferes with all aspects of maternal behavior, one could argue that the damage caused a global disruption of maternal motivation, resulting in females that were simply disinterested in their pups, and this outcome would be consistent with an appetitive role for preoptic neural circuits. Second, in Chapter 4 we suggested that retrieving might be considered an appetitive component of maternal behavior, with nursing being its goal or consummatory component. In those cases when preoptic damage interferes more strongly and permanently with retrieving, we might still conclude that preoptic neurons are more heavily involved in the appetitive aspects of maternal behavior. However, we also noted in Chapter 4 that some investigators might refer to retrieval behavior as a species-typic consummatory maternal response. If such were the case, then preoptic damage might have its most long-lasting and drastic effects on the ability to actually perform retrieval behavior in response to the appropriate pup stimuli (perioral tactile stimulation).

A recent study by Lee et al. (2000) has provided support for the view that the MPOA is involved in the appetitive aspects of maternal behavior in rats. They found that electrical lesions of the MPOA not only disrupted retrieving, nursing, and nest building, but also disrupted an operant bar-press response when pups were used as a reward, but not when a palatable food was used as reward. They suggest that MPOA-lesioned females do not find pups rewarding (but still find food rewarding) and therefore do not engage in an instrumental response to gain access to pups. It is worth reemphasizing that during the early postpartum period of rats, pups are more reinforcing than is cocaine (as tested in the conditioned place preference paradigm), while in the later postpartum period, cocaine is more reinforcing (Mattson et al., 2001). Interestingly, the tendency to engage in retrieving behavior is high during the early postpartum period, and then declines afterward (Rosenblatt & Lehrman, 1963). One interpretation of these results is that retrieving reflects appetitive maternal processes, and that the disruption of retrieval behavior after preoptic damage is a reflection of a decrease in such processes.

One final point needs to be tentatively offered. It appears that preoptic damage, when produced in postpartum females after maternal behavior has become established, does not result in the reemergence of avoidant responses that are typical of the virgin naive female. Such postpartum females approach pups, sniff them, and lick them; they don't show repetitve approach-avoidance responses or the stretched attention posture; and they don't bury pups under the bedding material. Therefore, although specific tests, such as placing pups in the female's preferred quadrant to determine if the female would switch her resting area to another quadrant, have not been performed, casual observations of post-partum preoptic-lesioned females do not suggest that they are afraid of pups. When all components of maternal behavior are disrupted in such females, the best descriptor of the female's behavior is that she appears disinterested in the pups.

To conclude this analysis we offer the following tentative hypothesis: When performed acutely during the postpartum period, preoptic damage may primarily interfere with the appetitive aspects of maternal behavior, although we cannot rule out an additional deficit in the ability to perform the requisite sensory-motor integration necessary for retrieval behavior. Postpartum lesioned females, however, are not afraid of the pups, and this lack of avoidance may result in the occurrence of some nursing behavior because of ventral tactile stimulation from pups. More specifically, on those occasions when a preoptic-lesioned female is in close proximity to pups, nursing and a crouch-like posture may be induced as a reflex in response to suckling and other ventral tactile stimulation even though appetitive maternal motivation is relatively low (Lonstein & Stern, 1997b; Stern, 1991).

The role of the MPOA/vBST in general maternal motivational processes is further strengthened by findings that show that stimulation of this region can activate all components of maternal behavior in rats. First, kindling stimulation of the MPOA has been found to facilitate sensitization in virgin female rats exposed to pups (Morgan et al., 1999). Second, as described below, MPOA/vBST neurons are one of the sites where estradiol and lactogenic hormones act to facilitate the onset of maternal behavior in rats.

Estradiol produces most of its cellular/neural effects by binding to intracellular estrogen receptors (ERs), which act as hormone-dependent transcription factors (McEwen & Alves, 1999; Paech et al., 1997). This estradiol-activated process has been referred to as a genomic mechanism of action because the estradiol-ER complex binds to DNA elements in the regulatory region of certain genes, in this way affecting transcription and protein synthesis, which ultimately results in an altered neuronal phenotype, with concomitant changes in neural function. Intracellular ERs are located in neurons in diverse brain regions that include MPOA, BST, MeA, AHA, VMN, lateral septum (LS), and paraventricular nucleus of the hypothalamus (PVN) (Pfaff & Keiner, 1973; Shugrue, Lane, & Merchenthaler, 1997; Simerly, Chang, Muramatsu, & Swanson, 1990). Importantly, biochemical and immunocytochemical studies on rats and mice have shown that the number of ERs increases within the MPOA as pregnancy advances, and that such an increase is preceded by increases in ER mRNA (Giordano, Siegel, & Rosenblatt, 1989, 1991; Koch & Ehret, 1989b; Wagner & Morrell, 1995, 1996). One interpretation of these findings is that as pregnancy advances, the MPOA becomes more sensitive to estradiol, and the subsequent transcriptional effects of estradiol modify the function of MPOA neurons in such a way as to promote the onset of maternal behavior at parturition. It should be noted, however, that the number of ERs also increases in other brain regions as pregnancy advances.

Studies employing implants of small amounts of estradiol into the MPOA have provided more direct evidence that estrogen acts on the neurons within this region to stimulate maternal behavior. Numan et al. (1977) found that implants of estradiol into the MPOA facilitated the onset of maternal behavior (retrieving, nursing, nest building) in female rats that were hysterectomized and ovariectomized on day 16 of pregnancy, whereas estradiol implants into other brain regions (VMN, mammillary bodies) or subcutaneously, or cholesterol implants into the MPOA, were ineffective. These findings have been replicated by others (Fahrbach & Pfaff, 1986; Matthews Felton, Linton, Rosenblatt, & Morrell, 1998a). Importantly, implants of an antiestrogen (4-hydroxytamoxifen) into the MPOA of rats on day 20 of pregnancy delays the onset of maternal behavior (Ahdieh, Mayer, & Rosenblatt, 1987). Implants of the antiestrogen into the VMN did not have this inhibitory effect. Tamoxifen is known to block some of the transcriptional effects of estradiol (Kuiper et al., 1997; Paech et al., 1997).

The above studies should not be viewed as showing that the MPOA is the only site where estrogen acts to facilitate maternal behavior. For example, estradiol is secreted by the ovary prior to day 16 of pregnancy (Bridges, 1984), and such circulating estrogen might act on extra-MPOA sites to influence maternal responsiveness. Also, it cannot be ruled out that the implanted estradiol (or tamoxifen) diffused to nearby neural sites bordering on the MPOA, such as the BST, LS, or PVN, to influence maternal behavior. The Matthews Felton et al. (1998a) study is noteworthy in this regard as they were able to facilitate maternal behavior in female rats hysterectomized and ovariectomized on day 16 of pregnancy after the implantation of diluted estradiol (1:10 dilution of estradiol in cholesterol) into the MPOA, which was used to decrease the diffusion of estradiol from the implant site. At a minimum, however, estrogen receptor–containing neurons in the vBST were undoubtedly affected by the MPOA implants, along with MPOA neurons.

Unlike estradiol, the lactogenic hormones (prolactin and placental lactogens) act on cell membrane receptors to produce their cellular effects, and two major types of prolactin receptors have been identified, a long form and a short form (Shirota et al., 1990). Importantly, both forms are located in the brain, but it is the long form that is dominant, and its synthesis within the brain increases as pregnancy advances (Nagano & Kelly, 1994; Sugiyama, Minoura, Kawabe, Tanaka, & Nakashima, 1994). In an excellent in situ hybridization anatomical study, Bakowska and Morrell (1997) mapped the location of rat forebrain neurons that express mRNA for the long form of the prolactin receptor. They found such expression in several regions, which included MPOA, BST, LS, MeA,

VMN, and PVN. High levels of expression were also found in the choroid plexus. Additionally, Bakowska and Morrell report that within the MPOA the number of neurons that contain mRNA for the long form of the prolactin receptor, and the amount of mRNA per MPOA neuron, increases between day 2 and day 21 of pregnancy (also see Grattan, 2001). In support of these general findings, immunocytochemical studies have also located prolactin receptor–immunoreactive neurons within the MPOA/BST and other brain regions (Roky et al., 1996).

As indicated in Chapter 2, circulating lactogens can gain access to the brain through a receptor-mediated transport system within the choroid plexus that transports lactogens into the CSF. Indeed, high levels of lactogenic hormones can be detected in the CSF between days 12 and 21 of pregnancy in the rat (Bridges et al., 1996). From the CSF, lactogens can presumably reach central neural sites where they can act on prolactin receptors. It should also be pointed out here that prolactin is also produced by neurons in the brain (Emanuele et al., 1992; Paut-Pagano, Roky, Valatx, Kitahama, & Jouvet, 1993), so that prolactin receptors can presumably be affected by both hormonal lactogens, and by a brain prolactin system. The possible role of a brain prolactin system in maternal behavior will be discussed in the next chapter.

There is good evidence that the MPOA/vBST region is one of the sites where lactogens act to stimulate the onset of maternal behavior in rats. Bridges, Numan, Ronsheim, Mann, and Lupini (1990) showed that bilateral injections of prolactin (40 ng/injection) into the MPOA facilitated the onset of maternal behavior in virgin female rats who were treated with a sequential steroid hormone regimen of progesterone followed by estradiol (see Chapter 2). Importantly, such females were also treated with bromocriptine to prevent estradiol-induced endogenous prolactin release. Critically, 80-ng injections of prolactin into the CSF were not effective in stimulating maternal behavior, although higher doses were (400 ng). Some of these results are shown in Figure 5.9. Subsequent research has shown that microinjections of placental lactogens in the MPOA region are just as effective as prolactin injections in facilitating maternal behavior (Bridges & Freemark, 1995; Bridges et al., 1996, 1997). These studies also show that the stimulatory effect on maternal behavior of lactogen injections into the MPOA is only effective if the virgin females are also exposed to progesterone withdrawal and estradiol. Finally, and importantly, microinjections of a prolactin receptor antagonist into the MPOA have recently been shown to disrupt the onset of maternal behavior in virgin female rats treated with a sequential steroid hormone regimen of progesterone followed by estradiol (Bridges, Rigero, Byrnes, Yang, & Walker, 2001). One shortcoming of these studies is that prolactin, placental lactogens, or the prolactin receptor antagonist were not injected into control sites within the hypothalamus, and therefore the possibility exists that the injected compounds may have spread to extra-MPOA sites where they exerted their effects.

Although we know that high levels of progesterone inhibit maternal behavior, and that its withdrawal facilitates maternal behavior, at present we do not know

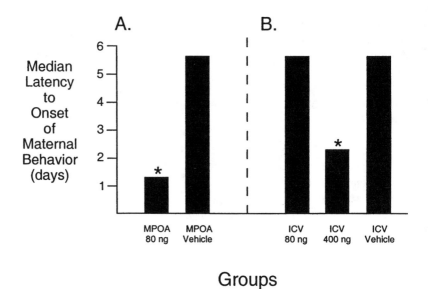

Figure 5.9. The effects of infusions of ovine prolactin into either the medial preoptic area (MPOA; shown in **A**) or intracerebroventricularly (ICV; shown in **B**) on sensitization latencies in nulliparous female rats treated with steroids. *80-ng injections into the MPOA or 400-ng injections ICV stimulate maternal behavior when compared to the remaining groups. Modified from Bridges et al. (1990).

where progesterone acts to produce these effects. Progesterone can have cellular/neural effects either by binding to intracellular progesterone receptors (PRs) to exert a genomic mechanism of action, or by acting on cell membrane receptors (Schumacher, Coirini, Robert, Guennoun, & El-Etr, 1999). Recent evidence suggests that a genomic mechanism of action is involved in progesterone's inhibition of maternal behavior. Numan et al. (1999) found that systemic treatment with RU 486 interfered with the ability of systemic progesterone injections to inhibit maternal behavior in pregnancy-terminated rats. RU 486 acts as an antagonist to the intracellular PR, blocking the transcriptional effects of progesterone while having no known antagonistic influence on the membrane effects of progesterone (Bitran et al., 1995; Gass, Leonhardt, Nordeen, & Edwards, 1998; Ramirez & Zheng, 1996; Spitz, Croxatto, & Robbins, 1996). Numan et al. (1999), in an immunocytochemical anatomical study, also located intracellular PR-immunoreactive neurons within several brain regions of late pregnant female rats, which included neurons in the MPOA, BST, anterior part of the PVN, and VMN. These results suggest that progesterone might act at one of these sites to influence maternal behavior. However, Numan (1978) implanted progesterone into diverse brain regions, which included the MPOA/BST and VMN, but was not able to inhibit estradiol-facilitated maternal behavior in hysterectomized and ovariectomized pregnancy-terminated rats, although systemic injections of progesterone were very effective in this regard. Therefore, the neural site of progesterone action on maternal neural systems remains a major unanswered question.

In summary, the facts that neuron-specific lesions of the MPOA/vBST disrupt maternal behavior and that estradiol and lactogens act on the MPOA/vBST to stimulate maternal behavior provide powerful evidence that this region gives rise to efferent neural projections that play a positive excitatory role in the maternal behavior of rats. Additional evidence indicates that it is the efferent lateral MPOA/vBST projections that are most important. Although much more comparative work needs to be done, the evidence that does exist suggests that the MPOA also plays a role in the parental behavior of species other than the rat (Ball & Balthazart, 2002; Buntin, 1996; Calamandrei & Keverne, 1994; Gubernick, Sengelaub, & Kurz, 1993a; Kendrick, Keverne, Hinton, & Goode, 1992a; Miceli & Malsbury, 1982).

3.3. The Effects of MPOA Lesions or Corticomedial Amygdala Lesions on the Maternal Behavior of Juvenile Female Rats

Recall from Chapters 2 and 4 that juvenile female rats can be induced to show maternal behavior through pup stimulation and that the sensitization latencies of such females are shorter than those of adult females, presumably because fearful and defensive responses have not completely developed in the juveniles. Interestingly, however, Oxley and Fleming (2000) report that corticomedial amygdala lesions shorten even further the sensitization latencies of juvenile females, suggesting that amygdaloid inhibition of maternal behavior already exists to some extent in these rats. Importantly, MPOA lesions have been found to disrupt pup-induced maternal behavior in juvenile females (Kalinichev et al., 2000a; Oxley & Fleming, 2000). Therefore, the same opposing neural mechanisms that influence the occurrence of maternal behavior in adult females also operate to regulate the degree of maternal responsiveness in juvenile female rats.

Two points are worth noting when comparing effects of MPOA lesions on the *pup-induced* maternal behavior in juvenile and adult female rats: (a) Larger lesions are needed to disrupt the onset of such behavior in juveniles than in adults (Kalinichev et al., 2000a); and (b) when MPOA lesions are performed prior to pup exposure, such lesions disrupt all components of maternal behavior in adults (retrieving, nest building, crouching), while in juveniles, although retrieving and nest building are eliminated, crouching or nursing-like behavior is unaffected (Kalinichev et al., 2000a; Numan et al., 1977; Oxley & Fleming, 2000). Perhaps crouching/nursing-like behavior occurs in juveniles with preoptic lesions for the same reason it does in postpartum females who have received such lesions during the postpartum period *after* maternal behavior has become established: Such females are less likely to avoid pups and, therefore, the nursing may simply be induced as a reflexive-type response to ventral tactile stimulation.

It is hypothesized that postpartum females are not afraid of the pups because they have already habituated to the novel pup odors during the maternal behavior that occurred prior to lesion production, while juvenile females may show a low level of fear toward novel pup odors even on their first exposure to such odors because their fear systems are only partially developed.

4. The Larger Neural Circuitry within Which the MPOA/ BST and MeA Operate to Influence Maternal Behavior

One way to gain an understanding of the function of MeA and MPOA/BST neurons with respect to maternal behavior that goes beyond saying that the former is inhibitory and the latter is excitatory is to uncover the larger neural circuitry within which these neurons are embedded. Such an analysis involves uncovering the afferents and efferents of these nuclei that are involved in regulating maternal behavior. This section will review the important research that has taken this approach toward delineating the neural underpinnings of maternal behavior.

The analysis of *fos* gene activation within neurons, measured by the increased production of various Fos proteins, has played a significant role in outlining the larger neural circuitries that may be involved in either inhibiting or facilitating maternal behavior. The *fos* family of genes, which are members of a class of genes referred to as immediate early genes, consists of *cfos, fos B, fra-1,* and *fra-2,* and their protein products are referred to as cFos, Fos B, Fra-1, and Fra-2 (Sheng & Greenberg, 1990). With respect to the function of Fos proteins within neurons, the following should be noted (Chaudhuri, 1997; Herrera & Robertson, 1996; Morgan & Curran, 1991; Sheng & Greenberg, 1990): When a neuron is affected by extracellular signals (for example, neurotransmitters, neuromodulators, hormones), in many cases its biosynthetic machinery is activated, which includes the stimulation of immediate early genes followed by late-responding genes. Basically, extracellular signals produce intracellular second-messenger cascades that, in turn, activate immediate early genes, the protein products of which serve as transcription factors that influence the expression of late-responding structural genes, the protein products of which are presumed to alter neural function (for example, by possibly altering the neuron's neurotransmitter levels or receptors). It is not surprising, therefore, that the production of Fos proteins within neurons has been used as evidence that those neurons have been activated, and this is supported by the facts that there is a close (but not perfect) correlation between cFos production and glucose uptake within neurons, and that neuronal depolarization can activate cFos production within neurons (Morgan & Curran, 1991; Sharp, Gonzalez, Sharp, & Sagar, 1989; Sheng & Greenberg, 1990). It should be noted, however, that not all neurons express Fos when they are activated, so an absence of Fos production within neurons does not necessarily mean that the neuron has not been active (Wan, Liang, Moret, Wiesendanger, & Rouiller, 1992). It is worth pointing out that located within the promoter regions of *fos* genes are several regulatory elements that enable Fos protein production to be influenced by a variety of extracellular factors (Bozas, Tritos, Phillipidis, & Stylianopoulou, 1997; Robertson et al., 1995; Xia, Dudek, Miranti, & Greenberg, 1996). Therefore, diverse types of cellular stimulation can activate the production of Fos proteins, allowing such induction to serve as a relatively ubiquitous marker of neuronal activation.

With this brief background, one should be able to see that the analysis of Fos production within neurons can be useful in a variety of ways. One way would be to use Fos immunocytochemistry as an *anatomical mapping procedure*, analogous to a PET scan, to locate neurons that have been activated in the brain after exposing an animal to certain types of stimulation. Since Fos proteins are nuclear proteins, their immunocytochemical detection provides excellent anatomical resolution of individual cells. Another use of Fos production within neurons would be as a *molecular neurobiological tool*. Within a particular neuron or nuclear group, one could investigate the particular extracellular factors that trigger Fos production, and one could also investigate the particular late-responding genes that are affected by Fos production. For the most part, the current chapter will discuss research that has used Fos as an anatomical mapping tool, and the next chapter will discuss the significance of Fos production in the brain of the maternal animal from a molecular neurobiological perspective.

If one were interested in using cFos immunocytochemistry to map neural regions involved in the excitation or inhibition of maternal behavior, one might examine the neural cFos response to the presentation of pups in two groups of naive females who have had no prior experience with pups: One group would be hormonally primed and therefore would show maternal behavior, and the other group would not be hormonally primed and therefore would avoid pups rather than engage in maternal behavior. Additional control groups would not be exposed to pups. This procedure was used by Sheehan et al. (2000) and Sheehan (2000). The basic experimental design is shown in Figure 5.10. Primigravid female rats were hysterectomized and ovariectomized on day 12 of pregnancy, and half of these females were injected subcutaneously with 20 µg/kg of estradiol benzoate (EB) immediately after the surgery while the remaining females were injected with oil. Forty-eight hours later, half of the EB and oil-treated females were presented with pups for 2 hours, while the remaining females were not exposed to pups. The EB-PUP females showed maternal behavior during the 2-hour period while the OIL-PUP females did not. Following the pup exposure period, all females were sacrificed and their brains were immunocytochemically processed for the localization of cFos in selected forebrain regions. Sheehan et al. (2000) and Sheehan (2000) were able to classify four types of neural responsiveness to pup presentation: (a) Some neural regions were most active, as evidenced by the number of cells that contained cFos immunoreactivity, in females showing maternal behavior, and it is possible that some of these regions are playing a positive role in maternal behavior; (b) some neural regions were most active in those females that were exposed to pups but did

A. Preparation

B. Groups

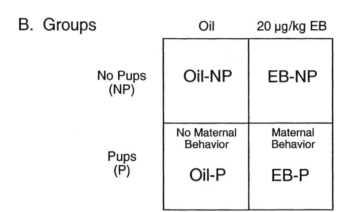

Figure 5.10. The experimental design of Sheehan, Cirrito, Numan, and Numan (2000) that utilized cFos immunocytochemistry to detect candidate neural sites that may exert either inhibitory or excitatory effects over maternal behavior in rats. **A** shows the procedure used: Primigravid female rats were hysterectomized (H) and ovariectomized (O) on day 12 of pregnancy and injected subcutaneously with either 20 μg/kg estradiol benzoate (EB) or oil. Forty-eight hours later, half of the EB-treated females and half of the oil-treated females were exposed to pups (P) for 2 hours while the remaining females were not exposed to pups (NP). For the pup-exposed females, those treated with EB showed maternal behavior during the 2-hour exposure phase and those treated with oil did not. The brains of all females were subsequently immunocytochemically processed to detect cFos immunoreactive neurons in selected brain regions. **B** shows the four groups that were formed in this 2×2 experimental design.

not show maternal behavior, suggesting brain areas that may be involved in inhibiting maternal behavior; (c) some neural regions were activated by pup stimulation, when compared to the females not exposed to pups, irrespective of whether the pup-exposed females showed maternal behavior or not; and (d) some neural regions were not affected by pup stimulation.

The putative inhibitory regions within the forebrain included the anterior hypothalamic nucleus (AHN, the major nucleus within AHA), the ventral part of the lateral septum (LSv), the dorsal premammillary nucleus (PMd), and the parvocellular part of the paraventricular hypothalamic nucleus (PVNp). These regions are shown with tightly spaced diagonal lines on the left side of

Figure 5.11B. The putative excitatory forebrain regions, which are shown with widely spaced diagonal lines on the right side of Figure 5.11B, included the dorsal part of the MPOA (MPOAd), vBST, and the intermediate part of the lateral septum (LSi). Figure 5.12 shows the actual number of Fos-positive neurons in each of these regions in the various experimental groups. The following neural regions, which are indicated in dark gray on both the left and right sides of Figure 5.11B, were activated by pup presentation, but such activation was equally intense in maternal and nonmaternal females: MeA (some experiments showed more intense activation in nonmaternal females, but others showed equal activation in maternal and nonmaternal females exposed to pups), lateral habenula (LHb), and the ventrolateral part of the ventromedial nucleus of the hypothalamus (VMNvl).

Several other studies have attempted to identify neural regions that play a positive role in maternal behavior by measuring the cFos neural response in the forebrain of postpartum females that are exposed to either pups or control stimuli such as candy. In this situation, of course, all females that are exposed to pups show maternal behavior. If certain forebrain neural regions show selective Fos activation in response to pup presentation, this suggests their positive involvement in maternal behavior. Conforming with the findings of Sheehan et al. (2000), these studies have indicated that the MPOA, vBST, MeA, and LSi are activated during pup presentation and the occurrence of maternal behavior in postpartum females (Fleming, Suh, Korsmit, & Rusak, 1994b; Lonstein, Simmons, Swann, & Stern, 1998b; Numan & Numan, 1994, 1995; Stack & Numan, 2000; Walsh et al., 1996). These studies on postpartum animals have also detected additional brain regions, which were not examined by Sheehan et al. (2000) but which showed cFos induction in response to pup presentation and the occurrence of maternal behavior: the shell region of the nucleus accumbens (NAs) (Lonstein et al., 1998b; Stack, Balakrishnan, Numan, & Numan, 2002), the anterior magnocellular part of the paraventricular hypothalamic nucleus (PVNam: Stack & Numan, 2000), and the somatosensory neocortex (Lonstein et al., 1998b). These additional, possibly positive neural regions are shown in solid black on the right side of Figure 5.11B (except that the PVNam is not shown). Interestingly, in postpartum animals there is no indication that the VMN is activated by pup stimulation and the occurrence of maternal behavior, and there is conflict over whether cFos is induced in the LHb in these animals (Fleming et al., 1994b; Lonstein et al., 1998b; Numan & Numan, 1994; Stack & Numan, 2000; also see Kalinichev, Rosenblatt, Nakabeppu, & Morrell, 2000b).

Although this work is strictly correlational, it does suggest candidate neural regions within the forebrain that may be involved in either the facilitation or inhibition of maternal behavior in rats. With respect to the Sheehan et al. (2000) study, since the AHN, PMd, LSv, and PVNp are most active in females who are exposed to pups but who are not maternally responsive, it is possible that these regions are activated by pup stimuli and that such activation inhibits maternal behavior, perhaps by promoting withdrawal and avoidance of pups. In

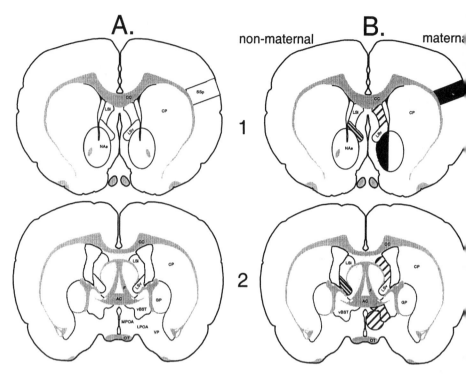

Figure 5.11. The distribution of cFos immunoreactive cells in selected regions of the rat brain for females rats who are exposed to pups but do not show maternal behavior (nonmaternal) and for female rats who are exposed to pups and show maternal behavior (maternal). Column **A** shows frontal sections of the relevant brain regions, with all abbreviations for reference. In column **B**, the same brain regions are shown, along with an indication of the conditions under which a high degree of cFos expression occurs in particular regions. The brain sections in **B** are divided into nonmaternal (left) and maternal (right) halves. Some brain regions show a high level of activity, as indicated by the number of cells expressing cFos immunoreactivity, only under conditions in which the female is exposed to pups and shows maternal behavior. These brain regions, which are putative excitatory regions for maternal behavior, are shown on the right side of the column **B** sections using widely spaced diagonal lines. Other brain regions show a high level of activity, as indicated by the number of cells expressing cFos immunoreactivity, only under conditions in which the female is exposed to pups and does not show maternal behavior. These brain regions, which are putative inhibitory regions for maternal behavior, are shown on the left side of the column **B** sections using tightly spaced diagonal lines. Additional brain regions show high levels of activity when females are exposed to pups, irrespective of whether they do or do not react maternally. These brain regions are shown in dark gray on *both* sides of the relevant sections in column **B**. Finally, some areas are indicated in black on the right side of certain column **B** sections. These areas show high levels of activity in postpartum females showing maternal

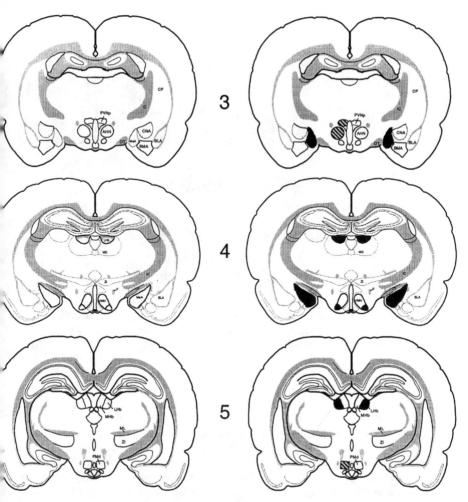

Figure 5.11 *Continued.* behavior, but the response of these regions to pup stimuli in nonmaternal females has not yet been measured. Abbreviations: AC = anterior commissure; AHN = anterior hypothalamic nucleus; BLA = basolateral amygdaloid nucleus; BMA = basomedial amygdaloid nucleus; CC = corpus callosum; CNA = central nucleus of amygdala; CP = caudate-putamen; F = fornix; GP = globus pallidus; IC = internal capsule; LHb = lateral habenula; LPOA = lateral preoptic area; LSi = intermediate nucleus of the lateral septum; LSv = ventral nucleus of the lateral septum; MD = medial dorsal thalamic nucleus; MeA = medial amygdaloid nucleus; MHb = medial habenula; ML = medial lemniscus; MPOA = medial preoptic area; NAs = shell region of the nucleus accumbens; OT = optic tract; PMd = dorsal premammillary nucleus; PVNp = parvocellular region of the paraventricular hypothalamic nucleus; SSp = primary somatic sensory cortex; vBST = ventral bed nucleus of stria terminalis; VMN = ventromedial hypothalamic nucleus; VP = ventral pallidum; ZI = zona incerta.

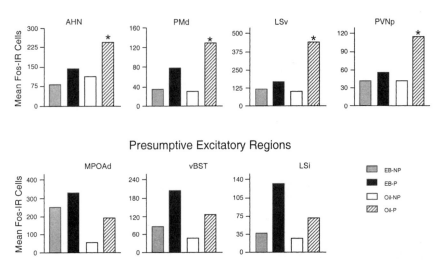

Figure 5.12. Presumptive inhibitory and excitatory neural regions regulating maternal behavior in rats. cFos immunocytochemistry was used to map these areas based on the experimental protocol outlined in Figure 5.10. The potential inhibitory regions included the anterior hypothalamic nucleus (AHN), dorsal premammillary nucleus (PMd), ventral part of the lateral septum (LSv), and the parvocellular part of the paraventricular hypothalamic nucleus (PVNp). These regions were most active, as indicated by the mean number of Fos immunoreactive (IR) neurons, in the pup-exposed, oil-treated females (Oil-P) who did not show maternal behavior. The potential excitatory areas included the dorsal part of the medial preoptic area (MPOAd), the ventral part of the bed nucleus of the stria terminalis (vBST), and the intermediate part of the lateral septum (LSi). These regions were most active in pup-exposed females that were injected with estradiol benzoate and showed maternal behavior (EB-P). NP = not exposed to pups. *Significantly greater than the remaining groups. Modified from Sheehan et al. (2000).

contrast, the fact that the MPOA, vBST, and LSi are most active in females who show maternal behavior when exposed to pups suggests their positive involvement in maternal behavior control mechanisms. Fos activation within the MPOA/vBST, of course, fits with the experimental evidence (reviewed in Section 3.2 of this chapter) indicating a central role for these neurons in maternal behavior. How is one to interpret the response of neural regions that are activated by pup stimuli irrespective of the occurrence of maternal behavior? One possibility is that different, although spatially overlapping, populations of neurons within the selected region are activated under maternal and nonmaternal conditions. A good case for this point might be the MeA. MeA projects to both MPOA/BST and to AHN (Canteras et al., 1995). Also recall that MeA receives olfactory and vomeronasal input and that pup odors inhibit maternal behavior in naive virgin females but are attractive, rather than aversive, to maternal females. Perhaps in nonmaternal females, olfactory and other inputs from pups activate an MeA-to-AHN circuit to inhibit maternal behavior, while in maternal females, olfactory and other inputs from pups activate an MeA-to-MPOA/BST circuit that promotes approach toward pups. Other possibilities to explain the equivalent activation of neural regions by pup stimuli in maternal and nonmaternal states are that the activated region contains cells that are either inhibitory or excitatory (or both) for maternal behavior, but that the downstream effects of activation of inhibitory neurons are prevented in maternal animals, and conversely, the downstream effects of activation of excitatory neurons are prevented in nonmaternal animals.

Additional possibilities exist to explain the findings of the Fos studies described above, including the possibility that the activated regions are not directly relevant to maternal behavior control mechanisms. In particular, it is possible that some aspects of Fos activation observed in maternal and nonmaternal females are related to activation of neuroendocrine mechanisms rather than neurobehavioral mechanisms. For example, the high Fos response in the PVNp of nonmaternal females exposed to pups may be indicative of stress-induced activation of the hypothalamic-pituitary-adrenal axis (see Figure 4.11), since the PVNp contains CRF neurons that project to the median eminence (Van de Kar & Blair, 1999; also see Dayas, Buller, & Day, 1999), while the high expression of Fos in the PVNam of postpartum females that are nursing their young may be related to the milk-ejection reflex, since the magnocellular part of the PVN contains oxytocin neurons that project to the neural lobe of the pituitary (Swanson & Sawchenko, 1983). Therefore, it is the purpose of the next two sections to present research that expands on the Fos mapping results, presenting experimental evidence that outlines the neural circuits through which MeA may inhibit and MPOA/vBST may facilitate maternal behavior.

4.1. Corticomedial Amygdala Inhibitory Circuits

The AHN, PMd, LSv, and PVNp all show increased Fos expression in nonmaternal female rats who avoid pups that they are exposed to for the first time, and this Fos response is much greater than that shown by females who are also exposed to pups for the first time but who, instead, act maternally (see Figure 5.12). Since MeA projects to AHN and LSv (Canteras et al., 1995), LSv projects to AHN (Swanson & Cowan, 1979), and AHN projects strongly to PMd (Risold, Canteras, & Swanson, 1994), it is possible that novel olfactory stimuli from pups activate these regions in suboptimally hormone-primed females, and that activation of this integrated network prevents maternal behavior by promoting withdrawal and avoidance of pups. In support of such a view, research has shown that a variety of stressful or threatening situations, such as immobilization stress or exposure to an elevated plus maze, a dominant conspecific, a predator, or a conditioned aversive stimulus, activate Fos expression in one or more of the following regions: corticomedial amygdala, AHN, LSv, PMd, and PVNp (Canteras, Chiavegatto, Valle, & Swanson, 1997; Da Costa et al., 1996b;

Dielenberg, Hunt, & McGregor, 2001; Duncan, Knapp, & Breese, 1996; Kollack-Walker, Watson, & Akil, 1997; Pezzone, Lee, Hoffman, & Rabin, 1992; Silveira, Sandner, & Graeff, 1993). These results suggest that these neural structures may participate in a "general aversion system" that can be accessed by a variety of threatening stimuli. The following findings provide additional support for such a view:

1. Lesions of PMd decrease the escape/flight responses shown by a rat in the presence of a predator (Canteras et al., 1997).
2. Electrical and chemical stimulation at the level of the medial hypothalamus caudal to the MPOA, which includes the AHN and VMN, produce defensive aggression and escape/flight behaviors in a variety of mammals, including rats (Bandler, 1988; Fuchs, Edinger, & Siegel, 1985; Silveira & Graeff, 1992). In this regard, note that the VMN receives input from MeA, and that VMN and AHN are reciprocally interconnected (Canteras, Simerly, & Swanson, 1994; Canteras et al., 1995; Risold et al., 1994). Also recall that VMNvl showed a Fos response to pup exposure in naive females irrespective of whether the females showed maternal behavior or not (Sheehan et al., 2000).
3. There is evidence for a direct spinohypothalamic tract, which projects to the anterior hypothalamus, primarily relaying nociceptive somatic sensory inputs to this region (Giesler, Katter, & Dado, 1994). Presumably, such input activates appropriate defensive responses.
4. AHN, VMN, and PMd project strongly to the periaqueductal gray (PAG) in the midbrain (Canteras & Goto, 1999; Canteras & Swanson, 1992; Canteras et al., 1994; Risold et al., 1994), a region that is critically involved in regulating diverse somatic and autonomic reactions related to fear, anxiety, escape/flight responses, and defensive aggression (Bandler, 1988; Bandler & Shipley, 1994; Behbehani, 1995; Fanselow, 1991; Fuchs et al., 1985). In this regard, it should be noted that others have suggested that a MeA-to-AHN/VMN/PMd-to-PAG circuit may underlie defensive responses to threatening situations under certain circumstances (Canteras et al., 1997; Han, Shaikh, & Siegel, 1996).

In nonmaternal female rats, if novel olfactory stimuli from pups activate the MeA, which, in turn, activates a "central aversion system" that promotes withdrawal and avoidance of pups (and perhaps infanticide, which might be considered an example of defensive aggression), then one would expect that MeA lesions would prevent the activation of this system. This hypothesis was tested by Sheehan, Paul, Amaral, Numan, and Numan (2001) in the following way, and was based on the fact that MeA efferents are primarily ipsilateral (Canteras et al., 1995): Primigravid female rats were hysterectomized and ovariectomized on day 12 of pregnancy, but were not injected with estradiol (and therefore should not show immediate maternal responsiveness). Immediately following this surgery, half of the females received a unilateral NMDA lesion of the corticomedial amygdala, and the remaining females received control injections of NMLA. Finally, half the females in the NMDA and NMLA groups were exposed to pups for 2 hours beginning 48 hours postoperatively, while the other half were not. None of the females exposed to pups showed maternal behavior during the 2-hour pup exposure period. The brains of all of these females were subsequently immunocytochemically processed to detect Fos-ir neurons in selected forebrain regions. On the side of the brain contralateral to the lesion, females exposed to pups had more Fos-ir neurons than the non-pup-exposed females in MeA, LSv, AHN, PMd, VMNvl, and PVNp, and this confirms previous results (Sheehan et al., 2000). Importantly, the NMDA lesions of the amygdala, which were centered in MeA, caused a significant reduction in the number of pup-induced Fos-ir neurons on the side of the brain ipsilateral to the lesion in AHN, PMd, VMNvl, and LSv, but not in PVNp. These results suggest that LSv, AHN, VMNvl, and PMd may lie downstream from MeA in a neural circuit that inhibits maternal behavior in the rat.

Experimental support for the view that some of these regions that are downstream from MeA are inhibitory for maternal behavior has recently been provided by Bridges, Mann, and Coppeta (1999). They found that lesions to the AHN/VMN region facilitated maternal behavior in virgin female rats that were suboptimally primed with hormones. Based on these findings and the Fos data just described, Sheehan et al. (2001) wanted to test whether MeA projects to AHN/VMN to inhibit maternal behavior. First, they confirmed the results of Bridges et al. (1999) by showing that NMDA lesions of the AHN/VMN region potentiate maternal behavior (Sheehan et al., 2001). Primigravid female rats were hysterectomized and ovariectomized on day 12 of pregnancy and injected subcutaneously with 1.25 μg/kg of estradiol benzoate (recall that this is suboptimal hormone priming, and that MeA lesions facilitated maternal behavior under this paradigm). Immediately following this surgery, females received either NMDA lesions of AHN/VMN or control injections of NMLA. Two days later, these females were exposed daily to pups until they showed maternal behavior or until 5 days had elapsed. Figure 5.13 shows the behavioral results and the lesion location. Females with NMDA lesions of AHN/VMN became maternal after a median latency of 1.5 days while the NMLA-injected females had a median latency to onset of maternal behavior of 6 days. The NMDA lesions destroyed the posterior part of the AHN and the rostral part of VMN, including part of VMNvl.

In a final experiment, Sheehan et al. (2001) employed an asymmetrical lesion design. Day 12 pregnancy-terminated females injected with 1.25 μg/kg of estradiol benzoate received one of the following treatments: unilateral NMDA lesion of MeA paired with a contralateral NMDA lesion of AHN/VMN, unilateral injection of NMLA into MeA paired with a contralateral injection of NMLA into AHN/VMN, or unilateral NMDA lesions of MeA and AHN/VMN that were placed ipsilateral to one another. Since MeA projections to AHN/VMN are primarily ipsilateral, it was predicted that the contralateral NMDA lesion group would show the greatest facilitation of maternal behavior because the hypothesized inhibitory circuit would be destroyed on both sides of the brain. The results indicated, however, that both the ipsilateral and contralateral NMDA le-

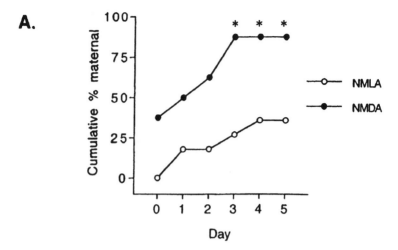

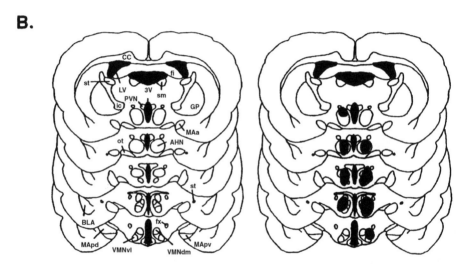

Figure 5.13. The effects of excitotoxic amino acid lesions (produced with N-methyl-D-aspartic acid [NMDA]) of the anterior hypothalamic nucleus and the ventrodmedial nucleus (AHN/VMN) on the onset of maternal behavior in female rats who were exposed to a suboptimal hormone-priming regimen. Primigravid female rats were hysterectomized and ovariectomized on day 12 of pregnancy and injected subcutaneously with 1.25 µg/kg of estradiol benzoate. On day 12, approximately half of the females also received NMDA lesions of AHN/VMN, while the remaining females received control injections of the inactive N-methyl-L-aspartic acid (NMLA). Forty-eight hours later, all females were exposed to pups and the latency to onset of maternal behavior was observed. **A** shows the cumulative percentage of females showing maternal behavior over the test days, and the right side of **B** shows the location of a typical lesion (drawn in black). An unlesioned brain is shown on the left side of **B** to indicate anatomical landmarks. AHN = anterior hypothalamic nucleus; BLA = basolateral amygdaloid nucleus; CC = corpus callosum; fi = fimbria; fx = fornix; GP = globus pallidus; ic = internal capsule; LV = lateral ventricle; MAa = anterior part of the medial amygdaloid nucleus; MApd = posterodorsal part of the medial amygdaloid nucleus; MApv = posteroventral part of the

sions facilitated maternal behavior (median latency to onset of maternal behavior of 1 day) in comparison to the NMLA contralateral group (median latency to onset of maternal behavior of 3 days).

How can we explain the finding that both unilateral and bilateral damage to a putative central aversion system was capable of facilitating maternal behavior in females rats who were exposed to a hormone profile that was suboptimal with respect to stimulating maternal behavior? One possibility is antagonistic to the view that we are dealing with a neural system that inhibits maternal behavior: Perhaps, instead, the NMDA injections into MeA and AHN/VMN had stimulatory neuroendocrine effects. In particular, perhaps such injections promoted prolactin secretion, which in turn facilitated maternal behavior, and ipsilateral injections were just as effective as contralateral injections in this regard. MeA and AHN/VMN are involved in the regulation of prolactin secretion (Erskine, 1995; Gunnet & Freeman, 1983), and there is evidence that the *initial* effects of NMDA infusions into MeA are to stimulate prolactin release (Numan et al., 1993; Polston & Erskine, 2001). Two points argue against this neuroendocrine hypothesis. First, bilateral NMDA lesions of MeA facilitate maternal behavior at a time when stimulatory effects on prolactin release are no longer evident (Numan et al., 1993). Second, damage to AHN/VMN stimulates maternal behavior in female rats whose endogenous prolactin release has been inhibited with bromocriptine (Bridges et al., 1999).

Two other explanations for the facilitatory effects of both ipsilateral and contralateral lesions of MeA and AHN/VMN on maternal behavior are more in line with the view that MeA projections to AHN/VMN represents a neural circuit that inhibits maternal behavior. The first explanation argues that this circuit inhibits maternal behavior because when it is activated by novel pup odors it stimulates defensiveness, and therefore withdrawal and avoidance of pups, which, of course, is antagonistic to positive maternal responses. Importantly, although MeA projections to AHN/VMN are primarily ipsilateral (Canteras et al., 1995), AHN shows a strong *bilateral* projection (with an ipsilateral dominance) to PMd (Risold et al., 1994). Given this anatomy, it is possible that a unilateral NMDA lesion of MeA paired with an ipsilateral lesion of AHN/VMN caused enough *bilateral* depression to a MeA-to-AHN/VMN-to-PMd aversion system to be capable of stimulating maternal behavior in females treated with a suboptimal hormone regimen. The second explanation, which may occur in conjunction with the first, is based on the fact that elements of our putative inhibitory system project to MPOA/vBST. In particular, both AHN and VMN

Figure 5.13 *Continued.* medial amygdaloid nucleus; ot = optic tract; PVN = paraventricular nucleus; sm = stria medullaris; st = stria terminalis; VMNdm = dorsomedial part of the VMN; VMNvl = ventrolateral part of the VMN; 3V = third ventricle. *Significantly greater than NMLA group. (Reproduced with permission from Sheehan T, Paul M, Amaral E, Numan MJ, Numan M [2001] Evidence that the medial amygdala projects to the anterior/ventromedial hypothalamic nuclei to inhibit maternal behavior in rats. Neuroscience 106:341–356. Copyright 2001 by Elsevier Science Ltd.)

project to MPOA (Canteras et al., 1994; Risold et al., 1994), and it is certainly possible that unilateral damage to AHN/VMN caused the MPOA on one side of the brain to be released from active inhibition. Since bilateral, but not unilateral, damage to the MPOA disrupts maternal behavior (Numan & Numan, 1991; Numan & Smith, 1984), and since unilateral implants of estradiol into MPOA stimulate maternal behavior (Numan et al., 1977), it appears that only one active MPOA is necessary for maternal behavior to occur. Therefore, disinhibition of the MPOA on one side of the brain by our ipsilateral lesions may have been sufficient to stimulate maternal behavior.

What we are arguing is that a neural sytem that originates in MeA and projects to AHN/VMN may inhibit maternal behavior by directly activating a neural system that promotes withdrawal and avoidance of pups, by directly inhibiting a neural circuit that promotes approach toward pups and positive maternal responses, or both. Furthermore, it is possible that unilateral damage to the MeA-to-AHN/VMN circuit is sufficient to promote maternal behavior in animals that have some degree of hormone priming. Please note that some degree of hormone priming appears necessary for MeA or AHN/VMN lesions to exert facilitatory effects (Bridges et al., 1999; Numan et al., 1993). Additional work is needed to determine whether unilateral MeA lesions alone, or unilateral AHN/VMN lesions alone, would be sufficient to stimulate maternal behavior in our rats, or whether, instead, the individual effects of each unilateral lesion must summate together in our ipsilateral lesion paradigm in order for a stimulatory effect on maternal behavior to manifest itself.

The working hypothesis that we are developing, with some additions (to incorporate the involvement of LSv, and to deal with MeA projections to MPOA), is depicted in Figure 5.14. In female rats that are not fully primed with a hormone profile that stimulates maternal behavior, olfactory input from pups activates a neural system that includes LSv and AHN/VMN. One effect of such activation may be to promote activity in PMd and PAG circuits that causes withdrawal responses from the threatening olfactory stimuli. Another effect of MeA activation of LSv and AHN/VMN may be to directly inhibit the neural circuit that promotes maternal behavior, since LSv (Simerly & Swanson, 1986) and AHN/VMN (Canteras et al., 1994; Risold et al., 1994) project strongly to MPOA/vBST, and we are hypothesizing that such projections inhibit MPOA/vBST output. Clearly, much more work needs to be done to validate this nascent hypothesis. To begin with, it would be important to examine whether LSv, PMd, and PAG exert inhibitory influences on maternal behavior. Interestingly, since PAG projects strongly to MPOA (Rizvi, Ennis, & Shipley, 1992), this could be another pathway through which the activation of fear systems might depress maternal behavior.

In Figure 5.14 we show two populations of MeA neurons, one projecting to MPOA/vBST and the other projecting to LSv and AHN/VMN. We have already presented the evidence that MeA projections to AHN/VMN (and possibly LSv) may be inhibitory with respect to maternal behavior, and we are now suggesting that MeA projections to MPOA/vBST may actually play a positive, though not

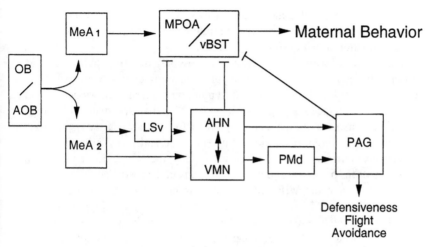

Figure 5.14. Hypothetical model to explain the neural inhibition of maternal behavior. Olfactory input from the main olfactory bulb (OB) and accessory olfacotory bulb (AOB) is shown as reaching two populations of neurons in the medial amygdaloid nucleus (MeA1 and MeA2). MeA1 is shown as sending chemosensory input to the medial preoptic area and the ventral part of the bed nucleus of the stria terminalis (MPOA/vBST), and such input is hypothesized to play a positive role in maternal behavior. In suboptimally hormone-primed nonmaternal females, however, we propose that MeA2 output is dominant over MeA1 output, and that MeA2 relays novel chemosensory inputs to a central aversion system that depresses maternal behavior by directly inhibiting MPOA/vBST and by also activating defensive behavior (escape and avoidance). The central aversion network is shown as consisting of the ventral part of the lateral septum (LSv), anterior and ventromedial hypothalamic nuclei (AHN and VMN, respectively), the dorsal premammillary nucleus (PMd), and the midbrain periaqueductal gray (PAG). Lines ending in arrows signify excitatory projections, and those ending in a vertical bar are inhibitory. Adapted from Sheehan et al. (2000).

essential, role in maternal behavior. Recall that when female rats are fully primed with maternal hormones (so that they would show maternal behavior) they are actually attracted to pup odors. Perhaps this occurs because the hormonal events of late pregnancy and parturition shut down the inhibitory MeA output system (depicted as MeA2 in Figure 5.14) either at the level of MeA or at the level of its downstream projections, while promoting activity in the MeA output system that stimulates interest in and approach toward pup odors (MeA1 system). If such were the case, it would explain why Fos activation in MeA is essentially equivalent in maternal and nonmaternal rats. Although equal numbers of neurons may be activated, the effects of such activation would be different because different neural systems would be affected. Recall, also, that since anosmia, or MeA lesions, do not disrupt maternal behavior in rats, the important process is that the MeA2 inhibitory system be shut down in parturient rats, while the switch to the MeA1 positive output, although facilitatory, is not essential. In support of some of these views, in a later section evidence will be presented that in fully maternal postpartum females, olfactory input facilitates neural activation in MPOA/vBST.

Some support for the general idea that there may be two populations of MeA neurons that respond to olfactory input from pups comes from work on the vomeronasal organ. There are two different vomeronasal sensory neurons, each with a different receptor type (V1R and V2R), that are spatially segregated within the vomeronasal organ, and each receptor/sensory neuron system appears to respond to different pheromone types (volatile and nonvolatile pheromones, respectively) (Krieger et al., 1999). Importantly, there is evidence that the V1R sensory neurons project to and activate neurons in the rostral accessory olfactory bulb (AOB) while the V2R sensory neurons project to the caudal AOB (see Kumar, Dudley, & Moss, 1999). In the context of our developing hypothesis, it would be interesting to determine the differential projections of the rostral and caudal AOB to MeA. One can conceive of a scenario, for example, where pup odors are composed of multiple pheromones, and that perhaps in a non-hormone-primed female a projection route from AOB to MeA to AHN/VMN is dominant, while in a female that is appropriately primed with pregnancy hormones a projection route from AOB to MeA to MPOA/vBST is dominant.

In Chapter 4 we argued that the physiological events of late pregnancy, parturition, and lactation modify the brain of the mother so that she is attracted to, rather than avoidant of, infant-related stimuli, and that she also displays a reduction in general fearfulness to a wide variety of stimuli beyond those presented by infants. The neural model outlined in Figure 5.14 is most appropriate for understanding how the valence of pup-related olfactory stimuli might be switched from negative to positive in the puerperal female. What might be the neural underpinnings of the reduction in general fearfulness? As we will see, maternal physiological factors may not only depress the reactivity of the proposed central aversion system to pup-related olfactory inputs, but might also depress the reactivity of this system to a variety of other sensory inputs. In this context, and because the current section is focussing on the amygdala, we would like to introduce the idea that depression of the output of the central nucleus of the amygdala (CNA; see Figure 5.6 for the location of this nucleus) might be one mechanism that contributes to the reduction in general fearfulness by postpartum females. The CNA is involved in the neural regulation of fear and anxiety, it receives a variety of sensory inputs (visual, auditory, somatic) from the lateral amygdala, and it projects to the PAG and other regions that are involved in fear-related behaviors (Davis, 1992; LeDoux, 1993; Maren & Fanselow, 1996). Evidence that CNA activity may influence the level of emotional reactivity in postpartum females will be presented in Chapter 8.

4.2. MPOA/vBST Facilitatory Circuits

All of the work reviewed so far indicates that the MPOA and adjoining vBST play a critical facilitatory role in the regulation of maternal behavior in rodents.

One might even argue that these regions play key integrative roles by translating hormonal and sensory inputs into maternal responsiveness. It is the purpose of this section to explore the larger neural circuitry within which the MPOA and vBST operate to influence the occurrence of maternal behavior. It will begin by exploring MPOA/vBST efferents relevant to maternal behavior control, since this is where most work has been done. Then it will turn its attention to afferent input to MPOA/vBST, which may induce and/or modulate activity in the critical efferent circuits.

4.2.1. MPOA/vBST Efferents Regulating Maternal Behavior: Functional Neuroanatomy

One approach that has been taken to explore the efferent projections of MPOA and vBST neurons relevant to maternal behavior control is to examine the sites of termination of neurons within these regions that also express Fos proteins during maternal behavior. The rationale for this approach is that if Fos is marking MPOA/vBST neurons that are active during maternal behavior, then the efferents of these neurons are presumably regulating maternal responsiveness. At this juncture it will be worthwhile to review the evidence that supports the view that Fos proteins are labeling neurons relevant to maternal behavior control rather than simply marking neurons that are involved in some other function that covaries with the occurrence of maternal behavior. First, the occurrence of maternal behavior in response to pup presentation is much more effective in activating cFos expression in the MPOA and vBST of postpartum females than is the presentation of control stimuli, such as novel foods or adult conspecifics, which elicit nonmaternal responses (Fleming et al., 1994b; Lonstein et al., 1998b; Numan & Numan, 1994). Second, Fos expression is activated in the MPOA and vBST of virgin female rats who have been sensitized to show maternal behavior through continuous pup stimulation (Kalinichev et al., 2000b; Numan & Numan, 1994). Therefore, such Fos expression is related more closely to maternal behavior than it is to a lactational phenomenon. Third, pup stimuli are much more effective in activating cFos expression in MPOA and vBST of rats who show maternal behavior toward pups than in nonmaternal rats who ignore and avoid the pups (Kalinichev et al., 2000b; Numan & Numan, 1994; Sheehan et al., 2000). Fourth, as suggested by the previous point, Fos expression in MPOA and vBST is closely tied to the performance of maternal behavior, and is not simply triggered by sensory stimulation from pups independent of the occurrence of maternal behavior. For example, exteroceptive stimulation from pups, which is not associated with the performance of typical maternal responses, is much less effective in activating cFos expression in these critical regions than is proximal stimulation from pups that covaries with the occurrence of maternal behavior (Calamandrei & Keverne, 1994; Lonstein et al., 1998; Numan & Numan, 1995). Also, postpartum rats who are thelectomized, olfactory bulbectomized, or both are still capable of showing maternal behavior when exposed to pups, and although cFos expression in MPOA/vBST is reduced in such females in comparison to intact females, the number of cells expressing Fos within these regions in the deafferented females is still significantly above that shown by control females who are exposed to stimuli (candy, for example) that do not induce maternal responses (Lonstein et al., 1998b; Numan & Numan, 1995; Walsh et al., 1996). We are not arguing that a critical sensory-motor linkage does not exist within the MPOA/vBST, which regulates maternal output. Clearly, some sensory inputs must be driving the Fos response in MPOA/vBST of the deafferented maternal females, and it is possible that perioral and general ventral tactile inputs are serving this function (see Stern, 1996a). What we are arguing, however, is that the critical sensory input that is driving the Fos response in MPOA/vBST is also activating cells that regulate the expression of maternal behavior. Importantly, Komisaruk et al. (2000) have reported that the expression of cFos in MPOA during maternal behavior is associated with increased uptake of radioactive 2-deoxyglucose in this region, which supports the view that excitatory input to the MPOA is driving the Fos response there.

If, as the above evidence suggests, Fos proteins are marking neurons in MPOA/vBST that are involved in the regulation of maternal responsiveness, then one might conclude that Fos proteins, in their role as transcription factors, are maintaining the functional integrity of the critical neurons. Although more will be said about this issue in the next chapter, in its simplest form we would argue that essential proteins may be used up or turned over in the MPOA/vBST during the occurrence of maternal behavior, and that Fos proteins may play the role of reactivating the synthesis of these proteins so that maternal behavior can continue. If such were the case, one might hypothesize that the expression of Fos proteins within MPOA/vBST would persist for as long as females were showing maternal behavior. In confirmation of this hypothesis, Stack and Numan (2000) have shown that both cFos and Fos B are persistently expressed in MPOA and vBST neurons during episodes of maternal behavior lasting as long as 48 hours. The persistent expression of Fos B within MPOA/vBST is particularly significant in light of the finding that a mouse line with a knockout mutation of the *fos B* gene shows a relatively specific disruption of maternal behavior (Brown, Ye, Bronson, Dikkes, & Greenberg, 1996). Although such a knockout mutation would remove Fos B proteins from all tissues, these results are certainly consistent with the view that Fos proteins are labeling MPOA/vBST neurons regulating maternal behavior, and that such proteins are necessary for the normal functioning of the critical neurons. The next chapter will discuss these findings in more detail, along with other studies that utilize transgenic mouse lines to study the molecular neurobiology of maternal behavior.

One final study adds strong support for the view that MPOA/vBST neurons that express Fos during maternal behavior are part of neural circuits relevant to maternal behavior control. Recall that knife cuts that are lateral to MPOA/vBST, but not those that are dorsal, anterior, or posterior to MPOA/vBST, selectively disrupt maternal behavior (Numan & Callahan, 1980). This finding suggests that the critical efferents and afferents of the MPOA/vBST necessary for maternal behavior must exit or enter this region from a lateral direction. Therefore, if Fos is marking neurons regulating maternal behavior, one would predict that knife cuts lateral to the MPOA/vBST would decrease the number of neurons that express Fos in this region during maternal behavior. In order to test this idea, Stack et al. (2002) took advantage of the fact that unilateral damage to the MPOA/vBST region, unlike bilateral damage, does not disrupt maternal behavior (Numan & Numan, 1991; Numan & Smith, 1984). They found that the occurrence of maternal behavior in postpartum female rats that received unilateral knife cuts severing the lateral connections of the MPOA/vBST was associated with a major decrease in the number of neurons that expressed either cFos or Fos B in the MPOA and vBST on the side of the brain ipsilateral to the knife cut, while the contralateral MPOA and vBST showed normal levels of Fos expression (see Figure 5.15).

Based on the assumption that Fos expression during maternal behavior is marking MPOA/vBST neurons involved in maternal behavior, two anatomical studies were performed to determine some of the regions to which these neurons project. In the first study (Numan & Numan, 1996), an anterograde tracer, *Phaseolis vulgaris* leucoagglutinin (PHAL), was iontophoretically injected into the dorsal MPOA and adjoining vBST of postpartum rats. This region was chosen as the injection site because of the high density of Fos expressing neurons that occur there during maternal behavior (see Figures 5.11 and 5.15). The results of this experiment are shown in Figure 5.16. Ascending projections from the injection site terminated in the lateral septum (LSi and LSv) and the shell region of the nucleus accumbens (NAs), while descending projections entered the medial forebrain bundle and projected to the following regions: the medial hypothalamus caudal to the MPOA, which included projections to AHN and VMNvl; and brainstem projections that included PAG, ventral tegmental area (VTA), retrorubral field (RRF), and locus coeruleus.

Although all of these sites to which MPOA/vBST neurons project may be involved in maternal behavior control, it should be realized that PHAL would be taken up by a heterogeneous population of neurons, and some of these neurons may not be involved in maternal behavior. Therefore, in a second anatomical study, which employed a double-labeling procedure to detect cFos immunoreactivity and a retrograde tracer, the attempt was made to determine some of the projection sites of MPOA and vBST neurons that express cFos during maternal behavior (Numan & Numan, 1997). Figure 5.17A shows the design of this study. Fully maternal primiparous rats were separated from their pups on day 4 postpartum. On day 5 postpartum, a female received an iontophoretic injection of wheat germ agglutinin (WGA; a retrograde tracer) into one of the following brain regions: lateral septum (LS; including LSi and LSv), medial hypothalamus (MH) posterior to the MPOA and centered in the region of VMN, VTA, RRF, PAG, and lateral habenula (LHb; we included this region as a control since our PHAL results did not find significant projections to LHb from MPOA/vBST). On day 7 postpartum, a female was exposed to either pups, during which maternal behavior occurred, or a control stimulus (candy) for 2 hours. After the 2-hour exposure period, each female was sacrificed and her brain was immunocytochemically processed to detect both cFos and WGA. This procedure

A.

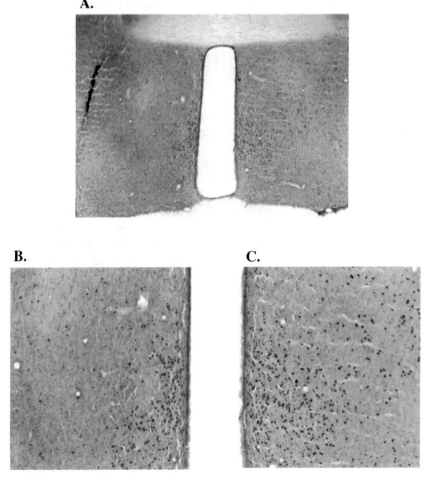

B.

C.

Figure 5.15. Photomicrographs showing the effect of a unilateral knife cut severing the lateral connections of the medial preoptic area (MPOA) and adjoining ventral bed nucleus of the stria terminalis (vBST) on the expression of cFos within cells in the ipsilateral and contralateral MPOA and vBST of maternal rats. **A**. A low-power magnification showing both the ipsilateral and contralateral regions. The knife cut is visible on the left side, the anterior commissure is located dorsally, and the third ventricle is in the middle of the photomicrograph. **B** and **C** show high-power magnifications of the ipsilateral and contralateral regions, respectively. The knife cut clearly decreases cFos expression on the ipsilateral side. Modified from Stack et al. (2002).

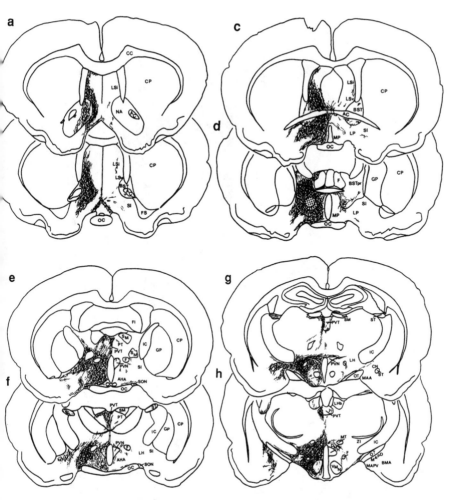

Figure 5.16. A series of frontal sections from rostral to caudal (**a** though **o**) through the rat brain showing the distribution of PHAL-labeled fibers after a unilateral iontophoretic injection of PHAL into the dorsolateral medial preoptic area and adjoining ventral bed nucleus of the stria terminalis. The injection site is indicated by stippling in section **d**. Brain sections were examined from the caudal nucleus accumbens through the locus coeruleus. Abbreviations: AC = anterior commissure; AHA = anterior hypothalamic area; B = Barrington's nucleus; BMA = basomedial amygdaloid nucleus; BST = bed nucleus of stria terminalis; BSTpr = principal subnucleus of the bed nucleus of the stria terminalis; CC = corpus callosum; CLi = central linear nucleus raphe; CN = central nucleus of amygdala; CP = caudate-putamen; CPD = cerebral peduncle; DMH = dorsomedial hypothalamic nucleus; DTN = dorsal tegmental nucleus; F = fornix; FI = fimbria; FR = fasiculus retroflexus; FS = fundus of the striatum; GP = globus pallidus; IC = internal capsule; IP and IPN = interpeduncular nucleus; LH = lateral hypothalamus; LHb = lateral habenula; LP = lateral preoptic area; LSi = intermediate nucleus of the lateral septum; LSv = ventral nucleus of the lateral septum; MAA = anterior medial amygdaloid nucleus; MAPD = posterodorsal medial amygdaloid nucleus;

159

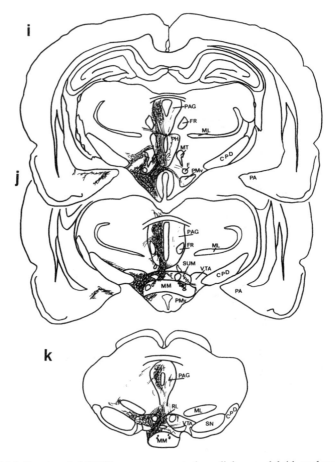

Figure 5.16 *Continued.* MAPV = posteroventral medial amygdaloid nucleus; ML = medial lemniscus; MM = medial mammillary nucleus; MP = medial preoptic area; MRN = mesencephalic reticular nucleus; MT = mammillothalamic tract; NA = nucleus accumbens; OC = optic chiasm; OT = optic tract; PA = posterior amygdala; PAG = periaqueductal gray; PBL = lateral parabrachial nucleus; PBM = medial parabrachial nucleus; PH = posterior hypothalamic nucleus; PLC = perilocus coeruleus region; PMv = ventral premammillary nucleus; PT = parataenial thalamic nucleus; PVN = paraventricular hypothalamic nucleus; PVT = paraventricular thalamic nucleus; PY = pyramidal tract; RL = rostral linear raphe nucleus; RM = raphe magnus nucleus; RN = red nucleus; RRF = retrorubral field; SCP = superior cerebellar peduncle; SI = substantia innominata (ventral pallidum); SM = stria medullaris; SN = substantia nigra; SON = supraoptic nucleus; ST = stria terminalis; SUM = supramammillary nucleus; TPF = transverse pontine fibers; VMN = ventromedial hypothalamic nucleus; VTA = ventral tegmental area; XSCP = decussation of the superior cerebellar peduncle; ZI = zona incerta. (Reproduced with permission from Numan M, Numan MJ [1996] A lesion and neuroanatomical tract-tracing analysis of the role of the bed nucleus of the stria terminalis in retrieval behavior and other aspects of maternal responsiveness in rats. Dev Psychobiol 29:23–51. Copyright 1996 by John Wiley & Sons, Inc.)

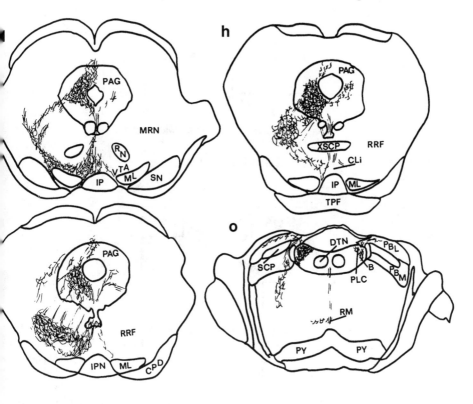

allowed us to determine the degree to which MPOA and vBST neurons that express cFos during maternal behavior project to particular brain regions. The basic findings are schematically presented in Figure 5.17B. Neurons in the MPOA that expressed cFos during maternal behavior projected to LS, MH, PAG, and VTA. Their vBST counterparts projected to MH, PAG, VTA, and RRF. Neurons in the MPOA and vBST that expressed cFos during maternal behavior did not show a significant projection to LHb.

Importantly, this double-labeling anatomical analysis only accounted for the projection sites of about 50% of MPOA and vBST neurons that express cFos during maternal behavior, indicating that the other Fos-expressing neurons may be interneurons that remain within the boundaries of MPOA/vBST, or that they may project to brain regions that were not among the sites we selected for our WGA injections. It would have been interesting to know the degree to which MPOA/vBST neurons that are active during maternal behavior project to NAs (see below).

These two anatomical studies have outlined candidate projection routes through which MPOA/vBST output may influence maternal behavior. In the succeeding sections we will examine each of these circuits in the context of other evidence that suggests their involvement in maternal behavior.

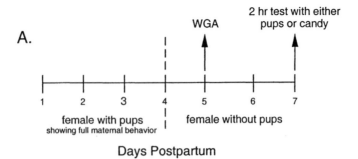

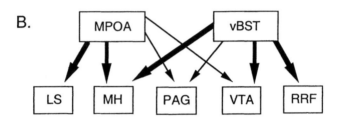

Figure 5.17. The use of a double-labeling immunocytochemical procedure to explore the neural projection sites of medial preoptic area (MPOA) and ventral bed nucleus of the stria terminalis (vBST) neurons that are active during the occurrence of maternal behavior in primiparous postpartum rats. The experimental procedure is shown in **A**. Fully maternal postpartum rats were separated from their pups on day 4 postpartum. On day 5 postpartum, the retrograde tracer, wheat germ agglutinin (WGA), was iontophoretically injected into one of several selected brain regions: the lateral habenula, lateral septum (LS), medial hypothalamus (MH) centered in the region of the ventromedial nucleus, periaqueductal gray (PAG), retrorubral field (RRF), or ventral tegmental area (VTA). On day 7 postpartum, some of these females were reunited with pups and were allowed to show maternal behavior for 2 hours. The remaining females were exposed to candy. Following the 2-hour exposure period, the brain of each female was immunocytochemically processed in order to detect the number of double-labeled cells containing both cFos and WGA in the MPOA and vBST. After the injection of WGA in a particular region, if the pup-exposed females had more double-labeled cells in the MPOA and/or vBST than did the candy-exposed females, we concluded that MPOA and/or vBST neurons that are active during maternal behavior project to that region. MPOA and vBST neurons that were activated during maternal behavior did not project to the lateral habenula. The positive results are schematically shown in panel **B**. MPOA neurons that were active during maternal behavior projected to LS, MH, PAG, and VTA. vBST neurons that were active during maternal behavior projected to MH, PAG, VTA, and RRF. The strength of each projection (number of double-labeled cells) is indicated by the thickness of the line and arrow representing a particular projection route. Panel **A** is modified from Numan and Numan (1997).

4.2.2. MPOA/vBST Projections to NAs, VTA, and RRF: Possible Role in Regulating Maternal Motivation

The evidence reviewed above indicates that MPOA/vBST neurons that are involved in maternal behavior project to the VTA, RRF, and NAs. We want to make the case that this projection is involved in regulating maternal motivation. It is important to note that the VTA and RRF give rise to a major ascending mesolimbic and mesocortical dopamine (DA) projection to the forebrain, and that the nucleus accumbens is among the structures that is innervated by the mesolimbic branch (Deutch, Goldstein, Baldino, & Roth, 1988; Dunnett & Robbins, 1992; Simon, Le Moal, & Calas, 1979; Swanson, 1982). Therefore, it is possible for MPOA/vBST neurons to influence the NA in at least two ways—through a direct projection to NAs (Numan & Numan, 1996; also see Brog, Salyapongse, Deutch, & Zahm, 1993), or indirectly via synapses in VTA/RRF. This anatomy is significant because the NA and the input it receives from the mesolimbic DA system have long been considered to be involved in *general* appetitive motivational processes (Blackburn, Pfaus, & Phillips, 1992; Dunnett & Robbins, 1992; Ikemoto & Panksepp, 1999; Mogenson, 1987; Salamone, 1996). A broad definition of motivation is that it is an internal process that regulates changes in responsiveness to a constant stimulus (Hinde, 1970): When a particular motivational state is high an animal responds appropriately to the relevant stimuli, and when the motivation is low the animal does not respond, or responds less vigorously, to the same stimuli. For example, a hungry animal is interested in food and engages in responses to gain access to food, while a satiated animal does not. Likewise, a maternal organism is interested in infant stimuli and behaves appropriately, while a nonmaternal female does not. A similar logic can be applied to other motivational states, such as thirst and water acquisition and intake, or sexual motivation and the occurrence of sexual behavior. It has been suggested that mesolimbic DA input to the NA can be activated by a variety of specific motivational states, and that such input modulates the ability of the organism to react appropriately to biologically significant stimuli (Blackburn et al., 1992; Mogenson, 1987; Numan, 1985; Salamone, 1996). In other words, the output of the NA may be involved in regulating a variety of appetitive responses, but just which responses occur is not only dependent upon the stimuli present in the external environment, but is also dependent upon the input the NA receives from *specific* motivational systems. Therefore, for the maternal case, it can be hypothesized that the mediation by the NA of appropriate responsiveness to infant stimuli may occur only if the NA, mesolimbic DA system, or both also receive input from neurons in the MPOA/vBST that are specifically related to maternal motivation.

The nucleus accumbens is part of the extrapyramidal motor system (see De Olmos & Heimer [1999], and Groenewegen, Wright, Beijer, & Voorn [1999] for an overview of NA anatomy), and it is also referred to as a component of the ventral striatum (nucleus accumbens and olfactory tubercle) to distinguish it from dorsal striatal stuctures (caudate nucleus and putamen). Although the output of the dorsal striatum projects to the dorsal pallidum (globus pallidus), ventral striatal structures project to the ventral pallidum, which has also been called the substantia innominata (Heimer, Zahm, Churchill, Kalivas, & Wohltmann, 1991). The output of the ventral pallidum, in turn, projects either directly or indirectly to cortical and brainstem systems involved in movement control and other functions (De Olmos & Heimer, 1999; Kalivas, Churchill, & Romanides, 1999; Swanson, Mogenson, Gerfen, & Robinson, 1984; Winn, Brown, & Inglis, 1997). An important distinction between the dorsal striatum and the ventral striatum is that although the former receives a major direct input from several neocortical regions, the ventral striatum gets many of its major inputs from cortical and subcortical limbic structures (De Olmos & Heimer, 1999). In particular, the nucleus accumbens has been shown to receive strong inputs from the lateral and basolateral amygdala, hippocampus, and hypothalamus (Brog et al., 1993; Groenewegen et al., 1999). Based on this anatomy, Mogenson (1987) was a strong proponent of the idea that the NA serves as a link between the limbic system and the motor system, and that it functions to translate motivational and emotional states into action. He further postulated that mesolimbic DA input to NA has a modulating or gating function on the transfer of information from the limbic system to the ventral pallidum (motor system) via the nucleus accumbens. Recent physiological support for this latter concept has been reviewed by Nicola, Surmeier, and Malenka (2000). They indicate that the major efferent neurons of the nucleus accumbens can exist in two states: a down-state (hyperpolarized membrane potential) and an up-state (depolarized membrane potential close to the threshold for action potential generation). Importantly, DA input to such neurons (acting on the D1 DA receptor) appears to further depress neurons that are in the down-state while stimulating neural activity in NA neurons that are in an up-state. They suggest that when convergent excitatory input to the NA from the limbic system brings particular NA neurons into an up-state, then concurrent DA input will drive such neurons above threshold, in this way stimulating the output of particular neural circuits. Importantly, DA input to NA is also conceived of as increasing the signal-to-noise ratio in the accumbens, as those NA neurons that are not receiving strong excitatory input from the limbic system would actually be inhibited by mesolimbic DA input to the striatum.

What kind of input can be relayed to the NA via the limbic system? We would like to focus attention on the strong projection of the basolateral amygdala (BLA; see Figure 5.6 for the location of this nucleus within the amygdala) to the NA (Brog et al., 1993; Krettek & Price, 1978; Pitkanen, 2000), which presumably utilizes glutamate as an excitatory neurotransmitter (Swanson & Petrovich, 1998). The BLA receives strong input from the lateral amygdala (LA), and this is noteworthy because LA is a recipient of diverse forms of sensory input from the neocortex and thalamus (Pitkanen, 2000; Pitkanen et al., 1997). Through such connections, BLA can presumably relay visual, auditory, gustatory, and somatic sensory input to the NA. Importantly, BLA input to the NA does not occur via the stria terminalis (Swanson & Petrovich, 1998). Olfactory input can also presumably reach the NA via BLA because the medial amygdala projects to LA (Canteras et al., 1995). With respect to additional olfactory inputs, the olfactory cortex (piriform cortex) and the nucleus of the lateral olfactory tract (NLOT) project directly to NA (Brog et al., 1993).

The anatomy we have just presented is incomplete and simplified, but it will suffice for the hypothetical neural model that is presented in Figure 5.18, which is meant to explain how MPOA/vBST input to NAs and VTA/RRF might influence maternal responsiveness (see Numan, 1985, 1988, for precursor versions of this model). The model shows that diverse forms of pup-related sensory inputs are capable of reaching the nucleus accumbens. However, the model further stipulates that such sensory input to the accumbens cannot be translated into appropriate maternal reactivity unless the NA also receives concurrent input from MPOA/vBST, either directly or via the mesolimbic DA system. The model further indicates that pup-related sensory inputs can also reach the MPOA/vBST (we will explore the evidence for this statement in a later section), and it is these inputs that drive the output of MPOA/vBST if these neurons have been appropriately primed with hormones and/or affected by previous maternal experience. This hypothetical model, therefore, postulates an interaction between a specific motivational system (MPOA/vBST and maternal motivation) and a nonspecific motivational system (nucleus accumbens) so that the reactivity of an organism to infant-related stimuli is regulated by the activity level of the MPOA/vBST.

In support of the model shown in Figure 5.18, Giordano, Johnson, and Rosenblatt (1990), and Stern and Taylor (1991) have found that the systemic treatment of postpartum rats with haloperidol (a DA receptor antagonist that preferentially blocks the D2 DA receptor) interferes with the active components of maternal behavior, retrieving and nest building, while having little effect on the occurrence of nursing. Interestingly, systemic treatment of postpartum rats with DA agonists (apomorphine, a mixed D1/D2 DA receptor agonist [Stern & Protomastro, 2000], and cocaine, which blocks the reuptake of DA, norepinephrine, and serotonin [Kinsley et al., 1994; Vernotica, Lisciotto, Rosenblatt, & Morrell, 1996]) is also associated with a significant disruption of maternal behavior. It has been suggested that an optimal level of central neural DA transmission is necessary for the normal occurrence of maternal behavior, and if this level is either too low or too high disruptions will be observed (Stern & Protomastro, 2000). A problem with these studies is that systemic drug treatment was applied, and therefore all DA systems would be affected, not just mesolimbic DA input to NA. However, additional findings do indicate the importance of DA input from the ventral mesencephalon to NA for maternal behavior:

1. Electrical lesions of the VTA, which would destroy DA and non-DA neurons in the VTA, and fibers of passage, have been found to disrupt maternal behavior in rats (Gaffori & Le Moal, 1979; Numan & Smith, 1984).

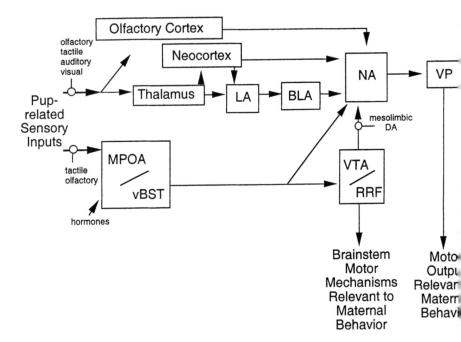

Figure 5.18. Hypothetical neural model to explain how the convergence of medial preoptic area (MPOA) and ventral bed nucleus of the stria terminalis (vBST) neural inputs to the nucleus accumbens (NA) with other neural inputs to NA may regulate maternal reactivity to pup-related stimuli in rats. The output of NA to the ventral pallidum (VP) is hypothesized to regulate the occurrence of proactive voluntary maternal responses to pup-related stimuli. A variety of pup-related sensory inputs (olfactory, tactile, auditory, visual) are capable of reaching NA through neural connections that include the thalamus, neocortex, olfactory cortex, lateral amygdala (LA), and basolateral amygdala (BLA). It is proposed that such inputs, by themselves, are not capable of activating NA-to-VP output regulating maternal responsiveness. Tactile, olfactory, and hormonal inputs are shown as affecting MPOA/vBST neurons. When MPOA/vBST neurons are sufficiently activated by tactile and olfactory inputs, they are hypothesized to send facilitatory projections to NA, either directly or via the mesolimbic dopamine (DA) system that originates in the ventral tegmental area (VTA) and retrorubral field (RRF). We propose that it is the MPOA/vBST input to NA that allows the NA to relay the effects of pup-related sensory inputs to the VP so that appropriate maternal reactivity occurs. We also suggest that MPOA/vBST neurons may influence the more reflexive aspects of maternal behavior by affecting the brainstem projections of the VTA, RRF, and periaqueductal gray (the latter area is not shown in the diagram).

2. Electrical lesions of the NA also disrupt maternal behavior in rats (Lee et al., 2000; Smith & Holland, 1975).
3. As determined through microdialysis measurement, maternal interaction with pups is associated with increases in the release of DA into the mother's NA (Hansen, Bergvall, & Nyiredi, 1993).
4. 6-hydroxydopamine (a dopaminergic neurotoxin) lesions of either the VTA or the NA (where DA terminals would be destoyed) disrupt retrieval in post-partum rats without affecting nursing (Hansen, 1994; Hansen, Harthon, Wallin, Lofberg, & Svensson, 1991a, 1991b). Similarly, microinjection of flupenthixol (a mixed D1/D2 DA receptor antagonist) directly into the NA of postpartum rats disrupts retrieval behavior without disrupting nursing (Keer & Stern, 1999). Importantly, 6-hydroxydopamine lesions or flupen-thixol injections into the dorsal striatum did not interfere with maternal behavior (Hansen et al., 1991b; Keer & Stern, 1999).
5. Microinjection of cocaine directly into NA of postpartum rats disrupts retrieving and nest building (Vernotica, Rosenblatt, & Morrell, 1999).

These studies clearly suggest that mesolimbic DA input to NA is involved in regulating maternal behavior. In many of the above studies, a depression in DA input to NA was associated with an interference with the active components of maternal behavior, particularly retrieving behavior, while nursing behavior was not disrupted. It should be noted that in most of these studies, if pups were not retrieved, the experimenter placed the pups back into the female's nest area at the conclusion of the retrieving test, and in some cases the female was actually placed over the pups at that time (see Keer & Stern, 1999; Stern & Taylor, 1991). As we have indicated previously, as long as a female does not avoid the pups (this is usually the case with postpartum females who have initiated maternal behavior prior to drug/neurotoxin treatment), nursing behavior may simply be elicited as a reflexive response to suckling and other ventral tactile stimulation (cf. Stern, 1991). It is interesting to point out, in this regard, that the deficits in maternal behavior seen after DA disruption in postpartum rats are similar to the disruptions observed after MPOA/vBST lesions: Retrieval and grouping of pups at a nest site are more severely disrupted than is nursing behavior (see Section 3.2 of this chapter). It is as if both systems are more concerned with regulating active, voluntary responses directed toward pups, rather than with regulating reflexive maternal responses, suggesting that the two systems may be linked together in this regulatory function.

What is the evidence that MPOA/vBST input to the mesolimbic DA projection to NA is one of the ways through which MPOA/vBST efferents influence maternal behavior? First, we should note that bilateral electrical lesions, or transverse knife cuts, of the lateral hypothalamus (LH) disrupt maternal behavior in postpartum rats (Avar & Monos, 1969; Numan, Morrell, & Pfaff, 1985), while excitotoxic amino acid lesions of LH, which would spare axons of passage, do not have this disruptive effect (Numan et al., 1988). Note in Figure 5.16 that the descending efferents of MPOA/vBST must pass through LH on their way to VTA and RRF. Importantly, Numan et al. (1985) provide anatomical evidence that LH knife cuts that disrupt maternal behavior damage the axons of MPOA and vBST neurons. Second, Numan and Smith (1984) showed that bilateral damage to a neural system that extends from the MPOA/vBST to the VTA disrupts the maternal behavior of postpartum rats. They found that a unilateral knife cut of the lateral MPOA/vBST connections paired with a contralateral electrical lesion of VTA disrupted maternal behavior to a much greater extent than did a variety of control lesions. Importantly, retrieving and nest building were more severely disrupted by the effective lesions than was nursing behavior. The various lesions are shown in Figure 5.19, and it should be noted that none of the control lesions was capable of bilaterally damaging MPOA/vBST-to-VTA neural connections (compare with Figure 5.16). In a study similar to that of Numan and Smith (1984), Numan and Numan (1991) have provided both behavioral and anatomical evidence that MPOA/vBST efferents may not only terminate in VTA to influence the occurrence of active, voluntary maternal responses, but may also pass through the VTA to affect RRF in order to exert such effects.

Although the work of Numan et al. (1985), Numan and Numan (1991), and Numan and Smith (1984), in the context of the other work reviewed in this section, strongly supports the view that MPOA/vBST projections to VTA/RRF influence maternal behavior, presumably by affecting mesolimbic DA neurons, it needs to be emphasized that the lesions employed in these studies (electrical lesions and knife cuts) were nonspecific in nature, and would destroy not only neuronal cell bodies and fibers of passage, but also would destroy ascending as well as descending systems that passed through the areas of neural damage. A recent study, however, provides very powerful evidence that the lateral efferent output of MPOA/vBST neurons is capable of influencing the functional activation of NA during maternal behavior.

Stack et al. (2002) asked whether *unilateral* damage to MPOA/vBST lateral efferents would disrupt Fos activation in the ipsilateral NA during maternal behavior. This experiment was based on the facts that unilateral damage to MPOA/vBST, unlike bilateral damage, does not disrupt maternal behavior (Numan & Smith, 1984; Numan & Numan, 1991), and that the efferent projections of MPOA/vBST are primarily ipsilateral (Numan & Numan 1996; Simerly & Swanson, 1988). Fully maternal primiparous rats received one of the following on day 3 postpartum: a unilateral knife cut severing the lateral connections of MPOA/vBST; a unilateral sham knife cut that left the lateral MPOA/vBST connections intact; a unilateral NMDA lesion of MPOA/vBST; or a unilateral NMLA injection into MPOA/vBST, which would not cause an excitotoxic lesion. As expected, none of these interventions disrupted maternal behavior. On day 5 postpartum all females were separated from their pups. Forty-eight hours later, on day 7 postpartum, half the females in each of the groups were reunited with pups for 6 hours (PUP females), while the remaining females were not (NO PUP females). All females that were reunited with pups showed maternal behavior. At the end of the 6-hour period, all females were anesthetized, perfused, and their brains were immunocytochemically processed to detect Fos proteins in various central neural sites. With respect to the effects of NMDA lesions of MPOA/vBST on cFos expression in NAs, the results are shown in Figure 5.20. In the first part of this figure, we show the mean number of cFos positive cells in NAs on the side of the brain contralateral to the NMDA or NMLA injection. Comparisons of these counts between PUP and NO PUP females allowed us to determine whether maternal behavior was associated with an increase in cFos expression in NAs. Subsequent sections of this figure examine the side of the brain ipsilateral to the preoptic neural intervention, and comparisons here allowed us to determine whether the preoptic damage decreased any cFos expression that might have been associated with maternal behavior. As can be clearly seen, maternal behavior was associated with an increase in cFos expression in the contralateral NAs (also see Lonstein et al., 1998b). Importantly, NMDA lesions of MPOA/vBST reduced the number of cFos positive neurons in the NAs on the side of the brain ipsilateral to the lesion to the level of the NO PUP females, while NMLA injections were ineffective. Basically, the same results were obtained when we examined the effects of unilateral knife cuts that severed the lateral MPOA/vBST connections. These data offer solid evidence that MPOA/vBST efferents positively influence Fos activation in NAs during maternal behavior in the rat. Additional findings from this study are (a) the unilateral NMDA lesions and knife cuts also reduced maternal behavior-induced Fos expression in the ipsilateral MPOA/vBST (see Figure 5.15 for the knife cut data), supporting the idea that MPOA/vBST Fos-containing neurons are driving the neural Fos response in NAs during maternal behavior; (b) although maternal behavior was associated with an increase in cFos expression in NAs, such an increase did not occur in the core of NA (NAc), suggesting that NAs is more important for maternal behavior than is NAc (also see Keer & Stern, 1999); and (c) no evidence was found for maternal behavior-induced Fos expression in VTA or RRF.

With respect to the neural route over which MPOA/vBST efferents influence activity in NAs during maternal behavior, the work of Numan and Smith (1984) and Numan and Numan (1991) argues for an MPOA/vBST-to-VTA/RRF projection. A problem with this interpretation, as just noted, is that Stack et al. (2002) found no evidence for an increase in Fos expression in VTA/RRF during maternal behavior. However, since not all activated neurons express Fos proteins (Morgan & Curran, 1991), it is possible that MPOA/vBST output induced activity in VTA/RRF neurons that remained undetectable with our Fos immunocytochemistry procedure. Alternatively, since MPOA/vBST neurons project directly to NAs (Brog et al., 1993; Numan & Numan, 1996), it is certainly possible that this is the significant projection route. In other words, perhaps a MPOA/vBST-to-NAs circuit potentiates the responsiveness of NAs neurons to input from other sources, which might include input from tonically active VTA/RRF mesolimbic DA neurons.

The nucleus accumbens is divided into a lateral core region (NAc) and a medial shell region (NAs), and the results just reviewed suggest the importance

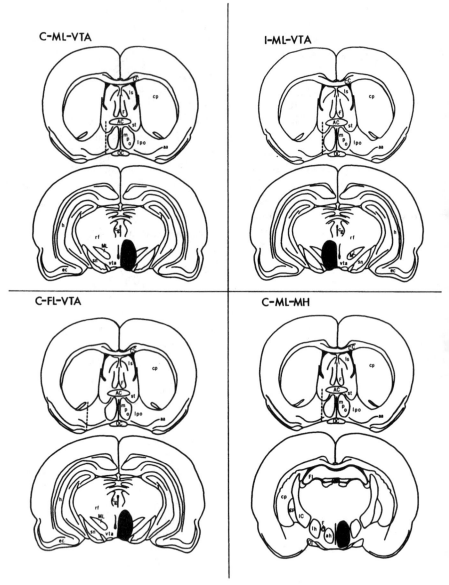

Figure 5.19. Frontal sections showing the asymmetric lesions produced by Numan and Smith (1984): C-ML-VTA, a unilateral cut (dashed line) severing the lateral connections of the medial preoptic area (mpo) and ventral part of the bed nucleus of the stria terminalis (st) paired with a contralateral lesion of the ventral tegmental area (vta; lesion shown in solid black); I-ML-VTA, same as above, exept that the two lesions are ipsilateral to one another; C-FL-VTA, a unilateral knife cut lateral to the lateral preoptic area (lpo) paired with a contralateral lesion of VTA; and C-ML-MH, a unilateral knife cut lateral to medial preoptic area and ventral bed nucleus of stria terminalis paired with a contra-

of NAs for maternal behavior. Although the NAc and NAs have similar afferents and efferents, there are also important anatomical differences between these two territories (Groenewegen et al., 1999). Future research should capitalize on this knowledge in order to further explore the neural regulation of maternal behavior. Since the NAs, but not NAc, receives afferents from vBST and from the hypothalamus (Brog et al., 1993), perhaps the NAs is particularly involved in regulating the occurrence of species-typic behaviors that are also governed by hypothalamic mechanisms (see Kelley, 1999).

The research we have presented is roughly supportive of the neural model presented in Figure 5.18. Additional support comes from the findings that electrical lesions of BLA and electrical or excitotoxic lesions of ventral pallidum (substantia innominata) have been found to disrupt maternal behavior in postpartum rats (Lee et al., 2000; Numan et al., 1988). Indeed, excitotoxic amino acid lesions of the ventral pallidum and adjoining lateral preoptic area disrupt maternal behavior as effectively as do lesions of MPOA (Numan et al., 1988). Our hypothetical model proposes that NA may receive important pup-related sensory inputs from both the BLA and MPOA/vBST, and that the output of NA may influence maternal responses through projections to ventral pallidum. However, since the major efferent neurons of the NA are GABAergic, and therefore inhibitory (Tzschentke & Schmidt, 2000), and since lesions of the ventral pallidum disrupt maternal behavior, it is clear that more research needs to be done to validate our model while also specifying its mechanistic details. In particular, research should focus on the exact contributions of the NA to maternal behavior, and the possibility should not be excluded that the direct and indirect projections of the preoptic area to NA may ultimately influence maternal behavior by *inhibiting* the activity of certain GABAergic NA neurons that project to the ventral pallidum. Relevantly, Mogenson (1987) has argued that DA input to NA potentiates movement by inhibiting the GABAergic efferents of NA to ventral pallidum. Finally, in accord with the undeveloped state of our knowledge, in Chapter 8 we will present a perspective on NA function that diverges significantly from the views presented in our model (Figure 5.18).

The lateral habenula (LHb) is intricately interconnected with the neurocircuitry of the mesolimbic DA system (Ellison, 1994; Matthews Felton, Linton, Rosenblatt, & Morrell, 1998b), and either electrical or excitotoxic lesions of LHb have been found to disrupt all the major components of maternal behavior

---

Figure 5.19 *Continued*. lateral lesion of the medial hypothalamus posterior to the medial preoptic area. Other abbreviations: aa = anterior amygdala; ah = anterior hypothalamus; CC = corpus callosum; cg = central gray (= periaqueductal gray); cp = caudate-putamen; ec = entorhinal cortex; F = fornix; FI = fimbria; gp = globus pallidus; h = hippocampus; IC = internal capsule; lh = lateral hypothalamus; ls = lateral septum; ML = medial lemniscus; OT = optic tract; rf = reticular formation; sn = substantia nigra. (Reproduced with permission from Numan M [1994] Maternal Behavior. In: The Physiology of Reproduction, Vol 2 [Knobil E, Neill JD, eds] pp 221–302. New York, Raven Press. Copyright 1994 by Raven Press.)

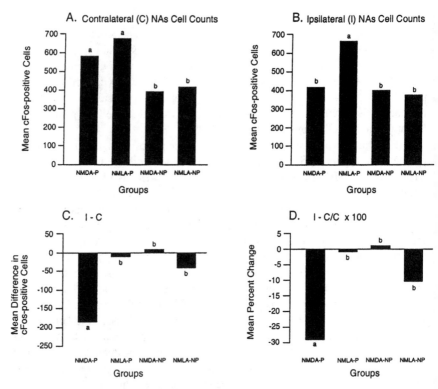

Figure 5.20. Mean cFos immunoreactive cells in the shell region of the nucleus accumbens (NAs) after various treatments. Fully maternal primiparous female rats received *unilateral* N-methyl-D-aspartic acid lesions (NMDA) of the medial preoptic area and adjoining ventral bed nucleus of the stria terminalis (MPOA/vBST), or control injections of N-methyl-L-aspartic acid (NMLA), on day 3 postpartum. The unilateral lesions did not disrupt maternal behavior. On day 5 postpartum, all females were separated from their pups. On day 7 postpartum, approximately half of the females in the NMDA and NMLA treatment groups were reunited with pups, and these females showed maternal behavior. The brains of all females were subsequently immunocytochemically processed to detect Fos proteins in various brain regions. Panel **A** shows the mean number of cFos positive cells in the NAs contralateral to the MPOA/vBST injection site. Females exposed to pups (P) had more cFos positive cells than did the non-pup-exposed females (NMDA-NP and NMLA-NP). Panel **B** shows the same data for the ipsilateral NAs, where it can be seen that a unilateral NMDA lesion of MPOA/vBST reduced the number of cFos positive cells in the NMDA-P females to the level displayed by females that were not exposed to pups. Panels **C** and **D** present the same data in terms of the mean difference in cFos positive cells in NAs when the ipsilateral counts are subtracted from the contralateral counts, or in terms of the mean percent change in cFos positive cells when the ipsilateral side is compared to the contralateral side. Groups that do not share the same letter are significantly different from one another. Panel **D** is modified from Stack et al. (2002).

in rats under a variety of experimental conditions (Corodimas, Rosenblatt, Canfield, & Morrell, 1993; Matthews Felton, Corodimas, Rosenblatt, & Morrell, 1995). In addition, anatomical studies have found that although MPOA/vBST neurons do not have a significant direct projection to LHb (Numan & Numan, 1996, 1997), they do project indirectly to the LHb, presumably via a synapse in the lateral preoptic area (Matthews Felton, Linton, Rosenblatt, & Morrell, 1998b). The question we ask here is whether MPOA/vBST output to LHb is involved in regulating maternal behavior. First, we should note that there is controversy over whether Fos protein expression increases in LHb during maternal behavior in rats. Although some laboratories have reported a maternal behavior-induced cFos expression in LHb (Kalinichev et al., 2000b; Lonstein et al., 1998b), other laboratories have consistently reported negative results (Sheehan et al., 2000; Stack et al., 2002; Stack & Numan, 2000). In addition, Stack et al. (2002) have reported that neither unilateral MPOA/vBST knife cuts nor NMDA lesions reduced the number of cells that expressed maternal behavior-independent cFos in the ipsilateral LHb. The results from the Numan lab suggest, therefore, that maternal behavior is not reliably associated with increases in cFos expression in LHb, and that MPOA/vBST efferents probably do not influence maternal behavior by affecting the functional activity of LHb. More work, however, is needed to uncover the exact way in which the LHb fits into the overall neural circuitry of maternal behavior, and to determine the functional role of this important structure.

In conclusion, we have argued that MPOA/vBST projections to VTA/RRF and/or NAs play a role in regulating maternal motivation. We have taken a mechanistic view to try to explain how activity in MPOA/vBST efferents may regulate the occurrence of maternal behavior by influencing the reactivity of NA neurons to pup-related sensory stimulation. It is interesting to point out, though, that RRF neurons also send descending projections directly to pontomedullary reticular formation neurons, which in turn project to premotor trigeminal system neurons (Von Krosigk & Smith, 1991). Therefore, it is possible that an MPOA/vBST projection to RRF (arising predominantly from vBST; see Numan & Numan, 1996, 1997) plays a role in regulating oral motor maternal responses to perioral sensory input from pups at the level of the brainstem. Note how both of these influences (MPOA/vBST-to-NAs or vBST-to-reticular formation) would fit our broad definition of maternal motivation: Both influences could result in an increase in responsiveness to a constant pup stimulus. In a forthcoming chapter we will more deeply explore the distinction between appetitive and consummatory maternal responses (as defined in Chapter 4) and how such a distinction relates to these two putative projection routes that may influence maternal behavior. For the time being, we suggest that the MPOA/vBST-to-NAs route regulates appetitive maternal responses while the vBST-to-reticular formation route might regulate aspects of consummatory maternal responses.

4.2.3. MPOA/vBST Projections to LS and MH: Possible Role in Regulating Fear Reduction during Maternal Behavior

MPOA and/or vBST neurons, including those that express cFos during maternal behavior, project to the lateral septum (LSi and LSv), and to the medial hypothalamic area (MH) caudal to the MPOA, which includes the AHN and VMN (see Figures 5.16, 5.17; Numan & Numan, 1996, 1997). We want to make the case that some of these projections may be involved in reducing fearfulness and defensiveness, and, therefore, may function to facilitate the onset of maternal behavior at parturition in primiparous females, and may also allow a postpartum female to take acceptable risks in caring for her young.

Not only do MPOA neurons that express cFos during maternal behavior project strongly to LSi, but LSi neurons also express cFos during maternal behavior (Lonstein et al., 1998b; Stack & Numan, 2000). These data suggest that LSi may be playing a positive role with respect to maternal behavior and that MPOA Fos-containing neurons may actually be driving the Fos response in LSi. In support of this view, Stack et al. (2002) found that both unilateral knife cuts of the lateral MPOA/vBST connections and unilateral NMDA lesions of MPOA/vBST caused an ipsilateral decrease in maternal behavior-induced cFos expression in LSi, reducing such expression to the level observed in postpartum females who were not exposed to pups. Clearly, MPOA neural activity during maternal behavior appears to stimulate neural activity in LSi.

Bilateral electrical lesions of the septal area, which include damage to both the lateral and medial septal area, have been found to severely disrupt maternal behavior in rats, mice, and rabbits, supporting the positive involvement of this area in maternal behavior (Fleischer & Slotnick, 1978; Numan, 1985, 1994). Interestingly, in rodents such lesions do not appear to eliminate maternal motivation, since septal-lesioned females are interested in pups and retrieve them. The primary disturbance is that aberrant retrieval responses occur: Septal mothers persistently carry their pups around the cage and drop them at random locations. The retrieving is so disorganized and persistent that the females are unable to proceed to other maternal activites, such as nest building and nursing.

What function(s) might be impaired by septal lesions to result in such erratic retrieving? One possibility is an interference with the well-known role of the septohippocampal system in spatial perception and spatial memory (Olton, Walker, & Wolf, 1982). Clearly, a deficit in spatial perception/memory could result in an inability to perceive or remember where the nest site is and whether a pup had been retrieved there, in this way causing persistent retrieval responses. It should be noted, however, that although direct damage to the hippocampus proper does cause some deficits in maternal behavior in rats (Kimble, Rodgers, & Hendrickson, 1967; Terlecki & Sainsbury, 1978), such deficits do not match the persistent retrieval responses observed in females with septal lesions, suggesting that different functions are being affected by the different lesions. As an alternate possibility, since the septal area has been proposed to regulate behavioral inhibition, and since animals with septal lesions typically show perseverative responding (see Fleischer & Slotnick, 1978), perhaps the persistent retrieval observed in females with septal lesions is simply the result of the females' inability to inhibit a prepotent response.

We would like to emphasize a third possibility (see Sheehan & Numan, 2000). Although there is some disagreement on this point, many investigators have suggested that the output of the septal area is related to anxiety or fear reduction, and that septal lesions are associated with increases in anxiety, fear, and defensiveness (Albert & Chew, 1980; Gavioli, Canteras, & De Lima, 1999; Melia, Sananes, & Davis, 1992; Risold & Swanson, 1997; Sparks & Le Doux, 2000; see Treit & Menard, 2000, for a contradictory point of view). Therefore, perhaps septal females show persistent retrieval responses because of heightened fearfulness. Stamm (1955) found that normal postpartum rats will relocate their litters to new nest site locations in response to stressful situations, such as exposure to air blasts. Perhaps septal lesions exacerbate such tendencies, resulting in persistent retrieving under relatively mild stressful situations. This possibility is significant in light of the evidence presented in Chapter 4, which showed that lactating postpartum rats are generally less fearful than are their nonlactating counterparts. Perhaps the decreased fearfulness that occurs during lactation allows for the occurrence of efficient and organized maternal responses under threatening or demanding environmental situations, and perhaps MPOA-to-LSi projections play a role in maintaining such responsiveness.

Similarly, we would like to suggest the possibility that MPOA/vBST projections to MH (including AHN and VMN) may serve similar fear-reducing functions, except that we would argue that such projections are inhibitory in nature. Recall that MH is involved in potentiating defensive responses to threatening situations (see Section 4.2 of the current chapter). Therefore, our neural hypothesis for fear reduction during the postpartum period would include two possibilities: (a) MPOA stimulates LSi or (b) MPOA/vBST inhibits MH. In regard to the latter proposal, Lonstein and De Vries (2000c) have reported that a proportion of MPOA and vBST neurons that express cFos during maternal behavior also contain glutamate decarboxylase, an enzyme involved in the synthesis of the inhibitory neurotransmitter GABA. This finding is consistent with the involvement of inhibitory MPOA/vBST projections to MH in the regulation maternal behavior.

In comparing the roles of MPOA/vBST projections to NAs and to LSi and MH, we would like to make some distinctions. As already indicated, we have argued that MPOA/vBST projections to NAs may be directly involved in stimulating maternal motivation. In contrast, MPOA/vBST projections to LSi and MH are conceived of as being involved in fear reduction. For postpartum rats, in which maternal behavior has already become established, such projections may allow for the occurrence of efficient maternal responses under stressful and demanding situations. With respect to the initiation of maternal behavior at parturition in naive or inexperienced females, however, such projections may also be indirectly linked to maternal motivation, in that they may play a role in decreasing fear responses to novel pup stimuli, in this way allowing maternal motivation levels to remain higher than maternal avoidance levels (see Figure 4.1).

In Chapter 4 we reviewed the evidence that fear reduction in postpartum females may be important for the occurrence of maternal aggression. The current analysis would suggest, therefore, that the hypothesized MPOA/vBST excitatory projections to LSi and inhibitory projections to MH may be important for the display of maternal aggression. The only evidence we have on this issue is from Hansen (1989), and it is contradictory. Hansen found that excitotoxic amino acid lesions of MH actually decreased maternal aggression in postpartum rats, while our overall analysis would suggest that such lesions would either have no effect or might actually increase such aggression (because we would hypothesize that such females would be even less fearful than the typical postpartum female). It is interesting to note, however, that Hansen tested for maternal aggression by using a female intruder at the nest site rather than a male intruder (males are more likely to attack the pups). Perhaps a certain level of fear is necessary for maternal aggression to occur, and when a relatively harmless intruder is coupled with an MH lesion, fear is so low that aggression does not occur. More work, of course, will be needed to resolve these issues, including the possibilities that MH may contain a heterogeneous group of neurons, and that different MH neurons may influence maternal aggression in different ways.

4.2.4. MPOA/vBST Projections to PAG: Possible Role in Regulating Nursing Reflexes and Fear-Related Functions

Recall that Stern and Johnson (1990) have categorized rat nursing behavior into several different postures. The work of Lonstein and Stern has shown that the ventrolateral portion of the caudal periaqueductal gray (cPAGvl) in the midbrain (at the border of the superior and inferior colliculi) is involved in regulating the occurrence of the quiescent-crouch nursing posture (low-crouch and high-crouch nursing, also referred to as kyphosis), while not being essential for the regulation of other aspects of maternal behavior. First, Lonstein and Stern (1997a) have reported that suckling stimulation from pups, which is an important stimulus in the elicitation and maintenance of kyphosis (see Section 2.2.2 of the current chapter), is also a potent stimulus for the induction of cFos expression in cPAGvl. Second, electrical lesions specifically aimed at the cPAG result in a severe disruption of kyphosis (Lonstein & Stern, 1997a, 1998; Lonstein et al., 1998a), while not disrupting the active and voluntary aspects of maternal behavior, such as retrieving, licking, and nest building. Also, total nursing time was not diminished by such lesions, and the decrease in kyphosis was compensated for by increases in other types of nursing (nursing while in a hover, prone, or supine position). Additional data support the view that the cPAG is involved in the sensorimotor integration underlying the kyphosis brainstem reflex: The cPAG receives somatic sensory inputs from the spinal cord (Bandler & Shipley, 1994; Yezierski, 1991), and in turn sends descending projections to the medullary reticular formation, whose axons contribute to the reticulospinal tract (Cameron, Khan, Westlund, & Willis, 1995; Rizvi, Murphy, Ennis, Bebehani, & Shipley, 1996). In addition, some of the descending PAG projections regulate the occurrence of motoric quiescence and immobility (Bandler & Shipley, 1994; Monassi, Leite-Panissi, & Menescal-de-Oliveira, 1999).

Given that components of the MPOA/vBST appear critical for voluntary, rather than reflexive, maternal behaviors, such as retrieving and grouping young at a nest site, one could hypothesize that MPOA/vBST efferents might occassionally inhibit activity in cPAGvl in order to temporarily disrupt the quiescent crouching posture, in this way allowing the female to more voluntarily interact with her litter through licking/grooming and readjusting the position of the individual pups. In other words, kyphosis does not go on indefinitely, and MPOA/vBST projections to PAG may be one way of terminating the reflex. It is also possible that PAG projections to MPOA serve to inhibit active maternal behavior, in this way facilitating quiescence (Lonstein & Stern, 1997b). In regard to these hypotheses, note that the MPOA/vBST and PAG are reciprocally interconnected (Numan & Numan, 1996; Rizvi et al., 1992), and that MPOA/vBST neurons that express cFos during maternal behavior project to cPAG (Numan & Numan, 1997). (See Figure 5.16, which shows that MPOA/vBST neurons project to cPAGvl.)

Recent findings from Stack et al. (2002) offer partial support for the hypothesis that MPOA/vBST projections might inhibit cPAGvl and quiescent nursing. They found that either unilateral MPOA/vBST knife cuts or NMDA lesions were associated with increases in the incidence of high crouches. Importantly, Lonstein and Stern (1997a) report that *unilateral* cPAG lesions decrease kyphosis, but, of course, not nearly to the same extent as do bilateral lesions, suggesting that unilateral disinhibition of PAG should potentiate kyphosis. In addition, Stack et al. (2002) found that unilateral knife cuts severing the lateral connections of MPOA/vBST were associated with *increases* in the maternal behavior-induced expression of cFos in the cPAG on the side of the midbrain ipsilateral to the knife cut. Unfortunately, unilateral NMDA lesions of MPOA/vBST did not replicate this effect. Interestingly, some of the NMDA-lesioned females did show an ipsilateral increase in cPAG cFos, while others did not. More research needs to be done on this problem, but the results are suggestive of an inhibitory effect of preoptic efferents on cPAG activity.

Recall that we have previously argued that aspects of the PAG may be a final output locus for a hypothesized central aversion system, and that there is strong evidence for a positive role for certain PAG neurons in the regulation of withdrawal, avoidance, and other defensive responses to threatening situations. In this context, then, it is also possible that some MPOA/vBST projections to PAG serve a function similar to the suggested role of MPOA/vBST projections to MH (see previous section). In support of this possibility, Lonstein and Stern (1997a, 1998), and Lonstein et al. (1998b) argue that there are additional neurons in the cPAG that overlap with those regulating kyphosis, but that serve another function, that of regulating the level of fearfulness. In support, they find that postpartum rats with electrical lesions of cPAG show decreased fearfulness, as measured by the amount of time spent in the open arms of an elevated plus maze, and that this decrease in fearfulness is associated with *increases* in maternal aggression directed toward a *male* rat intruder at the nest site. Importantly, such females actually showed less aggression than control females toward a relatively harmless weanling rat intruder. These findings fit nicely with the ideas we developed in the previous section with respect to MPOA/vBST projections to MH, which were hypothesized to be inhibitory. Perhaps MPOA/vBST projections to PAG also serve to inhibit neural systems that mediate fearful responses.

4.2.5. Afferents to MPOA/vBST that May Regulate Maternal Responsiveness

What are the positive afferent inputs to MPOA/vBST that activate this region, allowing for the occurrence of maternal behavior? Surprisingly, in comparison to the work that has explored MPOA/vBST efferents regulating maternal behavior, very little work has examined the involvement of inputs to this region. The Fos data clearly show that MPOA/vBST neurons are activated during maternal behavior, suggesting that some sources of afferent input are driving the response of MPOA/vBST output neurons that regulate maternal behavior. The most significant source of sensory input to MPOA/vBST is olfactory/vomeronasal input, which can reach the MPOA/vBST via MeA (Krettek & Price, 1978; Scalia & Winans, 1975; Simerly & Swanson, 1986). In this regard, it should be noted that olfactory bulbectomy (main and accessory olfactory bulb removal) eliminates the cFos response in MeA that occurs during maternal behavior in postpartum rats (Numan & Numan, 1995; Walsh et al., 1996), indicating that it is indeed olfactory input that activates MeA during maternal behavior. Importantly, olfactory bulbectomy slightly and significantly decreases, but does not eliminate, the cFos response that occurs in vBST during maternal behavior in postpartum rats (Numan & Numan, 1995). These data indicate that olfactory input to vBST is one source of input that plays a role in the neural activation of this region during maternal behavior. The fact that olfactory bulbectomized females show relatively normal maternal behavior indicates that this input to vBST, acting alone, is not essential for maternal behavior. However, such input may play a role in arousing maternal interest in pups through activation of a vBST-to-VTA/RRF pathway (see Numan & Numan, 1997), and such a possibility fits with the fact that postpartum females, unlike naive virgins, are attracted to pup-related odors.

Ventral tactile stimuli or olfactory stimuli appear capable of affecting certain MPOA neurons during maternal behavior. First, note that the peripeduncular nucleus (PPN) in the lateral midbrain tegmentum, which relays suckling-related and other ventral trunk somatic sensory inputs to other brain regions (Factor, Mayer, & Rosenblatt, 1993; Hansen & Kohler, 1984), projects to the MPOA (Chiba & Murata, 1985; Simerly & Swanson, 1986). Second, Numan and Numan (1995) found that thelectomy, which would eliminate suckling stimulation but not other sources of ventral tactile stimulation, did not affect the cFos response that occurs in MPOA/vBST during maternal behavior. In accordance with the work of Stern and Johnson (1990), maternal behavior was normal in the thelectomized females except for a decrease in the occurrence of high-crouch nursing. Importantly, Numan and Numan (1995) found that when thelectomy was coupled with olfactory bulbectomy, although maternal behavior was still normal except for a decrease in high-crouch nursing, the cFos response in the MPOA that occurs during maternal behavior was slightly and significantly reduced, but not eliminated (also see Walsh et al., 1996). This result shows that either olfactory input or suckling input is capable of activating a population of cells in MPOA during maternal behavior, and that only when *both* sources of input are removed will a significant reduction in Fos expression occur. The cFos response that does occur in MPOA (and vBST) of the thelectomized and bulbectomized females is still much higher than baseline, however, and this presumably accounts for the occurrence of maternal behavior.

Lonstein et al. (1998b) have suggested that perioral tactile inputs from pups may provide a significant source of afferent input to MPOA/vBST that activates this region during maternal behavior. Although there is no direct evidence to prove this point, we should note that there is a direct trigeminohypothalamic tract to the rostral hypothalamus, as well as several indirect neural pathways through which perioral tactile inputs can reach the MPOA/vBST (Lonstein et al., 1998b; Malick & Burstein, 1998). Such input is important in light of the findings, previously reviewed, that perioral tactile inputs play an important role in regulating retrieval behavior and other aspects of maternal motivation.

In Section 2.2.3 of this chapter we reviewed the evidence that suggested that ventral tactile and perioral tactile inputs may play a role in regulating maternal motivation, and as long as one of these sensory pathways is intact maternal behavior will occur. Females that received both perioral anesthesia and wore nylon jackets that prevented ventral tactile stimulation showed a low level of maternal responsiveness. Perhaps we can add to this the possible importance of olfactory/vomeronasal input for aspects of maternal attraction toward pups. In other words, the postpartum female appears to receive three major sources of afferent input that drive neural activation in MPOA/vBST: olfactory/vomeronasal, ventral tactile, and perioral tactile. Perhaps if a female receives a significant amount of input from two or more of these sources, the MPOA/vBST will be activated, and this will allow for the occurrence of maternal behavior. As far as we are aware, no one has examined the maternal responsiveness of females with complete olfactory bulbectomies coupled with either perioral or ventral tactile desensitizations (except for the work of Beach and Jaynes [1956b], but we have questioned the effectiveness of their perioral deafferentations). Perhaps such females would show severe deficits in maternal responsiveness, analogous to those that appear to be shown by females who receive both perioral and ventral tactile desensitizations.

In addition to the above sources of sensory input to MPOA/vBST, other sources of afferent input to MPOA/vBST may play a role in regulating maternal behavior. Numan et al. (1990) examined possible sources of brainstem input to MPOA/vBST that may be involved in maternal behavior. They severed the dorsolateral connections of the MPOA/vBST region with a wire knife that was coated with horseradish peroxidase (HRP, a retrograde tracer). The cuts disrupted maternal behavior, and the HRP was taken up by the cut axons and retrogradely transported to their cell bodies of origin. This study found that the MPOA/vBST knife cuts severed the axons of neurons in the locus coeruleus (LC) of the pons and in the caudal nucleus of the solitary tract (cNST) of the medulla. The LC and cNST provide sources of norepinephrine (NE) input to the MPOA/vBST region (see Numan et al., 1990).

Finally, we should note that the MPOA/vBST region receives significant input from both the magnocellular and parvocellular cells of the paraventricular nucleus of the hypothalamus (PVN) (Ingram & Moos, 1992; Simerly & Swanson, 1986). The role of the PVN and its associated oxytocin-containing neurons in maternal behavior control will be discussed in the next section, and in the next chapter.

5. Other Brain Regions Implicated in Maternal Behavior Control

In addition to the neural regions already discussed, two other regions have been shown to be important for maternal behavior: the paraventricular nucleus (PVN) of the hypothalamus and the neocortex. The role of the PVN in maternal behavior is interesting because lesions to this structure interfere with the initiation of the behavior, but not its maintenance once it has become established. More specifically, if PVN lesions are performed during pregnancy in rats, then the onset of maternal behavior at parturition is severely disrupted (Insel & Harbaugh, 1989). Damage to the anterior parts of the PVN appears to be critical in producing this disruptive effect. In contrast, when similar lesions are performed during the postpartum period, after maternal behavior has become established, the maternal behavior of such rats remains intact (Consiglio & Lucion, 1996; Insel & Harbaugh, 1989; Numan & Corodimas, 1985). What these results suggest is that the output of the PVN may have some initial modulatory influence on neural circuits that play either an inhibitory or excitatory role in maternal behavior, but that as maternal behavior becomes established this modulatory influence wanes in importance. In this regard, it should be noted that the PVN is the main source of oxytocinergic neural projections to other parts of the brain (although the PVN and supraoptic nucleus of the hypothalamus both contain oxytocin neurosecretory cells that project to the neural lobe of the pituitary, evidence suggests that it is PVN neurons that supply oxytocin to the rest of the brain: De Vries & Buijs, 1983; Lang et al., 1983). In the next chapter we will detail the evidence that oxytocinergic central neural circuits are critical for the onset of maternal behavior in a variety of species. For the present, we should note that oxytocin, presumably arising from the PVN, can reach several brain regions implicated in the control of maternal behavior, including the ol-factory bulb (Yu et al., 1996b), central nucleus of the amygdala (Bale et al., 2001), lateral septum (Caldwell, Greer, Johnson, Prange, & Pedersen, 1987; Landgraf, Neumann, & Pittman, 1991), MPOA/BST region (Ingram & Moos, 1992; Insel, 1990c; Kremarik, Freund-Mercier, & Stoeckel, 1991, 1995; Ped-ersen, 1997), the medial hypothalamus/VMN region (Johnson, Coirini, Ball, & McEwen, 1989), and the VTA (Pedersen, 1997; Swanson, 1977). It is tempting to speculate that these oxytocinergic projections play an important modulatory role in the regulation of the onset of maternal behavior at parturition by de-pressing the output of inhibitory neural areas and promoting the output of neural regions that play an excitatory role in maternal behavior (Da Costa, Guevara-Guzman, Ohkura, Goode, & Kendrick, 1996a; also see Da Costa, de la Riva, Guevara-Guzman, & Kendrick, 1999). Although PVN lesions that disrupt maternal behavior would damage oxytocin and nonoxytocin projections from the PVN, the overall evidence suggests that a disruption of oxytocin pathways is the likely cause of the lesion's interference with the onset of maternal behav-ior. In further support for a role of the PVN in maternal behavior, research has found increased expression of Fos proteins in certain parts of this area during maternal behavior in both rats and sheep (Da Costa, Broad, & Kendrick, 1997a; Lin et al., 1998; Stack & Numan, 2000). Although such expression may be indicative of neural activation associated with uterine contractions, the milk-ejection reflex, or ACTH release, it may also be marking neurons that influence maternal responsiveness under certain conditions. (See Stack & Numan [2000] for a discussion of some of these issues.)

Evidence for a role of the neocortex in the control of maternal behavior in rats was initially provided by Beach (1937). Neocortical lesions of various sizes were produced in adult virgin female rats. These females were subsequently mated, and their maternal behavior (retrieving and grouping of pups to a single location, nest building, number of surviving young) was studied over the first 4 postpartum days. Although lesions involving less than 20% of the neocortex left maternal behavior intact, larger lesions were associated with clear deficits, and when more than 40% of the cortex was destroyed maternal behavior was almost completely abolished. Examination of the location and extent of cortical damage suggested that the degree of disruption of maternal behavior was not significantly related to the site of the lesion, but was directly related to the amount of neocortical damage. These findings have been replicated by others (Davis, 1939; Stamm, 1955; Stone, 1938).

It is possible that these maternal behavior deficits derive from the fact that as the amount of neocortex destroyed increases, more cortical projection sites of the various sensory modalities are included within the lesioned area. Indeed, Beach and Jaynes (1956c) implied that a relationship existed between their work suggesting that maternal behavior in the rat is under multisensory control and their findings on the effects of neocortical damage. Such a view would also fit with the model we have presented in Figure 5.18, where it can be seen that the neocortex provides one of the routes through which pup-related sensory inputs can reach the nucleus accumbens. In addition to causing multiple sensory deficits, large neocortical lesions may also cause severe deficits in maternal behavior because they interfere with memory functions, spatial perception, and neocortical motor mechanisms. Although no single neocortical area may be essential for maternal behavior in rats, the interference with several neocortical areas and their associated functions after large lesions appears to have drastic consequences for maternal behavior.

A recent study by Xerri, Stern, and Merzenich (1994) has clearly shown that adaptive modifications in neocortical sensory mechanisms are associated with the maternal condition in rats. These investigators recorded neural activity within the primary somatic sensory cortex, at approximately 2 weeks postpartum, of rats who either remained with their pups or were separated from their pups at birth. They found that the representation of the ventral trunk region within the somatic sensory cortex was approximately 1.6 times larger in females that remained with their pups, and that the receptive fields of individual cells within this cortical region in these females were about one-third the sizes of the receptive fields recorded from females that were separated from their pups. These results indicate that postpartum females who are caring for their young are more sensitive to ventral tactile inputs than are their control counterparts. Perhaps this increased tactile sensitivity results in subtle maternal postural adjustments during nursing bouts, which, in turn, facilitate each pup's access to the nipple region. Although the cortical lesion studies described above suggest that the primary somatic sensory cortex by itself is not essential for maternal behavior in rats and for the survival of the mother's young, the work of Xerri et al. (1994) does indicate that this cortical region plays a role in ongoing maternal behavior, and that damage to this area, when added to damage to additional cortical regions, could cause more profound deficits in maternal behavior.

In contrast to the work on rats, research on rhesus monkeys (*Macaca mulatta*) has indicated that circumscribed regions of the neocortex are involved in primate maternal responsiveness (Franzen & Myers, 1973; Myers, Swett, & Miller, 1973). Ablations of either the prefrontal cortex or the anterior temporal cortex at approximately 2 months postpartum have been found to result in severe and long-term deficits in maternal behavior, which included a loss of protective retrieval of infants in threatening situations, increases in maternal withdrawal from their infant, increases in infant punishment by the mother, and decreases in the amount of active infant cuddling. Interestingly, and in some ways similar to rats with preoptic lesions, rhesus females with prefrontal cortical lesions would occasionally tolerate the presence of their infants, allowing them to nurse at their own initiative, with the mother playing a completely passive role. It is worth pointing out that lesions to the prefrontal cortex or anterior temporal cortex not only disrupted maternal behavior, but also seemed to generally interfere with social behavior in male and female rhesus monkeys. Such lesioned individuals simply withdrew from most social interactions within their troops. Based on these findings, and the fact that the prefrontal and anterior temporal neocortex are anatomically interconnected, Myers et al. (1973) suggested that these two neocortical regions act together to regulate social behavior in nonhuman primates. Importantly, there is some preliminary evidence that the prefrontal cortex is activated during maternal behavior in humans (Loberman et al., 2002).

The prefrontal cortex lesions that disrupted maternal behavior in rhesus females destroyed the neocortical tissue lying anterior to the frontal eye fields, and involved the orbitofrontal cortex in its entirety. In the context of the neural model we have presented for the control of maternal behavior in rats (Figure 5.18), the prefrontal cortex is relevant because it projects to the nucleus accumbens, and it also receives direct projections from dopaminergic neurons in the ventral tegmental region and indirect projections from the nucleus accumbens (nucleus accumbens–to–ventral pallidum–to–mediodorsal thalamus–to–prefrontal cortex) (Fallon & Loughlin, 1995; Kalivas et al., 1999; Zahm, 1999). It is interesting to speculate that in primates the interaction of prefrontal cortical mechanisms with subcortical mechanisms has evolved to play a greater role in maternal behavior than it does in rats. It should be noted that in addition to the older studies by Beach (1937) and Stone (1938), more recent studies have also indicated that lesions restricted to the prefrontal neocortex do not appear to disrupt maternal behavior in rats, and therefore do not replicate the severe effects observed in primates (Ferreira et al., 1987; Lee, Li, Watchus, & Fleming, 1999; Numan, 1985). The only caveat we have about this conclusion is to suggest that the involvement of the orbitofrontal cortex in the maternal behavior of rats be investigated more thoroughly than was done by Ferreira et al. (1987). (In the rat, the prefrontal cortex can be divided into a dorsomedial part and an orbito-frontal/insular part [Zilles & Wree, 1995].) We will return to the prefrontal cortex's role in maternal behavior in Chapter 9.

6. Conclusions

Figure 5.21 presents a composite diagram derived from previous figures and information in order to summarize much of the data presented in this chapter. The figure is a hypothetical neural model of the regulation of maternal responsiveness that is based on both scientific evidence and educated speculation. It should be obvious from the material presented in this chapter that much more evidence will be needed to verify elements of the model and to expand its detail. Since the model is derived mainly from work on rats, more comparative research will be necessary to examine its generality. We argue that olfactory projections to MeA2 are dominant in naive nonmaternal females, and this leads to avoidance of pups. As a result of the hormonal and other physiological conditions associated with late pregnancy and parturition, MPOA/vBST output systems become dominant while the MeA2-to-MH-to-PAG aversion system is depressed, and this results in approach behavior and the occurrence of voluntary maternal responses. Notice how this model maps onto the motivational models presented in Chapter 4: Maternal behavior occurs when maternal motivation and interest in pups becomes dominant over the tendency to withdraw from and avoid pup-related stimuli. Note also that this model is primarily concerned with the occurrence of

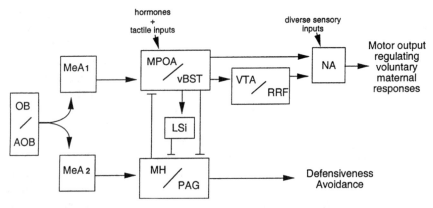

Figure 5.21. A composite neural model of the regulation of maternal behavior in rats that combines the models shown in Figures 5.14 and 5.18. Projection lines ending in arrows signify excitatory connections and those ending in a vertical bar indicate inhibitory connections. Voluntary proactive maternal responses occur when the medial preoptic area (MPOA)/ventral bed nucleus of the stria terminalis (vBST) system is dominant over the medial hypothalamus (MH = anterior and ventromedial nuclei)/periaqueductal gray (PAG) system. The latter system promotes avoidance of novel pup-related stimuli. LSi = intermediate part of the lateral septum. See Figures 5.14 and 5.18 for other abbreviations. Although not shown in this diagram, oxytocinergic projections from the paraventricular hypothalamic nucleus are proposed to play a critical coordinating role in biasing the hypothesized neural circuitry to favor maternal responsiveness. Oxytocin facilitates maternal behavior and may do so by acting at multiple sites, which include MPOA, BST, lateral septum, olfactory bulbs, ventral tegmental area (VTA), and MH.

active, voluntary maternal responses, and therefore the brainstem mechanisms regulating the sensorimotor trigeminal integration necessary for the performance of retrieval, and the sensorimotor integration within the PAG necessary for kyphosis, are not shown.

How do hormones act to affect these various systems so that maternal behavior occurs immediately at parturition in first-time mothers? As we reviewed in Section 3.2 of this chapter, evidence indicates that MPOA/vBST is one of the sites at which both estradiol and lactogens act to facilitate maternal behavior. Therefore, a simple explanation of the mechanism of hormone action might be that hormone action on MPOA/vBST neurons modifies the receptors present in these neurons and also modifies the neurotransmitters/neuromodulators they contain. Such modifications may allow MPOA/vBST to be stimulated by both tactile inputs and olfactory inputs from pups, which in turn activate MPOA/vBST efferents that inhibit the putative aversion system while stimulating neural systems that promote attraction, approach, and interaction with pup-related stimuli. As reviewed in Section 3.2, however, the available evidence does not preclude the possibility that hormones act at additional sites outside the MPOA/vBST to influence maternal behavior. Indeed, estradiol and prolactin receptors are located in MH, LS, MeA, PVN, and PAG in addition to MPOA/vBST (Bak-owska & Morrell, 1997; Pfaff & Keiner, 1973; Shughrue et al., 1997). It is possible, therefore, that these hormones change the phenotype of several neural systems toward the end of pregnancy, with the net result that the responsiveness of the aversion system to novel pup stimuli (and other stressful situations) is downregulated, while the responsiveness of neural systems that play a facilitatory role in maternal behavior is potentiated. Again, we should emphasize that the downregulation of our putative aversion system (either through the direct action of hormones or through the inhibitory effects of afferent input from other regions) is believed to mediate the general reduction in fear associated with the postpartum condition, and therefore not only depresses fearful responses to novel pup stimuli, but also allows the postpartum female to care for her young under stressful and demanding environmental conditions. In Chapter 8, we will elaborate on the model presented here in order to incorporate postpartum modifications in other neural regions (such as the central nucleus of the amygdala) to explain how fearfulness might be downregulated to a variety of external stimuli in addition to olfactory inputs. At this juncture, it is also worth emphasizing the important role of the PVN in the onset of maternal behavior at parturition, a point that will be more fully developed in the next chapter. As we will see, hormone action at the PVN can increase the synthesis of oxytocin, and hormone action at several PVN target sites affects the synthesis of oxytocin receptors. As indicated in the caption of Figure 5.21, the PVN oxytocin system is positioned to affect many neural regions that influence maternal reactivity, in this way suggesting its important coordinating role.

With respect to hormones acting at the level of MPOA/vBST to influence the onset of maternal behavior, the following information is instructive. Figure 5.22 shows the spatial pattern of cFos and Fos B expression in the MPOA/vBST during maternal behavior in postpartum rats (Stack & Numan, 2000). This pattern of expression matches the location of neurons that contain mRNA for the long form of the prolactin receptor (Bakowska & Morrell, 1997), and also overlaps with the location of MPOA/vBST neurons that both bind estradiol and project to the VTA (Fahrbach, Morrell, & Pfaff, 1986b). The high degree of spatial overlap in the neurons that contain these compounds suggests that estradiol and prolactin may act on MPOA/vBST neurons that express Fos proteins during maternal behavior and also project to the VTA (also see Numan & Numan, 1997). In further support of this view, Lonstein, Greco, De Vries, Stern, and Blaustein (2000) have recently reported that cFos is colocalized with the estrogen receptor within MPOA/vBST neurons during maternal behavior.

How might hormones prepare the MPOA/vBST so that at parturition this system is active in potentiating maternal behavior? One posssibility is that the hormonal events of late pregnancy activate Fos protein synthesis in MPOA and/or vBST, and Fos, in turn, has activational effects on the transcription of late-responding genes that ultimately alter the phenotype of the affected neurons. Indeed, there is evidence that the hormonal events that trigger immediate maternal responsiveness also participate in the activation of cFos synthesis in

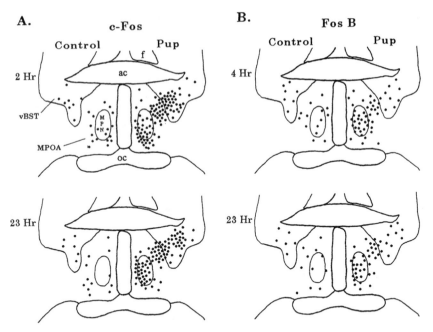

Figure 5.22. Representative frontal sections through the level of the medial preoptic area (MPOA) on which is plotted the location of cFos (panel **A**) and Fos B immunoreactive cells (panel **B**) in postpartum primiparous female rats who were exposed to pups and showed maternal behavior for either 2, 4, or 23 hours, and for control females who were not exposed to pups. ac = anterior commissure; f = fornix; MPN = medial preoptic nucleus; oc = optic chiasm; vBST = ventral part of the bed nucleus of the stria terminalis. (Reproduced with permission from Stack EC, Numan M [2000] The temporal course of expression of c-Fos and Fos B within the medial preoptic area and other brain regions of postpartum female rats during prolonged mother-young interactions. Behav Neurosci 114:609–622. Copyright 2000 by the American Psychological Association, Inc.)

MPOA neurons of rats and sheep and BST neurons of sheep *prior to any exposure to infant-related stimuli* (Da Costa et al., 1997a; Sheehan et al., 2000; Sheehan & Numan, 2002). The work of Sheehan and Numan (2002) is particularly instructive. Previous studies had shown that systemic estradiol treatment can increase cFos expression in MPOA (Auger & Blaustein, 1995; Insel, 1990b), and this fits with data that show that the *cfos* gene contains an estrogen response element in its promoter region (Schuchard, Landers, Sandhu, & Spelsberg, 1993). However, Sheehan and Numan (2002) showed that systemic estradiol treatment was more effective in activating cFos synthesis in the MPOA (in terms of number of cFos-containing cells) of pregnancy-terminated females (primiparous females that were hysterectomized and ovariectomized on day 12 of pregnancy) than it was in hysterectomized and ovariectomized virgin rats. This finding is important because the dose of estradiol used (20 μg/kg) would induce maternal behavior in the pregnancy-terminated rats, but not in the virgins

(Sheehan, 2000). Additionally, when the pregnancy-terminated females were administered both estradiol and a dose of progesterone that would have inhibited estradiol-induced maternal behavior if the females had been exposed to pups, the systemic progesterone treatment inhibited the estradiol-induced increase in cFos expression within MPOA. One interpretation of these results is that exposure to some of the hormonal events of late pregnancy, which would include withdrawal from progesterone dominance, sensitizes the MPOA to the effects of estradiol so that more neurons synthesize cFos. Activation of high levels of cFos synthesis may, in turn, alter the phenotype of the affected cells in ways that would be facilitatory for the display of maternal behavior. In this regard, it is interesting to note that Fos proteins can have transcriptional effects on both the oxytocin receptor gene and the prolactin receptor gene (Hu, Zhuang, & Dufau, 1996; Rozen, Russo, Banville, & Zingg, 1995). Perhaps some of the hormonal events of late pregnancy prime the MPOA so that it becomes more sensitive to oxytocin and prolactin, and this is one of the ways in which Fos proteins exert an influence over maternal behavior.

We have just shown that the hormonal events of late pregnancy can induce the expression of cFos in MPOA, and the work we presented earlier in this chapter showed that during the postpartum maintenance phase of maternal behavior, the continued occurrence of maternal behavior is associated with the continued expression of Fos proteins within MPOA (Stack & Numan, 2000). These data are consistent with the ideas that the hormonal events of late pregnancy prepare the phenotype of MPOA neurons so that they can facilitate maternal reactivity to pup stimuli, that Fos proteins are one of the mediators of this process, and that once maternal behavior occurs the continued expression of Fos maintains the continued functional integrity of MPOA neurons so that maternal behavior continues. This argument suggests that the action of Fos proteins within MPOA is essential for both the onset and maintenance of maternal behavior. At the time of this writing, however, there is not definitive proof that this is the case. If it could be shown that disruption of Fos synthesis within MPOA neurons, through antisense treatment procedures, for example, would disrupt maternal behavior, then the case would be strengthened for our argument. Please recall, however, that a mouse line with a Fos B knockout mutation (which would prevent Fos B production in all tissues) shows an absence of maternal behavior (Brown et al., 1996). Finally, although we have been emphasizing the MPOA, it should be clear that hormonal, sensory, and experiential factors are likely to operate on many other neural regions to influence the occurrence of maternal behavior.

A recent study by Kalinichev et al. (2000b) has raised some questions concerning the issue of whether Fos protein expression in MPOA/vBST during maternal behavior is actually labeling neurons that participate in neural circuits controlling maternal behavior. These researchers studied pup-stimulated or sensitized maternal behavior in both juvenile and adult female rats. As has been reported previously (Numan & Numan, 1994), sensitized maternal behavior in adult females was found to be associated with an increase in Fos protein expression in both MPOA and vBST. Surprisingly, however, the pup-stimulated maternal behavior of juvenile female rats was not associated with increased Fos protein expression in these regions. Much more work will be needed before we will be able to fully appreciate the significance of these findings. Kalinichev et al. (2000b) speculate that these differences between adults and juveniles may be related to the fact that MPOA/vBST neural mechanisms regulating maternal behavior may not be completely mature in juveniles, and they argue that this idea is supported by another one of their findings that we discussed in Section 3.3 of this chapter, that larger lesions of MPOA/vBST are needed to disrupt sensitized maternal behavior in juveniles than in adults (Kalinichev et al., 2000a). Kalinichev et al. (2000b) argue that this lack of maturity in MPOA/vBST neural control mechanisms may be related to the fact that juvenile females are less avoidant of novel pup stimuli than are adults (see Chapters 2 and 4). In reference to Figure 5.21, perhaps the activation of MPOA neurons that we have hypothesized to be involved in inhibiting defensiveness and avoidance of novel stimuli is less needed in the juvenile female than in the adult.

We would like to relate these speculations to the recent finding of Lonstein and De Vries (2000c), who showed that approximately 50% of the neurons in the *adult* MPOA and vBST that express cFos during postpartum maternal behavior also contain glutamate decarboxylase, an enzyme needed for GABA synthesis. This finding suggests that a large proportion of Fos-expressing neurons in the maternal MPOA/vBST of adults are inhibitory. Figure 5.23 shows some models of how GABAergic neurons in MPOA might be involved in adult ma-

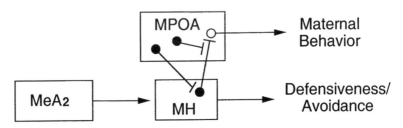

Figure 5.23. Hypothetical neural model that attempts to explain the possible role of medial preoptic area (MPOA) GABAergic neurons in maternal behavior control, and why fewer MPOA neurons may be needed to regulate maternal behavior in juvenile rats than in adults. In the adult rats, some MPOA efferents may directly promote maternal responsiveness, for example, by possibly projecting to the nucleus accumbens. Such MPOA neurons are shown with clear cell bodies. Other MPOA neurons, which are presumed to be GABAergic, may inhibit neurons in our hypothesized central aversion system, in this way decreasing the likelihood of avoidance responses to novel pup stimuli. Such MPOA neurons are shown with black cell bodies, and they are depicted as exerting either postsynaptic or presynaptic inhibition. In juvenile rats, where the central aversion system may not be fully developed, the need for GABAergic inhibition of this system may not be required to the same degree as in the adult. Neuronal projections ending in an arrow represent excitatory connections, and those ending in a vertical bar indicate inhibition. MeA2 = a particular population of neurons in the medial amygdaloid nucleus that projects to MH; MH = medial hypothalamus (anterior and ventromedial nuclei).

ternal behavior, based on the evidence presented throughout this chapter and summarized in Figure 5.21. Although all 3 neurons shown in the MPOA of Figure 5.23 express Fos during maternal behavior, each is involved in different functions, and we show 2 of the 3 neurons as being GABAergic. One of the GABA neurons is shown as an output neuron that inhibits MH, while the other is an MPOA local circuit interneuron that is shown as exerting presynaptic inhibition on incoming inhibitory axons originating in MH. The third neuron, which is hypothesized as not being GABAergic (mainly for clarity in the presentation of this model), is shown as projecting out of the MPOA to stimulate neural regions (the VTA or NAs, for example) that play positive roles in maternal behavior. Based on this model and on the speculations and data discussed above, perhaps the GABAergic neurons need not be activated for maternal behavior to occur in juveniles, while such activity would be necessary in adults. In contrast, our model would argue that for maternal behavior to occur in both juveniles and adults, the third, non-GABAergic neuron would have to be active. This model would result in fewer Fos-expressing neurons in MPOA during juvenile maternal behavior when compared to adult maternal behavior, as was found by Kalinichev et al. (2000b). Our model, however, would require that more Fos-expressing neurons be contained within the MPOA of maternal juveniles than in nonmaternal control juveniles, but this was not found by Kalinichev et al. (2000b). Perhaps if they did a more fine-grained analysis of Fos protein expression within MPOA they would have detected an increase in Fos expression during juvenile maternal behavior. A recent finding by Sheehan et al. (2000) has shown that cFos expression within the dorsolateral MPOA is strongly associated with the occurrence of maternal behavior.

As the above analysis suggests, research should be aimed at determining the neurochemical makeup of many of the neural circuits that have been proposed to be involved in the regulation of maternal responsiveness. Neuropharmacological manipulations could then be applied to the relevant synapses in order to examine the involvement of the neurochemically identified circuits in the regulation of maternal responsiveness. Our current state of knowledge with respect to this important issue will be reviewed in the next chapter.

6

Neurochemistry and Molecular Biology of Maternal Behavior

1. Introduction

In previous chapters we examined the role of gonadal steroids and the activation of select neural circuits associated with maternal behavior. Gonadal steroids alter cellular function by binding to receptors that function as transcription factors. Or, in other words, estrogen and progesterone work by indirectly turning on or turning off select genes. In this chapter we examine some of the likely targets of steroid hormone action. We will describe three classes of neurotransmitters: neuropeptides, monoamines, and amino acids, reviewing findings linking members of each class to maternal behavior. Later in this chapter, we will consider how one might find new genes and proteins important for maternal behavior. We will end this review with the unlikely discovery that some of the genes for maternal behavior appear to be of paternal origin.

There are three basic approaches to investigating these aspects of neurotransmitter action: (1) descriptive studies look at correlated changes between behavior and either neurotransmitters or their receptors; (2) to go beyond correlations, pharmacological studies ask how maternal behavior changes after an increase or decrease in neurotransmitter function; and (3) finally, lesion studies, either neurotoxic or transgenic, investigate maternal behavior in the absence of a given neurotransmitter or receptor. Our review of various neurochemicals and maternal behavior will follow the following general approach: We will provide a general introduction to the neurotransmitter system, then look at how each system changes during pregnancy and lactation; consider the evidence from pharmacological studies implicating each system in maternal behavior; and, whenever possible, summarize the findings with lesion studies.

In the previous chapter, we described lesion studies for interrupting neural pathways. In this chapter, we will describe lesion studies for interrupting genetic pathways, using genetic knockouts or null mutations. Because results with this powerful technique contribute to much of what we will be discussing in the remainder of this chapter, we will take a moment here to consider the promises and problems of this approach. Gene knockouts involve replacing a native gene with a defective or null version. This is done by microinjecting cultured embryonic stem cells with a vector containing a mutated version of the native gene sequence. In a small fraction of stem cells, this mutated version will be incorporated during recombination. Stem cells with the mutation are then inserted into normal blastocysts, which are then implanted in the reproductive tract of a surrogate female. The resulting offspring will be chimeras, consisting of mutant cells that arose from the stem cell and nonaffected cells that arose from the normal blastocyst. If the mutation is incorporated into the germ line (cells that will become sperm and ova), then these chimeric offspring can be bred to each other to produce an F2 generation with homozygous $(-/-)$ knockout mice that completely lack the gene, heterozygous $(+/-)$ mice that have reduced amounts of the gene, or wild-type $(+/+)$ mice that have the full complement of the gene. These steps, which require the success of several low-probability events, are shown in Figure 6.1. The appeal of this approach is its specificity: Only a single gene has been targeted for disruption.

Note that, as with anatomic lesions, the behavioral results from genetic lesions can be difficult to interpret. Specifically, a loss of maternal behavior following knockout of gene X does not mean that gene X is a gene for maternal behavior, because (a) the gene is deleted throughout development so that behavioral differences can result from an embryonic event and not the absence of the gene in adulthood; (b) the gene may have multiple nonspecific effects with maternal care inhibited indirectly due to general changes in physiology or behavior; (c) genes do not work in isolation so that, even at the level of transcription, hundreds of genes may be altered by single gene disruption; and (d) adjacent genes can be carried along with the targeted mutant during recombination in the stem cell (Gerlai, 1996). Moreover, the preservation of maternal behavior in a knockout animal cannot be taken as evidence that the gene has no physiologic role because other genes may compensate for the absent gene. Finally, knockout studies have thus far been limited to mice (*Mus musculus*). As we have seen, not only are mice remarkably different from rats in the onset of maternal behavior, we know much less about the neural systems underlying maternal behavior in mice. Unfortunately, maternal behavior has not been a major focus of efforts to phenotype knockout mice (for an excellent review, see Leckman & Herman, 2001). Indeed, most mutant mice are generated by breeding heterozygous rather than homozygous pairs, so that deficits in maternal behavior might never be detected. Nevertheless, a few interesting observations have been reported, summarized in Table 6.1, and described throughout this chapter.

Estrogen receptor knockout mice provide a good illustration of these points. There are two genes for the estrogen receptor: ER-α and ER-β. The ER-α–KO mouse has a deficit in maternal behavior with over 50% of homozygous females exhibiting infanticide and noninfanticidal mice showing reduced retrieval behavior (Ogawa et al., 1998; Ogawa, Taylor, Lubahn, Korach, & Pfaff, 1996). By contrast, the ER-β–KO mouse shows no deficit in maternal care (Ogawa et al., 1999). However, the changes in the ER-α–KO mouse are not specific for maternal care; these females are hyperaggressive in many situations (Ogawa et al., 1998). And there are many compensatory changes in these mice that might

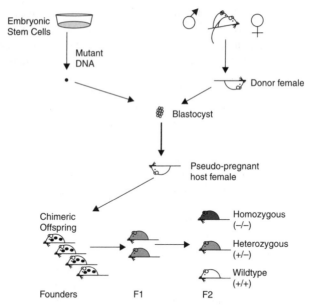

Figure 6.1. Knockout mice are created to eliminate a single gene. The gene is initially eliminated in embryonic stem cells. A stem cell with the null mutation is inserted into a blastocyst, which then is implanted into a pseudo-pregnant female. If the cell with the null mutation gets incorporated into the offspring developing from this blastocyst, chimeric mice will be produced ("chimeric" meaning that these mice are a mixture of mutant and nonmutant cells). If the null mutation becomes carried in the germ line (precursor cells to sperm or ova) of these chimeric mice, they can be bred to produce homozygous (–/–) or knockout mice. If homozygous mice are bred with mice lacking the mutation (wild-type or +/+) then heterozygous (+/–) mice result. Knockout mice have a selective mutation of a single gene, but behavioral abnormalities in these mice can have multiple causes that may be only indirectly related to the missing gene.

contribute to the observed phenotype, including 10-fold increases in circulating concentrations of estrogen (Couse et al., 1995) and testosterone (Lindzey & Korach, 1997). Although the absence of the ER-α receptor in development does not affect gross sexual differentiation (Lubahn et al., 1993), there is a profound loss of the normal sexual dimorphism in dopamine cells in the mouse preoptic area, possibly contributing to the behavioral phenotype (Simerly, Zee, Pendleton, Lubahn, & Korach, 1997). As we describe other knockout studies in this chapter, it is important to remember that, as with any lesion approach, one cannot easily assume that a deficit observed after a lesion of a candidate gene or structure indicates a specific physiological role for this candidate in the intact animal.

Table 6.1. Maternal behavior in knockout mice.

Gene	Background	Phenotype	Maternal behavior	Reference
ER-alpha	C57BL/6J + 129	aggressive, nonfertile	high infanticide, slow retrieval	(Ogawa et al., 1998)
Oxytocin	C57BL6/129SvE	social amnesia, decr. anxiety	unremarkable	(Nishimori et al., 1996)
				(Insel et al., 2001)
Prolactin	not stated	infertile females, mammopoeisis	full maternal behavior in virgin fem.	(Horsemann et al., 1997)
Prolactin rec	C57BL/6 × 129/S	Nl fear conditioning, water maze	failure to retrieve (homoz and heteroz)	(Lucas et al., 1998)
				(Ormandy et al., 1997)
DA Transporter	not stated	hyperactive, hyperresponsive	decr. retrieval	(Spielewoy et al., 2000)
Dbh	129/SvCPJ-C57BL	mod. cognitive def., incr. DA	deficit in first preg., deficits in virgins	(Thomas & Palmiter, 1997)
5HT-1b	129sv	hyperactive, impulsive	decr. time in nest	(Brunner et al., 1999)
nNOS	129/SvCPJ-C57BL	no deficit in nurturing	decr. maternal aggression	(Gammie & Nelson, 1999)
FosB	129/Sv × BALB/	no deficit in lactation, olf. fcn.	no retrieval/crouch postpartum or virgin	(Brown et al., 1996)
Peg1 (Mest)	129/sv	small litters, placental defic.	decr. retrieval/crouch postpartum or virgin	(Lefebvre et al., 1998)
Peg-3	129/sv	small litters, placental defic.	decr. retrieval/crouch postpartum or virgin	(Li et al., 1999)

2. Neuropeptides

Neuropeptides are particularly good candidates for the mediation of maternal behavior. Of the roughly 100 neuropeptides known in mammals, several are localized to brain regions implicated in maternal behavior; many are neuromodulators that have slow and integrative effects on behavior; and some, such as oxytocin and prolactin, are known to be important in peripheral aspects of maternal care (e.g., lactation, parturition). Our review will focus predominantly on oxytocin and prolactin because these remain the best candidates for central mediators of maternal behavior in rats. With the abundant literature on oxytocin and prolactin in maternal behavior, one might presume that these neuropeptides are major neurotransmitters in the brain. It is important to remember that neuropeptides, although critical for behavior, are relatively sparse in the brain. Whereas most neuropeptides and their receptors function in the picomolar range (10^{-12}), the monoamines are in the nanomolar range (10^{-9}), and the amino acids are in the micromolar (10^{-6}) range. Although neuropeptides may be sparse relative to other neurotransmitters, they may be abundant in local circuits within the hypothalamus. One other important feature of neuropeptides is that they may be released in a paracrine fashion rather than via classical axonal neurotransmission. As a result, neuropeptide function is often regulated by the number and location of receptors, as one finds with peripheral endocrine systems.

Much of our review will focus on neuropeptides that do not lend themselves to pharmacological studies. Neuropeptides do not cross the blood-brain barrier, so they must be given intracerebroventricularly (ICV) or directly into brain tissue to access central receptors. In contrast to monoamines or amino acid neurotransmitters, for which we have large numbers of powerful and selective drugs, there are relatively few compounds that can be given systemically to influence brain neuropeptide receptors. There are a few nonpeptide antagonists, such as naloxone, and even fewer nonpeptide agonists, such as morphine, which can be given systemically to influence opioid receptors. But the study of most other neuropeptide systems has been constrained by the lack of effective or selective drugs.

2.1. Oxytocin

2.1.1. Oxytocin—Background

Oxytocin is a 9-amino-acid neuropeptide found almost exclusively in placental mammals. Oxytocin belongs to a family of 9-amino-acid peptides that has been recently traced back to invertebrates (Goodson & Bass, 2001; Moore & Lowry, 1998). The members of this family occurring in snails (conopressin) (van Kesteren et al., 1992) and earthworms (annetocin) (Satake, Takuwa, Minakata, & Matsushima, 1999) have a unique distribution in neural and gonadal tissues and have been shown to influence reproductive behaviors (De Lange, De Boer, Ter Maat, Tensen, & Van Minnen, 1998; Fujino et al., 1999). Early in vertebrate evolution, a gene duplication event resulted in two 9-amino-acid neuropeptides in each species. One of these genes codes for vasotocin in nonmammalian vertebrates and vasopressin in mammals. The other codes for isotocin in fish, mesotocin in tetrapods and birds, and finally oxytocin in mammals. All of these various 9-amino-acid peptides differ by substitutions at one or two positions. This story of biochemical evolution is important because these various peptides have been implicated in the mediation of reproductive behaviors across phylogeny, from the mating calls of midshipman fish (Goodson & Bass, 2000) to the copulatory clasping of rough-skinned newts (Moore, Richardson, & Lowry, 2000).

In mammals, oxytocin is synthesized primarily in two nuclei of the hypothalamus—the paraventricular (PVN) and the supraoptic (SON) nuclei—and is transported by large neurosecretory axons to the posterior pituitary. Oxytocin is important for two prototypically mammalian functions: milk ejection during nursing and uterine contraction during labor. During suckling or with vaginocervical stimulation during labor, oxytocin is released from the posterior pituitary into the bloodstream where it finds its way to receptors on mammary myoepithelium (for milk ejection) or uterine myometrium (for uterine contraction). Oxytocin release in these circumstances is associated with synchronous bursts of electrical firing of cells in the PVN and SON (Wakerley, Jiang, Housham, Terenzi, & Ingram, 1995). In fact, oxytocin cells in these nuclei undergo a morphological change at parturition, marked by retraction of the glial processes that normally separate neurons so that cells are in direct contact and can become electrically coupled (Theodosis & Poulain, 1987; Yang & Hatton, 1988). But the increases in oxytocin's effects at parturition and lactation are not limited to release. In both mammary and uterine contractile cells, receptors for oxytocin increase several-fold at parturition, amplifying the effects of oxytocin specifically in these tissues (Soloff, Alexandrova, & Fernstrom, 1979). The concept here is important: The endocrine effect of oxytocin depends on the amount of hormone released, the location of receptors, and the number of receptors in a specific tissue. Below we will consider whether a similar process in the brain facilitates maternal behavior.

2.1.2. Oxytocin—Physiology

Given that oxytocin as an endocrine hormone influences labor and lactation, two cardinal features of mammalian motherhood, one might suspect that the peptide is also important for the central integration of maternal motivation and/or maternal behavior. Several neuropeptides, such as corticotropin releasing hormone and angiotensin, have been shown to have central effects related to their peripheral endocrine actions. In the case of oxytocin, pituitary stores of the hormone are increased just prior to parturition and plasma levels are increased during labor and nursing (reviewed in Russell & Leng, 1998). But changes in the pituitary or plasma may not be relevant to what is happening in the brain (Kendrick, Keverne, Baldwin, & Sharman, 1986). To investigate oxytocin's role in maternal behavior (as opposed to labor and nursing), we will need to know about changes in the paraventricular nucleus, which supplies both the pituitary for peripheral targets and nonpituitary projections for targets within the brain. Even more to the point, we will need to know about release of the peptide at select terminal sites.

Landgraf et al. (1991) have shown with microdialysis that oxytocin release in the ventral septal area increases by roughly 100% during parturition relative to the virgin state. Although this seems strong evidence for increased central release of oxytocin coincident with the onset of maternal behavior, similar levels were found during pregnancy and even greater increases were found for vasopressin (Landgraf et al., 1991). Perhaps more important, there is increased release of oxytocin locally in the PVN (300% relative to late pregnancy) and SON (254% relative to late pregnancy) at parturition without a concurrent increase in vasopressin (Neumann, Russell, & Landgraf, 1993). Not surprisingly, oxytocin synthesis, as measured by mRNA in the paraventricular nucleus as well as the supraoptic nucleus, increases several-fold during lactation (Lightman & Young, 1987). No doubt, synthesis in these regions must increase to support labor and milk ejection. It is not clear, however, whether oxytocin mRNA increases specifically in those cells that are projecting to central sites.

Moreover, there has been considerable debate about the mechanism for the increased oxytocin mRNA following parturition. Based on what we described in Chapter 2, an obvious mechanism posits the physiological increase in estrogen at parturition as the stimulus for increasing oxytocin mRNA. In fact, the paraventricular nucleus, especially the region with central projections, is enriched with ER-β receptors (Alves, Lopez, McEwen, & Weiland, 1998; Hrabovszky, Kallo, Hajszan, Shughrue, & Merchenthaler, 1998) and there are multiple estrogen-response elements in the promoter region of the oxytocin gene (Burbach, van Schaik, de Bree, Lopes da Silva, & Adan, 1995). However, estrogen given to ovariectomized females conspicuously fails to increase oxytocin mRNA (Burbach et al., 1990). Amico and colleagues resolved this paradox by using an experimental regimen of steroid treatments that mimic the physiological changes of pregnancy (Amico, Crowley, Insel, Thomas, & O'Keefe, 1995). When estrogen and progesterone were administered for 13 days and then progesterone was withdrawn for two days, oxytocin mRNA increased. Treatments with estrogen alone or with estrogen and progesterone without progesterone withdrawal were ineffective (Amico, Thomas, & Hollingshead, 1997). Thus, the same steroid treatment regimen that induces maternal behavior in ovariectomized rats (*Rattus norvegicus*) (see Chapter 2) induces oxytocin synthesis. Increases in oxytocin synthesis are correlated temporally with the onset of lactation, but the critical question remains: Are brain pathways for oxytocin important for maternal behavior?

2.1.3. Oxytocin—Pharmacology

In a classic set of experiments, Pedersen and Prange (1979) were the first to test the hypothesis that oxytocin (400 ng ICV) in the brain facilitates maternal behavior in virgin rats. In their initial report, they described a very rapid onset of full maternal behavior (defined as retrieving, grouping, licking, and crouching over pups) in 42% of the females within 2 hours of injection of the peptide into the lateral ventricle. None of the females injected with saline or an equivalent dose of vasopressin exhibited full maternal behavior. The rapid onset of this effect was notable given the previous reports of several days of exposure for steroid treatments to facilitate "short-latency" maternal behavior. When Pedersen and Prange examined their data for an interaction with hormonal state, they found an important relationship. All but one of the females showing full maternal behavior in response to oxytocin were either in estrus, proestrus, or about to enter estrus, whereas only 1 of 12 females in diestrus showed full maternal behavior. This observation suggested that elevated estrogen was critical for oxytocin to induce full maternal behavior. Indeed, in an additional experiment, Pedersen and Prange showed that of 13 ovariectomized females primed with estradiol benzoate (100 µg/kg), 11 exhibited full maternal behavior within 2 hours of a central oxytocin injection, an effect not observed in any of 14 nonprimed ovariectomized females. Pedersen et al. subsequently described the specificity of oxytocin's effects (Pedersen, Ascher, Monroe, & Prange, 1982). Within 60 minutes of injecting 400 ng oxytocin into the lateral ventricle, 72% of females exhibited full maternal behavior whereas equimolar doses of prolactin, beta-endorphin, LHRH, prostaglandin F2alpha, substance P, or neurotensin were not significantly different from saline (18% fully maternal) in their behavioral effects. Arginine vasopressin, like oxytocin, facilitated the onset of maternal behavior, but the effects were only observed in the second and third hours after administration when respectively 42% and 55% of the females were fully maternal (Figure 6.2). It is possible that vasopressin at this dose (400 ng) was activating oxytocin receptors, as vasopressin's affinity for the oxytocin receptor is only an order of magnitude less than oxytocin's affinity for this receptor (Barberis & Tribollet, 1996).

Several subsequent studies could not replicate these initial results with oxytocin (Bolwerk & Swanson, 1984; Rubin, Menniti, & Bridges, 1983; Wamboldt & Insel, 1987) although others found similar effects (Fahrbach, Morrell, & Pfaff, 1984). In a series of studies designed to explain the differences in findings, oxytocin's effects on maternal behavior appeared to be dependent on the strain (Sprague-Dawley rats were best) and testing conditions. In the studies that failed to find an effect, rats were tested in their home cage, whereas the studies in which oxytocin appeared to induce maternal behavior removed females to a novel environment. Indeed, in a direct comparison, Fahrbach et al. found that oxytocin could facilitate maternal behavior, but only in females tested in a novel environment (Fahrbach, Morrell, & Pfaff, 1986a). In a related study, Wamboldt and Insel (1987) found that oxytocin had effects only in anosmic females. Females with zinc sulfate lesions of the olfactory epithelium showed rapid onset of maternal behavior after oxytocin administration, but nonlesioned females did not respond to the peptide.

Reviewing these various studies from nearly 2 decades ago, one is struck that oxytocin's central effects were neither consistent nor, in most cases, robust. So

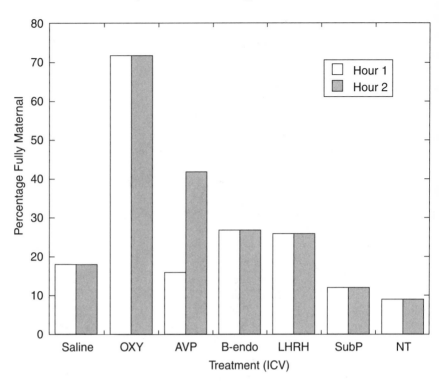

Figure 6.2. Oxytocin increases maternal behavior. Steroid-primed females received ICV injections of 400 ng of oxytocin (OT) or equimolar doses of vasopressin (AVP), β-endorphin, luteining hormone releasing hormone (LHRH), substance P (SubP), or neurotensin (NT). As shown, in the first hour after injection, 72% of females injected with oxytocin exhibited full maternal behavior (retrieval, grouping of pups, and crouching over pups). All other treatments were equivalent to the saline control injection. In the second hour, an increase (42%) in the percentage of females that were fully maternal was also observed in the group treated with vasopressin. Data adapted from Pedersen et al. (1982).

it comes as a particular surprise that a subsequent series of studies reducing oxytocin neurotransmission reported profound and consistent decreases in maternal behavior. In the first such study, Fahrbach, Morrell, and Pfaff (1985b) studied females who had undergone a 16-day pregnancy termination protocol: hysterectomy and ovariectomy followed by injections with EB (100 μg/kg). As expected, in the first hour of being presented with 3 foster pups, 75% of these females showed licking and grouping of pups as well as nest building. Of females injected ICV with the oxytocin antagonist (d(CH$_2$)$_5$–ornithine-vasotocin (800 ng × 2), only 50% showed licking, 16.7% grouping, and 0% nest building. Similarly, females receiving ICV infusions of an oxytocin antiserum (4 injections) showed reduced maternal behavior, with 37.5% licking, 12.5% grouping, and 12.5% nest building. Small groups (n = 3) treated with an anti-AVP antiserum or normal rabbit serum (n = 4) did not differ from uninjected controls (Fahrbach et al., 1985b). Similar results were subsequently reported by others using antisera (Pedersen, Caldwell, Fort, & Prange, 1985), antagonist injections prior to natural parturition (van Leengoed, Kerker, & Swanson, 1987), or lesions of the paraventricular nucleus (Insel & Harbaugh, 1989) (Figure 6.3A). The results across these various studies are clear: Reducing oxytocin neurotransmission inhibits the onset of maternal behavior either in hormonally primed or in naturally parturient female rats.

The consistent effects with these various inhibitors of oxytocin may be more significant than the contradictory results following administration of the peptide, as effects with an antagonist suggest a role for the endogenous peptide. A reasonable interpretation of the reduction of maternal behavior following antagonist administration is that oxytocin is necessary for maternal behavior. But more informative is the observation that similar treatments after a female has become maternal are without effect (Fahrbach et al., 1985b; Insel & Harbaugh, 1989; Numan & Corodimas, 1985) (Figure 6.3B). As shown in Figure 6.3, a lesion of the PVN on day 15 of pregnancy inhibits the onset of maternal behavior at parturition but a similar lesion has no effect on maternal behavior if performed on day 3 postpartum, after maternal behavior is established. The PVN is the source of limbic oxytocin. These lesions leave the SON-derived oxytocin, for pituitary and systemic release, unperturbed. Thus, PVN-lesioned females show no deficit in parturition or lactation. It appears, therefore, that oxytocin projecting from the PVN to central sites is critical for the initiation but not the maintenance of maternal behavior in rats. Thus, oxytocin may be especially important for maternal motivation or for overcoming the natural avoidance of pups at the time of parturition. Note that very recent data have demonstrated that an oxytocin antagonist given postpartum may decrease interaction with pups (Pedersen, personal communication), but this finding may reflect oxytocin's more general influence on affiliation rather than a specific effect on maternal behavior (Witt, Winslow, & Insel, 1992).

2.1.4. Oxytocin—Mechanisms

Above we have summarized the data for an increase in oxytocin synthesis at parturition as well as pharmacological evidence that oxytocin is necessary and sufficient for the onset of maternal care. How does oxytocin influence the onset of maternal behavior? To understand how and where oxytocin acts, we need to know more about oxytocin receptors. The oxytocin receptor is a membrane-bound, G-protein coupled receptor that along with the three vasopressin receptors (V1a, V1b, V2) forms the OT-AVP receptor family (Barberis & Tribollet, 1996). In rats, oxytocin receptors are abundant in the uterus, kidney, and brain. Two observations from the studies described above provide important clues to how this peptide may alter behavior. First, the facilitation of maternal behavior appears to require an increase in central rather than peripheral oxytocin. Al-

A. Prepartum Lesion

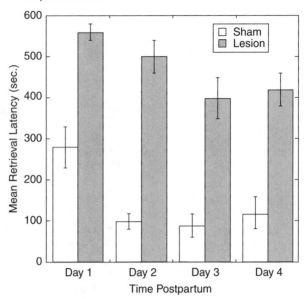

B. Postpartum Lesion

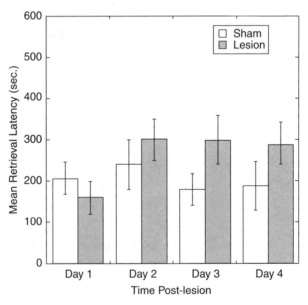

Figure 6.3. PVN lesions inhibit the onset but not the maintenance of maternal care. **A.** Female rats received electrolytic lesions of the PVN on day 15 of pregnancy. Lesions of the PVN should reduce central projections of oxytocin, leaving the SON oxytocin pathways to the pituitary still available for parturition and lactation. Relative to sham-lesioned controls, females with a PVN lesion exhibit delayed retrieval of 3 test pups on days 1–4 postpartum. **B.** Females receiving an equivalent lesion on day 3 postpartum exhibit no significant delay in retrieval. Data from Insel and Harbaugh (1989).

though there are no published studies of the effects of intravenous peptide on maternal behavior, based on the evidence that PVN-lesioned females have unperturbed peripheral oxytocin function and yet lack maternal behavior, one would surmise that maternal care involves central, not peripheral, oxytocin receptors. Second, oxytocin's effects require estrogen priming. Oxytocin receptors in specific brain sites are dependent on estrogen. Indeed, just at parturition, oxytocin receptor binding assessed by autoradiography increases (roughly 25%) in the bed nucleus of the stria terminalis (BST) and the ventromedial nucleus of the hypothalamus (VMN) (Insel, 1986, 1990c), but not in several other regions. Note that the part of the BST affected is specifically in the nucleus ovalis (BSTov), which is distinct from the ventral BST (BSTv) that was a focus in Chapter 5 (Figure 6.4). Studies of homogenates of microdissected brain regions from parturient females also report increased binding (roughly 40%) in the ventral tegmental area (VTA) and medial preoptic area (MPOA) (Pedersen, Caldwell, Walker, Ayers, & Mason, 1994). Although in these two regions specific binding is not observed consistently with autoradiographic studies in slices (Insel, 1986, 1990c), oxytocin receptor mRNA has been detected and appears to be steroid responsive in the MPOA (Young, Muns, Wang, & Insel, 1997a), lending support to the importance of oxytocin receptors in this region. The critical point is that oxytocin receptors in different regions show differential responses to estrogen, probably depending on the local presence of estrogen receptors. Remarkably, these receptors not only are regulated independently, but they also appear to be linked to different intracellular effectors depending on their location in the brain (Bale et al., 2001).

What is the functional importance of increasing oxytocin receptors in these various brain regions at parturition? In uterine myometrium and mammary myoepithelium, oxytocin receptors increase several-fold at parturition, amplifying the functional response to the peptide in these tissues, supporting labor and milk ejection (Soloff et al., 1979). Although receptor changes in the brain are much less dramatic, they may still be important for local changes in neurotransmission. For instance, oxytocin in the VMN appears critical for lordosis behavior (McCarthy, Kleopoulos, Mobbs, & Pfaff, 1994b). The increase in receptors in this region may thus reflect an increase in the local response to released oxytocin supporting postpartum estrus.

In previous chapters, we have described how hormones facilitate the onset of maternal behavior by decreasing pup avoidance, increasing pup attractiveness, and decreasing general fearfulness. How do the local changes in oxytocin neurotransmission contribute to these changes? The increase in oxytocin receptors in the BST, specifically in the nucleus ovalis, may be important in two ways. Moos et al. (1991) have shown that oxytocin microinjected into this region facilitates burst-firing of oxytocin cells in the hypothalamus. This region must have a powerful efferent connection to the oxytocin cells in the paraventricular nucleus, so that increasing responsiveness within the BST essentially amplifies a positive feedback loop. The BST, frequently identified as part of the extended amygdala (Alheid & Heimer, 1988) is, along with the central nucleus of the amyg-

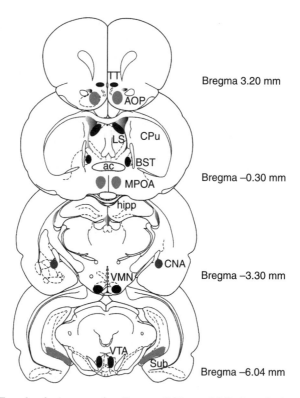

Figure 6.4. Four levels (representing Bregma 3.20 to –6.04) through the rat forebrain show major receptor fields for oxytocin. In the most anterior section, receptors are abundant in the anterior olfactory nucleus (AOP) and tenia tecta (TT). At the level of the anterior commissure (ac), receptors can be found in the bed nucleus of the stria terminalis (BST) as well as the shoulder of the lateral septum (LS). Homogenate binding and in situ hybridization studies have also identified receptors in the MPOA. At the level of the hippocampus (hipp), receptors are abundant in the ventromedial nucleus of the hypothalamus (VMN) and the central nucleus of the amygdala (CNA). Finally, oxytocin receptors are found in the ventral subiculum (Sub), the major outflow tract for the hippocampus. Homogenate binding and in situ hybridization studies have also identified receptors in the ventral tegmental area (VTA). Receptors in the BST, MPOA, and VMN increase in response to estrogen; receptors in the CNA increase in response to glucocorticoids or stress. Females with high levels of licking and grooming have higher levels of oxytocin receptor binding in MPOA, LS, BST, and CNA. (See text for details.)

dala, an important site for anxiety-related behaviors in the rat (Davis, Rannie, & Cassell, 1994), and oxytocin injected into the central nucleus has anxiolytic effects (Bale et al., 2001). Oxytocin's effects on anxiety might be mediated via a decrease in corticotropin releasing hormone (CRF) activity. The subregion of the BST, the nucleus ovalis, where oxytocin receptors are most abundant (Kremarik, Freund-Mercier, & Stoeckel, 1991), and the central nucleus of the amygdala (CNA) are both islands of CRF cells (Sakanaka, Shibasaki, & Lederis, 1987). CRF (possibly from the BST) activates caudal aspects of the dorsal raphe in what Lowry has considered a mesolimbic serotonergic pathway for anxiety in the rat brain (Lowry, Rodda, Lightman, & Ingram, 2000). It is possible, although not proven, that increased oxytocin responsiveness in the BST/CNA inhibits CRF activity and thereby reduces avoidance of pups and/or general fearfulness (see Chapter 8). Clearly, lactating females show a reduced response to fearful stimuli as we described in Chapter 4. They also show a reduced response to CRF (Da Costa et al., 1997b). What is missing from this story is clear evidence that an oxytocin antagonist injected into the BST or CNA blocks maternal behavior. Although the oxytocin receptors in the BST are responsive to estrogen, oxytocin receptors in the CNA appear more responsive to glucocorticoids or stress (Liberzon et al., 1994).

In an intriguing study of individual differences, Champagne et al. (2001) have reported that oxytocin receptor binding in the MPOA, lateral septum, CNA, PVN, and BST is higher in females exhibiting more maternal licking and grooming and that an oxytocin antagonist (ICV) eliminates the difference in licking and grooming behavior. Curiously, in this study, virgin female offspring of high- and low-licking and grooming dams differed in their response to estrogen (Champagne et al., 2001). Only the high-licking and grooming females responded to estrogen in the MPOA and lateral septum (response of the BST/CNA was not described). Thus, there is strong correlational evidence linking oxytocin receptors in the MPOA and, in some cases, the BST/CNA to maternal care. We still do not know if an oxytocin antagonist injected into these regions will inhibit the onset of maternal behavior, but, as noted above, oxytocin injected into the CNA has anxiolytic effects (Bale et al., 2001).

Site-specific injections of oxytocin antagonists into two other regions have been reported to reduce the onset of maternal behavior in primiparous females (Pedersen et al., 1994). Pedersen et al. (1994) infused the oxytocin antagonists 1.0 μg (Pen[1], Pen[2], Thr[4], delta[3], [4]Pro[7], Orn[8])-oxytocin (PPT) into the VTA or 0.25 μg (d[CH$_2$]$_5$ o-Me-Tyr[2], Thr[4], Tyr[9], Orn[8])-vasotocin (OTA) into the MPOA immediately after delivery of the first pup, 1 hour later, and 15 minutes and 55 minutes after delivery of the last pup. Pups were removed after delivery. Immediately after the last injection (approximately 1 hour after delivery of the last pup), 8 of the dam's pups were returned and latencies for retrieval and crouching were measured. Control females received either saline or a vasopressin antagonist d(CH$_2$)$_5$,O-Me-Tyr[2]-Arg[8]-vasopressin (V1A) in the same regions at the same times. Only 1 of 10 rats treated with the oxytocin antagonist in the VTA retrieved pups, compared with 7 of 10 receiving the V1a antagonist, and 8 of 9 receiving saline. Although females with oxytocin antagonist infusions in the VTA showed a longer latency to crouch, 5 of 10 of these females ultimately crouched over pups, compared to 8 of 10 with V1A, and 9 of 10 with saline infusions. In the MPOA, both the oxytocin and vasopressin antagonists delayed retrieval (0/9, 5/9, 8/8 retrieved 8 pups for OTA, V1A, and saline, respectively) and reduced crouching (3/9, 4/9, 8/8 crouched over pups in the OTA, V1A, and saline groups, respectively). These results are generally consistent with a previous study reporting short-latency maternal behavior in roughly half of the estrogen-primed females given oxytocin injections (200 ng/0.5ul) into the VTA or MPOA, but not in the medial amygdala (Fahrbach, Morrell, & Pfaff, 1985a). As Pedersen et al. (1994) used doses of antagonist roughly 3 orders of magnitude greater than other studies with site-specific injections of antagonists, the specificity of their results might be questioned. However, they find no effect with intracerebroventricular administration of the same doses and they note that females with cannula placements outside of the target areas fail to respond, providing two controls for specificity. Based on studies described in Chapter 5, one might hypothesize that oxytocin in the VTA and MPOA increases the dam's attraction to pups. These changes, coordinated with a decrease in fearfulness (mediated via the BST-amygdala pathway), would provide the requisite change in behavior for the onset of maternal behavior.

One other set of studies with site-specific injections of oxytocin deserves note here. Yu et al. (1996b) reported that oxytocin inhibited olfactory processing by reducing mitral cell firing and increasing granule cell (inhibitory interneuron) activity. Studies in ovariectomized virgin females found that high-frequency stimulation of the PVN inhibited 60 of 75 mitral cells (mean latency 32 seconds) and excited 17 of 23 granule cells (mean latency 35 seconds) and that both effects were blocked by intrabulbar infusion of the oxytocin antagonist $d(CH_2)_5,Tyr(Me)^2$ornithine-vasotocin (10pmol). This result is somewhat surprising as there are few oxytocin fibers or oxytocin receptors in the rat olfactory bulbs. In fact, Yu at al. (1996b) note that transections of fibers into the bulb or local injections of anesthetic do not block the response to PVN stimulation, so the decrease in olfactory processing may be due to oxytocin reaching the bulb via extracellular fluid. In a subsequent test of oxytocin and maternal behavior, Yu et al. infused 5fmol of the oxytocin antagonist $d(CH_2)_5$ o-Me-Tyr^2,Thr^4, Tyr^9,Orn^8-vasotocin or saline into the bulb just after birth of the first pup and 40 minutes after birth of the last pup (Yu, Kaba, Okutani, Takahashi, & Higuchi, 1996a). Wistar females were then tested in a novel cage 40 minutes after delivery of the last pup. Although saline-injected females showed full maternal behavior, OTA-infused females showed increased self-grooming and completely failed to build a nest, retrieve, lick, or crouch. Females receiving the same dose of OTA ICV also showed full maternal behavior and were not significantly different from saline-infused females on any measure. Conversely, when oxytocin (20pmol) was administered into the bulb of estrogen-primed, ovariectomized virgin females, 50% showed full maternal behavior, with 90% licking and

70% retrieving and crouching over 4 foster pups within 2 hours. None of the females infused with saline showed maternal behavior. Females receiving the same dose of oxytocin ICV failed to show maternal behavior, although half of this group cannibalized pups.

Taken together, what can we say about oxytocin and maternal behavior in rats? First, oxytocin is clearly important for peripheral aspects of maternal care, especially lactation. Second, oxytocin has central effects mediated by receptors in the brain that are identical to receptors found in peripheral target tissues. Third, central oxytocin is important for the onset but not the maintenance of maternal behavior. The available evidence suggests that oxytocin is important for the transition from avoidance to approach of pups, a critical event in the initiation of maternal behavior in rats. The steroid induction of oxytocin receptors by gonadal steroids in select regions of the brain may be critical for this transition, but at this point we cannot be certain which of the many sites identified are most important. Figure 6.5 proposes a model for integrating the available data. The BST, VTA, MPOA, and olfactory bulbs have all been implicated. What is required now is a careful dissection of these regions to determine how receptor activation in discrete circuits alters (a) avoidance behavior and (b) motivation for nurturing. Oxytocin given centrally has anxiolytic effects (Windle et al., 1997b) and an oxytocin antagonist appears to increase the neuroendocrine response to stressors (Neumann et al., 2000). Curiously, the effects of brain oxytocin on stress responses and anxiety appear to depend on the reproductive state of the female (Neumann et al., 2000). Site-specific injections of oxytocin antagonists point to the amygdala as an important site for anxiolytic effects (Bale et al., 2001). Although it seems likely that oxytocin inhibits the stress response in cycling females (Chapter 4), it probably does not serve this function during lactation (Neumann, Toschi, Ohl, Torner, & Kromer, 2001; Torner et al., 2001). The evidence that oxytocin increases motivation is also unclear. Although subcutaneous oxytocin induces a place preference (Liberzon, Trujillo, Akil, & Young, 1997) and facilitates cocaine sensitization (Kovaks, Sarnyai, & Szabo, 1998), there is no evidence that central oxytocin has reinforcing or hedonic properties. It seems likely that future studies will find that oxytocin facilitates maternal behavior both by reducing avoidance of pups and increasing the motivation to groom them, but where and how this happens remains to be demonstrated.

2.1.5. Oxytocin—Other Species

Although oxytocin's site-specific actions have not yet been tested in rats, a series of studies in sheep (*Ovis aries*) has already completed the marathon task of injecting oxytocin into multiple brain sites to determine where the peptide is effective and what behaviors are affected by infusions in different neural sites. Recall from Chapters 1 and 5 that sheep, unlike rats, form selective bonds with their young, rejecting alien lambs. Keverne et al. (1983) described the immediate onset of maternal care in estrogen-primed sheep by vaginocervical stim-

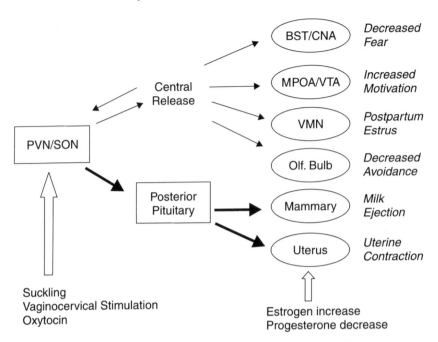

Figure 6.5. A model of how oxytocin might influence maternal behavior. Oxytocin release is stimulated by suckling, vaginocervical stimulation, or oxytocin feedback on to PVN neurons. The major pathway from the PVN and SON is via neurosecretory axons terminating in the pituitary from which oxytocin is released for labor (uterine contraction) and milk ejection (mammary myoepithelium). A parallel pathway is believed to influence central receptors, perhaps via paracrine (nonsynaptic) release. Each of the forebrain regions with oxytocin receptors may mediate selective aspects of oxytocin's effects on behavior, contributing in a coordinate fashion to the onset of maternal behavior. Studies in sheep suggest that none of these individual sites can support full maternal behavior, but that administration of oxytocin into the PVN increases activity in multiple receptor fields, leading to an integrated increase in maternal care. An important aspect of this model is the estrogen modulation of oxytocin receptors, not only in the uterine and mammary targets, but also in the BST, MPOA, and VMN.

ulation (VCS). Furthermore, VCS will induce acceptance of an alien lamb even 2–3 days after the ewe has bonded with her own lamb (Kendrick et al., 1991). Epidural anesthesia blocks these effects of VCS, demonstrating that central feedback is essential (Levy et al., 1996). As VCS is a potent stimulus for central as well as peripheral oxytocin release (Kendrick et al., 1986), it seemed likely that oxytocin might induce maternal behavior when given alone. Indeed, oxytocin can facilitate acceptance of an alien lamb, even in a nonpregnant ewe within *30 seconds* of ICV injection (Kendrick, Keverne, & Baldwin, 1987; Keverne & Kendrick, 1992). In postpartum females in which maternal behavior has been prevented by epidural anesthesia, ICV administration of oxytocin can induce a maternal response (Levy et al., 1992). Although the location of oxytocin's effects is not indicated by these studies, oxytocin mRNA and oxytocin receptor mRNA are increased regionally in the sheep brain postpartum (Broad, Kendrick, Sirinathsinghji, & Keverne, 1993b; Broad et al., 1999). In particular, oxytocin receptor mRNA increases in the anterior olfactory nucleus, medial preoptic area, ventromedial hypothalamus, lateral septum, medial amygdala, BST, and diagonal band of Broca (Broad et al., 1999). Maternal experience has an additional effect: Increases in oxytocin receptor mRNA were observed following parturition in the PVN but only in multiparous ewes (Broad et al., 1999).

A series of microdialysis and retrodialysis studies by Kendrick and colleagues (1997a) have exploited the relatively large brain of the ewe to map where oxytocin increases physiologically and where increases in oxytocin alter the ewe's behavior. For instance, exposure to a lamb increases oxytocin release in the MPOA and olfactory bulbs, and retrodialysis of oxytocin into these same regions reduces rejection of an alien lamb. Administration of oxytocin into the medial basal hypothalamus (roughly equivalent to the VMN in the rat brain) inhibits postpartum estrus (precisely opposite to the findings from studies in rats). What is important to understand about these elegant studies is that in none of these putative terminal areas does oxytocin induce full maternal behavior. However, infusions of the peptide near the oxytocin-synthesizing cells of the paraventricular nucleus of the hypothalamus (PVN) induce the entire complement of sheep maternal behaviors. It seems likely that oxytocin receptors, probably autoreceptors, on these cells mediate a short-loop feedback to increase oxytocin cell firing and coordinate oxytocin release in several terminal fields (Da Costa et al., 1996a) similar to the response described for BST cells in the rat brain (Moos et al., 1991).

Before concluding that oxytocin is the key to sheep maternal behavior, we need to consider three details of these studies. First, all of the studies described above involve measuring oxytocin release or infusing oxytocin. What about oxytocin antagonists? We know from studies in rats that antagonists are a better test of the importance of endogenous peptide. Thus far, oxytocin antagonists have not been successful at preventing sheep maternal behavior (Kendrick, 2000). One reason for this may be that antagonists that are potent and selective in the rat appear ineffective in the sheep brain. This leads to the second point: Oxytocin receptors have not been adequately mapped in the sheep brain. The extant ligands, iodinated antagonists, apparently do not bind to sheep receptors, possibly explaining why these antagonists do not have behavioral effects. Although receptors are now being mapped with in situ hybridization and immunocytochemistry (Broad et al., 1999), the development of effective antagonists will be important for behavioral experiments. Finally, all of the effects of oxytocin described above are in experienced ewes with a history of parental care (Keverne & Kendrick, 1991). Oxytocin can induce lamb acceptance in an inexperienced ewe in which peridural anesthesia blocks maternal behavior (Levy et al., 1992), but oxytocin release (in the olfactory bulb) is less during parturition and virtually absent following VCS in inexperienced ewes compared with experienced females (Kendrick et al, 1997a; Levy, Kendrick, Goode, Guevara-Guzman, & Keverne, 1995a). Oxytocin receptor induction in the PVN following parturition was only observed in multiparous ewes (Broad et al., 1999). Following development of a bond, even within a few hours, VCS can stimulate oxytocin release and induce maternal acceptance for the rest of the ewe's life. This permanent effect of experience on a neurochemical and behavioral response to a simple sensory input serves to remind us that neural systems are dynamic; oxytocin's effects on behavior depend not only on amount of peptide and number of receptors, but also on the history of the female, as represented in complex neural circuits.

2.1.6. Oxytocin—Knockouts

Given the robust effects of oxytocin on maternal behavior in rats and sheep, one might expect that mice with a knockout of the oxytocin gene would completely lack maternal behavior. So it came as a complete surprise when two groups almost simultaneously reported that female laboratory mice with a null mutation of the oxytocin gene were excellent mothers (Nishimori et al., 1996; Young et al., 1996). Although these groups had used slightly different methods to delete the oxytocin gene, both produced mice lacking oxytocin and both found that these oxytocin-knockout (OT-KO) mice were unremarkable in terms of reproduction, parturition, or maternal care except for an inability to lactate. When given exogenous oxytocin, OT-KO females were able to nurse and care for young without any evident disability (Nishimori et al., 1996). Of course, it is possible that some other neuropeptide, such as vasopressin, compensates for the lack of oxytocin, permitting ostensibly normal maternal behavior. This explanation seems unlikely, as vasopressin synthesis is not higher in these females (Nishimori et al., 1996) and females injected centrally with high doses of a potent, long-acting oxytocin antagonist for 4 days to block brain receptors still showed excellent maternal behavior (Insel, Gingrich, & Young, 2001).

The most likely explanation is that oxytocin is not necessary for maternal behavior in the laboratory mouse. As pointed out by Russell and Leng (1998), this apparent contradiction may highlight an important insight into what oxytocin is doing in other species where it appears necessary for maternal behavior. Unlike rats and sheep, laboratory mice do not shift from avoidance to nurturance of young at parturition. Laboratory mice (at least the inbred strains such as 129 and C57B6J used for transgenic experiments) appear fully maternal prior to parturition. In a sense, they are like juvenile rats that find pups attractive. Although it is not known whether oxytocin is involved in juvenile rat "maternal" behavior, it seems likely that oxytocin's role in the maternal behavior of adult rats and sheep is essentially "anxiolytic", that is, it reduces the avoidance and facilitates the approach to neonates. Laboratory mice spontaneously approach neonates and therefore oxytocin is not required for maternal behavior. Oxytocin receptors in the mouse brain are found in different regions and are regulated differently than oxytocin receptors in the rat brain (Insel, Young, Witt, & Crews,

1993), so perhaps we should not be surprised that the peptide has different behavioral effects in the two species. Indeed, OT-KO mice manifest a remarkable behavioral deficit—a profound and selective social amnesia—due to the absence of oxytocin in the medial amygdala, a region in which oxytocin receptors are abundant in the mouse brain and lacking in the rat brain (Ferguson, Aldag, Insel, & Young, 2001; Ferguson, Young, Hearn, Insel, & Winslow, 2000). But this deficit in social recognition does not impair maternal care.

This is not to say that oxytocin is unimportant for maternal behavior in all mice. In contrast to laboratory mice, between 60% and 90% of virgin male and female wild mice spontaneously kill unrelated young. McCarthy (1990) has shown that oxytocin administered ICV (40 ng) or sc (15.2 µg) reduced the rate of infanticide from 90% to below 40%. This effect was not dependent on gonadal state, as oxytocin was equally effective in intact, ovariectomized, and estrogen-primed females. The site of this effect is not clear, nor is it clear how a reduction in infanticide in mice relates to the facilitation of maternal care in other species. However, there may be a conceptual parallel. Both male and female wild housemice are infanticidal. Unlike male mice in which infanticide is reduced by sexual experience, cohabitation with a pregnant female, parental experience, or castration, in female wild mice infanticide is reduced by parturition. It is possible that the release of oxytocin at parturition is responsible for this reduction in infanticide in female mice, analogous to the reduction of pup avoidance in rats or lamb rejection in sheep.

2.1.7. Oxytocin—Summary

Let's summarize what we know about oxytocin and maternal behavior. First, oxytocin and vasopressin are part of an ancient family of 9-amino-acid peptides involved in sociosexual behavior in invertebrates and virtually every class of vertebrates studied. A too-simplistic but heuristic view is that oxytocin is critical for female, estrogen-dependent behaviors while vasopressin is involved in male, androgen-dependent behaviors (this concept is developed further in Insel & Young, 2000). In Chapter 7 we discuss vasopressin's role in paternal care. Second, oxytocin is a centrally active neuropeptide with projections that extend to limbic and autonomic centers and receptors in discrete neural circuits. Brain oxytocin receptors are highly plastic, with profound species differences in regulation and distribution. As a result, one might expect that (a) oxytocin's behavioral effects will be species-specific and (b) oxytocin may be involved in species-typical behaviors. With oxytocin, we need to be cautious extrapolating from rats to other species and we need to study each species with a fresh eye. Preliminary data suggest that oxytocin increases rhesus monkey parental care (*Macaca mulatta*) (Holman & Goy, 1995), but there are no comparable data for humans. Third, in species in which maternal behavior requires overcoming avoidance of neonates, oxytocin facilitates the transition from avoidance to approach of the young. As a fascinating (and not fully proven) corollary of this point, oxytocin's evolutionary history appears to confer approach to select social stimuli and not novel stimuli in general. Note that OT-KO mice lose social memory but show no other cognitive deficits (Ferguson et al., 2000), and rat pups treated with an oxytocin antagonist fail to learn odors associated with mother but have no deficit learning odors associated with nonsocial cues (Nelson & Panksepp, 1996).

Finally, oxytocin should be thought of as a neuromodulator. Its behavioral effects are likely mediated via effects on classical neurotransmitters. An important area of future research will be the description of the cellular cascade that follows oxytocin receptor activation. Thus far, this cascade has been considered most carefully in the sheep brain, where oxytocin seems to require coordinated activation of multiple target pathways to activate maternal behavior. In at least one of these pathways, the MPOA, oxytocin increases noradrenergic release (Kendrick et al., 1992a). It is certainly likely that oxytocin's effects on maternal behavior are mediated via monoaminergic, GABA-ergic, or glutamatergic effectors, but these more mechanistic studies of how oxytocin alters behavior remain to be done.

2.2. Prolactin

2.2.1. Prolactin—Background

The other major neuropeptide implicated in maternal behavior is prolactin. Compared to oxytocin, prolactin is a much larger molecule (198 amino acids) belonging to a peptide family that includes growth hormone and placental lactogens (PLs). Prolactin is present in all vertebrates and has been long associated with various aspects of parental behavior in birds (Lehrman, 1961). In Chapter 2, we described the evidence that prolactin as an endocrine hormone is involved in mammalian maternal care. To review quickly, prolactin is synthesized primarily in the anterior pituitary, under the influence of several prolactin-releasing and prolactin-inhibiting factors. In rodents, prolactin increases early in pregnancy, then is suppressed through mid- and late gestation when PL1 and PL2 (placental lactogen 1 and placental lactogen 2) are high. Prolactin increases again just prior to parturition. In primates, prolactin and placental lactogen rise steadily across gestation. The most potent prolactin-inhibiting factor is dopamine via the tuberoinfundibular (TIDA) neurons of the arcuate nucleus of the hypothalamus. TIDA neurons project to the median eminence, releasing dopamine into the hypophyseal portal blood to reach D2 receptors on lactotrophs in the anterior pituitary, inhibiting prolactin release. Lactotrophs also have receptors for oxytocin (Breton, Pechoux, Morel, & Zingg, 1995) and many other factors that are considered prolactin-releasing or prolactin-inhibiting factors (for instance TSH, VIP).

There are several indications that prolactin may also be synthesized within the brain. Prolactin in the brain has been detected by immunoreactivity (Paut-Pagano et al., 1993; Shivers, Harlan, & Pfaff, 1989) and radioimmunoassay even after hypophysectomy (DeVito, 1988). But the amount of prolactin in the brain is several orders of magnitude below concentrations in the pituitary, with 30–

500 pg/mg protein found in the brain versus roughly 10 μg/mg protein reported in the pituitary (Emanuele et al., 1992). Perhaps more important, prolactin mRNA has been demonstrated by PCR in the hypothalamus, amygdala, and caudate (Emanuele et al., 1992). The level of message was so low in this report that it could not be detected by Northern blot analysis, but several controls for nonspecific amplification support the presence of low levels of prolactin synthesis in select brain regions (for a critical review of this issue see Dutt, Kaplitt, Kow, & Pfaff, 1994). Supporting a role for CNS prolactin, central injections of the peptide have been associated with a number of behaviorial effects, including increased feeding and grooming and decreased fertility and stress responsiveness (reviewed by Grattan, 2001).

Prolactin's peripheral and central effects are mediated via specific lactogenic receptors that belong to class I of the cytokine receptor superfamily. Prolactin receptors are abundant in liver, mammary, and ovary tissue, and have been found to a lesser extent in virtually every peripheral tissue examined. They are found in the brain, especially in the choroid plexus, where they mediate transport of circulating lactogens into the CNS (Walsh, Slaby, & Posner, 1987) (Figure 6.6). To a much lesser extent, prolactin receptors are evident in the hypothalamus and limbic sites. Either receptor mRNA (Bakowska & Morrell, 1997; Chiu & Wise, 1994) or receptor immunoreactivity (Pi & Grattan, 1998) can be detected in the MPOA, anterior periventricular, paraventricular, supraoptic, arcuate, and ventro-medial nuclei of the hypothalamus as well as the BST and medial nucleus of the amygdala. Prolactin receptors on TIDA neurons in the arcuate nucleus mediate a short-loop feedback system.

There are at least two isoforms of the prolactin receptor, considered splice variants of the same gene. A long form (roughly 600 amino acids) and a short form (roughly 300 amino acids) differ in their cytoplasmic but not their ligand-binding domains. Although both forms of the receptor bind prolactin, the long form is coupled to the Jax2/Stat pathway, while the less-characterized short form with its truncated cytoplasmic domain may be coupled to MAP kinase. The difference in signaling appears critical to function. For instance, only the long form will transduce the prolactin signal for milk protein transcription, whereas either isoform will mediate prolactin's mitogenic effects. Transgenic mice deficient in prolactin receptor, D2 receptor, or Stat5b all have markedly elevated levels of prolactin (30–100-fold elevations in the PR-knockout mice) presumably due to loss of negative feedback. These observations, consistent with anatomic and physiological studies of prolactin receptors on TIDA neurons, suggest that prolactin stimulation of TIDA neuron activity is mediated via the long form of the receptor coupled to the Jax2/Stat pathway (Grattan et al., 2001). A dopamine agonist normalizes serum prolactin in the PR-knockout mouse, probably by replacing the inhibitory signal at the lactotroph.

Recent studies have demonstrated that prolactin receptors, like oxytocin receptors in the brain, are surprisingly plastic. Not only is staining for prolactin receptors greater in ovariectomized and estrogen-treated female rats than in male rats in both the hypothalamus and ventral pallidum, but lactating females have

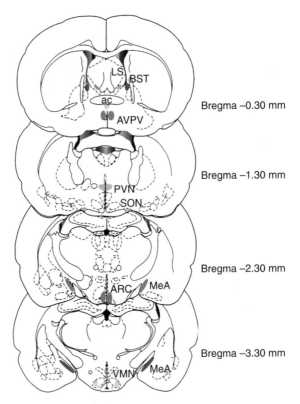

Figure 6.6. Prolactin receptors (long form) in the rat forebrain can be detected by in situ hybridization at four levels (representing Bregma –0.30 to –3.30). Prolactin receptors are most abundant in the choroid plexus, throughout the lateral ventricles. At a much lower level of expression, receptor mRNA can be detected in the anterior ventral periventricular nucleus (AVPV) of the MPOA, across much of the lateral BST, and in the ventral segment of the LS at the level of the anterior commissure (ac). Receptor mRNA is present in both the PVN and SON, the arcuate nucleus (ARC), and, at lower levels, in the VMN. Receptors in the ARC are believed to be localized to dopamine cells, regulating prolactin release via short-loop feedback. In the amygdala, prolactin receptor mRNA can be found in the medial (MeA) and, to a much lower extent, in the central nucleus. Compared to diestrous females, lactating females have more prolactin receptors in the choroid plexus, MPOA, PVN, SON, and VMN (Grattan et al., 2001). Figure adapted from Bakowska and Morrell (1998).

2–3-fold higher levels of immunostaining in the medial preoptic and arcuate nuclei relative to diestrous females (Pi & Grattan, 1999). In the PVN and VMN, staining was only detected in lactation and not in diestrous rats. In another study, using in situ hybridization, prolactin receptor mRNA (for the long form) was greater in the choroid plexus and the MPOA on day 21 of pregnancy compared to day 2 of pregnancy (Bakowska & Morrell, 1997).

Sugiyama et al. (1994) were the first to describe a preferential increase in mRNA for the long form of the receptor in the rat brain beginning on day 7 of pregnancy, increasing through gestation and remaining high through most of lactation. Similar increases could be induced by treatment with prolactin, growth hormone, or progesterone. In an interesting follow-up, Sugiyama et al. (1996) found that removal of the pups on day 3 of lactation eliminated this increase, but return of pups on day 8 for 5 additional days of lactation reinstated the increase in the mRNA for the long form of prolactin receptor. Nulliparous females induced into maternal behavior by pup exposure also showed a marked increase (roughly 8-fold) in the long form but not the short form of the receptor (Sugiyama et al., 1996). This effect was not observed in ovariectomized or hypophysectomized females even when they exhibited full maternal behavior, suggesting (along with the dependence on pup stimuli and lactation) that prolactin may be critical for the induction of the long form of the receptor. As Sugiyama et al. (1996) used mRNA extracted from whole brain, their effects most likely reflect prolactin receptors in the choroid plexus. However, Pi and Grattan have reported increased prolactin receptor mRNA (both long and short forms) in microdissected hypothalamic tissue during lactation (Grattan, 2001; Pi & Grattan, 1999).

2.2.2. Prolactin—Pharmacology

Aside from these changes in receptors, what is the experimental evidence that prolactin is involved in maternal behavior? In Chapter 2, we summarized the evidence that prolactin is an important component of the hormone-mediated onset of maternal behavior in the rat. Recall that these studies required simultaneous treatment with estrogen, involved several days of prolactin administration, and usually were manifested as a reversal of dopamine agonist effects with prolactin treatment. In contrast to oxytocin research, acute administration of prolactin has never been shown to induce rapid-onset maternal behavior. Moreover, we have lacked selective, high-affinity antagonists for prolactin receptors to decrease prolactin function. Over the past two decades, dopamine agonists and antagonists have been used to modulate prolactin levels, but these drugs have a range of neurochemical effects unrelated to prolactin. We described this approach in Chapter 2. In a prototype of these experiments, ovariectomized nulliparous females treated with progesterone on days 3–13, followed by estrogen on days 13–24, typically show full maternal behavior within 2–3 days of the initiation of estrogen treatment. Administration of the dopamine agonist bromocriptine (2 mg/kg, sc daily) beginning on day 13 reduces plasma prolactin by over 95% and increases the latency to maternal behavior to 5 days (Bridges & Ronsheim, 1990). Bridges and Ronsheim (1990) showed that prolactin (500 µg BID beginning on day 13) reversed the effects of bromocriptine on maternal behavior. This much we have discussed previously. The question that concerns us here is whether these effects are peripheral (mediated via pituitary prolactin acting on peripheral targets) or central (mediated via prolactin or other lactogens in the brain binding to central lactogen receptors).

Several observations support a central role for prolactin (or related placental lactogens). Recall from Chapter 5 that Bridges et al. (1990) showed that, relative to the systemic treatments described above, much lower doses of prolactin were effective when given centrally. In this study, only 40% of females receiving the sequential steroid protocol were fully maternal by day 6. In females receiving ICV injections of prolactin (dose range from 80 ng to 50 µg given twice the day before, twice the first day, and once the second day of behavioral testing), Bridges et al. observed a dose-dependent acceleration in the onset of maternal behavior. At the highest dose (50 µg), 90% of females were fully maternal at day 3. This same dose had no effect when administered systemically, demonstrating that the effects of central injections were mediated via central receptors. At the lowest effective dose (400 ng), 50% of females were fully maternal by day 3. A dose that was ineffective ICV (80 ng) was potent when given into the MPOA, with nearly 50% (6 of 13) showing full maternal behavior on day 2 and an overall group latency of 1.3 days compared to 5.6 days for vehicle-injected controls (Figure 6.7).

Bromocriptine is a nonspecific dopamine agonist that has a range of effects unrelated to its suppression of prolactin. As dopamine has been implicated in maternal behavior and dopamine fibers innervate the MPOA, it is certainly possible that the effects observed in these studies represent primarily the effects of bromocriptine on dopamine pathways with prolactin secondarily influencing dopamine release. Prolactin has been shown to alter dopaminergic activity with a latency of at least 17 hours. The slow effect of prolactin on dopamine cells may explain the delayed response seen in the studies of maternal behavior in which several days of prolactin treatment are required. On the other hand, other lactogenic hormones, such as rat placental lactogen I (but not growth hormone) mimic prolactin's effect on maternal behavior when infused into the MPOA (Bridges et al., 1997).

More direct evidence comes from recent studies that have compensated for the absence of a prolactin receptor antagonist by developing prolactin receptor antisense oligonucleotides to reduce prolactin transport into the CNS and neurotransmission selectively. Prolactin receptor antisense oligonucleotides given ICV for several days reduced choroid plexus receptors approximately 70% (Torner et al., 2001). When delivered for 3 days to postpartum females, pup retrieval latency increased over 100%, although there was no effect on lactation, as measured by pup weight gain or milk-ejection reflexes (Torner et al., 2002). This decrease in maternal behavior may be secondary to a general increase in fear-

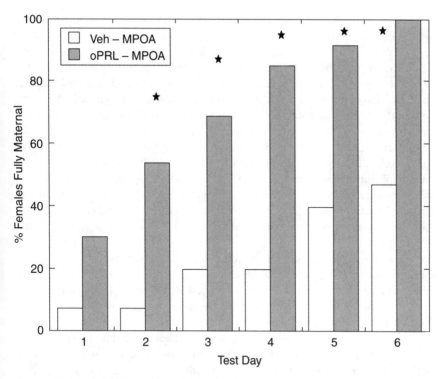

Figure 6.7. Prolactin injected into the MPOA facilitates the onset of maternal behavior in virgin female rats. In females injected with vehicle, fewer than 50% were maternal by day 6. In females with MPOA injections of 40 ng of ovine prolactin (a dose that is not effective ICV), more than 50% were fully maternal by day 2 and 100% were fully maternal by day 6. Differences (*) between vehicle and o-prolactin were significant on days 2–6 (Bridges et al., 1990).

fulness, as the antisense-treated females showed decreased exploration of the open arms of an elevated plus maze and showed markedly increased stress responses, as measured by ACTH release during the plus maze test (Torner et al., 2002). As most of the brain prolactin receptors are in the choroid plexus, ICV administration of antisense may work mostly by reducing transport of systemic hormone (prolactin and/or lactogens) into the CNS. It is also possible, since the pituitary is not essential for postpartum maternal behavior, that these antisense studies are revealing a role for a brain prolactin/lactogen system in maternal behavior.

These antisense studies along with the dopamine antagonist studies suggest that prolactin within the CNS is important for maternal behavior. Taken together with the increases in brain prolactin receptors, described above, we suspect that prolactin is stimulating maternal behavior by a two-stage process. The first stage involves the receptor in the choroid plexus that appears to serve as a transporter, moving plasma prolactin into the brain's extracellular fluid. Increasing receptors in the choroid plexus thus increases prolactin in the CNS, just as if the peptide were synthesized and released there. In the second stage, local hypothalamic increases in prolactin receptor may be important for specific circuits involved in maternal behavior, as suggested by the effects of prolactin injected into the MPOA. However, the MPOA is not the only potential target site. Prolactin receptors increase in a number of hypothalamic and limbic nuclei during lactation and any of these may be important for prolactin's behavioral and neuroendocrine effects (Figure 6.6). For instance, receptors in the AVPV in a region of GnRH cells may be important for the suckling inhibition of estrus, receptors (predominantly short-form) in the PVN and SON may influence oxytocin or CRF release, and receptors in the VMH may be important for prolactin stimulation of feeding. Torner et al. (2002) have demonstrated that prolactin is critical for the reduced stress response of lactating females, although the site for this effect is not yet clear.

What is the mechanism for inducing prolactin receptors? Above, we described that suckling is associated with increased prolactin receptor mRNA in the brain. Several lines of evidence suggest that it is prolactin, released by suckling, that induces its own receptor (reviewed by Grattan, 2001). But the story must be more complicated: Prolactin increases during pregnancy before suckling begins and there must be a braking mechanism for this autostimulation. Moreover, the reason for the slow onset of prolactin's effects remains a mystery. In contrast to oxytocin with effects in seconds (sheep) or minutes (rats), all of the studies either increasing or decreasing prolactin neurotransmission appear to require several days to influence maternal behavior. It appears that prolactin, like gonadal steroids, induces a number of downstream mechanisms to influence maternal care.

2.2.3. Prolactin—Other Species

What about prolactin and maternal behavior in other species? Although most of the research has been with rats using experimental induction of maternal behavior, there is an interesting literature on prolactin effects in mice. As early as 1973, Voci and Carlson reported that crystalline prolactin implanted near the MPOA enhanced pup retrieval and nest building (Voci & Carlson, 1973). However, others have not found effects with prolactin administration (Koch & Ehret, 1989a) or bromocriptine treatment (McCarthy, 1990).

There have not been the extensive investigations of either changes in prolactin receptors or pharmacological manipulations with multiple hormone priming conditions in other species, except in males (summarized in Chapter 7). In rabbits (*Oryctolagus cuniculus*), pharmacological reductions of prolactin late in gestation decrease maternal behavior (González-Mariscal et al., 2000). As noted in Chapter 2, this observation is particularly interesting because rabbits lack lactogens other than prolactin. In hamsters, bromocriptine (0.5 mg) disrupts maternal behavior when given on the day of parturition (McCarthy et al., 1994a).

In biparental prairie voles, prolactin receptor mRNA increases in the choroid plexus in females but not males postpartum (Khatib, Insel, & Young, 2001). In sheep, a single central injection of prolactin (50 or 100 µg ICV) did not facilitate maternal responsiveness of steroid-primed nonpregnant ewes (Levy et al., 1996) and administration of dibromoergocriptine to reduce prolactin release did not prevent maternal responsiveness in parturient ewes (Poindron et al., 1980). These results have led most investigators to conclude that prolactin, in contrast to oxytocin, is not important for sheep maternal behavior. It is important to remember that similar treatments have not been sufficient to influence maternal behavior in rats and that these treatments in sheep would not decrease placental lactogens. The evidence implicating prolactin in rat maternal behavior is based on multiple days of administration of prolactin and a very specific steroid-priming regimen paired with multiple days of dopamine agonist administration. Similar strategies have not been used in sheep. In marmosets (*Callithrix jacchus*) without parental experience, prolactin concentrations in plasma increase after infant carrying (Roberts et al., 2001b) and 3 days of bromocriptine (0.5 mg sc) decreases plasma prolactin and reduces parental responsiveness (Roberts et al., 2001a). In Chapter 7, we describe the evidence for prolactin involvement in paternal care.

2.2.4. Prolactin and Prolactin Receptor Knockouts

A null mutation of the prolactin receptor gene provides a powerful test of the role of prolactin and other lactogenic hormones on maternal behavior. Unfortunately, mice with a knockout of the prolactin receptor suffer a failure of embryonic implantation as well as preimplantation embryonic development, so they cannot carry a pregnancy and thus cannot be used for studying maternal behavior after parturition (Ormandy et al., 1997). Lucas et al. (1998) reported on maternal behavior in heterozygous primiparous females as well as homozygous null mutant virgin females (Figure 6.8). The tests of maternal behavior were similar to those used with rats: daily exposure to 3 foster pups (1–3 days old) to assess retrieval and crouching over pups. However, behavior was scored for only 30 minutes. On the first day of testing, heterozygous primiparous females showed a normal approach to and investigation of the pups, a slight but significant decrease in retrieval of the first and second pups, and a marked delay in retrieval of the third pup. Although testing in this group was not repeated over subsequent days, these results were consistent with an earlier observation that heterozygous females often left their pups scattered about the cage (Ormandy et al., 1997). However, the results were not entirely consistent as some heterozygous females showed no deficit in maternal behavior. Unfortunately, Lucas et al. (1998) failed to measure prolactin receptor gene expression to determine if the individual variation in behavior could be explained by variation in brain prolactin receptors within the heterozygous group. The homozygous virgin females also show a deficit in retrieval and crouching, failing to display full maternal behavior after 6 days of exposure while wild-type (+/+) virgin females

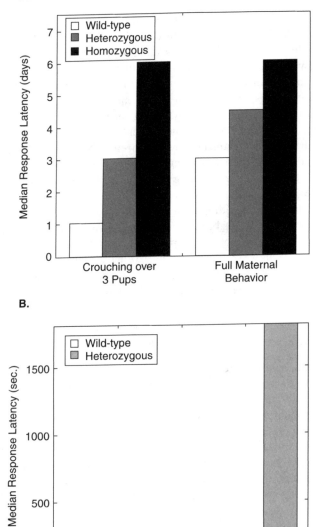

Figure 6.8. Mice with a null mutation of the prolactin receptor gene have a deficit in maternal behavior. **A.** Virgin females were tested for crouching over 3 pups or exhibiting full maternal behavior. Homozygous females have longer latency for crouching as well as showing other components of maternal behavior (retrieval, grooming) relative to wild-type females. Heterozygous females are intermediate on both measures. **B.** Relative to wild-types, heterozygous postpartum females exhibit no deficit when tested for pup retrieval of the first two pups, but are slower to retrieve a third pup. Females homozygous for the mutation of the prolactin receptor are not fertile and thus could not be tested postpartum (Lucas et al., 1998).

were fully maternal in 2 days and heterozygous (+/–) virgin females were fully maternal in roughly 4 days. Mice with a null mutation of the prolactin receptor were normal on the Morris water maze test, so their spatial memory appears intact (Lucas et al., 1998). Also they have no obvious olfactory deficit (Lucas et al., 1998). It remains to be seen whether their failure to retrieve and crouch over pups reflects a specific deficit in nurturing behavior or a more general abnormality in sensory or behavioral abilities following a knockout of the prolactin receptor.

This loss of maternal behavior in prolactin receptor knockout mice may seem consistent with the results of dopamine agonists and prolactin antisense administration in rats. But it was surprising nonetheless because a knockout of the prolactin gene (as opposed to the receptor gene) does not disrupt maternal behavior. Homozygous prolactin knockout females are infertile, but when given pups they show the full range of maternal behaviors (Horseman et al., 1997). How can this be? There are a number of cases in which a knockout of the receptor has a greater impact than a knockout of the ligand. The most likely possibility is that another ligand compensates for the missing gene. In this case, it seems likely that one of the other lactogenic hormones can substitute for prolactin. This has not been proven, but the evidence that hypophysectomized mice (lacking pituitary release of prolactin) are fully maternal is consistent with the possibility that a lactogenic hormone other than prolactin is involved. However, given the important role of prolactin for inducing prolactin receptors, one would expect that prolactin receptors would not increase in prolactin-knockout or hypophysectomized mice and thus the necessary neural changes associated with the hyperprolactinemia of parturition and lactation would not develop. The presence of full maternal behavior in prolactin-knockout mice remains to be explained.

Earlier in this chapter we described how mice with a null mutation of oxytocin, in contrast to the prolactin receptor knockout, show no deficit in maternal behavior. Can we conclude that prolactin (or some related lactogenic hormone) is necessary and oxytocin is not necessary for maternal behavior in mice? A comparison of the studies does not support such a simple explanation. For one thing, as just discussed, the prolactin knockout, like the oxytocin knockout, is also maternal. So some of the difference between the oxytocin and prolactin receptor knockout may reflect deletion of the ligand versus deletion of the receptor. But there are other issues as well. The prolactin receptor mutation was developed in a 129SV strain. This strain is frequently used for transgenesis because the mice are easy to superovulate and targeted mutations in this strain have a high rate of incorporation in the germ line. Relative to other mouse strains, their maternal behavior is weak. Studies of OT-KO mice were on a background of C57BL6, a strain that shows more robust maternal behavior in virgin females (Nishimori et al., 1996). To compare the behavioral effects of null mutations, it is important to start with the same background strain or one may be left with markedly different baseline rates of behavior precluding a comparison of the behavioral effects of the mutations.

2.2.5. Prolactin—Summary

There remain several contradictions and shortcomings in the literature on prolactin and maternal behavior in rats. The first involves changes in prolactin's effects in the hypothalamus during pregnancy. Prolactin is intimately connected to dopamine, particularly within the arcuate nucleus via the tuberoinfundibular system. These tuberoinfundibular dopamine neurons are part of a feedback loop: They are normally stimulated by prolactin and by releasing dopamine they maintain a tonic inhibition over prolactin secretion. This short-loop regulatory loop appears desensitized during pregnancy (Grattan & Averill, 1995) and lactation (Demarest, McKay, Riegle, & Moore, 1983), permitting high levels of prolactin to be maintained throughout late pregnancy and lactation without negative feedback. What is confusing is that prolactin receptor mRNA and prolactin receptor protein are both increased in this region during lactation, just the opposite of what one would expect for a desensitized system. This reminds us that an increase in receptor mRNA or receptor protein does not ensure an increase in functional responsiveness. Apparently, the receptor must be uncoupled from signal transduction during late pregnancy and lactation, so that prolactin no longer has an effect even in the presence of high-circulating concentrations of hormone and high concentrations of receptors.

A second major problem is the absence of direct evidence with selective antagonists or antisense approaches given specifically into regions enriched with prolactin receptors. The first study with a selective antagonist, S179D-PRL, looks promising in that it delays the onset of maternal behavior in steroid-primed, nulliparous females (Bridges et al., 2001). In addition to selective antagonists, the recent development of powerful antisense tools for the prolactin receptor may be useful for localizing prolactin effects (Torner et al., 2001).

The critical questions are: How and where will reducing prolactin receptor activation affect maternal behavior in parturient females? Is prolactin or another lactogenic hormone important for reducing fear, increasing motivation, or facilitating approach? Torner et al. (2002) have shown that prolactin is critical for the blunting of the stress response during lactation, as prolactin receptor antisense restores the ACTH response to stress in lactating females. In addition to apparent anxiolytic effects, we do not know if prolactin alters the stimulus value of pups. Whatever the mechanisms, prolactin or lactogenic hormones facilitate maternal care over several days rather than several minutes, so these neuromodulators might be best considered as necessary but not sufficient for maternal care.

2.3. Opioids

2.3.1. Opioids—Background

In Chapter 4, we discussed the evidence that maternal behavior involves a change in motivational state. In one powerful example of this change of state, rat mothers prefer a context associated with pups to a context associated with cocaine (Mattson et al., 2001). We discussed dopamine as a neurotransmitter associated with motivation. The other class of neurotransmitters associated with motivation are the opiate peptides. The opiate family, including β-endorphin, the enkephalins, dynorphin, the invertebrate peptide FMRFamide, and the recently discovered endogenous endomorphin are the prototype neuropeptides for the study of behavior, implicated in analgesia, stress, locomotion, reproduction, learning, and fear. Behavioral studies of this family of peptides have been facilitated by the availability of excellent nonpeptide agonists (e.g., morphine) and antagonists (e.g., naloxone), allowing systemic treatments to modify central pathways. There are four groups of receptors for opioid peptides in the brain: the μ, δ, κ, and σ. Although the endogenous peptides (endorphins, enkephalins, and dynorphins) have an imperfect selectivity for these receptors, a range of selective agonists or antagonists have helped to discriminate between these receptor subtypes, demonstrating somewhat different functional effects. For instance, although agonists at each of the subtypes have been associated with analgesia and stimulation of feeding, μ and κ but not δ agonists stimulate prolactin release and increase aggression. Moreover, both the endogenous peptides and these various opioid receptors have unique distributions in the mammalian brain. The MPOA is rich in both fibers and receptors for opioid peptides (Simerly, Gorski, & Swanson, 1986).

2.3.2. Opioids—Physiology

β-endorphin and its receptor, the μ-opiate receptor, have been of particular interest for the study of maternal behavior because of evidence that this peptide mediates affiliative behaviors such as grooming in nonhuman primates (Fabre-Nys, Meller, & Keverne, 1982; Keverne, Martensz, & Tuite, 1989) and attachment behaviors in guinea pigs (Panksepp, Herman, Conner, Bishop, & Scott, 1978). Although β-endorphin is largely synthesized from the precursor molecule POMC in the anterior pituitary, there are also two major sources for β-endorphin within the brain: the arcuate nucleus of the medial basal hypothalamus and the caudal region of the nucleus tractus solitarius in the brainstem. During pregnancy, β-endorphin concentrations increase in these regions as well as in terminal fields in the amygdala. Relative to pregnancy, β-endorphin concentrations and POMC gene expression decrease during lactation (Mann, Rubin, & Bridges, 1997; Smith, 1993; Wardlaw & Frantz, 1983). The major receptor for β-endorphin, the μ-opiate receptor, is found specifically in the MPOA. Both the content of β-endorphin in the MPOA and the density of 3H-naloxone binding are decreased during day-12 lactation relative to day 12 of pregnancy (Hammer & Bridges, 1987). In the same study, parallel increases were found in ovariectomized females treated with both estrogen (14 days) and progesterone (11 days) relative to ovariectomized nontreated females (Hammer & Bridges, 1987), suggesting that changes in gonadal steroids are driving the opioid system during pregnancy.

A series of studies have described a change in opiate peptide gene expression and opiate receptors just at the time of parturition. These changes are concurrent with the onset of lactation and the reduced stress response that accompanies lactation. Opioid peptides are potent inhibitors of oxytocin release, oxytocin cell firing, and milk ejection. This inhibition of lactation is blocked at parturition when the density of μ-receptors is reduced in the SON (Sumner, Douglas, & Russell, 1992). But in addition to losing the opiate brake on milk ejection, lactating rats are less sensitive to opioid agonist and antagonist effects on analgesia, catalepsy, and food intake, suggesting a widespread decrease in opioid sensitivity. On the other hand, opioid peptides are integral for the normal increase in prolactin during lactation, probably by decreasing dopamine by a combined inhibition of TIDA neurons and a decrease in dopamine synthesis locally in the arcuate nucleus.

2.3.3. Opioids—Pharmacology

In addition to these changes in pregnancy and lactation, are opioid peptides important for the onset of maternal behavior? An early study using a pregnancy termination model (day-17 termination with hysterectomy and ovariectomy) reported that morphine prevented the onset of maternal behavior and that the opiate receptor antagonist naloxone blocked morphine's effects (Bridges & Grimm, 1982). Although this study used high doses of morphine (10 mg/kg), morphine effects were not due to a general decrease in motor activity or sedation. In a follow-up experiment, morphine given as a crystal blocked maternal behavior in sensitized ovariectomized females when placed into the MPOA, but not in the VMH (Rubin & Bridges, 1984). These findings led to a premature assumption that opioid peptides prevented the onset of maternal care, but the actual story is considerably more complex. In some regions, such as the MPOA, opiates acting via μ-receptors appear to inhibit maternal care. In other regions, opiates may facilitate maternal behavior. For instance, injections of morphine directly into the VTA reduced the latency for the onset of maternal behavior in nulliparous females from 8 days to 4 days (Thompson & Kristal, 1996). Naloxone, given systemically, has complex effects as it antagonizes multiple receptors in regions with opposing actions—such as μ-receptors in the MPOA and δ-receptors in the VTA. When given systemically in late pregnancy, naloxone fails to facilitate maternal care (Bridges, 1996), whereas in juvenile rats, 9 days of naltrexone injections delayed the onset of parental care in both males and females (Zaias et al., 1996).

Aside from these effects on the onset of maternal behavior, what about opiate effects on ongoing maternal care? Several studies from the Bridges lab have shown that opioids that bind to the μ-receptor decrease pup grouping and retrieval in lactating females. In lactating females, morphine (0.5 μg) injected bilaterally into the MPOA reduced maternal behavior, an effect that was blocked by naloxone and was not mimicked by the inactive opiate dextrorphan (Rubin & Bridges, 1984). Mann, Kinsley, and Bridges (1991) demonstrated that morphine effects were likely to be mediated via the μ-receptor: β-endorphin or the μ-receptor agonist DAGO infused into the lateral ventricle reduced maternal care for 1–3 hours, whereas other opioid agonists, including DPDPE (δ), U50488H (κ), and SKF10047 (σ) failed to disrupt maternal behavior across a wide range of doses. These μ-receptor effects are likely dependent on receptors in the MPOA as β-endorphin (0.06–0.72 nm) injected into the MPOA also reduced maternal behavior (Mann & Bridges, 1992) and morphine reduced cFos activation in the MPOA of lactating primiparous females (Stafisso-Sandoz, Holt, Polley, Lambert, & Kinsley, 1998).

There is some evidence that opiate effects may be conferred on selective aspects of maternal care, such as the olfactory response to pups (Kinsley & Bridges, 1990), maternal aggression (Kinsley & Bridges, 1986), the duration of nursing bouts (Byrnes, Rigero, & Bridges, 2000), or placentophagia (Mayer, Faris, Komisaruk, & Rosenblatt, 1985), but a particularly interesting observation suggests that opiate receptors have a special role in the adaptive changes that accompany the onset of maternal behavior. Byrnes and Bridges (2000) reported that the long-acting μ-opiate receptor antagonist β-funaltrexamine, given ICV on day 21 of pregnancy, had no effect on postpartum maternal behavior but blocked maternal memory, measured in females removed from their pups postpartum and then tested for the onset of maternal behavior 7 days later. Females treated with β-funaltrexamine prepartum (but not postpartum) showed a delay in the onset of maternal behavior when tested 7 days after pup removal. Later, in Chapter 8, we will discuss more fully the evidence that opiates have some role in the adaptive changes that occur with parturition, manifested by this maternal memory.

Our interpretation of these results can be summarized by recognizing that various receptors in different brain regions may be mediating quite different effects. In the MPOA, β-endorphin acting via μ-receptors appears to inhibit the initiation and maintenance of maternal behavior. The absence of Fos activation in this region following morphine administration in lactating females supports an opiate role in MPOA activation. The decrease in receptors in this region in primiparous but not multiparous females suggests that opiates are most important for the initial experience of maternal care. Opiate neurotransmission also decreases in the neuroendocrine hypothalamus (SON and possibly PVN) following parturition, releasing a brake on oxytocin cells. In the VTA, where opiate treatment facilitates the onset of maternal behavior, enkephalin release may activate dopamine cells, leading to a more rapid sensitization to pups.

2.3.4. Opioids—Other Species

In contrast to these effects in rats, studies in sheep demonstrate that opioids potentiate the effects of vaginocervical stimulation to promote maternal responsiveness. Recall that VCS induces lamb acceptance in estrogen-primed ewes. Naltrexone, administered ICV, blocked this effect and inhibited the maternal responses of parturient ewes (Caba et al., 1995; Kendrick & Keverne, 1989;

Kendrick et al., 1997a). Morphine administered ICV with 5 minutes of VCS induced full maternal behavior, comparable to the responses of parturient ewes and greater than observed with oxytocin and VCS (Kendrick et al., 1991). However, morphine was not effective without VCS. Kendrick and Keverne (1989) suggest that the opioid effects may be mediated via oxytocin. Although there is, as yet, no proof of this explanation (Kendrick et al., 1991), naltrexone blocked the central (but not peripheral) release of oxytocin following VCS (Kendrick & Keverne, 1989). Broad et al. have reported that pre-proenkephalin mRNA increases in the PVN at parturition and in the VMN during lactation (Broad, Kendrick, Keverne, & Sirinathsinghji, 1993a), suggesting a mechanism for opiate facilitation of oxytocin central release. By contrast, POMC mRNA expression in the arcuate nucleus decreased at parturition and increased during lactation compared to late pregnant and ovariectomized animals (Broad et al., 1993a).

Further evidence for a role of opiates in maternal behavior comes from studies with rhesus monkeys. Martel and coworkers described a decrease in maternal care in 7 lactating females, housed in social groups, that were administered naloxone (0.5 mg/kg) when their infants were 4, 6, 8, and 10 weeks old (Martel, Nevison, Rayment, Simpson, & Keverne, 1993). Relative to females injected with saline, mothers receiving naloxone exhibited less allogrooming and less protection of their infants. These results in monkeys are clearly different from the studies noted above with rats in which opiate agonists decrease maternal grooming of pups, and they are countered by other studies finding a decrease in maternal behavior in macaques treated with opiates just prior to reunion (Kalin, Shelton, & Lynn, 1995) when morphine (0.1 mg/kg) decreased and naloxone (5 mg/kg) increased the mother's clinging to her infant (also see Misiti, Turillazzi, Zapponi, & Loizzo, 1991).

2.4. Other Neuropeptide Candidates

2.4.1. Cholecystokinin

Cholecystokinin (CCK) is an 8-amino-acid peptide synthesized in the gut and brain with receptors found in several brain regions, especially the MPOA. Thus far, there has been no study of physiological changes in either CCK release or CCK receptors with parturition and lactation, although CCK is abundant in the anterior hypothalamus. Linden et al. reported that systemic administration of CCK-8 into estrogen-primed nulliparous Wistar rats induced a more rapid onset of maternal behavior than vehicle-injected controls (Linden, Uvnas-Moberg, Eneroth, & Sodersten, 1989). This finding could not be replicated in Sprague-Dawley rats (Mann, Felicio, & Bridges, 1995). However, a CCK-8 antagonist, proglumide (125 µg ICV or 20 mg/kg ip), increased retrieval latencies and decreased crouching behavior on days 5–6 of lactation (Mann et al., 1995). This finding may be nonspecific; any intervention that causes illness or distress might disrupt maternal behavior in lactating females. CCK appears to have a more selective effect in its interaction with β-endorphin. CCK-8 administered into either the lateral ventricle (Felicio, Mann, & Bridges, 1991) or MPOA (Mann et al., 1995) blocks the disruption of maternal behavior observed after β-endorphin administration (see above). It seems likely that CCK modulates a system downstream from β-endorphin, but there is currently no evidence that this same neural system is involved in normal maternal behavior. Mice with null mutations of CCK-A (Feifel, Priebe, & Shilling, 2001) and CCK-B (Nagata et al., 1996) receptors have been developed. The papers describing these mice do not comment on maternal behavior.

2.4.2. Tachykinins

The tachykinin family includes substance P, a splice variant substance K or neurokinin A, the gut peptide bombesin, and a group of related peptides found in invertebrates. Substance K has been implicated in male rat sex behavior (Dornan et al., 1993). Interest in this family of peptides as inhibitors of maternal behavior arose from their anatomic location. As noted in Chapter 5, neurons in both the medial amygdala and the VMH appear to inhibit the onset of maternal behavior. Noting that tachykinin fibers and tachykinin receptors are present in the VMH and that a tachykinergic projection from medial amygdala to VMH has been described (in cats), Sheehan and Numan (1997) reasoned that tachykinins in the VMH would inhibit the onset of maternal behavior. In female rats primed to be maternal by pregnancy termination and estrogen injection, substance K at doses as low as 66pmol delayed the onset of maternal behavior when injected into the VMH but not into a comparison receptor-rich region, the mediodorsal thalamus (Sheehan & Numan, 1997). Females injected in the VMH were not impaired in their approach or investigation of pups, but showed, on average, a delay of about 1 day for the onset of maternal care. The mechanism for substance K's effects are not clear. Tachykinins alter prolactin, β-endorphin, and CRF secretion, but the peptide might also directly alter the processing of fear-related information as neurokinin antagonists decrease fear-related behaviors (Rupniak & Kramer, 1999). At this point, the tachykinins are candidate peptides for the inhibition of maternal behavior. The question of how normal pregnancy circumvents this inhibition may be a fruitful future line of inquiry.

Recent studies with substance P (NK1) receptor knockout mice have revealed an interesting phenotype, including decreased aggression (De Felipe et al., 1998), reduced pup ultrasonic vocalizations (Rupniak et al., 2000), and an absence of rewarding responses to opiates (Murtra, Sheasby, Hunt, & De Felipe, 2000), all of which suggest a possible change in maternal behavior. However, there is no evidence from any of these studies that mice with a deletion of the NK1 receptor have actually been tested for maternal behavior. Certainly, there is no obvious deficit that required cross-fostering and, given that substance K inhibits maternal behavior, one would not expect a maternal deficit in NK1 receptor knockouts. Also note that, as with oxytocin, the tachykinin receptors show striking species differences, so this neuropeptide system in mice may confer different functional outputs from what we would predict based on pharmacological studies in rats.

2.4.3. Corticotropin Releasing Hormone

Thus far we have described studies with anterior pituitary peptides (β-endorphin, prolactin), posterior pituitary peptides (oxytocin, vasopressin), and so-called gut peptides (CCK, neurokinin A). Corticotropin releasing hormone is a 41-amino-acid neuropeptide that is predominantly a brain peptide synthesized in the paraventricular nucleus and transported via the hypophyseal portal system to the anterior pituitary, where it acts via CRF R1 receptors on corticotrophs to release ACTH. In Chapter 4 we described the changes in the HPA axis, and specifically in the dynamics of CRF release following stress during lactation. As CRF also has central effects on stress-related behaviors, with CRF pathways and CRF receptors independent of the pituitary pathway, one might ask whether CRF might also influence maternal behavior.

Although CRF as an endocrine hormone appears critical for the onset of labor in humans (Smith, 1999), there is less evidence for CRF influencing neural mechanisms for maternal care. Both CRF mRNA and CRF receptor binding in the brain changes with parturition (Da Costa, Ma, Ingram, Lightman, & Aguilera, 2001). Pedersen et al. investigated the effects of CRF in ovariectomized virgin Sprague-Dawley rats primed with estradiol and progesterone, with the progesterone capsule removed 24 hours prior to testing (Pedersen, Caldwell, McGuire, & Evans, 1991). Under these conditions, 27/32 females treated with saline showed full maternal behavior within 2 hours. Of 32 females injected with CRF ICV (1.0, 2.0, 4.0 μg), only 9 showed full maternal behavior within 2 hours and 10 attacked and killed pups. In a separate experiment, females were given 3 days of pup experience before being tested with CRF. In experienced females, CRF (1.0 μg ICV) did not induce pup killing, but CRF-treated females were less likely to show full maternal behavior within 2 hours relative to saline-treated controls (5/12 CRF versus 14/14 saline fully maternal, respectively). A CRF antagonist was not used in this study, so the observed decrease in maternal behavior is best considered a pharmacological effect rather than indicating a physiological role for CRF in the mediation of this behavior.

Neither the mechanism nor the location of CRF effects on maternal behavior is known. These behavioral effects of CRF are most likely not mediated via an increase in ACTH, as ACTH by itself does not decrease maternal behavior (cited as unpublished data in Pedersen et al., 1991). The effects of corticosterone have not been studied, although adrenalectomy has been reported to facilitate, inhibit, or not affect maternal behavior (Hennessy, Harney, Smotherman, Coyle, & Levine, 1977; Leon, Numan, & Chan, 1975; Siegel & Rosenblatt, 1978b). Most likely, the observed effects of CRF involve central effects on pathways mediating fear responses (see discussion in Chapter 4). In particular, the increase in pup killing is what one might expect from fearful or stressed females. Note, however, that there is a general reduction in the response to CRF during lactation as we discussed in Chapter 4. Indeed, Da Costa et al.'s (1997) evidence that the maternal brain is much less responsive to CRF is consistent with the hypothesis that CRF is an inhibitor of maternal care.

Mice with null mutations of CRF R1 and R2 have been developed. Although R1 knockouts show reduced anxiety (Contarino et al., 1999) and R2 knockouts manifest increased anxiety (Bale et al., 2000), neither genotype has been reported to show altered maternal behavior. A double knockout has been produced by breeding the CRF-R1KO and CRF-R2KO mice (Bale et al., 2002). The resulting male offspring have a phenotype that depends on the mother's genotype. That is, double-mutant mice born to females either heterozygous or homozygous for the R2 null mutation are more anxious than mice born to an R1 mother, regardless of the pup's genotype. Apparently, the influence of the mother, in this case, is related to prenatal influences.

Paradoxically, in sheep, CRF appears to facilitate maternal acceptance, at least in experienced ewes. Keverne and Kendrick (1991) reported that CRF given ICV potentiated the effects of VCS, although like morphine, CRF had no significant effect on maternal behavior when given alone. The mechanism for this effect is unclear. Broad et al. reported a transient increase in CRF mRNA in the PVN and BST of the ewe just at parturition (Broad, Keverne, & Kendrick, 1995). Similar changes in CRF mRNA in the BST, but not the PVN, could be induced in ovariectomized ewes by estrogen or progesterone treatment (Broad et al., 1995).

2.5. Neuropeptides—Conclusion

Neuropeptides are important factors for the central mediation of maternal behavior. Some neuropeptides, such as oxytocin and prolactin, appear to have been selected through evolution for sociosexual behaviors, with effects on nurturing behavior in nonmammalian vertebrates. In mammals, oxytocin and prolactin are critical for lactation, so there is an intriguing possibility of a conservation of function, using the same peptide for feeding the young via mammary receptors and for supporting maternal motivation via brain pathways. These neuropeptide systems are also attractive candidates for maternal behavior because they are exquisitely sensitive to gonadal steroids and thus can be modified by the physiological changes of pregnancy and parturition. With the advent of nonpeptide antagonists for oxytocin (Pettibone et al., 1993), tachykinins (Rupniak & Kramer, 1999), CCK (Maselli et al., 2001), and CRF (Pournajafi Nazarloo et al., 2001; Zobel et al., 2000), we should be able to make rapid progress in the near future. But we need to be careful in describing what these peptide hormones do. They do not induce maternal behavior in the sense that they trigger milk production or milk ejection. Maternal behavior is a complex of responses mediated by many factors. Neuropeptides, in the brain, are modulators with slow, modifying effects on other neurotransmitter systems, especially the monoaminergic neurotransmitters. In the next section we will review how monoamines influence maternal behavior.

3. Monoamines

3.1. Monoamines—General Physiology

Given the abundant literature on norepinephrine, dopamine, and serotonin in virtually every behavior of interest to neuroscientists, there is a surprising paucity of research investigating these systems in maternal behavior. Descriptive studies of monoamines and their metabolites during pregnancy and lactation report profound changes in these systems in the maternal brain. For instance, Desan et al., while noting no changes across the estrous cycle, described roughly a 30% decrease in DA, a 25% increase in NE, and a 10% increase in the 5HT metabolite 5-HIAA in the anterior cortex of the rat brain between pregnancy day 19 and postpartum day 6 (Desan, Woodmansee, Ryan, Smock, & Maier, 1988). Similar changes were noted for NE and 5-HIAA in the hippocampus but not in the cerebellum. Somewhat different findings were noted by Glaser, Russell, and Taljaard (1991) for DA and NE in the hypothalamus of the rat comparing pregnancy day 20 and postpartum day 4, with median DA content increasing 14%, NE content increasing roughly 25%, and turnover of NE (MHPG/NE) decreasing 33% during this interval. In this latter study, no change in serotonin or 5-HIAA was detected in the hypothalamus, although much of the 5-HT in this region is *en passant* so a real change may be obscured by fibers on their way to other areas of the forebrain.

Alterations in monoamine turnover may prove significant for maternal care as *hubb/hubb* mice, a spontaneous mutation of a single autosomal recessive gene, have increased DA turnover and show clear, specific deficits in pup retrieval and grooming (Alston-Mills, Parker, Eisen, Wilson, & Fletcher, 1999). In fact, pup retrieval and grooming may be associated specifically with DA release as Hansen et al. (1993) reported a 23% increase in extracellular DA and a small but significant increase in extracellular 5-HIAA in the ventral striatum when lactating females were reunited with their pups following an overnight separation. As these females were nursing for most of the period of sampling, the increase in extracellular DA and 5-HIAA could reflect nursing more than other aspects of maternal care. However, Hansen et al. repeated this study using dirty pups, which reduced the time spent nursing and increased the time spent in nest building and grooming. Under these conditions, extracellular DA in the ventral striatum increased 50% over baseline. Adams et al. had previously reported that perioral stimulation increased striatal DA release (Adams, Schwarting, Boix, & Huston, 1991), so one might expect that many of the behaviors inherent in rat maternal care would release DA. Of course, these correlational studies do not demonstrate that DA release is necessary or sufficient for maternal care.

Ironically, we know more about monoaminergic changes associated with parturition in the sheep brain than in the rodent brain. A series of microdialysis studies by Kendrick and colleagues has demonstrated increased extracellular NE in the olfactory bulbs, BST, septum, MPOA, PVN, and substantia nigra measured throughout parturition (Kendrick et al., 1992a, 1997a). DA is released in many of the same regions, although no increase in DA was noted in the BST during parturition or suckling (Figure 6.9). These widespread changes in NE and DA suggest profound activation of catecholamine cell bodies in the brainstem. One must remember, however, that these changes are associated with the onset of parturition and lactation and may be temporally correlated with parturition but functionally unrelated to the onset of maternal behavior.

There are at least 9 receptors for NE, 5 receptors for DA, and 14 receptors for 5HT, but little is known about how any of these receptors change just before or just after parturition. Glaser et al. (1991) reported the number of β receptors in the hippocampus was reduced roughly 12% between day 20 of pregnancy and day 4 postpartum. However, changes in other adrenergic receptors have not been reported. Although there have been studies of estrogen effects on many of these receptors, the prolonged estrogen-priming protocols that have been used to induce maternal behavior have not been tested for their effects on adrenergic or serotonergic receptors.

3.2. Dopamine

3.2.1. Dopamine—Pharmacology

We have already presented the evidence (in Chapter 5) that dopamine is important for maternal motivation. Recall that dopamine antagonists interfere with retrieval, pup licking, and nest building (Giordano et al., 1990; Stern & Taylor, 1991) and that dopamine indirect agonists, such as cocaine, can also disrupt maternal behavior (Kinsley et al., 1994). Surprisingly, the effects of dopamine antagonists are analogous to lesions of the MPOA/vBST in that they inhibit the active components of maternal care without inhibiting more reflexive components, such as nursing. Also recall that several lines of evidence (see Chapter 5) with both pharmacological (Keer & Stern, 1999) and lesion (Hansen et al., 1991b) approaches implicate the dopaminergic innervation of the nucleus accumbens in these active components of maternal care.

Which dopamine receptor mediates these effects? Although most studies have used nonselective dopamine antagonists, such as haloperidol or flupenthixol, Silva et al. compared the effects of DA receptor selective blockers on maternal behavior and locomotion in lactating dams (Silva, Bernardi, & Felicio, 2001). The D2 receptor blocker pimozide (0.5 and 0.2 mg/kg) mildly disrupted pup retrieval but not locomotion, the putative D4 receptor blocker clozapine (1.0 mg/kg) disrupted nest building but reduced motor activity significantly, and the D1-like receptor blocker SKF-83566 (0.2 and 0.1 mg/kg) significantly reduced both pup retrieval and nest building but also decreased motor activity. It appears, as

A. Birth

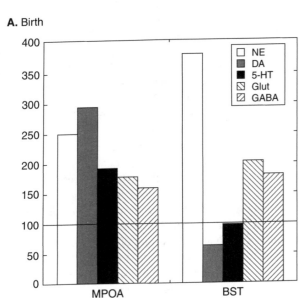

B. Suckling

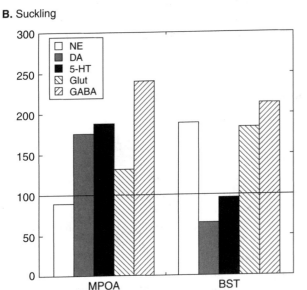

Figure 6.9. Neurotransmitter changes at birth and parturition in sheep. Pregnant ewes were fitted with microdialysis cannula aimed at either the MPOA or BST. Samples collected every 30 minutes were analyzed for norepinephrine (NE), dopamine (DA), serotonin (5-HT), glutamate (Glut), and GABA. Data shown are the percentage difference from baseline samples collected 5–10 hour prior to onset of labor; 100 represents no change from baseline. Note difference in Y-axis range between samples collected during birth and samples collected during suckling (5 minutes of suckling within the 30-minute sampling period) during the first 24 hours. In these same sheep, MPOA oxytocin during

with most behaviors, dopamine's effects are mediated via multiple receptor subtypes.

Although we have previously described evidence from lesion studies implicating the mesolimbic dopamine innervation to the nucleus accumbens, we thought it may be helpful to describe one set of studies in a bit more detail to provide an appreciation for the remarkable specificity of dopamine's effects. Hansen et al. (1991b) used 6-OHDA to reduce dopamine in either the dorsal or ventral striatum on day 2–3 postpartum in primiparous rats. In the ventral striatum, 6-OHDA reduced DA approximately 70% and NE approximately 30% relative to sham-injected controls. In the dorsal striatum, DA was reduced approximately 70% and NE, very low in the sham animals, was unaffected. When tested daily from 2 to 6 days after surgery, there were no group differences in nest building, nursing, or maternal aggression, but dams with ventral striatal DA depletion did not retrieve pups. In a test in which 3 pups were scattered in the home cage after a 2–3-minute dam-litter separation, only 28% of the dams from the ventral lesion group were able to retrieve all 3 pups within 10 minutes on any of the days of testing, whereas all of the dams in the dorsal lesion and sham groups were capable of retrieval.

What is the evidence for specificity? These same females were tested later for female sexual behavior and showed no deficits in either proceptive or receptive behavior. To understand better the nature of the retrieval deficit, Hansen videotaped another set of dams with ventral striatal 6-OHDA lesions (Hansen, 1994). In this experiment he noted that dams were more likely to feed (i.e., retrieve food) than retrieve pups. Neither denying dams the opportunity to feed nor coating the pups with sucrose increased retrieval. However, if pups were separated from the dams for 3–6 hours, dams showed rapid retrieval. These results indicate that the lesioned females are capable of retrieving but need some additional incentive that is not evoked by only a few minutes of separation. Remarkably, once lesioned females begin to retrieve pups following the 3–6-hour separation, they continue to retrieve on subsequent days following brief separations.

3.2.2. Dopamine—Cross-talk with Steroid Receptors

There is an apocryphal quote, often attributed to Einstein, that the secret to success in science is to make things as simple as possible, but not too simple. By studying one factor at a time, we are undoubtedly oversimplifying neurotransmitter action. As one example of the complexity of dopamine's effects on behavior, recent studies with dopamine have demonstrated that dopamine acting via D1 receptors can activate progesterone receptors in a ligand-independent fashion

---

Figure 6.9 *Continued*. birth was 398% and during suckling was 204% of baseline, and BST oxytocin during birth was 332% and during suckling was 462% of baseline. Oxytocin retrodialyzed into the MPOA increased NE release, but did not affect other neurotransmitters. Data adapted from Kendrick et al. (1992).

(Mani, Allen, Clark, Blaustein, & O'Malley, 1994). The evidence is as follows: (a) Both progesterone and a dopamine D1 receptor agonist facilitate female lordosis behavior (Mani et al., 1994); (b) neither progesterone nor a D1 receptor agonist facilitate lordosis behavior in progesterone receptor knockout (PR-KO) mice (Lydon, DeMayo, Conneely, & O'Malley, 1997; Mani, 2001); (c) PR antisense blocks the effects of both progesterone and a dopamine D1 agonist (Mani et al., 1994); (d) both progesterone and a D1 receptor agonist induce the second messenger cAMP and the downstream effectors protein kinase A and phosphorylated DARPP-32 in hypothalamus (Mani, 2001); and (e) a knockout of DARPP-32 prevents both dopamine and progesterone effects on lordosis (Mani et al., 2000).

Figure 6.10 provides a current working model of this cross-talk between dopamine and PR. The important point to note is that monoamines working via G-protein coupled receptors can modulate the activation or phosphorylation state of intracellular steroid receptors by activating either kinases that phosphorylate proteins (protein kinase A or C) or phosphatases that de-phosphorylate proteins (protein phosphatase-1 or its modulator DARPP-32). Kinases and phosphatases interact not only with the steroid receptor directly but also with the steroid receptor complex that binds to DNA response elements. Thus, a dopamine D1 receptor agonist can alter the activity of progesterone receptor-mediated gene transcription. This is a revolutionary discovery because it breaks down the divide between G-protein coupled receptors (including all of the peptide and monoaminergic receptors discussed in this chapter) and the steroid receptors discussed in Chapter 2. Although the dopamine D1 receptor is the only example of this effect thus far, it is likely that other examples will emerge in the future.

What does this mean for dopamine and maternal behavior? Above, we noted that dopamine appears to support rat maternal behavior by increasing maternal motivation. However, as we have noted elsewhere, dopamine antagonizes prolactin release. Now it appears that dopamine also increases progesterone receptor activation. For the onset of maternal behavior, prolactin increases and progesterone decreases. So, in contrast to the importance of increasing dopamine neurotransmission for maternal motivation, there must be a reduction of dopamine neurotransmission for prolactin and progesterone changes at parturition. How do we resolve this paradox? As D2 receptors mediate the prolactin effect, D1 receptors mediate the progesterone effect, and both have been implicated in maternal behavior, it is not likely that we can explain dopamine's opposing effects by receptor specificity. More likely, dopamine neurotransmission is modulated in a regional fashion, so that effects in the nucleus accumbens, influencing motivation, are regulated independently from effects in the hypothalamus, influencing endocrine function. As we learn more about regional differences in the intracellular events following dopamine receptor binding, the picture is likely to become even more complicated.

3.2.3. Dopamine—Knockout Studies

Mice with null mutations of each of the DA receptors as well as the DA transporter genes have been developed. Although DA receptor knockouts exhibit a

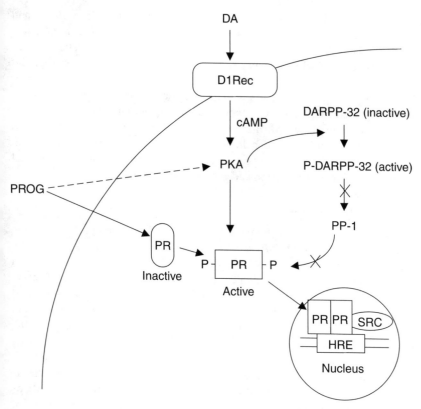

Figure 6.10. A model for cross-talk between dopamine and progesterone receptors. The classical model for progesterone action includes progesterone binding to a cytoplasmic receptor (PR), translocation of the activated receptor to the nucleus where the PR forms a heteromeric complex with another PR and several steroid response complex (SRC) proteins. This complex binds to a hormone response element (HRE), which is a consensus sequence within the promote region of genes influenced by PR. Activation of this response element drives transcription. The novel insight about PR action begins with the importance of phosphorylation for activation of the receptor. Phosphorylation is a dynamic event in the cell, promoted by kinases and reduced by phosphatases. Many intracellular signaling pathways alter phosphorylation. Dopamine (DA) binds to D1 receptors (D1Rec) to increase cAMP, which increases cAMP-dependent protein kinase (PKA). PKA phosphorylates several targets, including the PR and DARPP-32. DARPP-32, once phosphorylated, is a potent inhibitor of the major neuronal phosphatase, protein phosphatase-1 (PP-1). Remember that phosphatases reduce phosphorylation, so the increase in PKA and the decrease in PP-1 work together to lock the PR into an active state. Thus, dopamine can act with progesterone to drive PR-mediated gene expression. Adapted from Mani (2001).

number of abnormalities of motor behavior, response to psychostimulants, and neuroendocrine regulation, disrupted maternal behavior has not been noted in reports on their phenotype (reviewed in Glickstein & Schmauss, 2001). A notable exception are the DA transporter knockouts. In these mice, extracellular DA concentrations are elevated and hyperactivity is prominent. Although other social behaviors such as interaction with an unknown congener or aggressiveness were not modified relative to wild type mice, the maternal behavior of mutant females was severely disturbed (Spielewoy et al., 2000). Although these reports suggest that DA receptor deletions do not have specific effects on maternal care, the levels of extracellular DA appear to be an important regulator of maternal behavior via both neuroendocrine and mesolimbic pathways.

3.2.4. Dopamine—Summary

Taken together, the studies of dopamine release following reunion, dopamine antagonist effects, and 6-OHDA lesions all demonstrate that dopamine in the nucleus accumbens is important for pup retrieval and other active components of maternal care. As we discussed in Chapter 5, these results are best understood in the broader context of mesolimbic dopamine and motivated behaviors. Dopamine in the ventral striatum has been implicated in a number of motivated behaviors, including feeding, sex, pair bonding, and psychostimulant drug self-administration. Earlier, we suggested that the MPOA projection to the nucleus accumbens is critical for linking pup stimuli to this system in the context of maternal behavior. In a later chapter we will consider how cocaine, which hijacks this system at the level of the ventral striatum, is a potent disrupter of both the onset and maintenance phases of maternal behavior (Johns, Noonan, Zimmerman, Li, & Pedersen, 1994, 1997; Kinsley et al., 1994; Vernotica et al., 1999)—an observation that has important implications for human maternal care.

3.3. Norepinephrine

3.3.1. Norepinephrine—Pharmacology

In the same way that dopamine has been associated with motivated behaviors, norepinephrine has been linked to social cognition, especially olfactory memory. Studies in rat pups, mice, and sheep have demonstrated that NE release in the olfactory bulb is critical for learning about individual identity (Ferguson, Young, & Insel, 2002). As noted above, NE increases in diverse regions of the rat brain between pregnancy and lactation. In rats, there have been no published studies of noradrenergic agonist or antagonist effects on maternal behavior, but lesions that reduce NE in the forebrain have resulted in impaired maternal behavior. Rosenberg et al. injected 6-OHDA ICV either 2 days prior to parturition or 2 days postparturition (Rosenberg, Halaris, & Moltz, 1977). This treatment reduces both NE and DA and lacks regional specificity. Nevertheless, the study was of interest in that only the prepartum lesioned animals were affected, and the postpartum females showed no deficits in maternal behavior, analogous to the results with PVN lesions. Two other studies lesioning the dorsal bundle (Steele, Rowland, & Moltz, 1979) or both dorsal and ventral bundles (Bridges, Clifton, & Sawyer, 1982) during gestation resulted in roughly 70% depletions of NE and significant impairment in nest quality.

3.3.2. Norepinephrine—Other Species

One might not expect individual recognition to be critical for rat maternal behavior, but in sheep and other species with selective bonding to offspring, individual recognition is essential. And, not surprisingly, studies in sheep are more convincing for the importance of NE in maternal care. For instance, NE release in the olfactory bulb appears critical for the neural changes required in lamb recognition (Kendrick et al., 1997a) as the beta-adrenergic antagonist propranolol blocks maternal selectivity when injected into the olfactory bulb (Levy, Gervais, Kindermann, Orgeur, & Piketty, 1990). Remarkably, however, ICV infusions of the alpha adrenergic blocker phentolamine or the beta-adrenergic antagonist timolol fail to inhibit maternal responses following VCS in nonpregnant ewes (Kendrick & Keverne, 1989). Although it is possible that local injection of these antagonists into the MPOA would reduce maternal behavior, the available evidence suggests a role for NE primarily in individual recognition and not in other aspects of maternal care. We will have more to say about NE and maternal selectivity in Chapter 8.

3.3.3. Norepinephrine—Knockout Studies

In a transgenic experiment in mice, the gene for dopamine β-hydroxylase (*Dbh*) was deleted, eliminating both norepinephrine and epinephrine (Thomas & Palmiter, 1997). Homozygous *Dbh* null mutants die in utero of apparent cardiovascular failure. If adrenergic agonists are provided through prenatal development, these mice survive and, amazingly, develop into ostensibly normal adults without further treatments. Testing on a cognitive battery demonstrates some difficulty with acquisition of an active avoidance test and retention of spatial tasks, but on most cognitive, sensory, and motor tasks the *Dbh* knockouts cannot be distinguished from wild-type (129/SvCPJ X C57BL/6J) mice. Nevertheless, the offspring of these *Dbh* homozygous knockout mice fail to survive. These pups are not cleaned after birth, lie scattered throughout the cage or buried in the bedding, and are not fed. Several lines of evidence suggest a selective deficit in maternal behavior:

1. Heterozygous females show full maternal behavior and will successfully rear pups cross-fostered from homozygous females, demonstrating that the deficit is in the homozygous mothers and not in the pups.
2. Homozygous females treated with a synthetic precursor of norepinephrine before (but not the morning after) birth show full maternal behavior. Curiously, these females are able to raise subsequent litters without precursor treatment, suggesting not only that norepinephrine is necessary for the initial

triggering of maternal behavior but that norepinephrine or maternal experience induces neural changes with long-lasting effects, permitting maternal behavior after subsequent pregnancies in the absence of norepinephrine.

3. The *Dbh* knockout also impairs pup care in virgin females; however, the norepinephrine precursor did not rescue pup care in virgin females.

Although the evidence is consistent with a critical role for norepinephrine in the neural reorganization that takes place at parturition, Thomas and Palmiter do not investigate where this change takes place or how norepinephrine influences maternal care (Thomas & Palmiter, 1997). As the *Dbh* knockout mice produce excessive amounts of dopamine, some aspects of the observed phenotype might reflect increased dopamine (previously shown to inhibit maternal care in both rats and mice) rather than decreased norepinephrine.

3.4. Serotonin

3.4.1. Serotonin—Pharmacology and Lesions

Given that serotonin (5-HT) has been implicated in anxiety, affiliation, aggression, and reward, one might expect this to be the obvious candidate neurotransmitter for maternal care (Insel & Winslow, 1998). Yet, as noted above, changes in 5-HT in the maternal brain appear less dramatic than changes in DA or NE. Indeed, there is surprisingly little evidence that 5-HT is important for maternal behavior, with the possible exception of maternal aggression. Barofsky et al. reported that lesions of the median raphe 5-HT neurons disrupted maternal behavior when performed on day 1 of lactation, but overall the effects were transient, nonspecific, and possibly secondary to the surgical intervention rather than the loss of a defined 5-HT pathway (Barofsky, Taylor, Tizabi, Kumar, & Jones-Quartey, 1983). In a more careful pharmacological dissection, De Almeida & Lucion administered 20 μg 8-OH-DPAT (8-hydroxy-2-(di-n-propylamino) tetralin) for 5-HT1a receptors, 100 μg TFMPP (1-(3-trifluoromethyl-phenyl) piperazine hydrochloride) for 5-HT-1b/1d, and 100 μg DOI (1-(2,5-dimethoxy-4-iodophenyl)-2-aminopropane) for 5-HT2 receptors to lactating females (De Almeida & Lucion, 1994). None of the treatments altered pup care. The 5-HT1a and 5-HT2 agonists transiently decreased maternal aggression, but the 5-HT1b/d receptor agonist showed no significant difference as compared to saline. In a follow-up study, De Almeida & Lucion microinjected the 5-HT(1A) receptor agonist 8-OH-DPAT (0.2, 0.5, and 2.0 μg/ml) into the median raphe nucleus, activating the somatodendritic autoreceptor in lactating female rats on day 7 postpartum (De Almeida & Lucion, 1997). In the median raphe (as well as two regions with postsynaptic 5-HT1a receptors—the dorsal periaqueductal gray and in the corticomedial amygdaloid nucleus), 8-OH-DPAT decreased maternal aggression, again without evident effects on other aspects of maternal care. These results are consistent with a number of studies demonstrating that a generalized decrease in 5-HT, as expected after activation of the somatodendritic autoreceptors, decreases aggression. In that sense, the changes in maternal aggression are best considered in the broader context of agonistic behaviors and probably indicate little about other aspects of maternal care or nurturing behavior. Similarly, decreases in maternal nest time reported in 5-HT1b knockout mice are likely secondary to the hyperactivity and impulsivity noted in these mice in several domains (Brunner, Buhot, Hen, & Hofer, 1999). Mice with a null mutation of the 5-HT1a receptor are more fearful, yet show no evident decrease in maternal behavior (Gingrich & Hen, 2001).

3.5. Monoamines—Summary

We have discussed how neuropeptides are neuromodulators, influencing the activity of more rapid and abundant neurotransmitters, such as the monoamines. A simplistic summary of the monoaminergic role in maternal behavior would consider DA as the key to maternal motivation (pup retrieval), NE as critical for maternal memory (lamb recognition), and 5-HT as a mediator of maternal aggression. But how specific are these effects? For instance, mesolimbic DA has been linked to many motivated behaviors. Is there anything specific or unique about DA facilitating pup retrieval or is DA providing a general increase in motivation while other factors link this drive to pups? One answer to this question is that regional changes in neuropeptide modulators might provide this link (Insel & Young, 2001). For instance, estrogen/progesterone–driven increases in oxytocin receptor number in the BST or nucleus accumbens could facilitate local DA neurotransmission, so that a general increase in motivation occurring when pups are present becomes a specific increase in pup-directed behaviors. What we lack are clear studies that demonstrate how monoamines can facilitate maternal behavior. Nearly all of the literature summarized above describes how monoaminergic interventions impair rather than facilitate maternal behavior. In addition, no one has yet looked at how the prolonged steroid changes of pregnancy, specifically the decrease in progesterone following a several-day increase in estrogen and progesterone, influence monoamine content or receptors in the brain. Finally, monoamines are not only modulated by neuropeptides, they are also neuroendocrine factors, powerfully regulating oxytocin and prolactin release. Although we have noted the interaction of neurotransmitter and neuroendocrine functions above, the full complexity of these interactions remains to be investigated.

4. Amino Acids

4.1. Amino Acids—Background

The amino acid neurotransmitters are remarkably abundant in the brain, with 30% of all neurons making the major inhibitory neurotransmitter GABA and an even higher percentage making the major excitatory neurotransmitter glutamate. As noted at the outset, the relationship of neuropeptides to monoamines to amino acids is separated by roughly 3 orders of magnitude, so that there is roughly a million times more GABA or glutamate than oxytocin or prolactin in the brain. Accordingly, just as we asked whether a given neuropeptide synthesized by a few thousand cells in the hypothalamus and released at a handful of central targets is involved in maternal behavior, the question with amino acid neurotransmitters needs a different approach. These neurotransmitters are so abundant and so ubiquitous that it is difficult to imagine how they could not be involved in maternal behavior. The question is not whether they are involved, but where, when, and how. For instance, Lonstein and De Vries (2000c) looked at the number of Fos-immunoreactive cells in the MPOA, vBSTm, and PAG that were also GABAergic. In pup-stimulated lactating dams, more than half of the Fos-positive cells in each region were also positive for GAD_{67}, the synthesizing enzyme for GABA.

4.2. GABA—Pharmacology

In Chapter 4, we described a potential role for the GABAA receptor in the reduced fearfulness and increased aggression of maternal rats. We noted the behavioral characteristics of the lactating rat: less fear, more aggression, hyperphagic, and resembles a virgin female treated with benzodiazepines, a class of drugs that increases GABA inhibitory tone by enhancing anion transport into the cell (Hansen et al., 1985). Treating lactating females with the GABA antagonist bicuculline (60 ng) into the medial basal hypothalamus or amygdala decreases aggression and food intake so that lactating dams resemble virgin females (Hansen & Ferreira, 1986b). As we noted in Chapter 4, the endogenous ligand for the GABAA receptor may include metabolites of progesterone, such as allopregnenalone.

What we failed to mention in Chapter 4 was that there are local circuits where GABA may have rather selective effects on maternal behavior. Stern and Lonstein (2001) have reported one particularly interesting aspect of GABA function in the caudal PAG. One region within the caudal PAG, at the level of the trochlear nucleus, appears important for kyphosis. Cells in this region show an increase in Fos immunoreactivity during nursing. Bilateral lesions of this region reduce kyphosis by 85% (without disrupting pup retrieval, licking, or nest building) and also increase aggression and plus maze exploration (a measure of fearlessness). Reasoning that the PAG is tonically inhibited by GABA and that suckling may block this tonic inhibition to permit kyphosis, Stern and Lonstein administered several GABAergic compounds into the cPAG. After a dam-litter separation of 4 hours, infusions of the GABAA receptor antagonist bicuculline (15 ng/side) prolonged kyphosis over nonsuckling pups, whereas the GABAA receptor agonist muscimol (125 ng/side) reduced kyphosis relative to saline-injected nursing controls. These microinfusions did not affect other aspects of maternal care including retrieval and aggression.

4.3. Amino Acids—Summary

Although there are obvious parallels between the maternal rat and the benzodiazepine-treated virgin rat, we know very little about how or where GABA function is changing with the onset of maternal behavior. We know even less about changes in excitatory amino acid systems and rat maternal behavior. Because changes in GABA and glutamate are integral to neural plasticity and because the onset of maternal behavior involves long-term changes in neural organization, local adaptations in these systems will be important for future studies of the neurobiology of maternal care. We now have excellent tools for investigating both excitatory and inhibitory neurotransmission, so this is an exciting opportunity for studying the cellular changes in the MPOA or other brain regions that are associated with the onset of maternal care.

5. Finding Genes for Maternal Behavior

Is there a gene for parental care? In approaching this question, we need to consider two quite different meanings for the word "gene". Geneticists think of genes as DNA sequences unique to each species and to each individual within a species. Molecular biologists often think of genes as RNA, expressed sequences that are important for building proteins. The difference between these two concepts is easy to see when you consider that every cell in your body has the same DNA. Except for germ line cells (sperm and egg), each cell carries two copies of the full genetic program. However, in each cell only a small percentage, perhaps 20%, of this genetic information is converted into RNA and expressed as protein. The subset of DNA genes that is expressed as RNA and protein gives each cell its individual identity or phenotype. In other words, even though liver cells, brain cells, and blood cells all carry the full, identical DNA "text", each of these cells becomes specialized because different words are "read out" in each case.

Of the 30,000–40,000 genes in the mammalian genome, how would one find a gene for motherhood? As we have seen already, one could go after the known candidate expressed genes by measuring RNAs for oxytocin, prolactin, estrogen receptors, etc. Knockout studies of candidate genes have been informative, but they are limited to the very small fraction of genes that we know. Certainly one would be surprised if after investigating about 0.01% of the genome, we had identified all the key genes. Indeed, more than half of the 30,000–40,000 genes are expressed as RNAs in the brain, many of these will be important for behavior, and most have yet to be named or identified. The classic approach for finding genes (DNA) for behavior in molecular genetics is the linkage study. One studies behavioral variation and sequence variation in a population and looks for a particular pattern of association. The problem with this approach is that there is so much natural variation in both behavior and genetics that one is looking for the proverbial needle in a haystack. In human studies, we reduce the genetic variation by comparing pedigrees, families in which a specific behavioral variant occurs sporadically within a cohort that is closely related genetically. In animal studies, inbred mice are helpful as they have very limited genetic variation. Of course, all of these lines have intact parental care, but they differ predictably in the quality and quantity of their parental behavior.

Strain differences in mouse maternal behavior were recognized nearly 50 years ago (Thompson, 1953) but have only been quantified in recent years (Broida & Svare, 1982; Brown, Mathieson, Stapleton, & Neumann, 1999; Carlier et al., 1982; Cohen-Salmon et al., 1985). Similarly, the preoptic area shows predictable morphologic variation across inbred strains (Mathieson, Taylor, Marshall, & Neumann, 2000). For instance, cells in the medial region of the POA are nearly 50% larger in 129SvEv females compared to C57BL/6J females (Brown, Mani, & Tobet, 1999). What we do not know, at this point, is how much of either the behavioral or neuroanatomical variation can be accounted for by genetic differences in the inbred strains. This question can be answered by cross-fostering experiments, including embryo transfer studies that investigate the effects of prenatal environment on behavior. Such studies are underway (Francis, Szegda, Campbell, Martin, & Insel, in press). Although these will clarify the extent of the genetic versus epigenetic influence on maternal behavior, this approach will not identify which genes are involved.

To identify specific genes that account for strain differences in maternal behavior, one would want to find strains that are identical in most other behaviors and most other genes, but differ selectively in this behavior and in a very few genes. Inbred lines could be created that specifically differ in some aspect of maternal behavior. To our knowledge, this has not been done for maternal behavior, but attempts to breed selectively for related behaviors, such as pup responses to separation, have met with mixed success (Hofer, Shair, Masmela, & Brunelli, 2001). Frequently, selective breeding programs that target a behavioral variation end up selecting for a number of unwanted collateral features.

A more promising approach is to use mutagens that randomly influence only one or maybe two genes in a large number of mice. By careful behavioral screening, mutants with abnormal maternal behavior can be detected. Ultimately, the mutant genes can be identified with positional cloning. There are several large research programs currently using this approach. Although the data are too preliminary to make conclusions, abnormal maternal behavior has not been a key feature of the phenotypes identified thus far. Even if a "bad mother" mouse were identified, the task of positional cloning is formidable. So it is not clear that this approach will deliver a list of genes for maternal care in mice, genes that could be used to screen other species as well.

This brings us back to the candidate gene approach, focusing on RNA. Currently, the most interesting candidates have come from the hormones and neurotransmitters that we suspected to be important for maternal behavior. This "reverse genetics" approach goes from protein to RNA to DNA to search for mutations in gene structure. Earlier in this chapter, when we described the usual suspects and a few unusual candidates for maternal behavior, we neglected to note one candidate that is neither a hormone nor a neurotransmitter. Recall from Chapter 5 that pathways involved in maternal behavior have been mapped using the protein product of the immediate early gene *cfos*, called cFos. We described cFos as a marker of cellular activity, a marker that could identify which cells in the brain were activated by pups, suckling, or pup odors. For a variety of neural pathways, cFos is an excellent marker, but it also has an important function besides helping neuroscientists to map pathways. cFos is one of several proteins in the Fos family (along with fra-1, fra-2, and Fos B) that function as transcription factors, dimerizing with a member of the Jun family of proteins and then binding to AP-1 sites on DNA, short nucleotide segments that activate or repress other genes.

5.1. cFos

Could cFos serve as a marker for genes as well? If we use cFos as a beacon to identify AP-1 sites, wouldn't we identify the genes turned on or turned off by Fos and thus the genes involved in maternal behavior? It's a promising idea, but transcription is a combinatorial process, with many factors activating or repressing genes. So Fos may increase without a change in other genes in the cell. Conversely, not all genes have AP-1 sites and not all cells use cFos as a transcription factor, so there will be plenty of false negatives in following the cFos signal. Nevertheless, it might prove informative. One wholesale attempt is to wipe out Fos and see if maternal behavior disappears.

Surprisingly, mice with null mutations of *cfos* appear behaviorally relatively normal in terms of nonsocial cognitive tasks (Paylor, Johnson, Papaioannou, Spiegelman, & Wehner, 1994). However, to our knowledge, homozygous females have never been tested for maternal care. In a more targeted approach, antisense oligonucleotides against *cfos/c-jun* were administered into the PVN of sheep to decrease local cFos activity (Da Costa et al., 1999). The constructs were incorporated into 50–60% of cells with a resulting decrease in the birth-induced increase in oxytocin concentration in the PVN, but not in blood, a decrease in the birth-induced upregulation of cFos, oxytocin, CRF and preproenkephalin mRNA expression in the PVN. Although all the animals were fully maternal, the antisense treatment did reduce the peak expression of low-pitched bleats and lamb sniffing. These results suggest that Fos is functionally important for the cellular changes that accompany parturition.

5.2. Fos B

A null mutation of the *fos B* gene is of particular interest (Brown et al., 1996). As another member of the Fos family, *fos B* is an immediate early gene that codes for the Fos B protein as well as a truncated form of the protein called ΔfosB. Like cFos, both forms of Fos B function as transcription factors, dimerizing with a member of the Jun family of proteins and then binding to AP-1 sites. Mice with a knockout of the *fos* B gene fail to nurture their young (Brown et al., 1996). Although homozygous *fos* B knockout mice show normal reproductive behavior and unremarkable pregnancies, pups die within 1–2 days of parturition. With homozygous × heterozygous matings, pups continue to die if the mother is homozygous, but not if the father is homozygous, and the genotype of the pups appears unrelated to survival. Pup death thus appears to be due to an absence of maternal care by the mother. Homozygous females are capable of lactation, but they fail to retrieve or crouch over pups. The deficit in maternal care was found in both the 129Sv inbred and 129Sv × BALB/c mixed backgrounds. Brown et al. (1996) found that these homozygous knockout females show no deficits on tests of spatial memory or olfactory function, but appear to have a selective loss of nurturing behavior. In fact, juvenile males and females show a similar deficit in pup retrieval. Mice with deletion of the *fos* B gene have no alteration in plasma gonadal steroid levels, or oxytocin mRNA in the hypothalamus, or prolactin and growth hormone mRNAs in the pituitary. Brown et al. (1996) did not perform the critical experiment—rescuing maternal behavior by replacing the Fos B protein—but they discovered an increase in Fos B in the MPOA of wild-type mice after 6 hours of pup exposure. Based on this observation, they conclude that Fos B, possibly within the MPOA, is essential for maternal care. Of course, there are changes in Fos B (and cFos) in a number of brain regions with pup exposure. The proof that Fos B is functionally important in any of these pathways will require a demonstration that nurturing behavior can be reinstated by local injection of Fos B in the knockouts or that nurturing behavior is inhibited in the wild-type mice by local reductions of Fos B activity.

5.3. Fos Summary

Given that the antisense knockdown of cFos in sheep reduces oxytocin and the transgenic knockout of *fos* B in mice reduces maternal behavior, perhaps we should return to our notion that Fos proteins can be used not just as neuroanatomic markers but as beacons for gene discovery. We may have dismissed this too early. Although mapping AP-1 sites will deliver mostly false positives, we could use cFos in a different way. The microarray or gene chip can identify the profile of mRNA expressed in a given brain region or even in a collection of identified cells after pup exposure, nursing, or similar behavioral intervention. By using this technique in cells that are Fos positive in response to pup exposure, we should soon be able to identify the cohort of genes that are changed in neural circuits that are activated during maternal care.

5.4. Peg1 *and* Peg3

Thus far we have described deficits in maternal behavior following null mutations of a transcription factor and an enzyme for monoamine synthesis. All of these were candidate genes, in that they had been implicated in aspects of neural function and behavior. Deficits in maternal behavior have also been observed following knockouts of two genes that are more mysterious. *Peg1* (also known as *Mest*) and *Peg3* are imprinted genes. Nearly all autosomal genes are inherited as two copies, one from each parent. Only one of these copies is expressed in the offspring, with a 50–50 chance that any given gene will come from the mother and an equal chance that the expressed gene will be from the father. A small family of genes are imprinted, meaning that rather than a 50–50 chance of maternal versus paternal origin, they are only expressed if they come from the mother (for genes that are maternally imprinted) or father (for genes that are paternally imprinted). The gender of the offspring is unimportant in this process. For a maternally imprinted gene, both male and female offspring express only the mother's version (called an allele); the father's allele is silenced. *Peg1* and *Peg3* are paternally imprinted, so only the father's allele is expressed. Both are heavily expressed in the placenta and appear important for fetal growth. Both are also expressed in adults, almost exclusively in the brain.

Females with a knockout of either *Peg1* or *Peg3* have a normal pregnancy rate and deliver at term, but fail to care for their young (Lefebvre et al., 1998; Li et al., 1999). When provided with cross-fostered pups, *Peg3* mutant mothers required 11 times longer to retrieve and 8 times longer to build a nest than wild-type mothers, and the mutant mothers were never observed to crouch over the pups. Both *Peg1* and *Peg3* mutant virgin females also showed deficits in pup retrieval and nest building, so the deficit in maternal care was not restricted to the hormonally mediated changes of parturition. As with the mutant mice described above, these females have no obvious olfactory or motor impairment and show normal mammary gland development postpartum.

The mechanism for the deficit in maternal care is not clear. Keverne (2001) points out that *Peg3* mutant females have a reduced number of oxytocin cells in the hypothalamus, but this seems an unlikely explanation as females with a complete deletion of oxytocin exhibit full maternal behavior (Nishimori et al., 1996). Mutant mice have normal levels of Fos B, although the phenotype of these mice resembles the *FosB* knockout. As both *Peg1* and *Peg3* are expressed in the hypothalamus of adult mice, it seems possible that these genes influence pathways known to be critical for maternal behavior, but it is not known whether these genes are activated in wild-type mice during maternal behavior. As with Fos B, it will be important to perform rescue experiments in the knockout mice to determine where and how these genes are acting. For instance, it is not clear if *Peg1* or *Peg3* are required during adulthood for maternal behavior or if the absence of these genes during development renders the offspring unable to care for pups. Both gene deletions result in placental defects, smaller offspring, and decreased survival. But the effects on maternal behavior are not due completely to an abnormal intrauterine environment. Offspring of mutant males and wild-type females develop in a normal uterine environment, yet show a deficit in suckling behavior as pups and reduced maternal behavior as adults.

It may seem paradoxical that a paternally imprinted gene would be important for maternal behavior. Rather than a paradox, imprinting can be understood as a logical outcome of the conflict between paternal and maternal genomes (Haig & Graham, 1991). According to this "conflict theory", the paternal genes endeavor to obtain maximum resources from the mother, at the expense of future offspring who may have different fathers. The maternal genome wants to minimize investment in any given litter, because maternal genes will be equally apportioned to every litter irrespective of the father. The conflict gets played out most visibly in metabolically costly processes, such as growth and development of the placenta and postnatal feeding of the offspring. In line with this theory, null mutations of paternally imprinted genes (*Peg1*, *Peg 3*, *IGF*) reduce placental growth and result in smaller offspring. The paternal genome seems to code for a big placenta and robust offspring. By contrast, a null mutation of a maternally imprinted gene (*IGF2* receptor) results in increased placental and fetal growth; the maternal genome seems to code for genes that conserve resources.

Keverne has hypothesized that the results of the *Peg1* and *Peg 3* knockouts can be understood in this context. Placental defects, smaller offspring, and reduced suckling by the pups all suggest that these genes are normally important for the maternally costly process of producing offspring—what Keverne calls "maternalism" (2001). But what about the maternal behavior of these offspring? Certainly that doesn't cost their mother anything and the effects are observed in the grandsons and granddaughters, two generations removed from the original source of paternal imprinting. Keverne suggests that these effects of maternal behavior are secondary consequences of changes in maternalism that are first expressed during development. He argues for a curious complementarity in the effects on placental/pup development (largely consistent with the conflict theory) and the effects on maternal behavior, which may have led originally to the fixation of imprinted genes in the population. Specifically, based on the knockout results, *Peg1* and *Peg 3* appear important for suckling in the pups, milk letdown in the dam; placental priming of feeding in pups, and maternal behavior in the dam. Remember there is always a risk in extrapolating from lesion results to normal physiology; these ideas need to be treated as highly speculative. Nevertheless, these kinds of studies provide exciting new ways of thinking about the molecular evolution of parental care and remind us that genes can have complementary effects in mothers and offspring.

6. Conclusions

In the previous chapter, when describing the neuroanatomy of maternal behavior, we focussed on the circuitry that connects the sensory inputs for nurturing to the neural pathways for behavior by linking to centers for reward and for overcoming avoidance. The neurochemistry of maternal behavior appears to follow the same strategy, building on generic pathways with endocrine functions, and via the combined influences of gonadal steroids, pup stimuli, and experience, shaping these for maternal care. As we have seen, neuropeptides appear especially important in this process. Oxytocin and prolactin operate within discrete hypothalamic/limbic circuits and appear exquisitely sensitive to estrogen, suckling, and experience. These are neurochemical systems that are largely "target-regulated", meaning that local changes in receptors confer highly specific changes in the response to endogenous peptides. As estrogen appears to regulate these receptors, one can consider oxytocin (and perhaps prolactin) receptors as transducers of the steroid changes we described in Chapter 2. That is, we have two early steps in a multistep process: The steroid changes of pregnancy lead to local changes in brain neuropeptide receptors. The next steps are less clear. Increased oxytocin and prolactin neurotransmission appears necessary for maternal behavior, but what about increased dopamine release, changes in GABA or glutamate, or activation of *fos B*? Clearly, we are at the beginning of these studies with many leads but few links. Going forward, we will need to consider that (a) neuropeptide systems are notoriously species-specific (Insel & Young, 2000), (b) new candidates will likely emerge from gene discovery tools, and (c) neurotransmitters do not work in isolation, but via a combinatorial process that integrates thousands of signals in each neuron and perhaps millions of neurons for each circuit.

7

Paternal Behavior

1. Introduction

Male behavioral responses to offspring range from avoidance to infanticide to nurturing care. If this book were focussed on parental care among nonmammalian vertebrates, most of the preceding chapters would be about paternal rather than maternal care. In teleost fish, for example, males are more likely than females to guard the nest and transport newly hatched young (Blumer, 1979; Crawford & Balon, 1996). In some species, males have evolved specialized adaptations for parental care. Male seahorses (*Hippocampus* species) have a brood pouch for protecting and carrying infants and, in a piscine precursor to lactation, the males of many species exude a nutritional mucous from their skin on which the young feed until they become independent (Blumer, 1982). But even in species with less visible paternal anatomy, the male rather than the female fans and brushes the eggs while protecting them from predators. Among amphibia, there are extraordinary cases of males provisioning the young. For instance, in spotted poison frogs (*Dendrobates vanzolinii*), it is the male who transports the single tadpole to a tiny pool, then vocalizes to attract his mate to this pool where she releases unfertilized eggs for the tadpole to devour (Caldwell, 1997).

Paternal care is also the rule rather than the exception in avian species (reviewed by Ketterson & Nolan, 1994). Over 90% of bird species are biparental, with males sharing a range of responsibilities including nest building, brooding, and feeding the young. In many species, males as well as females produce crop milk, which in both genders is regulated by prolactin. In many species of seasonally breeding birds, parental care follows the breeding season. This is an important area of behavioral neuroendocrinology that has been reviewed elsewhere (Buntin, 1996). There are a number of species differences, but as a general rule, testosterone decreases at the end of the breeding season and prolactin increases with the onset of paternal care (Wingfield, Hegner, Dufty, & Ball, 1990). Testosterone may inhibit (Schoech, Ketterson, Nolan, Sharp, & Buntin, 1998) and prolactin facilitates paternal care, especially regurgitation feeding of the young (Buntin, Becker, & Ruzycki, 1991; Lehrman, 1961). Le-

246

sions of the preoptic area disrupt prolactin-induced regurgitation feeding in both male and female ring doves (*Streptopelia risoria*) (Slawski & Buntin, 1995).

As we noted in Chapter 1, mammals are defined by lactation. Other than male Dayak fruit bats (*Dyacopterus spadiceus*), which reportedly lactate (Francis, Anthony, Brunton, & Kunz, 1994), the provisioning of milk to neonates is the essential and exclusive role of the female. Nevertheless, in most mammalian taxa, there are males with extensive investment in the young, extensive enough to be considered paternal care. Kleiman and Malcolm define paternal care as "any increase in a pre-reproductive mammal's fitness attributable to the presence or action of a male" (Kleiman & Malcolm, 1981). Although this definition depends on the effect rather than the character of the behavior, Kleiman and Malcolm describe a broad range of activities as paternal, some direct (feeding, carrying, and grooming) and some indirect (nest building, defense, support of pregnant female). A commonly accepted earlier definition by Trivers required that paternal care also exact a cost to the male, such as a reduction in the ability to invest in other offspring (Trivers, 1972). Paternal care usually involves a cost, but we agree with Kleiman and Malcolm that Trivers's definition may prove too restrictive in practice. For the purpose of this review, which emphasizes neuroendocrine features of paternal care, we will adopt the Kleiman and Malcolm definition.

2. Evolutionary Mechanisms for Paternal Care

Why would male mammals provide parental care? Whereas female reproductive success is limited mainly by time and energy constraints and benefits directly from ensuring survival of offspring (Trivers, 1972), one might expect that a mammalian male's reproductive success would be maximized by mating with multiple females and leaving the lactating female to raise his young. In fact, this appears to be the strategy in most mammalian species. But in perhaps 6% of mammalian species, males shift from reproductive behavior to paternal care, investing in one female and her offspring (Kleiman & Malcolm, 1981). Direct paternal care is most common in primates, with nearly 40% of primate genera showing biparental support of young (Kleiman & Malcolm, 1981). Carnivores and rodents also include a number of species with altricial young and paternal care. Although there are exceptions, paternal care is generally associated with monogamous social organization. One might assume in pair-bonded, monogamous species that the male improves his reproductive success by caring for young that are his own. Perhaps this assumption is shared by the paternal male, but genotyping the pups in socially monogamous species often reveals that the offspring are the result of an extra-pair copulation. Indeed, from the last decade of studies on mating systems and social organization one might conclude that biparental care is more characteristic of monogamy than any specific mating system. Note, however, that mating systems, social organization, and parental care are labels for behaviors within populations. All of these vary not only across individuals but for the entire population with changes in season, population density, or ecological conditions. Although mammalian paternal care is generally associated with monogamy, males of nonmonogamous species will exhibit "facultative" paternal care following prolonged exposure to pups (Storey & Joyce, 1995) or under unfavorable breeding conditions (e.g., environmental stress) (Parker & Lee, 2001b). It may be useful therefore to consider mammalian paternal care within the broader rubric of cooperative breeding, with prereproductive males and females as well as a reproductive male providing alloparental care.

In general, paternal care is observed when two parents are required for provisioning and/or defending the young. Brown has called this "obligate" (as opposed to facultative) paternal care (Brown, 1993). For instance, in many New World monkeys that regularly deliver twins, the infants represent a metabolic demand that would be a considerable challenge to a single mother. For black tufted-ear marmosets (*Callithrix kuhlii*) or pygmy marmosets (*Cebuella pygmae*), both small, tree-dwelling primates, neonates comprise roughly 25% of the mother's body mass (Nunes, Fite, Patera, & French, 2001; Wamboldt, Gelhard, & Insel, 1988) (Figure 7.1). As they grow and continue to require fulltime carrying, both fathers and siblings become important for alloparental care. Although several studies in birds demonstrate that male presence increases offspring survival (reviewed in Bart & Tornes, 1989), there are relatively little data in biparental primates or rodents to prove the importance of paternal care. Studies in biparental Califonia mice (*Peromyscus californicus*) (Gubernick & Teferi, 2000) as well as Djungarian hamsters (*Phodopus campbelli*) (Wynne-Edwards & Lisk, 1989) have demonstrated that offspring survival is enhanced by the presence of fathers, but this effect in the laboratory appears to depend on environmental conditions and largely results from solo females attacking newborn pups (Gubernick, Wright, & Brown, 1993b). Some have suggested that male parental care is used to reduce female aggression (Maestripieri & Alleva, 1991). Curiously, one comparative study of nonhuman primates has reported greater longevity in males of biparental species compared to males of uniparental species (Allman, Rosin, Kumar, & Hasenstaub, 1998).

Whatever the adaptive value, rodent paternal care can be extensive. Although it is not observed in adult common laboratory rats (*Rattus norvegicus*) and mice (*Mus musculus*), in many naturally occurring rodents, such as prairie voles (*Microtus ochrogaster*), California mice, Djungarian hamsters, and mongolian gerbils (*Meriones unguiculatus*), males show virtually all of the species-typical behaviors associated with maternal care except lactation. Indeed, male prairie voles will crouch over pups in the kyphotic posture of nursing (Lonstein & De Vries, 1999b). And male Djungarian hamsters have been reported to assist in the delivery of pups by tearing away the membranes just after birth (Jones & Wynne-Edwards, 2000). It seems likely, however, that paternal care is not simply male maternal behavior. In biparental species, males may have a unique role that not only optimizes offspring survival but also is critical for maintenance of the pair bond.

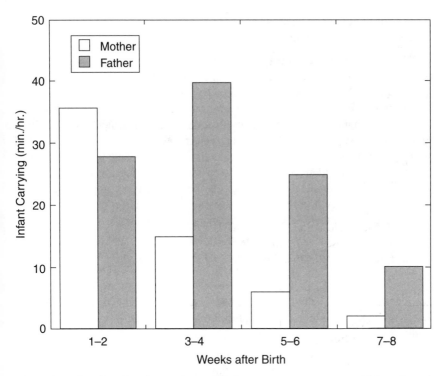

Figure 7.1. Carrying time by mother and father marmosets: rates at which male and female parents of black tufted-ear marmosets (*Callithrix kuhlii*) carry infants from twin litters during first 8 weeks postpartum. Marmoset family groups were studied for 1-hour observation periods. Not shown: Siblings also carried infants at a low rate in weeks 1–2, but at a rate equivalent to the mother during weeks 3–8. Data adapted from Nunes et al. (2001).

3. Proximate Mechanisms of Paternal Behavior

What factors regulate paternal care in mammals? Brown (1993) and Dewsbury (1985) have developed a list of factors. Here we will reorganize these into three major factors: social organization (described above), experience (with pups, females, and ecology), and neuroendocrine factors (gonadal hormones and neuropeptides). In light of earlier chapters, the two questions that will concern us here are:

1. Are the triggers for paternal care different from the triggers for maternal care? Clearly, males do not go through the endocrine changes of pregnancy and parturition, but other endocrine changes may be necessary or sufficient for paternal behavior.

2. Are there important neuroanatomic or neuroendocrine differences between species that are paternal and those that are not?

We will address these questions in three disparate forms of paternal care. First, we will review studies of biparental mammals to investigate whether there are experiential or neuroendocrine bases to paternal care and, if so, how these compare to what we have reviewed in earlier chapters for maternal care. A second line of inquiry will focus on juvenile male rats, which exhibit intense interest in pups. Although this may be more related to play than adult paternal care, the investigation of the onset and offset of this behavior should provide clues to factors regulating male interest in pups. Finally, there is a fascinating literature on infanticide. This subject has been reviewed comprehensively by Hrdy (1999) and will only be mentioned here in reference to studies in mice that lose their infanticidal behavior 21 days postcopulation.

3.1. Experiential and Neuroendocrine Factors in Biparental Species

To investigate experiential and neuroendocrine factors influencing paternal behavior in adults, we will look at five biparental mammalian species: prairie voles, California mice, Djungarian hamsters, Mongolian gerbils, and marmosets. Each of these is quite different, not only in the behaviors exhibited but in the neuroendocrine correlates of paternal care. The prolonged increases in gonadal steroids and lactogenic hormones and the vaginocervical stimulation of parturition described for females in earlier chapters appear critical for maternal behavior. Although there is no homologous experience in males, paternal care often occurs concurrently with mating behavior (recall that many rodent females exhibit postpartum estrus). Mating is not parturition, but these events share certain neuroendocrine responses (such as oxytocin and vasopressin release) (Murphy, Seckl, Burton, Checkley, & Lightman, 1987). Although it is possible that hormonal events associated with mating are important for paternal care, we have only a few studies that investigate a neuroendocrine basis for mammalian paternal care.

3.1.1. Prairie Voles

Paternal behavior in prairie voles has been investigated both in field and laboratory settings. Unlike males of many other biparental species, male prairie voles of all ages exhibit a keen interest in pups. As noted above, male and female care appear similar in this species except that inexperienced females are more likely than males to be infanticidal and only become parental late in gestation, before the onset of lactation (Lonstein & De Vries, 1999a). Both sexes participate in nest building, grooming, and licking, but curiously neither shows extensive retrieval (although also see Oliveras & Novak, 1986). For prairie voles in a laboratory setting, experience with pups seems surprisingly unimportant. Males are paternal without developmental or recent experience and even after prolonged periods of isolation (Lonstein & De Vries, 2000a), although in one study experience with pups increased the percentage of males showing paternal behavior (Roberts, Miller, Taymans, & Carter, 1998a).

As we have noted earlier, one attractive aspect of studying prairie voles is the availability of several congeneric species, such as montane voles (*Microtus montanus*) and meadow voles (*Microtus pennsylvanicus*), that are uniparental (McGuire & Novak, 1984; Olivera & Novak, 1986). Comparisons between these species may reveal important correlates of paternal care. Recall that by cross-fostering pups from uniparental meadow voles to biparental prairie voles, McGuire demonstrated the role of rearing conditions in the expression of parental care (1988). Meadow vole pups fostered to prairie vole parents receive considerably more paternal care than pups reared by meadow voles. These cross-fostered meadow voles exhibit increased parental care as adults relative to meadow voles reared by meadow vole parents. However, cross-fostering appears to affect females more than males (reviewed in Wang & Insel, 1996). (Figure 7.2). It appears that males are less subject to experiential or developmental influences on parental care.

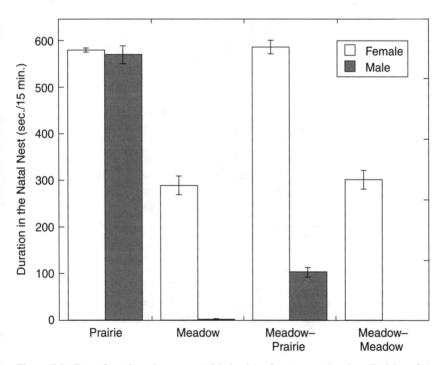

Figure 7.2. Cross-fostering alters parental behavior of nonparental voles. Prairie voles have higher levels of maternal and paternal behavior than meadow voles. When meadow voles are reared by prairie voles, both female and male parental care is higher in adulthood than meadow voles reared by meadow voles. Graph from Wang and Insel (1996).

What hormonal factors facilitate paternal behavior in prairie voles? Although an initial paper described a decrease in paternal care one month following castration (Wang & De Vries, 1993), subsequent studies (Lonstein & De Vries, 1999a) from the same group have found no evidence that castration either at weaning or during adulthood influences paternal care. The difference may be due to differences in the rate of paternal care in this colony, as later studies found nearly all males to be highly paternal (De Vries, personal communication). Administration of the glucocorticoid/progesterone receptor antagonist RU 486 (1 mg/day subcutaneously for 4 days) did not reduce paternal care. In addition, prenatal or postnatal treatment with an antiandrogen (flutamide) or with the aromatase inhibitor 1,4,6–androstatriene–3.17-dione (ATD) failed to alter subsequent male parental care (Lonstein & De Vries, 2000b). If paternal behavior were organized by early androgen or estrogen exposure, one might have expected these treatments to either decrease interest in pups or confer a feminized pattern of behavior, including infanticide.

To determine if prolactin influences paternal care, Lonstein and De Vries (2000a) administered the dopamine agonist bromocriptine (0.4 mg/day for 10 days) to adult male prairie voles to lower their prolactin levels. This treatment had no effect on paternal behavior, suggesting that prolactin was not essential for paternal care. Khatib et al. (2001) compared prolactin and prolactin receptors in adult virgin male and female prairie voles exposed daily to pups. Plasma prolactin was higher in males than females prior to exposure. There was no gender difference in the primary measure of paternal care (latency to contact pup). Following pup exposure, both prolactin and prolactin receptor mRNA increased in females; no change was evident in males.

In an earlier set of studies, De Vries and his colleagues investigated the role of vasopressin in male prairie voles. In many mammals, there is a dense plexus of vasopressin fibers in the lateral septum that is found only in males. In male prairie voles, this plexus becomes less densely stained and vasopressin mRNA in the cell bodies becomes more abundant just after mating. (Bamshad, Novak, & De Vries, 1994). The combination of reduced staining in terminals and increased AVP synthesis in cell bodies may indicate release of vasopressin with mating. Because of data suggesting that males become more paternal after mating (Bamshad et al., 1994), Wang et al. reasoned that vasopressin in the lateral septum might be important for paternal care (Wang, Ferris, & De Vries, 1994). Indeed, injections of vasopressin (0.1 ng, but not 0.01 or 3.0 ng) directly into the lateral septum increased while the vasopressin V1a receptor antagonist $d(CH_2)_5Tyr(Me)AVP$ (1.0 ng) reduced paternal care in virgin prairie voles tested for 10 minutes with a single pup in a novel cage. AVP increased the time spent grooming, contacting, and crouching over the pup, but had no effect on retrieval. Vasopressin gene expression increases in both males and females postpartum (Wang, Liu, Young, & Insel, 2000). Vasopressin even increases paternal interest in meadow voles that are not monogamous (Parker & Lee, 2001a). Indeed, relative to nonpaternal meadow voles, paternal meadow voles have more AVP (V1a) receptor binding in the anterior olfactory nucleus, less in the lateral septum, but no difference in the medial amygdala (Parker, Kinney, Phillips, & Lee, 2001). These results appear to parallel the evidence that oxytocin is important for maternal care—consistent with extensive evidence that vasopressin is more effective for male behavior and oxytocin for female behavior (Insel & Young, 2000). However, the story is not quite so simple. Recent studies demonstrating that castration markedly reduces vasopressin in the lateral septum without altering paternal care demonstrate that vasopressin in this region is not essential for paternal behavior (Lonstein & De Vries, 1999a). In another recent study, vasopressin mRNA was found to increase in both male and female prairie voles just after the birth of their litters (Wang et al., 2000). But this increase may be the result, not the cause, of paternal behavior. Vasopressin mRNA increases in response to an osmotic challenge. Wang has demonstrated that anogenital grooming of the pups results in both males and females ingesting hypertonic pup urine (Wang et al., 1997). It is certainly possible that the osmotic challenge of ingesting hypertonic urine contributes to the increase in vasopressin mRNA.

Are the same brain regions involved in paternal and maternal care? To investigate neural pathways activated by pup stimuli, Kirkpatrick et al. compared Fos protein staining in male prairie voles exposed for 3 hours to pups versus a nonsocial olfactory stimulus (Kirkpatrick, Kim, & Insel, 1994b). Pups activated Fos in the accessory olfactory bulb, medial nucleus of the amygdala, and medial preoptic area, among several other regions. Females showed a similar pattern of activation when exposed to pups, although activation in the medial preoptic area was greater in males than in females. In a separate study, Kirkpatrick et al. lesioned cells within the medial nucleus of the amygdala with n-methyl-d-aspartic acid and found a selective loss of paternal behavior without a reduction in other aspects of social interaction (Kirkpatrick, Carter, Newman, & Insel, 1994a). Lesions of the MPOA have not been described in prairie voles.

In summary, male prairie voles are highly parental, apparently unrelated to circulating or early exposure to gonadal steroids. Although there is some evidence for vasopressin involvement, the results are not consistent. Prolactin appears not to be essential in males. Whatever the neuroendocrine basis, if any, the medial amygdala and its projections (MPOA via the BST) appear critical.

3.1.2. California Mice

Peromyscus californicus, the California mouse, is one of the most carefully studied biparental rodents (Gubernick, 1988; Gubernick & Alberts, 1987). Unlike prairie voles, the virgin male California mouse is infanticidal rather than paternal. Some males (34%) exhibit increased paternal interest after copulation, but most continue to be infanticidal throughout the female's gestation (Gubernick, Schneider, & Jeanotte, 1994). Virtually all males become paternal after the female delivers. This process on the surface is reminiscent of the onset of maternal behavior in rats. However, the onset of paternal care is contingent on the presence of the female (Gubernick & Alberts, 1989). If the female and pups are removed, males revert to infanticidal behavior. It is likely that the female induces the male's cooperation via an olfactory cue, as even if she is kept behind a screen barrier in the cage, the male will continue to provide paternal care. In this species, male parental care resembles maternal behavior; however, females do most of the anogenital grooming, possibly an important adaptation to replace fluids lost from nursing (Gubernick, 1988). As noted above, field studies have demonstrated that nest sites lacking a father have reduced survival of pups (Gubernick & Teferi, 2000).

One might expect that the induction of paternal care in this species would provide an ideal opportunity to define the neural correlates of paternal behavior. In fact, relatively little is known. Testosterone does not change in males exposed to pups (Gubernick & Nelson, 1989). Castration reduces and androgen treatment increases paternal behavior without altering aggressive behavior (Trainor & Marler, 2001). Gubernick and Nelson reported a 35% increase in prolactin on postpartum day 2, but the baseline levels in this study were roughly 50-fold higher than in other rodents, raising a question about the specificity of the radioimmunoassay for prolactin (Gubernick & Nelson, 1989). Vasopressin has not been studied pharmacologically, but comparative studies have described more vasopressin fibers in the BST or more vasopressin receptors in the lateral septum of this species than male *Peromyscus leucopus* or *Peromyscus maniculatus*, which fail to provide paternal care (Bester-Meredith, Young, & Marler, 1999; Insel, Gelhard, & Shapiro, 1991). In one study, plasma oxytocin was described as increased during pregnancy but not during the postpartum period in male mice (Gubernick, Winslow, Jensen, Jeanotte, & Bowen, 1995). The MPOA is larger in virgin males relative to virgin females (Gubernick et al., 1993a). Following parturition, this difference is not evident due to an increase in the MPOA soma size in females (Gubernick et al., 1993a).

3.1.3. Djungarian Hamsters

The Djungarian hamster (*Phodopus campbelli*) is biparental and monogamous. As with the prairie vole, a closely related species, the Siberian hamster (*Phodopus sungorus*), is a useful uniparental congener for comparative studies (Wynne-Edwards & Lisk, 1989). Jones and Wynne-Edwards have recently described how males assist with the birth process, tearing fetal membranes from the pups, ingesting amniotic fluid, and eating the placenta (Jones & Wynne-Edwards, 2000). Males participate in later aspects of parental care, including both direct (grooming, carrying) and indirect (nest building, defense) behaviors. Females raising young without a male present are hyperthermic and may therefore spend less time on the nest (Walton & Wynne-Edwards, 1997).

There has been relatively little investigation of the neuroendocrine substrates of paternal care in this species. Reburn and Wynne-Edwards reported that testosterone is higher, prolactin lower, and cortisol equivalent in virgin male Djungarian hamsters relative to Siberian hamsters. In the former (but not the latter), testosterone increases in the male just prior to delivery of pups, then decreases the day after the pups are delivered, rising again 5 days later (Reburn & Wynne-

Edwards, 1999). At this latter time (but not before or after), prolactin is increased. Beyond these correlations, it is not clear if changes in either testosterone or prolactin are important for any aspect of paternal behavior.

3.1.4. Mongolian Gerbils

The development of paternal care in Mongolian gerbils (*Meriones unguiculatus*) is different from the examples cited above. Both virgin males and females are infanticidal, although male gerbils are less aggressive than females. Males become less infanticidal about 2 weeks prior to delivery, but this change requires direct exposure to the female. With the first litter, males leave the nest on the day of birth. Clark and Galef have shown that the male's absence from the nest is not due to the female's behavior but to an active avoidance of pups (Clark & Galef, 2000). The absence is transient, as males appear paternal by postpartum day 3 and continue to show paternal care thereafter. Curiously, although males are persistently parental, females revert to infanticidal behavior.

Brown et al. investigated the hormonal changes associated with reproduction and parental care in gerbils 1, 10, and 20 days after pairing and then 3, 10, and 20 days after pups were born (Brown, Murdoch, Murphy, & Moger, 1995). Androgen (testosterone and 17alpha-hydroxy-androgen) levels rose at 10 and 20 days postmating, then dropped and remained low postpartum. Prolactin rose significantly only at 20 days postpartum, relative to an unmated group of males. During the postpartum phase, pups were removed for 4 hours and then returned to either male or female parents. Plasma prolactin increased in females with pups returned, but neither prolactin nor androgen differed between males with or without pups. Although the data do not support a role for prolactin, several lines of evidence suggest that testosterone may inhibit gerbil paternal care. Castrated adult males are more attentive to young than are intact males, and replacing testosterone in castrated males reduces their interaction with young (Clark & Galef, 1999). In gerbils, Clark and Galef have described a naturally occurring increase in circulating testosterone peaking on day 75 (Clark & Galef, 2001). Males at this age show less paternal behavior than younger or older males (Clark & Galef, 2001).

3.1.5. Marmosets

Neuroendocrine studies of paternal care have been described in three New World biparental primates: common marmosets (*Callithrix jacchus*), cotton-top tamarins (*Sanguinus oedipus*), and black tufted-ear marmosets (*Callithrix kuhlii*). All of these species are small, arboreal primates frequently giving birth to twins that represent a considerable metabolic demand. Paternal behaviors differ in subtle ways between these species, but the general pattern includes direct paternal care (carrying infants, extensive grooming) from the day of birth. Cotton-top tamarin males carry as much as females in the first week (Ziegler & Snowdon, 2000) whereas in the other species males begin carrying more than females only after the first 2 weeks postpartum (Nunes et al., 2001; Wamboldt et al., 1988).

Does alloparental experience contribute to paternal care in nonhuman primates? As reviewed by Snowdon, the evidence supporting the importance of prior infant care in marmosets is slim (Snowdon, 1996). Indeed, Roth and Darms have argued that marmosets do not need direct exposure to infants as long as they can develop general social competence (Roth & Darms, 1993). In tamarins, with more immediate paternal care, males appear to benefit from prior experience (Johnson et al., 1991).

Dixson and George were the first to describe an increase in plasma prolactin in 5 male common marmosets housed with their mates and infants (10–30 days old) compared to males housed with either pregnant (n = 5) or nonpregnant (n = 5) females (Dixson & George, 1982). Daily blood samples were taken, with paternal males carrying infants just prior to at least 60% of the samples. Mean plasma prolactin levels for paternal males were 5.3 ± 0.5 ng/ml versus 0.84 ± 0.07 ng/ml for males with pregnant females and 1.08 ± 0.08 ng/ml for males with nonpregnant females (Figure 7.3). There was no overlap between the means of the males with infants and the means of males housed without infants. In paternal males, values were higher on days when infants were being carried immediately before the blood draw (6.9 ± 0.6 ng/ml) compared to days when the infants were not being carried (2.3 ± 0.3 ng/ml). The high values in the paternal males were comparable to prolactin levels in females (means of 3.9,

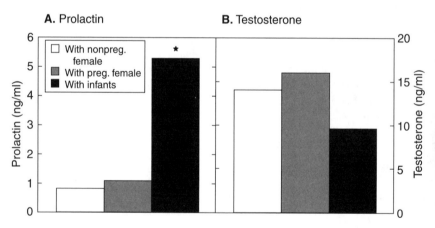

Figure 7.3. Prolactin is increased in male marmosets with infants. Male marmosets (*Callithrix jacchus*) were studied in captive family groups with nonpregnant females, pregnant females, or females with 10–20-day-old infants. Plasma samples were collected from 5 males in each group. Males with infants had significantly higher prolactin but no difference in testosterone relative to other groups. In a further analysis, plasma prolactin was 3-fold higher on days when males were carrying infants than when infants were not in contact with males. Data adapted from Dixson and George (1982).

6.3, and 30.7 ng/ml for nonpregnant, pregnant, and lactating, respectively). Testosterone did not differ in any of the male groups and there was no correlation of testosterone and prolactin. Prolactin may have been influenced by sexual behavior in males housed with postpartum females, but in 5 males sampled before and after mating, Dixson and George found no change in prolactin. Because prolactin increases with stress, one explanation for this finding was that males with offspring would be more stressed by capture and venipuncture than males housed without offspring. Dixson and George had habituated their subjects to the blood draw procedure, but it is possible that males with offspring were more responsive. This confound appears to be eliminated by a subsequent study that used urinary prolactin, collected without capture or stress. In this case, prolactin increased during the prepartum period, but only in males with parental experience (Ziegler & Snowdon, 2000). This second study also found no change in testosterone in males with offspring. Finally, bromocriptine (0.5 mg daily for 3 days) reduced prolactin and decreased retrieval or carrying in both male and female marmosets (Roberts et al., 2001a).

In black tufted-ear marmosets, urinary testosterone and estrogen decrease during the prepartum period, then return to normal levels following parturition. Nunes et al. (2001) reported that the postpartum testosterone level was inversely correlated to carrying time. Prolactin has not yet been assayed in these fathers.

Cotton-top tamarins differ from marmosets in that males are involved from the first day postpartum. In spite of this extensive immediate paternal care, tamarins show little evidence of a change in either prolactin or testosterone during the early postpartum period (Ziegler & Snowdon, 1997, 2000). Although prolactin increased in females postpartum, in males there was no significant change in prolactin, testosterone, dihydrotestosterone, or cortisol across the first 15 days postpartum relative to prepartum levels. The absence of an effect could be attributed to altered hormone levels prepartum. Indeed, in this study, just as in the earlier study with common marmosets, experienced males had higher prolactin levels prepartum compared to inexperienced males matched for age (Ziegler & Snowdon, 2000). This seems an insufficient explanation of the negative findings postpartum, however, as there was no difference in postpartum levels between males exposed to infants and those who lost their infants (Ziegler & Snowdon, 2000).

3.1.6. Summary of Experiential and Neuroendocrine Factors in Biparental Species

Based on the foregoing, what can be said about the experiential and neuroendocrine factors for paternal care in biparental species? How do biparental males compare to uniparental females? In terms of experience, we noted in Chapter 3 that female rats benefit from either juvenile exposure to pups or earlier adult experiences of maternal care. In biparental males, there is no single generalization for the role of experience. Male prairie voles appear paternal without experience. Although juvenile alloparenting has been reported to increase the number of prairie voles with paternal care (Roberts et al., 1998a), at least two-thirds show this behavior without previous experience. Male California mice, gerbils, and hamsters become less infanticidal either after copulation or later after the female delivers. In the latter case, the onset of paternal care may depend on cues from the lactating dam. The role of juvenile alloparental experience is not clear in these species. In biparental primates, the importance of juvenile experience appears to vary from critical in tamarins with immediate paternal care postpartum to useful in marmosets in which females provide the bulk of infant care for the first 2 weeks postpartum (reviewed in Snowdon, 1996).

Are the same hormonal factors influencing males and females to support nurturing behavior? Clearly, changes in gonadal steroids are not critical factors for males (see Table 7.1). Testosterone is not a factor in prairie voles, California mice, or (in most studies) marmosets. Modest decreases in testosterone have been reported postpartum in Djungarian hamsters and Mongolian gerbils, but there is only correlational evidence that this change in testosterone is related to paternal behavior. The results with prolactin are underwhelming, with the singular exception of the marmoset study by Dixson and George (see Table 7.2). After 20 years, this finding remains the strongest evidence that prolactin is increased when males are with offspring. Studies of prolactin changes in biparental rodents have found very modest or negative effects and there is no evidence that decreasing prolactin reduces paternal behavior (see, for instance, Lonstein

Table 7.1. Testosterone (T) in biparental males.

Species	Findings	References
Prairie Vole	No effect of castration, prenatal, postnatal testosterone, antiandrogen, aromatase inhibitor. In early study, castration reduces and T reinstates paternal care.	Lonstein & DeVries, 1999, 2000; Wang & DeVries, 1993
California Mouse	No difference in T between paternal and nonpaternal males. Castration decreases and T increases paternal behavior.	Gubernick & Nelson, 1989; Trainor & Marler, 2001
Djungarian Hamster	T higher in biparental versus uniparental hamsters. T increases just prior to delivery, decreases postpartum day 1 and increases postpartum day 5.	Reburn & Wynne-Edwards, 1999
Mongolian Gerbil	T and 17alpha-hydroxy-androgen increase prepartum then decrease postpartum. Castration increases and T decreases paternal care. Paternal behavior decreased at age 75 days when T peaks.	Brown et al., 1995; Clark & Galef, 1999, 2001
Marmosets	In *C. jacchus*, no difference in T between males carrying infants and males without infants. In *C. kuhlii*, T inversely correlated to carrying time.	Dixson & George, 1982; Nunes et al., 2001
Tamarins	No change in T or dihydroT postpartum.	Ziegler et al., 2000

Table 7.2. Prolactin (Pro) and Prolactin Receptor (PR) in biparental males.

Species	Findings	References
Prairie Vole	No effect of bromocriptine on paternal care. P and PR higher in virgin males than females; PR increases in females but not males with pup exposure.	Lonstein & DeVries, 2000; Khatib et al., 2001
California Mouse	35% increase in Pro on postpartum day 2.	Gubernick & Nelson, 1989
Djungarian Hamster	Pro lower in biparental versus uniparental hamsters. Pro increases postpartum day 5.	Reburn & Wynne-Edwards, 1999
Mongolian Gerbil	Pro increases only on postpartum day 20. Pup exposure increases Pro in females but not males.	Brown et al., 1995
Marmosets	In *C. jacchus*, Pro higher in males carrying infants relative to males without infants. Bromocriptine decreases carrying time.	Dixson & George, 1982; Roberts et al., 2001
Tamarins	Pro increases in females but not males; however, experienced males have higher Pro than inexperienced males prepartum.	Ziegler et al., 2000

& De Vries, 2000a). Given the evidence for oxytocin and maternal behavior in rats, vasopressin seemed an excellent candidate for the mediation of paternal behavior. Although early results support a role for vasopressin in the lateral septum in prairie vole paternal behavior, more recent results from the same lab have shown that castrated males with no detectable vasopressin in the lateral septum remain highly paternal (Lonstein & De Vries, 1999a).

What about the neural circuitry for paternal care? Here we have little to work with, but the available evidence from biparental species suggests that the neurobiology of paternal care does not resemble the neurobiology of maternal care. A lesion of the medial amygdala in male prairie voles decreases paternal care (Kirkpatrick et al., 1994a). Recall that in female rats, lesions of this region facilitate the onset of maternal care (Fleming et al., 1980). Why do lesions of this nucleus facilitate maternal behavior in rats while inhibiting paternal behavior in voles? The answer is not known. It is best to consider the medial nucleus in the context of related nuclei in a circuit for processing social information, such as the BST and the MPOA. Unfortunately, we do not have any data on the role of the MPOA in paternal care (although see below for relevant studies in rats). Following pup exposure to adult prairie voles, Fos staining increases in this region in males more than females (Kirkpatrick et al., 1994b), but there is, as yet, no evidence that this region is critical for paternal care in mammals, as previously shown in birds (Slawski & Buntin, 1995).

In summary, in biparental species paternal behavior resembles maternal behavior, but we know very little about its neuroendocrine mechanisms. This should be a rich area for investigation, particularly in species like the California mouse with a clear transition from infanticide to paternal care.

3.2. Nurturing Behavior of Male Rats

Rattus is not an ideal genus for the study of paternal care. In contrast to the monogamous species described thus far, adult rats do not form pair bonds and do not participate in care of the young. Nevertheless, rats may be instructive for studying paternal care in two ways. First, by suggesting the factors necessary to induce males to exhibit parental care, rats provide an important study in the sexual dimorphism of parental behavior. Second, juvenile rats are intensely interested in pups (male juveniles more than female juveniles). Although this behavior may not be entirely homologous to adult parental care, defining the factors that transform this interest into an avoidance of pups may prove useful for understanding the physiological requirements for parental care.

Paternal behavior in rats requires an inhibition of infanticide and an initiation of nurturing behaviors, such as nest building, retrieving, grooming, and crouching over pups. Obviously, the latter requires the former. Both can be accomplished in adult male rats by prolonged exposure to pups. In one study of Long-Evans rats, infanticide was reduced after copulation and cohabitation with a female (Brown, 1986). Is infanticide related to testosterone? Castration before sexual maturity reduces infanticide and is associated with more rapid acquisition and higher levels of paternal behavior (Rosenberg, 1974; Rosenberg, Denenberg, Zarrow, & Frank, 1971; Rosenberg, & Herrenkohl, 1976). Neonatal castration appears most effective and castration after day 70 has weak effects. Consistent with the effects of neonatal castration, prenatal stress, which has several "demasculinizing effects," also facilitates parental behavior in males (with opposite effects in females) (Kinsley & Bridges, 1988b). Taken together, the effects of neonatal castration and prenatal stress suggest that developmental processes that masculinize the male rat brain increase infanticide and decrease the likelihood for observing male parental care.

The role of circulating hormones in adulthood is much less clear. Although castration of adult males generally has weak effects, testosterone treatment increases infanticide (Rosenberg, 1974; Rosenberg & Herrenkohl, 1976). The effects of testosterone in adult male rats are modified by experience. That is, once males have accepted pups, testosterone does not induce pup killing (Rosenberg & Sherman, 1975).

Beyond the inhibition of infanticide, what factors stimulate the onset of paternal behavior in adult rats? There have been at least 8 studies comparing the latencies of adult males and females to be sensitized to pups (reviewed in Lonstein & De Vries, 2000a) with most of these showing either fewer males than females responding, females responding more quickly than males, or inconsistent responses in males (see for instance Samuels & Bridges, 1983). Administration of gonadal steroids can facilitate paternal responses in males, but these treatments require higher doses in males than in females (Lubin, Leon, Moltz, & Numan, 1972; Rosenblatt, Hazelwood, & Poole, 1996). For instance, Rosenblatt et al. reported that castrated males given estradiol 17-*B* (days 1–16) and progesterone (days 3–15) responded to estradiol benzoate 100 µg/kg with latencies of 2.5 days, whereas females given 5 μg/kg had a latency of 1 day (Mayer et al., 1990b; Rosenblatt et al., 1996). Is there a difference in the observed behavior between males and females? Males may not exhibit the classic arched-back nursing posture, but they demonstrate retrieval, grouping, crouching, and aggression toward intruders (Rosenblatt et al., 1996). In a careful study of hormonal correlates of the onset of paternal behavior, Brown and Moger provided 4–6 pups daily for 22 days to Long-Evans hooded rats (Brown & Moger, 1983). Paternal behaviors were recorded daily and groups of males were sacrificed at 1, 7, 14, or 22 days to measure hormones (Figure 7.4.). Although virtually all the males were fully parental (defined as displaying sniffing, licking, carrying, and crouching over pups on the same day) by day 7, there were no changes in testosterone, LH, FSH, or prolactin across the 22 days. Curiously, prolactin was lower at each time point in the group of males with pups compared to a control group housed without pups, and, in the pup-exposed group, paternal care was correlated with both prolactin and testosterone.

If hormones can facilitate the onset of paternal care in castrated males, are they working on the same neural targets as previously suggested in females (Chapters 2 and 5)? This question has been answered in two ways. Lesions of the MPOA prevent sensitization in naive males (Rosenblatt et al., 1996) and, if performed after pup care has been established, disrupt retrieval, grouping, and crouching over pups (Sturgis & Bridges, 1997). Rosenblatt and Ceus have shown that estrogen implants in the MPOA reduce the time required for parental care in castrated males from 11 days (cholesterol implants in the MPOA) to 6 days (10% estradiol in the MPOA) (Rosenblatt & Ceus, 1998). These findings provide strong evidence for the MPOA as the site of action for estrogen's effects on paternal care. However, this same experiment found even shorter latencies for males injected systemically with 100 μg/kg estradiol benzoate, so another estrogen-responsive site may also be important (Rosenblatt & Ceus, 1998). Taken together, the data with MPOA lesions and estrogen implants are consistent with the hypothesis that the paternal behavior of steroid-treated, castrated adult male rats resembles maternal behavior not only in terms of the components of care (retrieval, grouping, and crouching) but also in terms of neural mechanisms. The role of prolactin or other peptide hormones has not been investigated.

Paradoxically, juvenile male rats appear more parental than juvenile females (Mayer & Rosenblatt, 1979; Gray & Chesley, 1984). In direct contrast to the results with adults, juvenile males are more likely to be immediately parental and, if not immediately parental, acquire parental behavior more quickly than age-matched females. Interest in pups declines at weaning and declines further around day 30 (puberty), ultimately showing the female bias described above. Although this juvenile parental care has been used as a model for adult behavior, careful study of the structure and organization of this behavior reveals important differences from adult parental care (reviewed in Lonstein & De Vries, 2000a). Indeed, some have suggested that juvenile interaction with pups may be more closely related to play rather than parental care (Brunelli & Hofer, 1990).

In Chapter 5 we described differences in regional brain cFos and Fos B

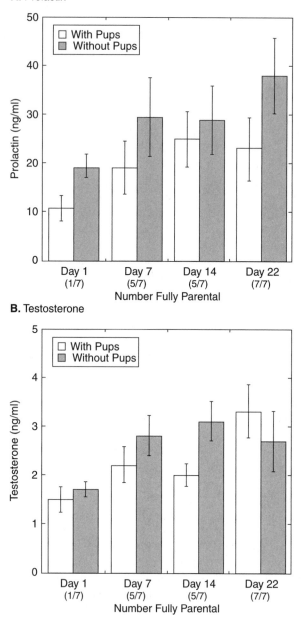

A. Prolactin

B. Testosterone

Figure 7.4. Hormone changes in male rats exposed to pups. Intact Long-Evans male rats were housed with pups for 1, 7, 14, or 22 days. Control rats were not exposed to pups. After day 1, at least 5 of the 7 males in each pup-exposed group exhibited paternal behavior (sniffing, licking, crouching over, and carrying). Relative to the control group, males with pups had consistently lower plasma levels of prolactin although overlap is evident at each point. There were no treatment differences in testosterone. Figure adapted from Brown and Moger (1983).

expression in juvenile versus adult maternal care (Kalinichev et al., 2000b). Although infusions of morphine into the MPOA are able to disrupt juvenile parental care (Wellman et al., 1997), small lesions of the MPOA that abolish maternal care in adults are not effective in juveniles, and large lesions (72% of the MPOA) do not reduce crouching behavior in juveniles whereas much smaller lesions reduce this behavior in adults (Kalinichev et al., 2000a).

We know relatively little about the neuroendocrine regulation of juvenile interactions with pups. As this behavior wanes with the onset of puberty, one might assume that gonadal steroids, especially androgens, inhibit the male's keen interest in pups. In fact, there has been little study of the role of perinatal or circulating hormones on juvenile parental care, perhaps because juveniles are usually studied in a sensitization paradigm that has been assumed to be hormone-independent. Kinsley and Bridges examined pup responsiveness in 25-day-old rats given three pups (1–6 days old) continuously (Kinsley & Bridges, 1988a). Full parental behavior (defined as grouping and crouching over pups within 60 minutes) was observed by the second day in males and the third day in females. Plasma prolactin concentration was roughly double in males (approximately 21 ng/ml) compared to females (approximately 10 ng/ml). Treatment with a dopamine agonist, CB-154, which decreases prolactin, shifted the latency for full parental behavior to 5 days in males, an effect that could be reversed by administration of prolactin. It was not clear from this study if prolactin alone could influence juvenile interest in pups, but the gender differences in the concentration of prolactin in juvenile rats (although also see Dohler & Wuttke, 1975) is an intriguing correlate to the gender difference in pup interest. In a separate study, Kinsley et al. showed that morphine reduced both the onset and the performance of juvenile parental care (Kinsley, Wellman, Carr, & Graham, 1993).

3.3. Infanticide in Laboratory Mice

Just as with maternal behaviors (Chapter 2), paternal behaviors of laboratory mice (*Mus musculus*) are highly variable depending on the strain. For instance, C57BL/6J males are more likely to be infanticidal than DBA/2J males (Svare, Kinsley, Mann, & Broida, 1984). In the CF-1 strain, about half of virgin males are infanticidal, and the other half exhibit pup acceptance. Following copulation, however, virtually all males of this strain will kill or attack pups (vom Saal & Howard, 1982). One "explanation" for this increase in infanticide is that when a virgin male finds an estrous female, she is likely to be postpartum and any associated litter will be from another male. By killing these pups, the male can impregnate this female, reduce the success of an unrelated male, and increase his own reproductive success. Of course, it is important that he not kill his own young. Remarkably, infanticide disappears across the course of the female's pregnancy, so that approximately 18–20 days later after the mate delivers pups, the male behaves paternally. Infanticide reemerges spontaneously about 50–60 days after copulation, at about the time of weaning (vom Saal & Howard, 1982). This pattern of inhibition of infanticide is triggered by copulation, but it does not require subsequent exposure to the mate, as the same behavioral sequence is observed in males that are socially isolated following copulation (vom Saal, 1985). In some strains of mice, the female's presence does appear critical for inhibiting infanticide, perhaps by subjugating the male (Elwood & Kennedy, 1991). Moreover, the delay of 18–20 days appears to involve an extraordinary chronobiological mechanism. By shifting the light cycle after mating, Perrigo et al. were able to demonstrate that the male's brain is tracking day-night transitions rather than the real time elapsed since copulation (Perrigo, Belvin, & vom Saal, 1992). It also appears that pup exposure can induce estrogen receptors in specific regions of the mouse brain, especially the arcuate nucleus and MPOA (Ehret, Jurgens, & Koch, 1993).

Unlike rats, there is no evident sex difference in the parental behavior of juvenile mice. Both males and females ignore pups. In the Rockland-Swiss (R-S) strain, infanticide emerges around day 32 in males (Gandelman, 1973c). This is not associated with a surge in testosterone, and injections of testosterone on day 22 do not induce pup killing. The decrease in infanticide after copulation also does not require a change in testosterone. Testosterone influences pup killing in adults, but the effects are complex. Castration in adulthood reduces infanticide in males but only if they do not have prior pup-killing experience (Gandelman & vom Saal, 1975). Most of all, the effects of testosterone in adulthood interact with a history of androgen exposure during development. Specifically, mice castrated neonatally are more likely to respond to testosterone in adulthood with an increase in pup killing compared to mice castrated in adulthood (Gandelman & vom Saal, 1977). Paradoxically, it appears that perinatal exposure to testosterone primes the developing mouse brain to be *less* responsive to the activational effects of androgens in adulthood. Similarly, the likelihood that virgin CF-1 mice will be infanticidal versus paternal may be related to androgen exposure during development. Vom Saal reported that intrauterine position could predict infanticidal behavior (vom Saal, Grant, McMullen, & Laves, 1983). The pregnant mouse uterus contains several contiguous embryos. An embryo positioned between two males receives more androgen exposure than an embryo positioned between two females. High prenatal androgen exposure is associated with increased likelihood of paternal care and decreased likelihood of infanticide in CF-1 adult virgin mice (Perrigo, Bryant, & vom Saal, 1989). As one further suggestion that androgens may reduce rather than increase infanticide, C57BL/6J males, which are more infanticidal than DBA/2J males and most other strains, have lower levels of testosterone at all ages (Svare et al., 1984).

One of the most surprising findings in this field has been the chance observation that male progesterone receptor knockout (PRKO) mice are highly paternal (Schneider et al., 2003). In this study, wild-type mice following mating had infanticide rates of 80%, whereas infanticide was not observed under the same conditions in the PRKO males. In a pup-retrieval test with sexually naive males caged alone, wild-type males attacked pups 16%, retrieved pups 20%, and ignored pups 64% of the trials. In contrast, PRKO males attacked pups

7.3%, retrieved pups 63%, and ignored pups 17% of the trials. What could be the mechanism for the apparent increase in paternal behavior in PRKO males? In Chapter 2 we described how progesterone in female mice may be important for nest building but, unlike rats, the hormones of pregnancy appear to have very modest effects on lab mouse maternal behavior, effects that are mostly detected when females are tested outside of their home cages. Also recall (from Chapter 6) that the progesterone receptor responds not only to progesterone but also to dopamine and that both ligands activate intracellular pathways involving DARPP-32. One possibility is that PRKO males have reduced dopaminergic tone in the hypothalamus and therefore develop high concentrations of prolactin, which facilitates paternal behavior. In addition, PRKO mice have elevated progesterone, suggesting the alternative possibility that increased allopregnanolone increases male nurturing behavior or decreases infanticide. PRKO mice also have elevated testosterone, demonstrating that testosterone does not antagonize paternal care or increase infanticide. At this point, we do not know the mechanism for the increase in paternal care in PRKO males, but this initial unexpected observation may prove instructive for defining a role for progesterone, a progesterone metabolite, or another effector in the neurobiology of mouse social behavior. It will be important to determine if PRKO females show equivalent changes.

4. Human Paternal Care

Humans, like many primates with altricial young, are a biparental species. Ethnographic studies report that at least 40% of human cultures exhibit paternal care. The degree of paternal care appears to be related to social organization, with fathers more involved in those societies with a monogamous social organization.

Storey et al. provided perhaps the first survey of hormonal changes in fathers during the course of their wives' pregnancies and postpartum periods (Storey, Walsh, Quinton, & Wynne-Edwards, 2000). In this study, 35 couples were monitored (all but three for their first birth), with blood drawn in a low-stress environment during pregnancy (either early, mid, or late) and during the postpartum period (early or late). At each time point, two samples were collected: one at baseline and a second 30 minutes later after exposure to a series of infant cues. This study was mostly cross-sectional (1 couple was studied longitudinally with 10 samples) and sampling times were variable between subjects, but a few patterns emerged. In fathers, testosterone decreased postpartum relative to late prepartum. In the early postpartum period when circulating testosterone was low, infant cues resulted in a significant increase. Prolactin did not change across the time points, although prolactin and cortisol were slightly higher in the group sampled in late pregnancy relative to the group sampled in early pregnancy. Plasma prolactin was higher in 9 male subjects who described couvade symptoms. The couvade syndrome describes the common response of men experiencing the pregnancy symptoms of their mates. Although this has been mostly considered a sympathetic response (see, for instance, a charming description by Churchwell, 2000), there is certainly a possibility that the male's subjective response has an endocrine correlate (Elwood & Mason, 1994). The data of Storey are an interesting first foray into the study of a hormonal basis for paternal behavior in humans. Although the changes in hormones and the levels of prolactin were much lower in men than in women, some of the changes in men resembled an independent report of changes in women (Fleming et al., 1997a), and changes were significantly correlated within couples. Wynne-Edwards has recently reported decreased testosterone and cortisol in expectant fathers relative to age-matched men who were not fathers (Wynne-Edwards, 2001).

5. Summary

At first glance, paternal behavior in mammals may seem unnecessary or even counterproductive. However, in monogamous species in which males are investing in a single female and her young, the male optimizes his investment by defending and caring for the young. In birds, the transition to paternal care usually follows mating and it is likely that the associated endocrine changes, decreasing testosterone and increasing prolactin, are important for the behavioral shift from copulation to incubating and brooding. In many rodent species, females enter postpartum estrus. Thus, a paternal male will need to copulate and care for young during the postpartum period. The relationship of testosterone and prolactin to paternal behavior is less clear in mammals than in birds. Taking the results from all of the species described above, there is no consistent evidence that high- or low-circulating testosterone is correlated with paternal care across all species, although there are specific examples of note (see Table 7.1). Similarly, although there is remarkable evidence of an increase in prolactin in marmosets housed with young, there is no evidence at this point that the presentation of young increases prolactin or prolactin receptors in males of any other species or that manipulating prolactin alters male parental care (Table 7.2). Other neuropeptides may be involved: Vasopressin facilitates paternal care in prairie and meadow voles and opiates reduce the parental-like care of juvenile rats, but much more study will be required to determine the physiological significance or the general applicability of these relationships.

Is paternal care neurobiologically equivalent to maternal care? We know that many of the same behaviors are involved, with lactation an obvious exception. But it is not yet clear that the same neural pathways are activated. Other than the study of Kirkpatrick et al. (1994a) in prairie voles, there has not been a systematic attempt to study regions of neural activation in biparental species, comparing males that are paternal and males that are not paternal. Both Fos activation and excitotoxic lesions in the prairie vole have focussed attention on the medial nucleus of the amygdala, a region that is critical for processing olfactory and pheromonal information with projections to the BST and MPOA. It is intriguing that male voles exposed to pups had even greater Fos activation in the MPOA than females exposed to pups. However, we remain far short of the circuitry analysis that has been used to investigate the maternal brain. Perhaps the most significant data come from rats induced into paternal care. Both lesion and implant studies support the importance of the MPOA for male parental care, at least as a site for estrogen action.

Finally, given the predominance of paternal care in other vertebrates, we might consider that it is the relative absence, not its occasional presence, in mammals that requires explanation. Perhaps we should be asking: What inhibits paternal care in mammals? The foregoing summary suggests that the answer is not a singular presence of testosterone or a chronic absence of prolactin. In the future, we might profit by considering the loss of paternal behavior as a by-product of sexual differentiation. Sexual dimorphisms in brain regions, such as the accessory olfactory bulb, the MPOA, and the medial amygdala, may harbor the circuits that normally prevent parental care in male rats—unless males are immature or exposed to prolonged estrogen treatment. Note that in biparental mammals such as California mice and prairie voles, the MPOA is not sexually dimorphic when adults are caring for young (Gubernick et al., 1993a; Shapiro, Leonard, Sessions, Dewsbury, & Insel, 1991).

In a sense, the apparent independence of paternal care from hormonal factors suggests an opportunity to study the neurobiology of nurturing behavior in a relatively pure form. There has been remarkably little investigation of the neural basis of paternal care. Although only a few biparental mammals (*M. ochrogaster, P. campbelli, P. californicus*) have been studied in the laboratory, these species provide a valuable resource for understanding the neural basis of parental care, particularly because of the availability of congeners that are not biparental (*M. pennsylvanicus, P. sungorus, P. maniculatus*).

8

Neural Basis of Parental Behavior Revisited

This chapter is meant to expand on the knowledge of the neural basis of parental behavior that has been built up in the previous chapters. We have three main goals: (a) to discuss the role of the hypothalamus in the control of parental behavior in the context of the specificity of the underlying neural control mechanisms, (b) to discuss how experiential factors might operate on the neural circuitry of parental behavior so that subsequent parental responsiveness is modified, and (c) to discuss the neural basis of parental behavior in the context of what is known about the neural underpinnings of social affiliation and social attachment, with an eye toward uncovering points of overlap or intersection between the respective neural processes.

1. The Specificity of Hypothalamic Mechanisms That Influence the Occurrence of Parental Behavior

In our neural and motivational models of the control of maternal responsiveness, we have taken an approach-avoidance perspective. We have suggested that maternal behavior occurs when the tendency to approach and interact with infant-related stimuli is greater than the tendency to avoid such stimuli, and we have related these dual tendencies to opposing hypothalamic mechanisms: a positive neural mechanism descending from the preoptic region through the lateral hypothalamus to the ventral mesencephalon, and a negative neural system descending from the medial hypothalamus (AHN/VMN) to the periaqueductal gray. Clearly, approach-avoidance tendencies influence a variety of motivated behaviors. Ethologists have long emphasized that various species-typical social displays (for example, courtship displays and threat displays) can be analyzed into components suggestive of approach or avoidance (Hinde, 1970), and physiological psychologists have long emphasized the existence of dual hypothalamic mechanisms regulating the *general* processes of reward and aversion (Panksepp, 1981). The question arises, therefore, as to the degree of specificity of the neural mechanisms that have been shown to influence maternal behavior. In Chapter

5 we argued that the positive or excitatory neural system regulating maternal behavior can be broken down into a specific system (MPOA/vBST) and a non-specific or general system (the nucleus accumbens system). In this context, we want to critically evaluate the degree of specificity involved in the neural organization of preoptic mechanisms that facilitate maternal responsiveness. Because our understanding of the neural basis of parental behavior (and other social behaviors) is still undeveloped, what follows should be viewed as neural modeling, rather than as the presentation of definitive facts.

In an insightful review, Newman (1999) has emphasized the large amount of overlap in the neural circuits that regulate a variety of social behaviors in mammals (also see Newman, 2002). These social behaviors included male and female sexual behavior, offensive aggression, and parental behavior, and the neural regions she referred to included the MPOA/BST, MeA, and LS. She speculates that there may be a common central network that regulates a variety of social behaviors, and just which behavior occurs may be determined by the stimuli that have access to this social circuit. We want to build on these ideas with respect to MPOA function and social behavior, and we will argue that there may be neural elements in the MPOA that serve a *general* function that can be utilized by a variety of social behaviors, but that there are also probably neural elements in the MPOA that play a role specifically related to the control of maternal behavior.

At this time, it will be worthwhile to review the concepts of appetitive and consummatory components of species-typical behaviors. The appetitive (preparatory, anticipatory) component of a goal-directed behavioral sequence is made up of those behaviors that bring an organism in contact with a desired or attractive stimulus, while the consummatory component is made up of those behaviors that are performed once contact with the desired stimulus is attained (Ikemoto & Panksepp, 1999). Therefore, the appetitive component, if successful (the desired stimulus is obtained), allows the organism to perform the consummatory component. Generally, appetitive behavior is variable and can be influenced by learning and higher cognitive processes (the organism needs to search, and possibly manipulate, its environment to gain access to the desired stimulus), while consummatory or terminal responses are more reflexive in nature and are dependent upon proximal stimulation from the relevant stimulus. How can we relate this appetitive-consummatory distinction to our broad definition of motivation (see Chapter 5) as a process regulating changes in responsiveness to a constant stimulus? Traditionally, the occurrence of the appetitive component of a species-typical behavior has been taken as evidence for increases in motivation: If an organism acts on its environment to obtain a particular stimulus (pups, mate, food), then that stimulus is attractive, signifying an underlying increase in motivation. Note that the occurrence of consummatory behavior might also be taken as evidence for increases in motivation, based on our definition, if it could be shown that a particular proximal stimulus is able to elicit a particular consummatory response when an organism is under one internal state (induced by hormones, for example), but not when it is under another hormonal state. In this section, however, we want to concentrate on the appetitive-consummatory distinction by asking the following questions: Do MPOA mechanisms regulate the appetitive component, consummatory component, or both components of maternal behavior? And to what extent are such regulatory mechanisms specific to maternal behavior?

Hormones act on the MPOA to stimulate parental behavior, and MPOA lesions disrupt parental behavior. What aspects of parental behavior are being affected by these manipulations? Recall that MPOA lesions have their largest and longest-lasting effects on the active components of maternal behavior in rats (*Rattus norvegicus*), primarily retrieving. Recall also that retrieving might be considered an appetitive maternal response: It allows the female to return pups to a nest site where the consummatory response of nursing can take place; retrieving/transporting behavior can also be viewed as being more variable or flexible than nursing and can also be based on higher cognitive processes since the female must determine where the pups are, where the nest is, and what the best route to take is in order to get the pups back to an old nest or brought to a new nest site. Alternatively, aspects of retrieving behavior might also be considered a consummatory response because it is a maternal response that occurs after proximal contact with an attractive stimulus has occurred. One would be better able to determine whether preoptic lesions interfere with the appetitive component of maternal behavior if it could be shown that such lesions disrupt *nonmaternal* responses that are used to gain access to pups (instrumental responses), and such data have been provided by Lee et al. (2000). They showed that MPOA lesions disrupted a postpartum female rat's operant bar-press response to gain access to pups. Importantly, MPOA lesions did not disrupt the female's response when food was used as a reward.

Based on the anatomy of MPOA projections that have been shown to play a role in maternal behavior, we would like to suggest that certain elements in the MPOA (and adjoining BST) influence the appetitive aspects of maternal behavior, while other elements may influence consummatory aspects of retrieving. Most investigators have associated the nucleus accumbens and the mesolimbic dopamine system with appetitive behaviors (Blackburn et al., 1992; Everitt, 1990; Ikemoto & Panksepp, 1999), and therefore it makes sense that MPOA/BST projections to VTA/RRF, or directly to NA, play a role in regulating the flexible, voluntary, appetitive aspects of maternal behavior. (In this section we will be emphasizing MPOA/BST influences on NA. Please note, however, that if an appetitive maternal response were to rely heavily on higher cognitive processes, the prefrontal cortex and MPOA/BST influences on mesocortical DA input to the prefrontal cortex might become more important. We will explore this possibility further in Chapter 9.) In contrast, MPOA projections to brainstem regions controlling sensorimotor integration may be involved in the more reflexive, consummatory aspects of retrieving. For example, in Chapter 5 we noted that the RRF not only projects to the nucleus accumbens, but also projects to brainstem reticular formation neurons that, in turn, can influence the trigeminal motor system. In addition, MPOA/BST neurons also project to the PAG, which in turn can influence the trigeminal system (Chiang, Dostrovsky, & Sessle, 1991; Li, Takada, Shinonaga, & Mizuno, 1993), and some evidence exists that suggests that lesions of the rostral PAG cause a moderate disruption in the sensorimotor integrative mechanisms involved in retrieval behavior (Lonstein & Stern, 1998). Given the possibility of these two control elements within the MPOA region, we would like to suggest that the neural elements that regulate the appetitive components of maternal behavior may constitute a relatively general mechanism that is also accessible to systems related to other social behaviors, while the neural elements in the preoptic region that are conceived of as regulating the consummatory aspects of maternal retrieval behavior are probably specifically related to that behavior.

In two excellent reviews, Risold, Thompson, and Swanson (1997) and Swanson (2000) have attempted to understand hypothalamic function in the context of its connections with other brain regions. Swanson (2000) places hypothalamic nuclei at the top of a motor hierarchy that he calls the behavioral control column, and he notes the importance of hypothalamic mechanisms in the control of reproductive, aggressive (defensive), and ingestive behavior. Risold et al. (1997) emphasize the dual nature of hypothalamic efferent projections, with one output channel descending to the brainstem and spinal cord to ultimately influence motor neuron output, and an ascending feedback connection to the telencephalon. This view is schematically represented in Figure 8.1. We would like to suggest that the descending system that activates the motor mechanisms in the brainstem and spinal cord regulates the consummatory or reflexive aspects of the various behaviors controlled by the hypothalamus, and that specific neural elements and pathways are probably specifically tied to unique behaviors since the different behaviors utilize different motor neurons. Additionally, we suggest that the ascending hypothalamic projection to the telencephalon is involved in influencing the appetitive aspects of the various behaviors controlled by the hypothalamus since the involvement of telencephalic mechanisms would allow for more flexible, adaptive, voluntary responding through the use of higher cognitive processes. In a simplification of the neural model presented in Chapter 5, it is as if hypothalamic mechanisms can inform the telencephalon of a particular motivational state, and that such telencephalic biases would then allow the organism to "figure out" the best way to gain access to a desired stimulus.

Risold et al. (1997) indicate that the hypothalamus can project to the telencephalon in at least three different ways: directly, indirectly via projections to the thalamus, and indirectly through projections to the brainstem. In our analysis of the neural mechanisms influencing maternal behavior, we have emphasized two routes of access through which MPOA neurons could influence the telencephalic nucleus accumbens: a direct route (MPOA/BST-to-NAs), and an indirect route via the brainstem (MPOA-to-VTA/RRF-to-NA). We should also note that MPOA neurons can access the nucleus accumbens and other telencephalic structures (the prefrontal cortex, for example) via projections to the paraventricular nucleus of the thalamus (please distinguish from the hypothalamic PVN) (Risold et al., 1997). Indeed, Numan and Numan (1996) have shown that the

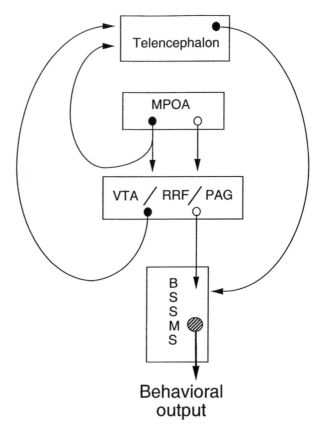

Figure 8.1. Diagrammatic representation of neural pathways through which medial preoptic area (MPOA) neurons might influence both the appetitive and consummatory aspects of maternal behavior. Certain MPOA neurons (shown as an open circle) may project, via a relay in the retrorubral field (RRF) or periaqueductal gray (PAG) (also shown as an open circle), to the brainstem and spinal motor system (BSSMS) in order to influence the reflex-like sensorimotor integration necessary for the consummatory aspects of certain maternal responses. In contrast, other MPOA neurons (shown as a closed circle) may project to the telencephalon, either directly or via a relay in the ventral tegmental area (VTA) and/or RRF (also shown as a closed circle), to influence the performance of proactive and voluntary appetitive maternal responses that are variable and may depend upon higher cognitive processes. The output of the telencephalon (also shown as a closed circle) ultimately influences the output of the BSSMS, whose motor neurons are shown as hatched. Some of the schematized pathways may be multisynaptic. See text for more details. Adapted from Risold et al. (1997).

MPOA/BST region that is highly active during maternal behavior projects strongly to the paraventricular thalamus.

An important question is whether the ascending MPOA projections to the nucleus accumbens and other telencephalic structures, which are conceived of as influencing the appetitive aspects of maternal behavior, are specific to maternal behavior control, or, instead, are these particular MPOA neuron pools used in the motivational/appetitive control of a variety of social behaviors? We want to outline the evidence that suggests that there may be a common population of MPOA neurons that influences the appetitive aspects of a variety of social behaviors in addition to maternal behavior. These additional behaviors include male sexual behavior, female sexual behavior, and offensive aggression.

The first point to emphasize is that Fos immunocytochemical studies in rodents have indicated that the MPOA is not only active during maternal behavior (see Chapter 5) but is also active during male sexual behavior, female sexual behavior, and during bouts of offensive aggression (Baum & Everitt, 1992; Coolen, Peters, & Veening, 1996; Fernandez-Fewell & Meredith, 1998; Heeb & Yahr, 1996; Joppa, Meisel, & Garber, 1995; Newman, 1999; Pfaus, Kleopoulos, Mobbs, Gibbs, & Pfaff, 1993; Wersinger, Baum, & Erskine, 1993). In accordance with the views expressed above, Fernandez-Fewell and Meredith (1998) note that the similar Fos induction pattern in the MPOA of female and male rodents during mating behavior suggests a similarity in the underlying neural control mechanisms. Second, we should note that it has been known for a long time that there is a remarkable similarity in the neural control mechanisms regulating maternal behavior and male sexual behavior (Numan, 1974, 1985). Most importantly, MPOA lesions abolish male sexual behavior in all species that have been studied (which includes members from several vertebrate classes) (Hart & Leedy 1985; Meisel & Sachs, 1994). Indeed, as for maternal behavior in rats, studies on the neural control of male sexual behavior in rodents have emphasized the importance of MPOA-to-midbrain tegmental connections, and some studies have even emphasized the importance of MPOA-to-RRF projections (Finn & Yahr, 1994; Maillard-Gutekunst & Edwards, 1994). Next, we should note that MPOA lesions depress intermale offensive aggression in rodents (Albert, Walsh, Gorzalka, Mendelson, & Zalys, 1986; Bermond, 1982; Edwards, Nahai, & Wright, 1993). Finally, what about female sexual behavior in rodents? The reflexive (consummatory) component of female sexual behavior in rodents is lordosis, which is not depressed by MPOA lesions, and may even be enhanced (Powers & Valenstein, 1972; Whitney, 1986). In contrast, however, the appetitive aspects of female sexual behavior in rodents and primates (marmosets) are depressed by MPOA lesions (Dixson & Hastings, 1992; Hoshina, Takeo, Nakano, Sato, & Sakuma, 1994; Kendrick & Dixson, 1986; Whitney, 1986; Yang & Clemens, 2000). For example, if a sexually active female and male rat are placed in a testing arena that allows the female to control the pace of sexual activity (for example, in a dual-chambered arena that allows the female to leave and return to the male), MPOA lesions do not depress lordosis per se, but the amount of time the female spends with the male is decreased (Whitney, 1986;

Yang & Clemens, 2000). That is, after a copulatory episode, females tend to leave the male, and females with preoptic lesions take a longer time to return to the male and resume copulation. These results suggest that the MPOA may promote a female's appoach to the male and the initiation of copulation. Interestingly, Whitney (1986) has implied that there may be a similarity in the role of the MPOA in female sexual behavior and maternal behavior. Using the terminology of Jacobson et al. (1980), he suggests that certain MPOA neurons may promote the active (rather than the passive) components of both maternal behavior and female sexual behavior.

The view that the MPOA may contain neurons that influence the appetitive aspects of parental behavior, male and female sexual behavior, and offensive aggression contrasts strongly with the work of Everitt (1990) and Everitt and Stacey (1987). Rather than functionally relating MPOA output to the mesolimbic DA system (as we have done for maternal behavior), these authors suggested that these two neural systems regulate different aspects of male sexual behavior in rats. In a very influential series of experiments, they presented evidence that the MPOA only regulates the consummatory aspects of male sexual behavior, while the mesolimbic DA system and the nucleus accumbens, acting independently of the MPOA, regulate the appetitive aspects of male sexual behavior. However, subsequent research has cast doubt on the functional dichotomy created by Everitt's work. That is, as we have been arguing, several studies have shown that MPOA output influences both the appetitive and consummatory aspects of male sexual behavior in rodents and other species (Balthazart, Absil, Gerard, Appeltants, & Ball, 1998; Edwards & Einhorn, 1986; Hull et al., 1999; Paredes, Highland, & Karam, 1993; Paredes, Tzschentke, & Nakach, 1998; Pfaus & Phillips, 1991; Shimura, Yamamoto, & Shimokochi, 1994). The study by Balthazart et al. (1998) on the effects of MPOA lesions on male sexual behavior in an avian species, Japanese quails (*Coturnix japonica*), provides a particularly clear example. They examined the consummatory aspects of male sexual behavior by observing males' copulatory activity. The appetitive aspect of male sexual behavior was analyzed through the use of a learned social proximity procedure that measured the time spent by the male in front of a window with a view of the female prior to the female's release into the male's cage. Importantly, they found that lesions of the rostral MPOA depressed both the appetitive and consummatory aspects of male sexual behavior, while lesions of the caudal MPOA disrupted only the consummatory aspects. Based on the model depicted in Figure 8.1, perhaps descending neurons that directly regulate the consummatory aspects of male sexual behavior are located more caudally, while neurons with ascending influences on appetitive behavior, or on the willingness to engage in sexual activity, are located more rostrally within the MPOA region (note that lesions that directly cause a severe disruption in appetitive aspects of a behavior should also indirectly disrupt the consummatory components by blocking the initiation of such behavior).

In rodents, if the same neuron population in the MPOA influences the appetitive aspects of both maternal behavior and male sexual behavior, then one would expect that similarly located MPOA lesions would disrupt both maternal behavior and male sexual behavior. Gray and Brooks (1984) examined this issue in rats. They found that lesions of the rostral or middle MPOA, but not of the very caudal MPOA, disrupted maternal behavior, while lesions to rostral, middle, or caudal levels of the MPOA disrupted male sexual behavior (their caudal lesions also damaged the anterior hypothalamus). These results would be consistent with a view that neurons in the rostral-to-middle parts of the MPOA may be involved in the appetitive aspects of maternal behavior and male sexual behavior. Consummatory aspects of maternal behavior might also be contolled by the rostral-to-middle MPOA, while the caudal MPOA may control the consummatory aspects of male sexual behavior in rats, as suggested by the work of Balthazart et al. (1998). (See Murphy, Rizvi, Ennis, and Shipley [1999] for an analysis of some MPOA efferent pathways that may underlie the consummatory aspects of male sexual behavior in rats.) The main point, however, is that the work of Gray and Brooks (1984) is consistent with the view that the same neurons in the MPOA may influence aspects of both maternal and sexual behavior.

If a common population of neurons in the MPOA influences the appetitive aspects of parental behavior, male sexual behavior, female sexual behavior, and offensive aggression, what is the evidence that they do so through ascending MPOA projections to the telencephalon? The best evidence for this proposal was outlined in Chapter 5 for maternal behavior, where we reviewed the importance of MPOA/BST projections to NA in the control of maternal motivation. For male sexual behavior and female sexual behavior there is no direct evidence to support our proposal. However, we want to emphasize the following logical progression of research data that provides indirect support. First, dopamine is released into the nucleus accumbens during male sexual behavior and female sexual behavior in rodents (Becker, Rudnick, & Jenkins, 2001; Damsma, Pfaus, Wenkstern, Phillips, & Fibiger, 1992; Kohlert & Meisel, 1999; Meisel, Camp, & Robinson, 1993; Pfaus, Damsma, Wenkstern, & Fibiger, 1995; Wenkstern, Pfaus, & Fibiger, 1993). Second, interference with nucleus accumbens function has been found to depress the appetitive aspects of male and female sexual behavior (Everitt, 1990; Jenkins & Becker, 2001; Pfaus & Phillips, 1991). For example, Pfaus and Phillips (1991) tested the sexual behavior of male rats in a bilevel apparatus where the male had to search through the different levels of the apparatus to find a sexually receptive female. They found that microinjection of a dopamine antagonist into the nucleus accumbens severely depressed the number of level changes (searching, anticipatory responses) shown by the male, but once the male was in the presence of the female, the number of copulatory responses was not affected. Similarly, Jenkins and Becker (2001) found that excitotoxic amino acid lesions of the nucleus accumbens, which included the shell region, resulted in female rats that would not approach a male to engage in sexual activity. However, when the male was allowed to approach the female, the actual number of lordosis responses was not affected. Third, since the MPOA contains neurons that appear to affect the appetitive aspects of male and female sexual behavior, and since neural pathways exist that would allow MPOA efferents to affect mesolimbic dopamine influences on the nucleus accumbens, it makes sense to conclude that there is a population of MPOA neurons that influences male and female sexual motivation through projections to the telencephalon, which would include influences on the nucleus accumbens.

In Chapter 5 and in the current chapter we have been emphasizing the neural mechanisms that may mediate the voluntary appetitive aspects of maternal behavior because we feel that an understanding of these mechanisms will be most relevant to understanding possible biological factors that underlie anomalies in human parenting. Child neglect and child abuse are best viewed as abnormalities in parental appetitive motivation, rather than as a result of an inability to perform particular acts of parental behavior. We will return to this issue in the final chapter. With respect to the current chapter, we have one final puzzle that needs to be discussed. If it is true that there is a common pool of neurons in the MPOA that influences the appetitive aspects of several types of social behavior, then how do we arrive at specificity in appetitive responding? For example, why should a postpartum female (outside the time of the postpartum estrus, which occurs in some species during the early postpartum period) be attracted to infant-related stimuli so that she can initiate maternal responding, while at the same time not be interested in male-related stimuli (Kinsley & Bridges, 1990), if the same pool of neurons is involved in both sorts of motivation? This enigma could be resolved if we argue that hormonal, genetic, and other factors may prepare the "appetitive MPOA neuron pool" to be activated by certain sorts of stimuli and not others at a particular point in time (Newman, 1999). Therefore, for a postpartum rat who is not in estrus, pup-related olfactory and tactile stimuli might be very effective in activating this pool, while male pheromones and other male-related stimuli would be ineffective. Through such a mechanism, a general or common pool of MPOA neurons could be modified to serve a specific function.

2. Possible Mechanisms through Which Experiential Factors Might Operate on the Neural Circuits Underlying Parental Behavior

In Chapter 3 we reviewed the influence of experiential factors on maternal responsiveness. Such factors undoubtedly produce their effects by modifying the nervous system. In the context of what we now know about the neural basis of maternal behavior, we want to present facts and hypotheses concerning how experiential factors might operate at the neural level to exert their effects. We will explore possible underlying neural mechanisms through which: (a) postpartum ewes (*Ovis aries*) form selective attachments to their lambs, (b) previous maternal experience decreases the need for hormonal stimulation in the onset of maternal behavior in rats, and (c) early life experiences influence the development of adult maternal behavior. Additional excellent reviews on some of these topics can be found in Carter and Keverne (2002), Insel and Young (2001), and Walker, Welberg, and Plotsky (2002).

2.1. Olfactory Learning: The Bruce Effect As a Neural Model

We will begin by outlining the olfactory learning mechanisms that underlie a modifiable neuroendocrine response referred to as the Bruce effect. Although this effect is not directly related to maternal behavior, an understanding of it will create a neural model through which we can then examine the role of olfactory learning in the maternal behavior of sheep and rats. Two excellent reviews (Brennan & Keverne, 1997; Kaba & Nakanishi, 1995) have analyzed the mechanisms of the Bruce effect, and these reviews will form the basis of the present section.

The Bruce effect is an example of pheromonal learning in mice (*Mus musculus*), and it was initially reported by Bruce (1959, 1960). A female mouse is mated with a male and remains with him for several hours, after which he is removed. If the female is subsequently presented with a novel male, or the urinary pheromones of such a male, during the first few days after mating, her pregnancy by the original male is aborted (pregnancy block effect). However, if the female is exposed to the original male or his pheromones, the pregnancy that was induced by this male continues undisturbed. It is important to note that the female must remain with the original male for several hours after mating, and if this procedure is not followed, then even the original male will cause the pregnancy he induced to be aborted. It is as if exposure to a male's pheromones for a critical period of time during and after mating allows the female to become familiar with the sire's scent (olfactory learning), so that the sire will not cause the block of the female's pregnancy, while strange pheromonal scents will. Subsequent research has indicated that the ability of the pheromones of a strange male to block pregnancy is dependent upon the presence of the vomeronasal organ (rather than the primary olfactory system), and that the locus of the synaptic plasticity that allows for the recognition of the familiar male's pheromones lies within the accessory olfactory bulb (AOB). It is also known that strange male pheromones block pregnancy by *activating* projections from AOB mitral cells that ultimately project to hypothalamic neurosecretory cells (tuberoinfundibular dopamine (TIDA) neurons), which inhibit anterior pituitary prolactin release. Since prolactin is necessary for the maintenance of early pregnancy (it is luteotropic), the pregnancy is terminated. With this understanding, one can conclude that the olfactory learning that prevents the pheromones of the familiar male from terminating pregnancy involves an experience-based modification of AOB neural networks that results in familiar pheromones losing the ability to activate mitral cell efferents that are involved in inhibiting prolactin release. In other words, the learning seems to involve the development of mechanisms that *prevent* the access of familiar urinary pheromones to hypothalamic TIDA neurons.

Before we can present the neural model that has been suggested to explain the Bruce effect, some additional information is needed. The olfactory bulbs (AOB and main olfactory bulb) are innervated by ascending noradrenergic inputs from the pontine locus coeruleus (Brennan & Keverne, 1997; Shipley & Ennis, 1996), and norepinephrine (NE) is elevated in the AOB for several hours following mating in the mouse as a result of mating-induced vaginocervical stimulation (Brennan, Kendrick, & Keverne, 1995). Importantly, 6-hydroxydopamine lesions of NE input to AOB, or infusions of α-NE antagonists into the AOB, prevent the formation of a memory for the mating male's pheromones (Brennan & Keverne, 1997). Note that although disruption of NE input to the AOB interferes with the formation of a new memory, this treatment has no effect on the recall of an already established memory.

Based on all the evidence that has been presented, and on additional findings, Brennan and Keverne (1997) have presented the following neural model of the Bruce effect, which is depicted in a very simplified form in Figure 8.2. The main output cells of the AOB are the mitral cells, which receive input from vomeronasal sensory neurons, and glutamate is the presumed excitatory neurotransmitter used by the mitral cells. Two groups of mitral cells are labeled M1 and M2, and we indicate that pheromones from different males activate different groups of mitral cells. We also show one of the major intrinsic neurons of the AOB, the granule cell, which contains GABA and exerts inhibition on the mitral cell. The mitral cell is shown as projecting (indirectly) to hypothalamic mechanisms that inhibit prolactin release, and to a neighboring granule cell. As we will see, the model suggests that olfactory learning is based on a strengthening of the mitral cell-to-granule cell synapse (we have circled this synapse between M1 and its granule cell (GC1) in Figure 8.2). Finally, NE from the locus coeruleus is released into the AOB as a result of the vaginocervical stimulation that accompanies mating, and this input is hypothesized to exert presynaptic inhibition on granule cell terminals, inhibiting the release of GABA, and the associated feedback inhibition onto the mitral cell. Assume that a female mates with male 1, and that his pheromones activate a population of mitral cells designated as M1. His pheromones will then activate the M1-GC1 synapse. Normally, GC1 would inhibit M1, but this is prevented by the presynaptic inhibitory effects of NE. Therefore, there is strong and continuous glutamatergic activity across the M1-GC1 synapse during the mating and postmating interval (as long as the pheromones are present and NE is elevated). Using a Hebbian presynaptic-postsynaptic coincidence model of synaptic facilitation, analogous to the development of long-term potentiation in the hippocampus (see Dudai, 1989), the M1-GC1 synapse is hypothesized to be strengthened if its activity persists for a critical amount of time. Subsequently, after NE release has subsided and the M1-GC1 synapse has been strengthened, the pheromones of male 1 exert such a strong activation of GC1 that M1 activity is almost immediately depressed by GABAergic feedback inhibition, very little activity is transmitted to the hypo-

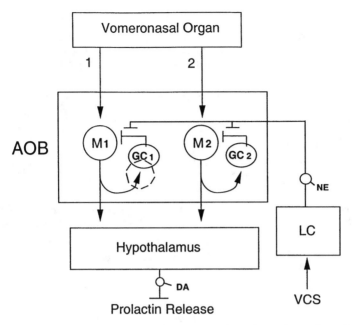

Figure 8.2. Neural model that outlines the neural circuits and experience-dependent synaptic strengthening that underlies the Bruce effect. Pheromones from two different male mice are shown as activating two different populations of mitral cells (M1 and M2) in the accessory olfactory bulb (AOB) of a female mouse as a result of neural input from the vomeronasal organ (shown as vomeronasal nerve pathway 1 and 2, respectively). Each mitral cell group is shown as exciting an associated granule cell (GC1 and GC2, respectively), and the granule cells, in turn, exert a feedback inhibitory effect on their associated mitral cell (for simplicity, we have depicted the synapses between mitral cells and granule cells and between granule cells and mitral cells as classical axo-somatic synapses; in actuality, such synapses are reciprocal dendrodendritic synapses). Mitral cells also project to the hypothalamus (via synapses in the amygdala) to inhibit prolactin release. Finally, a noradrenergic (NE) projection from the locus coeruleus (LC) to the AOB, which is activated by vaginocervical stimulation (VCS) during mating, is shown as exerting presynaptic inhibition on granule cell presynaptic terminals. If a female mates with male 1, then the M1-to-GC1 synapse is strengthened (this synapse is shown as a dashed circle). Subsequently, the pheromones of male 1 will not abort the female's pregnancy because the M1 projection to the hypothalamus will be strongly inhibited by the enhanced feedback inhibition exerted by GC1 onto M1. In contrast, an unfamiliar male (male 2 and his pheromones) will inhibit prolactin release and abort a female's pregnancy because M2 efferents will strongly promote hypothalamic dopamine (DA) release into the pituitary portal system (because the M2-GC2 synapse has not been strengthened). See text for details and for experimental support for the model. Pathways ending in an arrow are excitatory, and those ending in a vertical bar are inhibitory. Adapted from Brennan and Keverne (1997).

thalamus, prolactin release is not inhibited, and pregnancy continues normally. However, if male 2 is introduced to the female during early pregnancy, he disrupts pregnancy because the M2-GC2 synapse has not been strengthened, and therefore his pheromones gain access to the prolactin-inhibiting system in the hypothalamus.

This model has gained support from additional findings (see Brennan & Keverne, 1997; Kaba & Nakanishi, 1995). First, infusion of GABA antagonists into AOB of a female mouse can induce the formation of an olfactory memory upon exposure to male mouse pheromones in the absence of mating. Presumably, the GABA antagonist substituted for the mating-induced NE release by disinhibiting the mitral cells activated by the pheromones. Second, antagonism of ionotropic glutamate receptors in the AOB of a female mouse while she is mating with a male blocks the formation of a memory for that male's pheromones. The glutamate antagonists would presumably block the heightened activity across the mitral cell-to-granule cell synapses that are hypothesized to be necessary for their synaptic strengthening. Finally, structural evidence exists for the idea that a strengthening of particular mitral cell-to-granule cell synapses occurs during pheromonal learning (Matsuoka, Kaba, Mori, & Ichikawa, 1997). The main aspect of this model that we would like to reemphasize is that olfactory learning in the Bruce effect case is presumed to result in the inhibition of the transfer of olfactory information from the olfactory bulb to other parts of the brain. In what follows, we will explore the extent to which analogous olfactory learning mechanisms might mediate experiential influences on maternal responsiveness in sheep and rats.

2.2. Olfactory Learning Mechanisms and the Formation of a Selective Attachment between a Maternal Ewe and Its Lamb

Recall that in the immediate postpartum period a ewe will accept any lamb, but after several hours of interaction with a particular lamb, she forms a selective bond with that lamb, and will subsequently reject alien lambs (see Chapter 5). The formation of such a selective bond is dependent upon olfaction since olfactory bulbectomy results in a ewe that remains unselective, showing maternal behavior to all lambs. It should be noted that although olfaction is necessary for the development of a selective maternal attachment to a particular lamb, this does not mean that the synaptic modifications that underlie the development of the olfactory recognition occur in the olfactory bulb. For example, the necessary synaptic modifications could occur in downstream targets of the olfactory bulb. Therefore, to what extent is the development of selective lamb recognition based on modifications in olfactory bulb circuitry, and to what extent are the mechanisms leading to such recognition similar to those that occur during the Bruce effect in mice? As indicated in Chapter 5, it is the main olfactory bulb (MOB) system, rather than the AOB system, that is necessary for the development of maternal selectivity in sheep. Since the intrinsic circuitry of the MOB differs slightly from that of the AOB (see below), we might expect to see some differences in any olfactory learning mechanisms that occur in these two structures. In spite of this, there are remarkable similarities in the mechanisms underlying olfactory learning during the Bruce effect and during maternal attachment in sheep.

Recall that Kendrick et al. (1992b) recorded the neural activity of individual cells from the mitral cell layer of the MOB of prepartum and postpartum sheep. During the prepartum period, cells were not detected that showed a selective response to lamb odors, while during the postpartum period, after a selective maternal bond had been formed, many such cells were detected. Although most of these cells responded to both own and alien lamb odors, a population of cells was detected that only responded preferentially to the odor of the lamb to which the female had bonded (see Figure 5.4). These electrophysiological findings suggest that MOB circuitry is modified during the postpartum period. In this study, a microdialysis probe was also placed into the external plexiform layer of the MOB in order to measure neurotransmitter interactions between mitral and granule cells during the postpartum period. In ewes that had already formed a selective bond with their lambs, the odor of their lambs activated the release of both glutamate and GABA, while alien lamb odors did not. This finding suggested to Kendrick et al. (1992b) that the formation of a selective bond between a ewe and its lamb is associated with a stengthening of mitral cell-to-granule cell synapses, with a resultant increase in feedback inhibition onto mitral cells, analogous to what occurs during the Bruce effect in the AOB. Another similarity between the Bruce effect and the formation of a selective maternal bond in sheep pertains to the role of vaginocervical stimulation (VCS). Analogous to the olfactory learning situation in mice, VCS plays a role in the formation of a selective bond between a ewe and its lamb (see Chapter 5). Importantly, as in the mouse AOB, the VCS that occurs in sheep during parturition is associated with an increase in the release of norepinephrine at synapses in the MOB (Levy, Guevara-Guzman, Hinton, Kendrick, & Keverne, 1993; Levy et al., 1995a). The locus coeruleus is the source of norepinephrine in the MOB of sheep (Levy, Meurisse, Ferreira, Thibault, & Tillet, 1999).

All of the above evidence is correlational in nature. Experimental evidence also indicates that modifications within the MOB underlie olfactory learning in maternal sheep. First, 6-hydroxydopamine lesions of the noradrenergic input to the MOB, while not disrupting the onset of maternal behavior at parturition, and without causing anosmia, disrupt the formation of maternal selectivity in sheep: Such ewes act maternally toward their own or alien young and do not form selective bonds (Pissonnier, Thiery, Fabre-Nys, Poindron, & Keverne, 1985). Similar results are obtained when a β-noradrenergic antagonist is infused into the MOB (Levy et al., 1990). These results suggest, as in the mouse, that norepinephrine may be involved in disinhibiting mitral cells during olfactory learning so that specific mitral cell-to-granule cell synapses will be strengthened. Note, however, that in the mouse AOB α-NE receptors, rather than β-NE receptors, are presumed to mediate NE inhibition of granule cell GABA release onto mitral cells. Another similarity with the olfactory learning that occurs during the Bruce effect is that the infusion of ionotropic glutamate antagonists into the MOB of recently parturient sheep prevents the formation of maternal selectivity without interfering with maternal behavior per se (Kendrick et al., 1997b). Finally, we should note that there is evidence that the formation of nitric oxide is involved in the synaptic plasticity that occurs in the sheep MOB and the mouse AOB during olfactory learning (Kendrick et al., 1997b; Okere, Kaba, & Higuchi, 1996).

All of these results suggest that modifications in the synaptic circuitry of the main olfactory bulb underlie the development of maternal selectivity in sheep. Another interesting finding is that parturition and VCS also increase the release of oxytocin into the MOB of sheep, and that such oxytocin potentiates the release of NE into the MOB (Levy et al., 1995a). These results suggest that oxytocin may play a dual role in the maternal behavior of sheep. Acting outside the MOB, in the PVN, MPOA, BST, and VTA, for example, it may facilitate the onset of a general maternal responsiveness (see Chapters 5 and 6), while acting at the level of the MOB it may play a role in the development of maternal selectivity, perhaps by modulating the effects of NE and other neurotransmitters. Finally, there is evidence that cholinergic input to MOB may also be important for the development of maternal selectivity in sheep (Ferreira, Meurisse, Gervais, Ravel, & Levy, 2001).

Many similarities exist between the olfactory learning that occurs in mice and sheep, and a good case can be made for the hypothesis that a strengthening of mitral cell-to-granule cell synapses in the AOB and MOB, respectively, underlies both types of learning. A problem exists, however, in carrying this comparison too far. In examining Figure 8.2, one can see that the mechanisms that have been hypothesized to underlie the Bruce effect presumably result in a selective anosmia (with respect to affecting prolactin release): As a result of learning mechanisms that increase feedback inhibition onto specific mitral cells, the pheromone-induced neural activity of the stud male is prevented from leaving the AOB. Such a mechanism appears to be an unlikely process for mediating the selective attachment of a ewe to its lamb. Recall that in Chapter 5 we reviewed the behavioral evidence that indicated that the development of a maternal attachment toward a specific lamb is based primarily on the ewe's recognition of a familiar lamb odor rather than on her lack of detection of a foreign odor. These findings suggest that the scent of a familiar lamb should be able to gain access to those neural regions mediating positive maternal responses, while unfamiliar odors do not have such access (and may access aversion-promoting regions instead).

Although the strengthening of mitral cell-to-granule cell synapses may underlie olfactory learning during the Bruce effect and during maternal behavior in sheep, the differences in the anatomical organization of the AOB and the MOB may result in different sensory outcomes. As outlined by Brennan and Keverne (1997) and Mori, Nagao, and Yoshihara (1999), in the AOB a granule cell tends to exert feedback inhibition primarily on the mitral cell that activated it (see Figure 8.2), while in the MOB a granule cell that is activated by one mitral cell is capable of exerting inhibition on other mitral cells (lateral inhibition). Although this anatomical distinction is not absolute, but instead is one of degree, it may have important functional consequences. Figure 8.3 shows a very simplified view of synaptic circuitry in the MOB, which was adapted from Mori et al. (1999). Please compare this figure with Figure 8.2. Note that in the AOB a granule cell is shown as inhibiting the mitral cell that activated it, while in the MOB a granule cell is shown as inhibiting more than one mitral cell. Although the strengthening of mitral cell-to-granule cell synapses in the AOB may simply depress activity in single mitral cells, the strengthening of such synapses in the MOB may result in synchronized bursting activity in groups of mitral cells that respond to a particular complex odor because of synchronized and periodic cycles of activation followed by inhibition (Mori et al., 1999). Without becoming too speculative, we would like to propose a simple hypothetical model to explain how the development of such synchronous mitral cell activity in the MOB during olfactory learning might be able to promote maternally selective acceptance behavior (also see Keverne, 1995; Mori et al., 1999). Suppose that the odor of a specific lamb is complex and activates several groups of mitral cells. At parturition, because of the associated hormonal priming and other physiological events, a ewe will initially respond positively to the complex odor of any lamb. However, once she is exposed to a particular lamb, in conjunction with the VCS that occurs during parturition and that releases NE (and oxytocin) into the MOB, a specific group of mitral cell-to-granule cell synapses are strengthened, which subsequently results in a specific population of mitral cells firing in synchronized bursts of activity only upon exposure to the particular odor complex of the familiar lamb. Lamb odors that the ewe was not exposed to during the early postpartum period do not cause groups of mitral cells to fire in synchronous bursts. Let us also assume that some of the mitral cells that respond to lamb odors project downstream to *converge* on neurons within brain regions that mediate approach and acceptance behavior. If we hypothesize that these downstream acceptance neurons have high thresholds of activation, then familiar odors that cause synchronized bursts of activity in mitral cells may be able to activate these neural regions because of the associated temporal and spatial summation of excitatory postsynaptic potentials, while unfamiliar odors, which cause mitral cells to fire unsynchronously, are not capable of activating these downstream regions. We should note the development of maternal selectivity is obviously much more complicated than is depicted in the above model, and further elaborations would be needed to explain several aspects of maternal behavior in sheep. For example, why do females respond positively to all lambs at birth, and how do unfamiliar lamb odors lose their ability to activate positive responses? These events are likely tied to the waning of hormonal and other physiological influences that promote a generalized maternal responsiveness. It might also be that olfactory learning mechanisms not only promote the access of familiar lamb odors to positive neural regions, but also operate to depress the

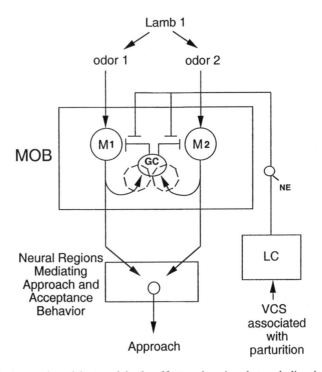

Figure 8.3. A neural model to explain the olfactory learning that underlies the development of maternal selectivity in sheep. As a result of the hormonal and neuroendocrine events associated with pregnancy termination, a ewe will accept any lamb at parturition. Later in the postpartum period, however, the ewe will only accept a lamb to which she was exposed at parturition, and she will now reject unfamiliar lambs. The model assumes that the odor characteristics of a newborn lamb are complex, composed of many odors (shown as odor 1 and odor 2). Each odor activates a different population of mitral cells (shown as M1 and M2) in the main olfactory bulb (MOB). Unlike mitral cells in the accessory olfactory bulb (see Figure 8.2), mitral cells in the MOB project to granule cells (GC) that exert feedback inhibition on more than one population of mitral cells (for simplicity, we have depicted the synapses between mitral cells and granule cells and between granule cells and mitral cells as classical axo-somatic synapses; in actuality, such synapses are reciprocal dendrodendritic synapses). At parturition, and under the influence of locus coeruleus (LC) norepinephrine (NE) release into the MOB, which is activated by vaginocervical stimulation (VCS), the M1-to-GC synapse and the M2-to-GC synapse are strengthened (shown as dashed circles) as a result of exposure to a particular lamb (shown as lamb 1). Later in the postpartum period, as a result of such synaptic strengthening, the complex odor of lamb 1, but not the complex odor of an alien lamb, is able to cause synchronized bursts of neural activity in the relevant population of mitral cells (M1 and M2). Such synchronized bursting activity is proposed to cause a temporal and spatial summation of synaptic inputs onto neurons in brain regions mediating maternal acceptance. If such acceptance neurons were to have high thresholds of activation, then only synchronized bursting inputs would fire them, while the unsynchronized inputs from mitral cells that respond to alien lamb odors would not be able to fire such "acceptance neurons". See text for further elaborations. Projections ending in an arrow are excitatory, and those ending in a vertical bar are inhibitory. Adapted from Brennan and Keverne (1997) and from Mori et al. (1999).

access of familiar lamb odors to negative response regions and/or facilitate the access of unfamiliar odors to such negative response regions. Finally, any neural model would need to explain why olfactory bulbectomized females, or females with olfactory bulb NE depletions, continue to show adequate maternal behavior in the absence of the development of maternal selectivity.

What might be the neural regions outside the olfactory bulb that are activated by familiar lamb stimuli? There is only one study that is relevant to this important issue. Da Costa et al. (1997a) treated one group of sheep with a hormone regimen coupled with VCS that is known to induce maternal behavior, but they did not expose such sheep to lambs. A second group of sheep was allowed to give birth naturally and interact with its lamb at birth for 30 minutes. The females in both groups were subsequently sacrificed and their brains were examined for the induction of cFos mRNA. Although both groups showed increases in cFos mRNA in neural regions known to promote maternal behavior, such as the MPOA, BST, and PVN, the group that gave birth naturally and was also exposed to its lamb showed additional activation of cFos mRNA synthesis in the piriform cortex, orbitofrontal cortex, and cingulate cortex. Therefore, it is possible that these latter regions play a role in the development of maternal selectivity. A study that is crying out to be done, but that has not been done yet, as far as we know, would be a comparison of the differences in neural activation that might occur when a maternally selective ewe is exposed to either its own lamb or an alien lamb.

2.3. Olfactory Learning Mechanisms and Experiential Influences on Maternal Behavior in Rats

Recall that in primiparous females, hormonal stimulation is essential for the activation of an immediate onset of maternal behavior at parturition. However, once a female interacts with her pups for a critical length of time, her future maternal responsiveness is less dependent upon hormonal stimulation (see Chapter 3). This effect has been referred to as the maternal experience effect, the long-term retention of maternal responsiveness, or maternal memory. In this section we want to explore whether olfactory learning mechanisms play a role in this experience-induced partial emancipation of maternal behavior mechanisms from hormonal control. It is interesting to note that the possible mediation of this maternal experience effect by mechanisms analogous to the Bruce effect would fit nicely with much (but not all) of what we already know about the control of maternal behavior in rats. In previous chapters we have argued that olfactory input from novel pups is initially aversive to the naive virgin female rat, and that the hormonal events of late pregnancy promote maternal behavior by modifying the impact of olfactory input, changing its valence from negative to positive. It could be argued, therefore, that once a primiparous female interacts with her young for a critical amount of time during the early postpartum period, olfactory learning mechanisms come into play that cause a long-term alteration in the synaptic circuitry of the MOB/AOB, so that on future occasions, when a female is no longer exposed to the facilitatory effects of hormones, pup odors are less effective in inhibiting maternal behavior or causing avoidance responses. In a very simple model, we could argue that full interaction with pups at parturition causes a Bruce effect–like process to occur in the olfactory bulb so that mitral cells that respond to aversive pup odors come under strong inhibition from their associated granule cells, in this way preventing such odors from accessing more centrally located regions involved in the inhibition of maternal behavior. The fact that the production of anosmia in naive virgin females facilitates maternal behavior is congruent with this idea. A problem, however, with this simple model is that postpartum females, who have presumably already been affected by the hypothesized olfactory bulb modifications, not only do not avoid pup odors, but are actually strongly attracted to such odors (Kinsley & Bridges, 1990). Therefore, a revised olfactory learning hypothesis might incorporate what we know about olfactory learning during the Bruce effect and what we know about olfactory learning during the formation of a selective bond between a ewe and its lamb. Assume that pup odors are complex, and are made up of at least two groups of pheromones/odors that activate spatially distinct groups of mitral cells, which, in turn, access brain areas that either promote attraction or promote aversion, respectively (see the following papers that indicate that such a proposal is possible: Krieger et al., 1999; Kumar et al., 1999; Mori et al., 1999). These odors would have to be rather general in nature across pups, as rats do not form selective bonds to specific pups. As a result of a postpartum maternal experience, perhaps a Bruce effect–like mechanism results in a long-term depression of the ability of inhibitory pheromones to strongly activate their associated mitral cells, while a synchronized bursting modification, analogous to what has been suggested to occur in the olfactory bulb of sheep, results in a long-term facilitation of the ability of positive pheromones to co-ordinate concurrent activity in groups of mitral cells whose effects can then summate on positive downstream regions.

Probably the best evidence in favor of olfactory learning mechanisms playing a role in the maternal experience effect comes from a study by Malenfant, Barry, and Fleming (1991a). Primigravid female rats were caesarean-sectioned on day 21 of pregnancy. Thirty-six hours later they were given a 1-hour period of full maternal interaction with pups who were scented with an artificial odorant (rum or vanilla). Ten days following the caesarean section, females were reexposed to pups in order to reinduce maternal behavior in the absence of hormonal facilitation, and two groups were formed. One group was exposed to pups that were scented with the same odor that was used during the experience phase, while the other group was exposed to pups that were scented with a different odor. The sensitization latencies to the onset of maternal behavior in the two groups differed, with the group exposed to pups with the same odor as that initially experienced showing shorter latencies.

Other findings are also supportive of olfactory learning mechanisms playing a role in the maternal experience effect. Full interaction with pups is necessary for the long-term retention of maternal responsiveness (see Chapter 3), and research has indicated a role for perioral and ventral tactile stimulation (Morgan et al., 1992). In addition to vaginocervical stimulation, general somatosensory stimulation is also capable of releasing norepinephrine into the olfactory bulb (Leon, 1992; Wilson & Sullivan, 1994), and Moffat, Suh, and Fleming (1993) have provided evidence for the involvement of noradrenergic mechansisms in the development of the long-term retention of maternal responsiveness. Females were caesarean-sectioned on day 21 of pregnancy, and 36 hours later they received either no maternal behavior experience, 15 minutes of maternal behavior experience, or 1 hour of maternal behavior experience. Immediately following the experience phase, the females that received 15 minutes of experience were injected with either a β-NE agonist or saline, the females that received 1 hour of maternal experience were injected with either a β-NE antagonist or saline, and the inexperienced females were injected with either a β-NE agonist, β-NE antagonist, or saline. All injections were administered systemically. Ten days following the caesarean section, all females were exposed to pups and sensitization latencies to the onset of maternal behavior were measured. Some of the results are shown in Figure 8.4: A β-NE agonist potentiated the ability of a

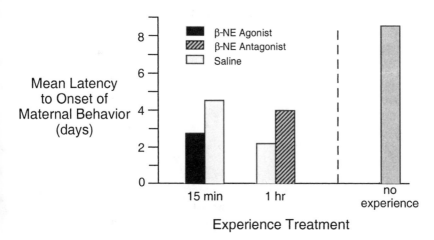

Figure 8.4. The effects of systemic treatment with a beta noradrenergic agonist (β-NE agonist, or isoproterenol) or a β-NE antagonist (propranolol) at the time of a maternal experience on the long-term retention of maternal responsiveness in rats. Female rats were caesarean-sectioned on day 21 of pregnancy, and 36 hours later they received either no experience with pups, 15 minutes, or 1 hour of maternal experience. Ten days following surgery, all females were exposed to pups and latencies to onset of maternal behavior were measured. Females with no experience showed the typical long latency to onset of maternal behavior (about 8.5 days). A β-NE agonist potentiated the effects of the 15 minute experience, and a β-NE antagonist disrupted the effects the 1-hour experience, on the long-term retention of maternal responsiveness. Modified from Moffet et al. (1993).

weak maternal experience (15 minutes) to promote the long-term retention of maternal responsiveness, while a β-NE antagonist depressed the ability of the 1-hour experience to promote the long-term retention of maternal responsiveness. Although these results were achieved with systemic injections, they are consistent with a view that NE release in the olfactory bulb, possibly triggered by somatic sensory stimulation, is playing a role in olfactory learning mechanisms that may underlie the maternal experience effect.

In an important experiment, Fleming et al. (1992) directly explored whether the maternal experience effect could be detected in females whose olfactory abilities had been compromised (also see Mayer & Rosenblatt, 1975). If one could detect a maternal experience effect in such animals, it would suggest that the experience is exerting some of its effects outside the olfactory system. Primigravid females received either bilateral transections through the MOB, sparing the vomeronasal nerves; transections of the vomeronasal nerves; or sham surgeries. Following a caesarean section on day 21 of pregnancy, females received either 2 hours of maternal experience or no maternal experience, and then 10 days after the caesarean section, all females were exposed to pups and latencies to onset of maternal behavior were measured (retention test). As expected, for inexperienced animals exposed to pups for the first time 10 days post–caesarean section, animals with damage to the olfactory system had shorter latencies to the onset of maternal behavior than did females receiving sham surgeries. These findings fit with the well-known fact that anosmia facilitates maternal behavior in inexperienced animals. However, at the retention test, when comparing across experience conditions, not only did experienced sham animals display shorter latencies than did their inexperienced counterparts (standard maternal experience effect), but the same relationship also held for those females that received damage to the olfactory system: Experienced females had shorter latencies than inexperienced females. These results suggested to Fleming et al. (1992) that at least a portion of the maternal experience effect occurred in neural areas outside the olfactory system. The only criticism we have of this conclusion is based on the fact that this experiment was not conducted on animals who had received damage to both olfactory systems. Therefore, the synaptic modifications that underlie the long-term retention of maternal responsiveness may occur in *both* the MOB and the AOB, and may be restricted to these systems. In other words, it is possible that experiential influences observed in the animals with MOB cuts occurred in the AOB, and the experience effects observed in animals with vomeronasal nerve cuts occurred in the MOB.

The analysis of olfactory learning mechanisms in the control of maternal behavior in rats is clearly undeveloped when compared to the work done on the Bruce effect and on maternal behavior in sheep. For example, although noradrenergic mechanisms appear to contribute to the development of the maternal experience effect in rats, this influence has not been shown to occur in the olfactory bulb, and may in fact occur in other regions of the nervous sytem, since systemic injections were utilized. We feel, however, that by developing the cases of olfactory learning during the Bruce effect in mice and during the formation of a selective maternal bond in sheep, we have highlighted the importance of this possibility as one of the ways through which maternal experience might modify future maternal responsiveness in rats.

In addition to the work of Fleming et al. (1992), other research exists that suggests that the long-term retention of maternal responsiveness might be influenced by maternal experience-induced modifications in neural systems outside the olfactory bulb. Indeed, it appears that an initial maternal experience may modify the nervous system at multiple levels so that future maternal responsiveness is potentiated. It is to this research that we now turn our attention.

2.4. The Involvement of Brain Regions and Systems Other than the Main and Accessory Olfactory Bulbs in the Long-Term Retention of Maternal Responsiveness in Rats

In addition to modifying the synaptic circuitry of the olfactory bulb, how could a maternal experience at the time of an initial parturition modify the brain so that in the future, maternal responsiveness to pups is potentiated even in the absence of hormonal facilitation? A reexamination of Figure 5.21 may be helpful, as we make some suggestions. We have argued that at the time of an initial parturition in primiparous rats, the genomic and nongenomic effects of hormones modify the brain so that defensiveness in response to pup stimuli is decreased while attraction to such stimuli is increased. A simple hypothesis to explain the maternal experience effect (Hypothesis 1) is that hormones and other physiological factors associated with late pregnancy and parturition modify the molecular architecture of particular brain regions so that the onset of maternal behavior is facilitated, and then, as a result of an episode of maternal experience, the brain is affected in such a way so that those hormone-induced changes are subsequently maintained in the absence of the hormonal influences. Of course, in the absence of an episode of maternal experience, these hormone-induced brain modifications would wane over time. A competing hypothesis would be that hormones induce one change, which wanes, but if a maternal experience occurs during the period of hormone stimulation it produces a change in the brain that is different from, but substitutes for, the hormone-induced change (Hypothesis 2). As examples of Hypothesis 1, suppose hormones modify neurotransmitter receptor sites in the MPOA, MeA1, and MeA2 so that at parturition MPOA is more excited by tactile inputs, while MeA2 is less excited, and MeA1 is more excited by olfactory inputs (see Figure 5.21); as a result of a maternal experience, these modifications are maintained in the absence of hormonal stimulation. As a proposal for Hypothesis 2, suppose hormones modify the MPOA so that it is more responsive to pup stimulation. As a result of a maternal experience, MPOA output drives the VTA/RRF input to the NAs. Perhaps this effect sensitizes the responsiveness of the mesolimbic DA system so that after hormonal effects on the MPOA have waned, the now less-than-optimal activation of the MPOA by pup stimulation is still capable of affecting a now hyperresponsive downstream mesolimbic DA system. More generally, irrespective of whether Hypothesis 1 or Hypothesis 2 is correct, it makes sense at a systems level of analysis to propose that a maternal experience might render a female more maternally responsive in the future because changes have occurred in brain systems so that the aversiveness of pup stimuli is decreased, the attractiveness of such stimuli is increased, or both. This could be accomplished through decreased excitation or increased inhibition of inhibitory neural systems and/or increased excitation or decreased inhibition of excitatory neural systems. In fact, from a behavioral perspective, one would predict that positive interactions and familiarity with infants should decrease their fear-eliciting characteristics and promote their attractiveness. It is within the context of such ideas that we would like to examine the remaining literature on the neural basis of the long-term retention of maternal responsiveness.

Research has clearly shown that for most types of learning, long-term memory is dependent upon experience-induced changes in protein synthesis (Davis & Squire, 1984; Dudai, 1989). In accordance with such findings, research has also indicated that protein synthesis is necessary for the long-term retention of maternal responsiveness. The critical experiments were performed by Fleming, Cheung, and Barry (1990), and some of their results are shown in Figure 8.5. Primigravid female rats were caesarean-sectioned on day 21 of pregnancy, and 36 hours later they received a 2-hour maternal experience. In one experiment, the females received an intracerebroventricular injection of cycloheximide (a drug that blocks protein synthesis at the ribosomal level) or saline at either 30 minutes before, 10 minutes after, or 24 hours after the maternal experience. The long-term retention of maternal responsiveness was measured at 10 days post–caesarean section. Two important points emerged from this study: First, when cycloheximide was administered at 30 minutes before or 10 minutes after the maternal experience, but not at 24 hours after the experience, it blocked the long-term retention of maternal responsiveness. Second, when cycloheximide was administered at 30 minutes before a maternal experience it did not block the initial immediate onset of maternal behavior at 36 hours post–caesarean section. To follow up on these findings, in a second experiment cycloheximide or saline was administered at 10 minutes after the 2-hour maternal experience, and retention tests were administered at 4, 6, or 10 days post–caesarean section. Interestingly, at the 4-day retention test both the cycloheximide and saline groups showed a short latency to onset of maternal behavior, but on the 6- and 10-day retention tests the long-term retention of maternal responsiveness was blocked in the cycloheximide groups (see Figure 8.5). Probably the best interpretation of these findings is that the hormonal and other physiological events associated with pregnancy termination alter the brain so that an immediate onset of maternal behavior occurs at 36 hours post–caesarean section. These hormone-induced brain changes appear to last up to day 4 post–caesarean section. Although steroid-induced genomic effects, and the accompanying changes in protein synthesis, are clearly involved in this hormonal facilitation, such changes have al-

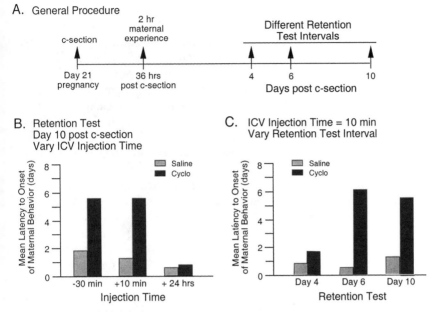

Figure 8.5. The effects of intracerebroventricular (ICV) injection of cycloheximide (Cyclo, a protein synthesis inhibitor) on the long-term retention of maternal responsiveness. **A** shows the general experimental methodology: Female rats were subjected to a caesarean-section (c-section) on day 21 of pregnancy, and 36 hours later they received 2 hours of maternal experience with pups. ICV Cyclo or saline was administered either 30 minutes prior to or 10 minutes or 24 hours after the maternal experience. A retention test to measure maternal responsiveness was given at either 4, 6, or 10 days post–c-section. **B** shows that when the retention test was 10 days post–c-section, Cyclo disrupted maternal responsiveness (as measured by increased sensitization latencies) if it had been administered 30 minutes prior to or 10 minutes after the 2-hour maternal experience. **C** shows that when ICV Cyclo was administered 10 minutes after the 2-hour maternal experience, it increased sensitization latencies on the day 6 and day 10 retention test, but not on the day 4 retention test. See text for interpretation. Modified from Fleming et al. (1990).

ready occurred at the times that cycloheximide was administered, and therefore a facilitation of maternal behavior is not disrupted when females are tested at either 36 hours or 4 days post–caesarean section. However, maternal experience does induce facilitatory brain changes that require protein synthesis (which would be blocked by the effective cycloheximide treatments), and these effects can be detected on days 6 and 10 post–caesarean section, after the changes induced by hormones have presumably waned. This interpretation assumes that cycloheximide would have been able to block the hormone-induced immediate onset of maternal behavior at 36 hours post–caesarean section if the drug had been administered earlier in order to disrupt hormone-induced protein synthesis, but as far as we are aware, no one has performed this particular study. If hormones induce changes in protein synthesis that facilitate the immediate onset of maternal behavior, and if a maternal experience episode also induces changes in protein synthesis that promote the long-term retention of maternal responsiveness, future research will be needed to determine whether there are any similarities in the induced proteins caused by either hormones or experience. The fact that the hormonal facilitation effect appears to be shorter lasting than the maternal experience-induced long-term retention of maternal behavior (see Chapter 3) suggests that there should be some differences in the proteins that are induced by the different procedures, although this point requires further study.

In the above study, since cycloheximide was injected into the brain ventricles, we are not informed with respect to where in the brain the critical changes in protein synthesis are occurring that mediate the long-term retention of maternal responsiveness. When considering brain regions outside the olfactory bulbs, because of the essential facilitatory role of the MPOA for maternal behavior of rats, and because of the inhibitory involvement of MeA, a simple hypothesis is that critical changes in protein synthesis induced by a maternal experience may occur in one or both of these regions, so that after a maternal experience the MPOA is more easily activated, the MeA exerts a weaker inhibition over maternal behavior, or both. There is some indirect evidence to support aspects of these ideas. Fleming and Korsmit (1996) allowed primigravid female rats to give birth normally, and for half of these females the pups were immediately removed at birth, while the remaining females were allowed 4 hours of maternal experience with their pups. Ten days later, females in each of these groups were exposed to either pups in a plastic perforated box, which would only allow the females to detect distal or exteroceptive pup stimuli, but not proximal stimuli, or simply to the plastic box. The results showed that exteroceptive pup stimuli were capable of activating cFos synthesis in the MPOA of the experienced females, but not in the inexperienced females. One interpretation of these data is that as a result of maternal experience the MPOA is more effectively activated by pup stimuli, including those stimuli that are ineffective in females that have minimal maternal experience. Although it is possible that protein synthesis changes, induced by maternal experience, alter the MPOA so that it is more responsive to pup stimulation, it is also possible that the critical changes are occurring in the brain regions that provide afferent input to the MPOA. Although these results by Fleming and Korsmit (1996) are interesting, the findings of other investigators have failed to confirm that exteroceptive pup stimulation is capable of inducing cFos expression in the MPOA of maternally experienced postpartum females (Lonstein et al., 1998b; Numan & Numan, 1995). Since there were methodological differences among these studies, more work will be needed to resolve this important issue.

With respect to the MeA as a site that might be altered by a maternal experience, Featherstone, Fleming, and Ivy (2000) have reported that more glial fibrillary acidic protein (GFAP) immunoreactive cells were found in this region of primiparous females when compared to multiparous females. Although GFAP is a marker of glial astrocytes, these result do suggest, in a very preliminary way, that previous breeding experience alters the structural organization of MeA. Whether such changes have implications for the facilitated maternal responsiveness observed in experienced females remains to be determined.

Another approach that has been taken to explore the neural basis of the maternal experience effect in rats has been to investigate the effects of brain lesions on this process. Before examining these studies, we should examine what such an approach can actually tell us. First, it appears most probable that the brain changes that underlie the experience effect should take place in those brain regions that play a role in the initial regulation of maternal behavior. In the context of what has already been reviewed, how can one use brain lesions to explore the possibility that the brain changes that underlie the maternal experience effect occur in the MPOA or MeA? For the MPOA such an approach could not be taken, since animals with such lesions do not show normal maternal behavior, irrespective of their previous experience. Exploring the effects of MeA lesions on the long-term retention of maternal behavior is just as problematical, since such animals will show a facilitation of maternal behavior irrespective of their previous experience. The lesion approach, therefore, is not useful for detecting modifications in the MPOA and/or MeA that might mediate the long-term retention of maternal responsiveness. There is a case, however, in which the lesioning method might provide useful information: If a brain region (area X) were not necessary for maternal behavior per se, but were necessary for the development of experience-induced long-term changes in some other region (like the MPOA, for example), then lesions to area X might prevent the maternal experience effect without disrupting maternal behavior per se. Area X might therefore serve a function similar to that ascribed to the hippocampus in the consolidation of declarative or relational memories (Eichenbaum, Otto, & Cohen, 1992; Squire, 1992).

With these issues in mind, we will present the results of a study by Lee et al. (1999). These investigators produced electrolytic lesions or sham lesions in one of the following structures during pregnancy in primigravid female rats: nucleus accumbens, MeA, basolateral amygdala, dorsal hippocampus, prefrontal cortex, piriform cortex, and dorsomedial thalamus. All females were allowed to give birth normally, and were allowed to remain with their pups for 24 hours. Importantly, all females, irrespective of lesion location, showed normal maternal behavior when examined during a 1-hour test at 24 hours postpartum. After this test, pups were removed and all females were given a retention test, to measure their maternal responsiveness, on day 10 postpartum. All females, except those in the nucleus accumbens lesion group, showed short sensitization latencies to the onset of maternal behavior when tested on day 10, with such latencies averaging from 1 to 3 days. In contrast, females that received nucleus accumbens lesions were not affected by the 24 hours of maternal experience, with a large proportion of these females not showing maternal behavior over the 11 days of sensitization testing beginning on day 10 postpartum. The results for the nucleus accumbens females are shown in Figure 8.6. These findings indicate that although these particular nucleus accumbens lesions did not block the initial onset of maternal behavior at parturition, they did block the long-term retention of maternal responsiveness induced by the 24 hours of postpartum maternal experience. One interpretation of these findings is that although the nucleus accumbens might not be critical for maternal behavior per se, it might be involved in the consolidation of a maternal experience by triggering an alteration in some other brain region that is part of the neural machinery that regulates maternal behavior. Such a modification would then allow the subsequent occurrence of maternal behavior to be less dependent upon hormonal stimulation. Additional findings are also consistent with this view of the NA as being important for the consolidation of the long-term retention of maternal responsiveness: Both Lee et al. (2000) and Smith and Holland (1975) performed nucleus accumbens lesions during pregnancy and then followed the course of maternal behavior through the postpartum period. Although relatively normal maternal behavior occurred over the first 2–3 days postpartum, the behavior declined drastically after that point. Refer to Figure 3.2 to see how such an NA lesion effect might be consistent with a role for the nucleus accumbens in consolidation processes.

Note how this consolidation view of NA function diverges sharply from the evidence presented in Chapter 5, which argued strongly for the idea that NA

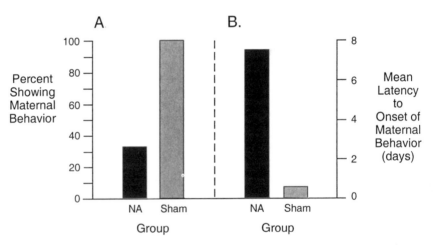

Figure 8.6. The effects of nucleus accumbens (NA) or sham lesions on the long-term retention of maternal responsiveness. Female rats received NA or sham lesions during pregnancy and were allowed to interact with their pups for 24 hours after parturition. All females were subsequently given a retention test to measure maternal responsiveness on day 10 postpartum. At the retention test, NA lesions decreased the percentage of females showing maternal behavior and increased the mean sensitization latency. Modified from Lee et al. (1999).

(and its mesolimbic DA input) is critical for the appetitive aspects of maternal behavior during all of its phases. That is, the evidence indicates that the NA system is an essential part of the neural machinery that regulates maternal behavior. Although it is possible that NA may only be involved in maternal consolidation processes, we feel that we can present an explanation of the above lesion findings that is more consistent with the data presented in Chapter 5. In each of the above studies, when NA lesions did not disrupt the hormonal onset of maternal behavior but did disrupt later occurrences of the behavior, the electrical lesions were partial in nature and left large portions of the NA intact. Perhaps at parturition, when maternal motivation is high, possibly as a result of hormonal stimulation, maternal behavior can occur as long as some of NA remains intact. However, assuming that maternal motivation declines as the postpartum period proceeds, partial lesions of NA might eventually become effective in causing maternal deficits during subsequent postpartum days. Clearly, much more research is needed to resolve these issues and to define the exact functional role of the nucleus accumbens in the regulation of processes relevant to maternal behavior control. These are extremely important issues.

Even though we feel that the overall data show that NA is part of the neural circuitry underlying maternal behavior, please note that such a role for NA does not rule out the possibility that the maternal experience effect is mediated by a molecular alteration within NA. Indeed, it is an attractive hypothesis that maternal experience alters the nucleus accumbens in some way so that it is more easily activated by the MPOA in the absence of hormonal mediation. However, the lesion approach is not the best methodology to test this hypothesis, which is distinct from the consolidation hypothesis outlined above.

Neurochemical approaches have also been taken to explore the neural basis of the maternal experience effect, and some evidence exists that suggests that experience-induced modifications in central oxytocinergic and central opioid systems may influence the long-term retention of maternal responsiveness. In Chapters 5 and 6 we reviewed the evidence that central oxytocin systems, mainly arising from the PVN and projecting to a variety of sites that include the MPOA/BST, LS, VTA, and olfactory bulbs, play a role in the immediate onset of maternal behavior at parturition in rats, sheep, and other species. Also recall that the physiological events associated with pregnancy and parturition cause an upregulation of oxytocin receptors in certain brain regions, such as the MPOA, VTA, and BST, and that such increases appear to be crucial for the immediate onset of maternal behavior. An obvious hypothesis is that an initial maternal experience modifies the brain in such a way so that central oxytocin systems become more excitable by pup stimuli even after the physiological influences associated with pregnancy termination have waned, and this increased excitability underlies the long-term retention of maternal responsiveness. One might initally reject this hypothesis because of the results that have shown that the PVN and oxytocin are essential for the onset, but not the maintenance, of maternal behavior in rats. That is, PVN lesions and oxytocin antagonists disrupt the onset of maternal behavior when administered prepartum or at parturition, but do not block maternal behavior when administered during the postpartum period, after the behavior has become established (see Chapters 5 and 6). However, a recent study has compared the effects of microinjections of an oxytocin antagonist into the VTA on the maternal behavior of postpartum female rats who had either been separated from their pups for several days (beginning on day 5 postpartum), or who had remained with their pups without a period of separation (Pedersen, 1997; Pedersen et al., 1995). When examined during a 2.5-hour test on day 10 postpartum, an oxytocin antagonist did not disrupt the maternal behavior of females who remained with their pups, but did disrupt the reemergence of the behavior in females who had been separated from their pups. One interpretation of these findings is that once maternal behavior is initiated, oxytocinergic systems are no longer required for the continuance of maternal behavior if females remain with their pups, but if the pups are removed for a period of time and then returned, oxytocin systems may be required for a short latency reemergence of the behavior.

These results suggest, but by no means prove, that a modification of oxytocin systems by maternal experience may contribute to the long-term retention of maternal responsiveness in rats. It is possible, of course, that it is other systems that are modified by a maternal experience, and that such modifications allow an oxytocin system that remains unaffected by the experience to exert its effects. In support of the idea that oxytocin systems may not be modified by maternal experience, the increases in oxytocin receptor density that are induced in several brain regions by the physiological events associated with pregnancy termination are *not* maintained into the postpartum period and after weaning (Insel, 1990b; Pedersen, 1997; Pedersen et al., 1994). However, in support of the idea that central oxytocin systems might be altered by maternal experience, several studies have shown that the synaptic architecture of hypothalamic oxytocin neurons in the supraoptic nucleus and in the PVN is modified in postpartum lactating rats, in comparison to virgin controls, and that such changes persist for variable amounts of time after a female is separated from her litter (Hatton & Ellisman, 1982; Hatton & Yang, 1994; Modney & Hatton, 1994; Modney, Yang, & Hatton, 1990; Montagnese, Poulain, Vincent, & Theodosis, 1987; Theodosis & Poulain, 1984, 1989). These changes include an increase in dendrodendritic and somasomatic contacts among oxytocin neurons, and an increase in the number of double synapses, in which an afferent axon terminal to the PVN and SON is seen to contact two postsynaptic neurons. These changes presumably result in oxytocinergic neurons that are more easily excitable, and are more capable of synchronized activity, which would cause the release of large amounts of oxytocin into terminal regions. Importantly, these structural changes disappear quickly (in a matter of days) if pups are removed at birth, but may persist for a month or more after pup removal if a female is first allowed to raise a litter to weaning (Montagnese et al., 1987; Theodosis & Poulain, 1984). Since these changes were detected in oxytocinergic magnocellular neurons, they may be more relevant to alterations in neuroendocrine, rather than neurobehavioral systems. However, it is possible that these changes, or related structural changes in parvocellular oxytocinergic neurons, may influence maternal responsiveness. It is interesting to speculate that a maternal experience modifies oxytocinergic systems relevant to maternal behavior so that nonsuckling pup stimuli are more capable of activating the release of large amounts of oxytocin into critical brain sites, and that such activation facilitates the rapid reemergence of maternal behavior in females that have been separated from their pups for a period of time.

With respect to the involvement of central opioid systems in the maternal experience effect, first recall the role that opioids play in maternal behavior. As reviewed in Chapter 6, opiates can have dual effects on maternal responsiveness in rats, and the specific effect is dependent upon the opioid system that is involved. In particular, β-endorphin acting on the MPOA has been found to have inhibitory effects on maternal behavior (Mann & Bridges, 1992), while an opioid system acting on the VTA has been found to facilitate maternal responsiveness (Thompson & Kristal, 1996). An interesting possibility is that as a result of a maternal experience, endogenous opioid inhibition on the MPOA is decreased and/or opioid facilitation at the level of the VTA is increased. Mann and Bridges (1992) have provided some evidence that is consistent with the first part of this hypothesis. They found that microinfusions of β-endorphin into the MPOA were much more effective in disrupting the maternal behavior of primiparous lactating females than they were in disrupting the behavior in age-matched multiparous females (one previous lactation).

Concerning the relevance of the positive opioid system to the maternal experience effect, Byrnes and Bridges (2000) performed the following experiment: Primigravid female rats received intracerebroventricular (ICV) injections of either a long-acting μ opioid antagonist, β-funaltrexamine, or sterile water, on day 21 of pregnancy. Such females were allowed to give birth normally and were allowed to remain with their pups for 3 hours after parturition. Maternal behavior tests indicated that all females showed normal maternal behavior at this time. The pups were then removed from all females, and 7 days later the females were tested for the retention of a high level of maternal responsiveness. Females treated with the vehicle solution showed a short latency to onset of maternal behavior at the retention test, while the females treated with the opiate antagonist showed a delayed onset to maternal behavior. One interpretation of these results is that the antagonist disrupted the maternal experience effect. Although the central site of action of the opioid antagonist cannot be determined from this study, we will consider one possibility. Since opioid action at the level of the VTA indirectly activates the mesolimbic dopamine system through a disinhibiton mechanism (Bontempi & Sharp, 1997; Kalivas, 1985; Kalivas & Stewart, 1991), it is possible that ICV administration of the opioid antagonist *partially* dampened the effects of mesolimbic DA activation during the immediate postpartum period, and although this effect was not strong enough to block the onset of maternal behavior (see Thompson & Kristal, 1996, for evidence that direct intra-VTA injections of opiate antagonists can depress the onset of maternal behavior in rats), it might have disrupted the mechanisms responsible for the development of the long-term retention of maternal behavior. Indeed, these findings are reminiscent of the work of Lee et al. (1999), described above, during which it was found that partial prepartum lesions of the nucleus accumbens blocked the effects of a parturient maternal experience on future maternal responsiveness. With respect to the opioid effects, it is interesting to speculate that opioid action at the level of the VTA during an initial parturition might sensitize the VTA system to opioid action (see Kalivas & Stewart, 1991), so that at a future time point smaller amounts of opioids acting on μ receptors in the VTA are capable of effectively activating the mesolimbic DA system, in this way facilitating the onset of maternal behavior (see Thompson & Kristal, 1996).

The preceding review on the involvement of various brain systems in the maternal experience effect in rats was based on the directly relevant literature. In what follows, we will suggest some additional possibilities about the neural basis of the long-term retention of maternal behavior that are based on much more indirect evidence. First, future research might consider the possibility that modifications in neural systems that include the lateral septum (LS) may be involved in the long-term retention of maternal responsiveness. Research on rodents has shown that the LS is involved in social memory processes related to olfactory-based individual recognition of conspecifics (see Sheehan and Numan [2000] for a review). Since the LS receives input from both MeA and MPOA (Chapter 5; Sheehan & Numan, 2000), it is in a position to integrate olfactory input with neural information about the occurrence of maternal behavior (recall from Chapter 5 that MPOA neurons that express Fos during maternal behavior project to LS). Finally, remember from Chapter 5 that the output of the ventral part of LS (LSv) may be involved in promoting aversive responses to pup stimuli, while the output of the intermediate part of LS (LSi) may be involved in inhibiting defensiveness. It is worth speculating that as a result of an initial maternal experience, the circuitry of LS systems is modified in some way so that olfactory stimuli from pups no longer activate LSv mechanisms underlying aversion and withdrawal, but instead may potentiate the output of LSi.

Second, cellular models of neural plasticity have shown that electrical stimulation of central neural circuits can result in either long-term potentiation (LTP) or long-term depression (LTD) of synapses within the circuit, and that the outcome (potentiation or depression) is dependent, in part, on the initial frequency of electrical stimulation and on the amount of Ca^{2+} ions that enter the excited neurons (Beggs et al., 1999; Luscher, Nicoll, Malenka, & Muller, 2000). The mechanisms underlying LTP and LTD may serve as useful models for the mechanisms that underlie the effects of a maternal experience on future maternal responsiveness. In particular, during an initial maternal experience, perhaps neural regions positively involved in maternal behavior (MPOA, VTA, for example) undergo changes analogous to LTP, while neural regions that inhibit maternal behavior (MeA, AHN/VMN, PAG, for example) undergo a process similar to LTD. If such were the case, pup stimuli would be more likely to strongly excite positive circuits relevant to maternal behavior and would be less likely to strongly excite neural circuits that serve inhibitory functions with respect to maternal behavior. In regard to such hypotheses, it is worth pointing out that mechanisms similar to LTD have been shown to occur in the amygdala and in the PAG, and that LTP has been shown to occur in the VTA (Adamec, 2001; Bonci & Malenka, 1999; Li, Weiss, Chuang, Post, & Rogawski, 1998; also see Watanabe, Ikegaya, Saito, & Abe, 1996). (Conversely, it should be noted that if processes similar to LTP were to occur in MeA, we would predict inhibitory effects on maternal behavior. As reviewed in Chapter 5, this is exactly what has been reported by Morgan et al. [1999], where it was found that kindling stimulation of MeA increased the sensitization latencies of female rats who were exposed to pups 9 days after the kindling procedure.)

If the above hypotheses are correct, one might expect glutamate receptors of the NMDA type to be involved in the maternal experience effect because such receptors play a positive role in many examples of LTP and LTD (Beggs et al., 1999; Luscher et al., 2000). Malenfant, O'Hearn, and Fleming (1991b) tested this idea by systemically injecting female rats with MK801, an NMDA antagonist, either immediately before or immediately after a 1-hour maternal experience. Such injections did not disrupt the long-term retention of maternal responsiveness induced by the maternal experience, suggesting that NMDA receptors are not involved in the maternal experience effect. It should be noted, however, that the systemic dosage of MK801 that was employed may not have effectively blocked central NMDA receptors. Even if NMDA receptors are not involved in the maternal experience effect, this does not mean that mechanisms analogous to LTP and LTD are not playing a role in the plasticity of neural mechanisms underlying maternal behavior since there are forms of LTP and LTD that do not require the activation of NMDA receptors (Beggs et al., 1999; Daniel, Levenes, & Crepel, 1998).

Continuing along this line of inquiry, it is interesting to speculate that the hormonal changes of late pregnancy set up a neural state that allows a maternal experience to cause changes in maternally relevant neural circuits that are akin to either LTP or LTD, depending on the particular circuit affected. Such a view is consistent with the fact that a maternal experience is more effective in causing long-term changes in maternal responsiveness if it occurs in hormonally primed rats than if it occurs in sensitized adult virgins (see Chapter 3). A heuristic model of how hormones might set up such a state comes from work on the effects of estradiol on hippocampal morphology and function. Estradiol increases the density of dendritic spines on hippocampal pyramidal cells (Murphy & Segal, 1996; Woolley, Weiland, McEwen, & Schwartzkroin, 1997). This increase in spine density is associated with an increased responsiveness of pyramidal cells to the excitatory effects of glutamate acting at NMDA receptors, and this, in turn, is related to an increase in the number of NMDA receptors on the pyramidal cells (Gazzaley, Weiland, McEwen, & Morrison, 1996; Weiland, 1992; Woolley et al., 1997). Additional work has provided a mechanism through which these changes may be induced by estradiol. Murphy, Cole, Greenberger, and Segal (1998) have indicated that estrogen receptors are not located within hippocampal pyramidal cells, but instead are located within GABAergic hippocampal interneurons, and that estradiol decreases GAD (glutamic acid decarboxylase, the enzyme needed for GABA synthesis) levels within these neurons through a genomic mechanism of action (also see Murphy & Segal, 1996). They hypothesize that the decrease in GAD within the inhibitory interneurons lowers the level of GABAergic inhibition on the pyramidal cells, which, in turn, causes an increase in excitatory drive on the pyramidal cells. The increased excitation on the pyramidal cells subsequently causes intracellular changes that induce the formation of increased dendritic spines (see Murphy & Segal, 1997). It is worth pointing out that the estradiol-induced increase in hippocampal spine density is transient. Indeed, the degree of spine density on hippocampal cells varies during the estrous cycle, being highest at proestrus, when estradiol levels are highest (Woolley & McEwen, 1992). Importantly, hippocampal LTP is enhanced when it is induced in proestrus rats compared to when it is induced in rats during other phases of the estrous cycle (Warren, Humphreys, Juraska, & Greenough, 1995).

Can this hippocampal model that we have just described be transferred to maternally relevant neural circuits? Let us use the MPOA as an example, although what we have to say can also be generalized to other areas. Perhaps the hormonal events of late pregnancy act on the MPOA to acutely increase spine density (or some other aspect that affects responsiveness to excitatory neural input), and perhaps this resultant increase in MPOA responsiveness to excitatory input plays a role in the initial stimulation of the onset of maternal behavior. More importantly, when maternal behavior occurs during this period of hormonal priming, perhaps the "prepared" MPOA allows the stimuli associated with maternal behavior to produce persistent changes in MPOA neurons, which can be described as a long-term potentiation or as a long-term increase in excitability. In this way, in the future, pup stimuli might be capable of quickly activating MPOA circuits in the absence of hormonal mediation. As far as we know, no one has directly measured spine density in the MPOA in periparturient rats (but see Anderson, 1982), so we have no definitive information as to whether MPOA spine density increases in association with the onset of maternal behavior. It is worth indicating that spine density in other hypothalamic nuclei has been shown to be positively affected by estradiol (Calizo & Flanagan-Cato, 2000). Very importantly, a recent study has shown that other aspects of the morphology of MPOA neurons are modified by the hormonal events that occur at the end of pregnancy (Keyser-Marcus et al., 2001). Although spine density was not measured, these authors report that the total basal dendritic length of MPOA neurons was increased in late pregnant (day 21) primigravid female rats. In a related preliminary report, Francis, Brake, McEwen, Allen, Greengard, and Insel (2001) found increased levels of spinophilin in the MPOA of late pregnant rats. Since spinophilin is a protein found predominantly in dendritic spines, increased levels are highly suggestive of increases in spine density. Could a depression of GABAergic mechanisms be involved in these effects? In this regard, there are contrasts between how estradiol acts on the MPOA and how it acts on the hippocampus. According to Herbison (1997), estradiol does not affect GAD levels within MPOA neurons, and therefore there is a limit to how far we can carry the analogy to the hippocampal model. However, estradiol is capable of influencing the subunit composition of GABA receptors within the MPOA, and it is also capable of increasing GABA transporter activity (Herbison, 1997; Herbison & Fenelon, 1995). Therefore, one might still be able to create a scenario where the hormonal events of late pregnancy, which include rising estradiol and declining progesterone, depress GABAergic inhibition on selected MPOA output neurons, in this way increasing their overall excitatory input, which, in turn, may lead to additional structural and other changes within these neurons that enable them to be even more receptive to excitatory neural inputs (also see Parducz, Perez, & Garcia-Segura, 1993). An additional relevant finding is that progesterone withdrawal leads to a hyposensitivity of GABA receptors in certain parts of the brain (Smith et al., 1998; also see Brussard & Herbison, 2000; Fenelon & Herbison, 1996). Finally, another potentially resolvable lack of congruence between the hippocampal and MPOA response to estradiol is that estradiol treatment increases NMDA glutamate receptors in the hippocampus, but not in the MPOA (Weiland, 1992). Importantly, however, estradiol does increase non-NMDA glutamate receptors in the MPOA (Diano, Naftolin, & Horvath, 1997; Weiland, 1992; also see Karlsson, Sundgren, Nasstrom, & Johansson, 1997). In conclusion, although the details may be different concerning the effects of estradiol on MPOA and hippocampus, it is certainly possible that at parturition there is an increase in excitation on MPOA neurons, and that if such excitation is coupled with the occurrence of maternal behavior, long-term changes may occur in such neurons that increase their future excitability to pup-related stimuli. Additional findings on alterations in the molecular architecture of MPOA neurons in periparturient rats are consistent with the ideas being presented here (O'Day, Payne, Drmic, & Fleming, 2001).

Another model of neural plasticity that may have relevance to the brain changes that underlie the maternal experience effect is that of the brain changes induced by repeated intake of drugs of abuse. The mesolimbic DA system is not only involved in the control of a variety of motivated behaviors, but is also involved in the rewarding effects of drugs of abuse such as cocaine, amphetamine, and morphine, all of which activate the system (Koob, 1999). Importantly, repeated administration of these drugs results in a process referred to as sensitization wherein the rewarding effects of the drugs are increased with their repeated use (Koob, 1999). In other words, sensitization increases the responsiveness of the mesolimbic DA system to the drugs. The analogy to the maternal experience effect is the possibility that as a result of a maternal experience, the mesolimbic DA system may become more reactive to pup stimulation, in this way allowing maternal behavior to occur more easily in the absence of hormonal priming. In particular, perhaps some of the changes that occur in the mesolimbic DA system during drug sensitization are also induced by a maternal experience. In this regard, it should be noted that sensitization to the rewarding effects of cocaine and/or morphine has been related to changes in the subunit composition of glutamate receptors in the nucleus accumbens and in the VTA (Carlezon et al., 1997, 2000; Fitzgerald, Ortiz, Hamedani, & Nestler, 1996; Kelz et al., 1999). Importantly, Fos proteins may mediate some of these changes (Kelz et al., 1999; Nestler, Kelz, & Chen, 1999). Interestingly, repeated administration of amphetamine to rats has also been found to produce structural alterations within the nucleus accumbens, increasing dendritic spine density on accumbens output neurons (Robinson & Kolb, 1997). Based on the involvement of the mesolimbic DA system in maternal behavior, we believe it will be worthwhile to explore whether any overlap exists between the effects of drug sensitization on this system and the effects of maternal experience on this system. The fact that a positive endogenous opioid system appears important for the maternal experience effect (see above) increases the attractiveness of such an investigatory approach. Significantly, a recent finding has shown that rats with previous maternal experience are more sensitive to the effects of apomorphine, a DA receptor agonist, than are nulliparous females (Byrnes, Byrnes, & Bridges, 2001).

We will conclude our analysis of the maternal experience effect by indicating that the evidence we have reviewed suggests that the long-term retention of a high level of maternal responsiveness in rats after an initial maternal experience is probably the result of modifications in several components of the neural networks that influence maternal behavior.

2.5. Possible Central Neural Mechanisms through Which Early Life Experiences Might Influence the Development of Adult Maternal Responsiveness

In this section we want to explore the data that examine the neural modifications that might mediate the effects of early life experiences on the development of adult maternal responsiveness. As reviewed in Chapter 3, the work of Harlow, Suomi, and colleagues found that motherless mother monkeys showed aberrant maternal responsiveness, and the research of Fairbanks and others showed that the level of protectiveness shown by a mother monkey toward its infants influences the level of protectiveness that these offspring show to their own infants once they reach maturity and reproduce. Also in Chapter 3, we noted the somewhat similar findings on rats from Fleming's group, who found that adult rats who were exposed to either complete maternal deprivation as neonates or to prolonged periods of maternal separation as neonates showed reductions in maternal behavior, which included decreases in nursing behavior and maternal licking and grooming. Finally, recall that the research of Meaney and colleagues showed that the level of licking, grooming, and arched-back nursing (LG-AGN) that a mother rat displays toward her pups influences the development of those maternal characteristics in her offspring. In this section we are concerned with how maternal deprivation or maternal treatment of offspring might influence the offsprings' brain development so that their adult maternal behavior is affected. In Chapter 3, our dominant hypothesis was that the kind of maternal treatment, or lack thereof, that offspring receive influences their emotional development, which in turn influences their adult maternal behavior. In particular, we argued that maternal deprivation, high maternal protectiveness, or exposure to low levels of LG-ABN result in offspring that are more fearful, and that this increased fearfulness has ramifications for the maternal behavior of the offspring. This hypothesis is compelling from both a behavioral and neural point of view. As reviewed in Chapter 4, fearfulness typically decreases during the postpartum period, suggesting that such reductions allow for more effective maternal behavior, particularly under stressful and demanding environmental conditions. We actually argued in Chapter 4 that early environmental manipulations that result in increases in adult fearfulness might actually lead to a major deterioration of maternal behavior under stressful environmental conditions. However, even under relatively benign environmental conditions, it is easy to conceive of how the increased wariness that is associated with fearfulness would lead to a female that is easily distractable, so that such a female would nurse and lick her pups less than a more confident female would. From a neural point of view, as shown in Figure 5.14, the central neural regions that regulate fear-related behaviors also project to neural regions such as the MPOA that regulate maternal responsiveness, and we have argued in Chapter 5 that such projections are likely to be inhibitory. It makes sense that if two motivational states are in conflict (maternal motivation with attention toward pups; general fearfulness with attention toward the external environment), when one state increases the other may show modifications of some sort. It needs to be emphasized, however, that the hypothesis that early environmental influences on adult maternal behavior are mediated by alterations in emotionality is, for the most part, based only on correlational evidence. For example, although rats that have been exposed to frequent periods of prolonged maternal separation as neonates are more fearful in adulthood and also show reductions in nursing and maternal licking of their own pups, it has not been proven that the alterations in emotionality are the cause of the alterations in maternal behavior. Indeed, some have argued against such a causal relationship (Gonzalez et al., 2001; Lovic et al., 2001). Therefore, from a neural point of view, additional possibilities should be considered, and we would like to propose the following. Adverse early environmental experiences, such as maternal deprivation or poor mothering, could decrease and/or modify the adult maternal behavior of the affected offspring through the induction of enduring developmental outcomes that cause one or more of the following effects: (a) an increase in the activity of neural regions regulating fearfulness and anxiety-related behavior; (b) a decrease or dysregulation in the activity of brain regions regulating *general* appetitive motivation: In this case, such effects should not only cause decreases in adult maternal behavior, but alterations in other sorts of appetitive motivations should also occur; and (c) a decrease in the activity of those brain circuits more closely related to maternal behavior. Using maternal deprivation as an example, perhaps such an early adverse experience alters the development of the nervous system so that in adulthood the medial hypothalamus/periaqueductal gray fear system is more active, which causes increases in anxiety and decreases in maternal behavior. Alternatively, perhaps activity in the mesolimbic dopamine system is depressed by maternal deprivation, in this way causing depressions in a variety of appetitive behaviors, including interest in pup-related stimuli. Finally, perhaps adult MPOA functioning could be directly affected by maternal deprivation such that it is more refractory to olfactory and tactile inputs from pups, and therefore less capable of activating the mesolimbic dopamine system (or brainstem sensorimotor mechanisms) in response to these inputs. In the sections that follow, we will explore the evidence favoring each of these possibilities. (For a related review see Fleming, Kraemer, Gonzalez, Lovic, Rees, and Melo [2002]).

2.5.1. The Effects of Maternal Deprivation on the Development of Neural Systems Regulating Fear and Anxiety

As outlined in Chapter 3, exposure of neonatal rats to prolonged periods of maternal separation results in adults who are more fearful, who show reductions in nursing and maternal licking of their offspring, and who also display increased hypothalamic-pituitary-adrenal (HPA) responses to stressful stimuli. With respect to the latter effect, adult rats who have previously been exposed to maternal separation have increased levels of hypothalamic corticotropin releasing factor (CRF) mRNA and immunoreactivity in comparison to controls (Ladd et al., 1996; Plotsky & Meaney, 1993). Recall that CRF produced by parvocellular cells in the hypothalamic PVN stimulates ACTH release from the anterior pituitary. Since CRF is not only a neurohormone that is released into the pituitary portal system, but is also a neurotransmitter/neuromodulator that is released at synapses in diverse parts of the brain (Morin, Ling, Liu, Kahl, & Gehlert, 1999; Sakanaka et al., 1987; Steckler & Holsboer, 1999), these results raise the possibility that maternal separation may cause an upregulation of CRF activity in diverse brain regions. Importantly, activity in central CRF neural systems has been related to the promotion of fearfulness and anxiety-related behaviors (Dunn & Berridge, 1990; Schulkin, 1998; Steckler & Holsboer, 1999), which would be consistent with the increased fearfulness shown by adult rats that had been exposed to periods of prolonged maternal separation. Also worth recalling is that intracerebroventricular administration of CRF has been found to disrupt maternal behavior in rats (Almeida, Yassouridis, & Forgas-Moya, 1994; Pedersen et al., 1991). Therefore, a proposal would be that maternal separation causes an upregulation of central CRF systems that would then increase fearfulness and reduce maternal behavior. There is evidence that supports one aspect of this proposal: Maternal separation does cause an upregulation of CRF systems outside the HPA system. It has been reported that maternal separation leads to the following changes in the adult rat brain: increases in CRF mRNA levels in the central nucleus of the amygdala (CNA), and increases in CRF protein levels and CRF receptors in the locus coeruleus (LC) of the pons (Francis et al., 1999a; Ladd et al., 2000). Since there is evidence for a CRF neural system that extends from the CNA to LC (Steckler & Holsboer, 1999; Valentino et al., 1998), these results suggest that maternal separation leads to increased activity in this pathway in the adult brain. Other research indicates that this pathway promotes aspects of anxiety-related behaviors (Caldji et al., 1998; Ladd et al., 2000). The CNA has long been known to play a role in the neural integration involved in the production of both behavioral and autonomic fear-related responses (Davis, 1992; LeDoux, 1993; Maren & Fanselow, 1996). It receives a variety of sensory inputs, relayed through the lateral amygdala, and in turn projects to several other relevant neural regions. For example, in addition to projecting to the LC, the CNA also projects to the periaqueductal gray (PAG), which we have previously emphasized as being an important regulator of fear-related responses (Bandler, 1988; Behbehani, 1995). Additional anatomical and functional points are that LC projects to PAG (Cameron et al., 1995), and that ascending noradrenergic projections from the LC to the neocortex and hippocampus are involved in regulating the level of behavioral arousal, vigilance, and selective attention (Valentino et al., 1998). Finally, a variety of stressful stimuli have been found to promote neural activity in LC norepinephrine neurons, and such activation is, in part, the result of CRF stimulation (Sauvage & Steckler, 2001; Valentino et al., 1998).

Additional evidence has been presented supporting the idea that maternal separation leads to increased activity in a CNA-to-LC circuit in the adult rat brain. Maternal separation has been associated with decreases in GABAA receptor density in LC, and decreases in benzodiazepine receptor density in LC and CNA (Caldji et al., 2000; Francis et al., 1999a; Ladd et al., 2000), which, as the authors indicate, should result in decreased inhibitory effects on the CNA-to-LC system. Furthermore, we should note that congruent data have been obtained from the offspring of mothers who showed either high LG-ABN or low LG-ABN. The adult offspring of low LG-ABN mothers, who display increases in fearfulness as adults, present neurochemical data consistent with increased activity in a CRF system from CNA to noradrenergic LC neurons, while the converse is true for the less fearful offspring of high LG-ABN mothers (Caldji et al., 1998; Francis et al., 1999b). Finally, relevant data from primates have been provided by Coplan et al. (1996) and by Higley et al. (1992). Coplan et al. (1996) found that adult bonnet macaques (*Macaca radiata*) who were reared by mothers who showed aberrant maternal behavior had higher levels of CRF in their cerebrospinal fluid then did normally reared monkeys (such monkeys are also more fearful: Rosenblum & Andrews, 1994). Higley et al. (1992) reported that peer-reared rhesus monkeys (*Macaca mulatta*) had higher cerebrospinal fluid levels of 3-methoxy-4-hydroxyphenolglycol, an NE metabolite, which would be consistent with a CRF-induced hyperactivation of LC NE neurons.

If adverse early experiences, such as maternal deprivation or separation, cause an upregulation of a CNA-to-LC CRF pathway in the adult so that the hypothesized resultant increase in fearfulness also interferes with maternal behavior, how could such an outcome be incorporated into the neural model we have presented in Chapter 5 for the regulation of maternal behavior? One possible model is shown in Figure 8.7. Although the MeA may relay aversive olfactory input to a central aversion system, as outlined previously, perhaps the CNA is the avenue through which other, nonolfactory stimuli gain access to a central aversion system. Projections of the CNA to the PAG and LC may activate both fearfulness and vigilance, and in an adult who had been exposed to early maternal separation, these systems may be hyperactive. Since the MPOA is reciprocally connected to the PAG (see Chapter 5), it is conceivable that increased PAG activity not only promotes fear-related responses, but also reduces maternal behavior through an hypothesized inhibitory projection from PAG to MPOA. Hyperactivation of an LC-to-PAG projection might also make the PAG hypersensitive to its other afferents, while hyperactivity in LC projections to the neocortex and hippocampus might cause a hypervigilant state that makes it difficult for the mother to focus her attention on her pups, making her easily distractable. All of these effects could result in a mother who nurses and grooms her young at lower than normal levels, and these effects might even cause a severe deterioration of maternal behavior under stressful environmental conditions.

With respect to this model, one can suggest that the decrease in general fear-

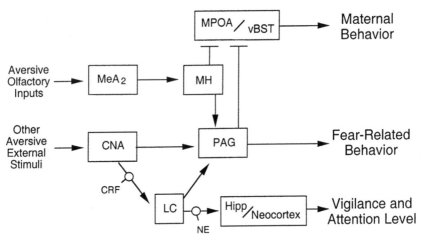

Figure 8.7. A neural model, based on known anatomical connections, which shows how stressful and aversive olfactory and nonolfactory external events might activate a central aversion system that could then result in a depression of maternal responsiveness and a concomitant increase in fear-related behavior and hypervigilance. This figure is an elaboration of Figure 5.14. See text for details. Abbreviations: CNA = central nucleus of the amygdala; CRF = corticotropin releasing factor; Hipp = hippocampus; LC = locus coeruleus; MeA$_2$ = a cell population in the medial amygdala that projects to the medial hypothalamus; MH = medial hypothalamus (includes anterior hypothalamic nucleus, ventromedial nucleus, and dorsal premammillary nucleus); MPOA = medial preoptic area; NE = norepinephrine; PAG = periaqueductal gray; vBST = ventral bed nucleus of the stria terminalis. Projections ending in an arrow are hypothesized excitatory connections, and those ending in a vertical bar are hypothesized to be inhibitory.

fulness that is shown by *normally reared* lactating rodents (see Chapter 4) may be caused by a downregulation of the CNA-to-PAG/LC pathway by the physiological events associated with pregnancy termination and lactation. Such a hypothesis is supported by the finding that CRF mRNA levels in CNA are lower in normally reared lactating rats than in their virgin counterparts (Walker et al., 2001), and is consistent with a decrease in central noradrenergic tone that has been reported to occur in normally reared pospartum females (Liu et al., 2000; Toufexis et al., 1998, 1999). Furthermore, such a hypothesis also fits with the fact that oxytocin receptors are located in the CNA (Young et al., 1997a), that central oxytocin systems are involved in the decreased level of fearfulness shown by lactating females (Neumann et al., 2000; Nissen et al., 1998; Uvnas-Moberg et al., 2000; Windle et al., 1997b), and that oxytocin action at the level of CNA causes an anxiolytic effect (Bale et al., 2001). It is tempting to speculate, as indicated in Chapter 6, that although oxytocin action at the level of the olfactory bulbs, MPOA, and VTA may promote the immediate onset of maternal behavior at parturition, the continued action of oxytocin at the level of CNA throughout the postpartum period may be involved in suppressing CNA output, in this way causing the decrease in general fearfulness associated with the postpartum period. Significantly, Champagne et al. (2001) have reported that lactating rats who display high levels of LG-ABN also have higher levels of oxytocin receptors in CNA than do their low LG-ABN counterparts, and that intracerebroventricular administration of an oxytocin receptor antagonist decreases the amount of LG-ABN shown by postpartum rats. Importantly, the adult female offspring of high LG-ABN dams have increased oxytocin receptor binding in CNA when compared to the adult female offspring of low LG-ABN dams (Francis, Young, Meaney, & Insel, 2002). Furthermore, in a preliminary report, Boccia and Pedersen (1999) indicate that early maternal separations are associated with decreased oxytocin binding in the adult CNA and increased fearfulness during the postpartum period. Finally, since we argued in Chapter 4 that a general decrease in fearfulness during the postpartum period may be necessary for the appropriate occurrence of maternal aggression, it is relevant to recall that maternally deprived rats have been found not only to be more fearful in adulthood, but also to show lower levels of maternal aggression when compared to control females (Boccia & Pedersen, 2001). Importantly, lesions of the paraventricular nucleus of the hypothalamus, which would presumably decrease oxytocin input to CNA, have been found to depress maternal aggression in postpartum rats (Consiglio & Lucion, 1996; cf. Giovenardi, Padoin, Cadore, & Lucion, 1998; Neumann et al., 2001) Also, there is evidence that increased CNA activity may cause decreases in maternal aggression (Hansen & Ferreira, 1986b; also see Qureshi, Hansen, & Sodersten, 1987).

   It cannot be emphasized too strongly that the model we have presented is highly speculative and is based on correlational and incomplete evidence. Although maternal separation of neonates does appear to cause an upregulation of CRF systems that project from CNA to LC, it has not been shown that such upregulation is the basis for either the increased fear or decreased maternal behavior shown by these females. For starters, it would be important to show that treatment of adult rats that had been exposed to prolonged maternal separations as neonates with either CRF antagonists, GABAA agonists, or CNA lesions would be capable of normalizing both their fearfulness and their maternal behavior. One other point is also worth considering. In addition to upregulating a CNA-to-LC CRF system, early maternal separations may upregulate other CRF systems, and it might be the increased activity in these systems that causes increased fearfulness and reductions in adult maternal behavior. In particular, there is a CRF projection from the medial amygdala to the ventromedial hypothalmus (Sakanaka, Shibasaki, & Lederis, 1986), and it appears that early maternal deprivation activates this system in neonates (Eghbal-Ahmadi, Avishai-Eliner, Hatalski, & Baram, 1999; Eghbal-Ahmadi, Hatalski, Avishai-Eliner, & Baram, 1997). Perhaps an upregulation of this system causes reductions in adult maternal behavior, particularly since we have argued in Chapter 5 that MeA projections to AHN/VMN are inhibitory for maternal behavior. A related consideration is that CRF input to the lateral septum, presumably originating from the AHN (Sakanaka, Magari, Shibasaki, & Lederis, 1988), has been found to promote anxiety-related behaviors (Radulovic Ruhmann, Liepold, & Spiess, 1999). In this regard, note that oxytocin receptors are located in LS, and that high LG-ABN dams have more oxytocin receptor binding in this region than do low LG-ABN dams (Champagne et al., 2001; also see McCarthy et al., 1996). Finally, as noted in the discussion of oxytocin receptors in Chapter 6, there is a projection from the oval nucleus of BST to the caudal region of the dorsal raphe, linking CRF to a mesolimbic serotonergic system. Although there is no evidence as yet that this system is a target of early adverse experience, this is one additional site where effects on CRF activity could influence maternal behavior and fearfulness in adulthood.

2.5.2. The Effects of Maternal Deprivation on the Development of Neural Systems Regulating General Appetitive Motivational Systems

Ladd et al. (2000) have reported that adult rats exposed to prolonged periods of maternal separation as neonates displayed what they referred to as anhedonia. Compared to controls, such rats showed decreased ingestion of a sweet sucrose solution. Other evidence also suggests that maternally separated rats show decreased responsiveness toward reward and reward-related stimuli in adulthood (Matthews, Wilkinson, & Robbins, 1996b). Since the mesolimbic DA system is involved in regulating general appetitive motivational processes, these results are consistent with the possibility that early maternal deprivation modifies the development of this system. In support of this view, additional findings have shown that maternally deprived rats are less behaviorally responsive in adulthood to systemic treatment with a presynaptic DA agonist (amphetamine), and more behaviorally responsive to systemic treatment with a postsynaptic DA antagonist (sulpiride) (Matthews, Hall, Wilkinson, & Robbins, 1996a; cf. Weiss, Domeney, Heidbreder, Moreau, & Feldon, 2001). For example, amphetamine-induced locomotion was reduced in maternally deprived female rats. These findings fit with the view that maternal deprivation may decrease the functional activity of the mesolimbic DA system. In order to test this idea directly, Hall, Wilkinson, Humby, and Robbins (1999), in an in vivo microdialysis study, measured the release of DA into the nucleus accumbens in response to systemic amphetamine treatment or in response to direct application of potassium ions (a depolarizing agent) to DA axon terminals in the nucleus accumbens. Surprisingly, they found that amphetamine and K^+ actually caused an enhanced release of DA into the nucleus accumbens of adult rats that had been maternally deprived as neonates. In order to reconcile the behavioral data with the physiological data, they suggest the maternal deprivation might operate to decrease the postsynaptic responsiveness of nucleus accumbens neurons to DA, and that the increased DA release may act as a partial compensatory mechanism. It is possible, of course, that maternal deprivation induces additional effects that alter the responsiveness of the adult rat to rewarding stimuli (Matthews, Dalley, Matthews, Tsai, & Robbins, 2001).

Work on other species also indicates that maternal deprivation can depress the development of certain dopaminergic systems in the brain. Martin, Spicer, Lewis, Gluck, and Cork (1991) have reported on the neuroanatomical effects of exposing rhesus monkeys to complete social isolation for the first 9 months of age. In comparison to socially reared monkeys, the isolates, at approximately 2 years of age, had reduced numbers of tyrosine hydroxylase (an enzyme necessary for DA synthesis) immunoreactive axon terminals in the dorsal (caudate nucleus and putamen) and ventral (nucleus accumbens) striatum, and decreased numbers of tryrosine hydroxylase immunoreactive cell bodies in the substantia nigra and ventral tegmental area. These results indicate that social deprivation severely depressed the development of the mesolimbic and nigrostriatal DA systems (also see Lewis, Gluck, Beauchamp, Keresztury, & Mailman, 1990). Also, Poeggel et al. (1999) have reported that prolonged periods of maternal separation coupled with postweaning social isolation of Octodon degus offspring (a lagomorph) are associated with a decrease in the number of NADPH-diaphorase (a marker for nitric oxide synthase) positive neurons in the nucleus accumbens (and some other brain regions) as measured on day 45 of life. In a related study, Braun, Lange, Metzger, and Poeggel (2000) found that a similar protocol of social deprivation also decreased the number of tyrosine hydroxylase immunoreactive fibers in the medial prefrontal cortex of Octodon degus offspring. Therefore, social deprivation in this species may not only affect the mesolimbic DA system and the nucleus accumbens, but may also retard the development of the mesocortical DA system. Importantly, isolation rearing does not appear to indiscriminately affect all DA systems. In rhesus monkeys, Ginsberg, Hof, McKinney, and Morrison (1993) report that isolation rearing does not affect the development of the tuberoinfundibular dopamine system.

Not all studies agree that early maternal deprivation depresses the development of the mesolimbic and mesocortical DA systems. Indeed, it has recently been argued that prolonged periods of maternal separation in neonatal rats result in a facilitation of activity within the adult mesolimbic DA system by reducing the levels of the DA transporter protein, in this way slowing down the DA reuptake mechanism (Meaney, Brake, & Gratton, 2002). More work needs to be done on this issue, but what is clear is that maternal deprivation causes a dysregulation of the mesolimbic DA system. It is worth recalling that adult maternal behavior can be disrupted by both decreases and supernormal increases in DA neural activity (see Chapter 5).

2.5.3. Conclusions

We have argued that adverse early experiences, such as the isolation or separation of infants from their mothers, influence the development of the brain so that the adult maternal behavior of the offspring is affected. Although we have focussed on research that suggests that increased activity in central fear systems or decreased activity in the mesolimbic DA system may mediate some of these developmental effects, it is likely that maternal deprivation has widespread central neural effects that would have an impact on a broad range of functions, which, in turn, could influence maternal responsiveness in adulthood (see Hall, 1998). For example, some research has shown that social rearing conditions are capable of influencing the adult anatomical organization and/or function of the medial amygdala and the medial preoptic area (Cooke, Chowanadisai, & Breedlove, 2000; Fleming et al., 2002; Ichikawa, Matsuoka, & Mori, 1993; Sanchez-Toscano, Sanchez, & Garzon, 1991). Therefore, it is possible that early life experiences could directly influence the sensory integration processes involved in maternal responsiveness and/or preoptic mechanisms controlling aspects of maternal behavior. It is also worth emphasizing that although we have separately discussed the effects of early social rearing conditions on the development of central fear systems and general appetitive motivational systems, it is possible that the effects on these two systems are interrelated. For example, early adverse experiences may increase activity in central fear systems, and such an increase may, in turn, disrupt the activity of the mesolimbic and mesocortical DA systems (this view is developed more fully in Chapter 9, where we discuss the relationships between anxiety, postpartum depression with its associated anhedonia, and disturbances in maternal behavior).

Before concluding this review, it is worth considering the mechanisms by which maternal behavior can have such profound effects on the emotional and neuroendocrine systems of the offspring. In one of the most insightful formulations of mother-infant attachment, Hofer suggested that we view the mother-infant dyad as a unit with multiple "hidden regulators" (Hofer, 1984). Hofer discouraged thinking of maternal care as a monolithic behavior with good or bad outcomes in the offspring. Instead, he hypothesized that each of the components of maternal behavior—nursing, grooming, thermal comfort, vestibular stimulation—matched with specific components of the infant's growth and development. In this context, maternal separation might result in multiple regulatory failures but variations in grooming might be expected to have highly specific physiological effects in the infant with predictable outcomes on development. For instance, maternal licking and grooming (but not maternal warmth or nursing) increase ornithine decarboxylase acutely and facilitate growth of the infant (Schanberg & Field, 1987). Ornithine decarboxylase is an enzyme important for protein synthesis, so low levels of this enzyme could result in a failure to thrive. Although it is possible that changes in ornithine decarboxylase contribute to the consequences of natural variations in licking and grooming, there are several other mechanisms by which early experience could program neural development. One target of early adverse experience that has received the most attention has been the glucocorticoid receptor (GR) in the hippocampus that is increased permanently by postnatal handling. McCormick and colleagues have demonstrated that the GR mRNA exists in 11 different splice variants with considerable tissue-specific variation in distribution (McCormick et al., 2000). Postnatal handling selectively elevates GR mRNA containing the hippocampus-specific splice variant (McCormick et al., 2000), possibly by influencing methylation of the GR gene. There is a CpG island, a target for methylation, in the region of the GR gene that regulates transcription. Meaney's group has shown that methylation of this gene is influenced by maternal licking and grooming, providing a mechanism by which short-term changes in maternal care could lead to long-term changes in the GR gene in the hippocampus. Although this mechanism could explain the changes in HPA axis reactivity associated with high versus low LG-ABN, one would have to consider analogous changes on other neural targets for the changes in emotional reactivity or maternal behavior described above.

In the context of our emphasis on early environmental rearing effects on the development of central aversion systems and general appetitive motivational systems, note how these possible outcomes map nicely onto the approach-avoidance models of maternal responsiveness that we developed in Chapter 4. Maternally deprived offspring may display decreases in adult maternal behavior because of increases in avoidance tendencies and/or decreases in approach/attraction tendencies. Finally, we should mention that the effects of early social isolation on adult maternal responsiveness appear to be much more severe in primates than in rats. Although maternally deprived primates may abuse or neglect their young, maternally deprived rats are maternally responsive, although they show significant reductions in nursing and maternal licking and grooming. Although this may represent an important species difference, we would emphasize that the maternal behavior of rats has been tested under relatively benign laboratory conditions. We would predict that under more natural conditions, with concomitant increases in environmental stressors, the maternal responsiveness of maternally deprived rats might show more drastic reductions (see Chapter 4).

3. Interactions between the Neural Systems Regulating Maternal Behavior and Those Regulating Social Attachment

Earlier in this chapter we referred to the review by Newman (1999) in which she describes the overlap in the neural circuits regulating a variety of social behaviors. The MPOA/BST, MeA, and LS appear to be involved in male and female sexual behavior, offensive aggression, and parental behavior. At the outset, we struggled with this question: How does the brain develop specificity if the same circuit conveys the information for aggression, sex, and parental care? We proposed that specificity arises at a cellular level. In this final section, we will address another approach to identifying these cellular circuits and to do this we will shift our focus from parental care to pair bonding.

Thus far we have described prairie voles (*Microtus ochrogaster*) as monogamous rodents, with a focus on their biparental care. Here we will focus on prairie voles as a species that pair bonds, that is, exhibits stable, selective social attachments. In field studies, pair-bonded prairie voles generally will not accept a new mate if either the male or female of the pair is removed (Getz, McGuire, Pizzuto, Hoffman, & Frase, 1993). In the lab, prairie voles predictably form a long-term preference for the mate versus a stranger in partner preference tests (Carter, DeVries, & Getz, 1995). In both field and lab, pair bonds form as a consequence of mating.

Like the VCS of parturition in sheep, one might assume that the VCS of mating induces neural changes that lead to pair bond formation. Indeed, there are a number of ostensible similarities between the social attachment of male-female prairie voles and the selective mother-offspring bonds of sheep. Both are stimulated by reproduction; both involve an enduring, selective bond; and in both cases aggression emerges toward strangers. Although there is still much research to be done, the current evidence suggests that in both voles and sheep, oxytocin and the related peptide vasopressin are important players in the bonding process. We have already discussed the evidence for oxytocin in sheep maternal behavior. What concerns us now is oxytocin (and vasopressin) in vole pair bond formation, because this research provides an important model for the problem of specificity.

Kendrick et al. (1986) have shown that oxytocin is released in the sheep brain with copulation, and several groups have described surges of oxytocin and/or vasopressin in plasma of several other mammalian species, including humans (reviewed by Witt, 1995). Although no one has shown that oxytocin is released centrally during mating in prairie voles, central (ICV) administration of an oxytocin antagonist to female prairie voles just before mating blocks partner preference formation (Insel & Hulihan, 1995). Conversely, ICV administration of oxytocin to females housed with males but not given an opportunity to mate is sufficient to induce a partner preference (Williams, Insel, Harbaugh, & Carter,

1994). Analogous to the studies of VCS, oxytocin, and selective parental care in sheep, one explanation for these results is that oxytocin released with mating activates the neural pathways for selective bonding in voles, in this case with a mate instead of a lamb.

The problem with this formulation, of course, is specificity. Rats, rabbits, and humans also release oxytocin with mating, but pair bonding does not follow mating in any of these species. How does oxytocin induce partner preference formation in prairie voles but not these other species? The answer is likely in the receptors. The pattern of brain oxytocin receptors in prairie voles is clearly different from the pattern observed in nonmonogamous species, including the closely related montane voles (*Microtus montanus*) and meadow voles (*Microtus pennsylvanicus*) (Insel & Shapiro, 1992). In prairie voles, oxytocin receptors are most abundant in the NA and prelimbic cortex (PLC). In montane and meadow voles, oxytocin receptors are virtually absent in these reward circuits but are abundant in LS. Several observations (Young, Lim, Gingrich, & Insel, 2001) suggest this anatomic difference contributes to partner preference formation: (a) an oxytocin antagonist injected into the NA or prelimbic cortex (but not other regions) blocks partner preference formation, (b) oxytocin (ICV) increases preproenkephalin mRNA in the NA of prairie voles but not montane voles, (c) the increase in preproenkephalin mRNA following oxytocin administration is correlated with oxytocin receptor binding in the NA, and (d) similar receptor distribution patterns have been described in other monogamous species.

Presumably, oxytocin is released in response to VCS but partner preference develops only in a species with receptors in reward pathways, just as a place preference develops by increasing dopaminergic neurotransmission in the NA (White & Carr, 1985). If this model were valid, then one might expect that a partner preference could be induced not only by oxytocin but also by dopamine injections into the NA, essentially entering the system downstream from oxytocin. Gingrich, Liu, Cascio, Wang, and Insel (2000) showed that the dopamine D2 agonist quinpirole (50 pg–10 ng) administered bilaterally into the NA induced a partner preference in female prairie voles in the absence of mating. Conversely, the dopamine D2 antagonist eticlopride (100 ng) administered bilaterally into the NA prevented a partner preference in prairie voles permitted to mate. Recall from Chapter 5 that dopamine antagonists in the NA inhibit the active components of rat maternal behavior. Similarly, it seems likely that oxytocin's effects on partner preference involve dopamine D2 receptors (and possibly enkephalin) pathways in the NA. In contrast to voles, in rat maternal care, oxytocin appears to influence the VTA, the site of dopamine cell bodies for the mesolimbic circuit (Pedersen et al., 1994).

Why are oxytocin receptors in different neural pathways in closely related species? Current evidence suggests that the promoter region of the prairie vole oxytocin receptor gene may lead to this species-typical pattern. This region varies between species (Young, Huot, Nilsen, Wang, & Insel, 1996) and when this region was coupled to a reporter gene and inserted into the mouse genome, the pattern of expression of the reporter in the mouse brain resembled the pattern of the prairie vole oxytocin receptor (Young et al., 1997b). But the best evidence for the molecular mechanism of species-typical receptor patterns comes from studies of the vole vasopressin receptor. Just as oxytocin facilitates partner preference formation in female prairie voles, vasopressin appears to be necessary and sufficient for partner preference formation in males (Winslow, Hastings, Carter, Harbaugh, & Insel, 1993). A vasopressin V1a receptor agonist induces, and an antagonist inhibits, partner preference formation in male prairie voles. And just as with the oxytocin receptor, there is a striking difference in the neuroanatomical pattern of vasopressin V1a receptors between monogamous and nonmonogamous voles (irrespective of gender) (Insel, Wang, & Ferris, 1994). There are no V1a receptors in the NA of the prairie vole, but the ventral pallidum, which is a major projection of the NA, has abundant receptors in the prairie vole but few in the montane or meadow vole (Young et al., 2001). Mating induces Fos in this region in prairie voles. In the one site-specific study of AVP reported in prairie voles, this region was not tested although injections into the overlying septum altered partner preference formation (Liu, Curtis, & Wang, 2001). Young, Nilsen, Waymire, MacGregor, and Insel (1999) discovered that the species differences in V1a receptor distribution were associated with variations in a microsatellite within the promoter of the V1a receptor gene. Microsatellites are repetitive DNA sequences that are common throughout the genome. They do not code for specific RNA or protein but, when inserted in a vulnerable region, they may alter the architecture of the genome and thereby change the pattern of transcription of adjacent genes. The microsatellite in the promoter region that flanks the V1a receptor gene may influence where and when this receptor is expressed in the brain. In the prairie vole this microsatellite is several hundred bases long, whereas in nonmonogamous voles it is barely detectable.

To determine if this variation in the promoter contributes to the pattern of receptors, Young et al. (1999) created a transgenic mouse, inserting the prairie vole V1a receptor gene along with 5kB of its promoter (including the microsatellite) into the mouse genome. The resulting transgenic mouse had a pattern of V1a receptor distribution, which was similar, but not identical, to the prairie vole. Remarkably, this transgenic mouse was not monogamous, but it responded to vasopressin with increased affiliation, similar to the response observed in prairie voles but clearly different from the response of wild-type mice. Of course, this experiment failed to identify which region with the V1a receptor contributed to the behavioral response. Pitkow et al. investigated the active site by overexpressing V1a receptors using a viral vector (Pitkow et al., 2001). Adeno-associated virus was engineered to express the V1a receptor instead of normal viral proteins. When injected into a specific brain region, the virus invades cells, hijacks the cells' genetic machinery, and begins transcribing the gene of interest. In addition to allowing spatial and temporal control of gene expression in the brain, this technique can be used in many species, including voles. When Pitkow et al. (2001) injected the virus into the ventral pallidum of the prairie vole, they found an increase in affiliation and a rapid development of a partner preference. Indeed, 13 of 17 males with V1a overexpression in the ventral pallidum developed partner preferences without mating, whereas injections into the striatum were without a significant effect (Pitkow et al., 2001).

These results with oxytocin and vasopressin receptors in voles provide one solution to the specificity problem. The location of receptors within MPOA/BST, MeA, LS or within projections from these regions may create a neurochemically mediated circuitry, providing unique cellular targets within broader cytoarchitecturally defined nuclei. In the vole studies, the presence of these receptors in brain regions associated with reward is found only in monogamous species that form partner preferences. The working model is quite simple: Oxytocin and vasopressin are released centrally in response to VCS and social interaction, receptors for these peptides form the link between social stimuli and vast information processing streams in the brain. In monogamous species, receptors in the NA and ventral pallidum link the sensory properties of the mate to the generic reward signals of the mesolimbic dopamine pathways. In nonmonogamous species, activation of the LS leads to a different set of behavioral outcomes, without a conditioned preference for the mate. The species differences in receptor distribution appear to result from hypervariable regions in the promoters of these receptor genes.

9

Human Implications

This chapter will discuss the implications of the findings we have presented in the previous chapters, which have been obtained primarily from nonhuman species, for human maternal behavior. First, we want to evaluate whether human maternal behavior is emancipated from hormonal control, and we will also explore the related issue of whether what we have learned about the neural basis of parental behavior can be applied to humans. Second, we will examine the etiology of postpartum mood disorders and depression in humans, and the impact of such disorders on maternal behavior. This examination will occur in the context of the perplexing data that suggest that periparturitional hormonal changes may play a role in precipitating postpartum depression, an effect that would appear to oppose adaptive maternal responding. Third, we will present findings related to the possible biological causes of child abuse and child neglect. In particular, we want to make the argument that such disruptions in maternal behavior control may be related to disruptions in the maternal behavior neural mechanisms that we have outlined in previous chapters. Fourth, we want to relate the animal data on the neurobiology of maternal behavior to a model that might explain the intergenerational continuity of faulty maternal responsiveness within human families. Finally, we will discuss the effects of drug addictions on human maternal behavior. Our main intent in this chapter is not to present an exhaustive overview of human parenting, but instead to try to relate the neural and hormonal models we have developed in previous chapters to an understanding of problems associated with the human maternal condition. In an attempt to link the human data to the animal data, many of the proposals we will present are speculative and are meant to encourage integrative thinking. Therefore, this chapter should not be read as a separate entity, but instead should be interpreted in the context of the other chapters in this book.

1. Hormonal and Neural Basis of Human Maternal Behavior

Is maternal behavior in humans completely emancipated from endocrine control? In Chapter 2 we argued that it is not, and we presented evidence that hormonal factors do indeed influence the level of appetitive maternal motivation in humans and other primates. However, high levels of alloparental behavior do occur in human populations, and adoption of infants by nulliparous females occurs frequently. Evolutionary arguments have been presented (Hrdy, 1999) that propose that the high level of alloparental behavior in human populations may be adaptive because the long period of infant dependence requires that the postpartum mother receive assistance in child rearing from nonlactating/nonpregnant females and other individuals. Indeed, the existence of paternal behavior in humans was undoubtedly influenced by these same evolutionary forces. Therefore, although the biological mother is still the infant's primary caregiver in human populations, the occurrence of alloparental and paternal behavior suggests that the hormonal events of pregnancy termination are not an absolute requirement for human parental behavior, even in first-time mothers. As we argue below, although hormonal influences may not be essential for the onset of parental care in humans, there is still a relevant neurobiology that may be conserved in evolution.

Pryce (1995) has probably presented the best theoretical model for the control of human maternal behavior. This model, which is strongly based on the animal data we have presented in this book, is not only relevant to the current section, but also has relevance for most of this chapter. Pryce suggests that maximal maternal behavior will occur in a primate female if she (a) has a genotype that results in a confident (low-anxiety) personality; (b) had a secure relationship with her mother and also had experience with play mothering; (c) was exposed to the normal hormonal milieu associated with pregnancy, pregnancy-termination, and the postpartum period; and (d) was exposed to low stress during pregnancy and the postpartum period in conjunction with good levels of social support. According to this model, changes in any one of these factors could cause decreases in maternal behavior, and, more importantly, multiple alterations in these factors might lead to infant neglect or abuse.

Comparisons of the parental behavior and attitudes of monozygotic and dizygotic twins in human populations have provided evidence that genetic factors contribute to differences in parental responsiveness in humans (Kendler, 1996; Perusse, Neale, Heath, & Eaves, 1994): Monozygotic twins are more similar in their parental behavior than are dizygotic twins. As with most complex traits, genes appear to account for only part of the variance. Scores on the Parental Bonding Instrument (which gives composite factors of care and overprotection) in 1,117 adult twins yield estimates of 39% heritability for maternal care. Therefore, although certain genes may be relatively fixed in human populations and may allow for a relatively high level of maternal responsiveness, allele differences may still function to promote differences in human maternal behavior. It is likely that such gene differences operate in part by influencing the temperamental characteristics of the mother.

In previous chapters we reviewed the evidence that hormones influence maternal responsiveness in primates. With respect to humans, we should restate the following: (a) Primiparous females reported a stronger attachment to their infants during the early postpartum period if their estradiol-to-progesterone ratio increased throughout pregnancy, while they reported a weaker attachment if that ratio showed a decline over pregnancy (Fleming et al., 1997a); (b) early post-partum primiparous women are more attracted to infant odors than are nulliparous women (Fleming et al., 1993); and (c) postpartum human mothers who are breastfeeding tend to have lower anxiety scores and stress reactivity than control females (for a recent review see Carter, Altemus, and Chrousos [2001]). Since suckling stimulation triggers the release of prolactin from the anterior pituitary (Grattan, 2001) and oxytocin from the posterior pituitary and also presumably from oxytocin neurons within the brain (Neumann, 2001), and given the evidence that we have previously reviewed (Chapters 4, 6, 8) that shows that both prolactin and oxytocin have anxiolytic effects (for a review, see Neumann, 2001), these results suggest that prolactin (or perhaps a brain prolactin system) and possibly a brain oxytocin system (since systemic oxytocin has low penetrance across the blood-brain barrier) may be involved in a general fear-reduction process in postpartum women. These ideas are attractive in terms of the neural model presented in Chapter 8, which suggested that oxytocin might act at the level of the amygdala to inhibit anxiety and stress-related responsiveness by possibly inhibiting CRF neurons located there. Beyond the reduction of anxiety, as we have discussed earlier, oxytocin may have specific rewarding or hedonic effects that would support the development of the mother-infant bond. It should also be considered that nonsuckling stimulation may similarly release prolactin and oxytocin in postpartum women with concomitant anxiolytic or hedonic effects.

With respect to hormonal control mechanisms, an ideal study would be to compare aspects of maternal behavior in two groups of women, a sample of nulliparous women who adopted a newborn infant and a matched sample of primiparous females. The attempt would be made to match these two samples in all respects except that one group would be exposed to the hormonal (and neural) events of pregnancy and the postpartum period while the other would not. If the maternal behavior of the postpartum females was superior in some respects, this would suggest the importance of hormonal variables. This ideal experiment is difficult to accomplish since adoptive parents are usually older, have more established careers, and are more financially secure than typical first-time natural mothers (Brodzinsky & Huffman, 1988). In other words, adoptive parents are usually exposed to fewer life stresses than are many first-time natural mothers. One study has approximated a match between adoptive and natural parents (Singer, Brodzinsky, Ramsay, Steir, & Waters, 1985). This study measured the security of an infant's attachment to its mother at 14 months of age in the Strange Situation paradigm (which is presumably reflective of the quality of maternal care received by the infant up to that point), and no significant differences were observed between a group of natural mother-infant pairs and a group of adoptive mother-infant pairs. Importantly, the groups were matched for socioeconomic status (middle class), maternal (30 years) and paternal (33 years) age, and maternal and paternal education. This study suggests that a hormonally primed mother is not essential for normal infant development, and we are sure that most adoptive parents would agree with this. A problem with this study, however, is that in order to match the two samples, the natural mother-infant dyads may not have been representative of typical first-time mothers and their infants. Importantly, other comparison studies have suggested that physiological factors may play a more important role in human maternal behavior. Brodzinsky and Huffman (1988), citing the work of Hoopes, indicate that adoptive mothers tend to be more protective of their infants, and tend to display higher levels of anxiety with regard to parenthood than do natural mothers. One could interpret this evidence as suggestive of the possibility that the hormonal events associated with natural motherhood play a role in decreasing anxiety-related processes. Therefore, if one could compare groups of adoptive and non-adoptive first-time mothers who were younger in age than in the Singer et al. (1985) study, and also exposed to more stressful or demanding life circumstances (lower socioeconomic status and poor social support, for example), perhaps one would be able to detect major differences between the groups in maternal responsiveness. But a matched sample would be hard to create with respect to the adoptive group. We suggest that the typical adoptive parent is under relatively low life stress and has strong social support, and that under these conditions processes akin to sensitization may eventually allow for the occurrence of perfectly normal maternal behavior. However, parturitional events may normally play a role in stabilizing the mother-infant relationship and preventing maternal abandonment of infants (by increasing maternal motivation and decreasing anxiety) in the early postpartum days in primiparous women with weak social support systems and high levels of stressful life events.

For the sake of argument, and in order to allow for the consideration of all possibilities, if we took the extreme position that human maternal behavior was truly emancipated from endocrine control, what would this mean with respect to the neuroanatomy of maternal behavior in humans? Would the same structures and circuits be involved? For example, would we predict that the MPOA, through the neural pathways described in Chapter 5, would be involved in enhancing maternal motivation in human mothers? In rats we know that estradiol, prolactin, and oxytocin all act on the MPOA to promote maternal behavior. If hormones were not needed for maternal behavior in humans, would the need for the MPOA also be eliminated? In Figure 2.1 we proposed that the neural mechanisms underlying maternal behavior are probably highly conserved across species, and what varies is the degree to which hormones are a requirement for infant stimuli to gain access to this conserved neural system. This model would predict, therefore, that the MPOA should be critically involved in primate maternal behavior. In support of this prediction, we know that MPOA lesions in rats disrupt those aspects of maternal behavior that no longer require hormonal mediation. That is, MPOA lesions disrupt maternal behavior in experienced postpartum females and in sensitized virgins (see Chapter 5). However, very little work has been done on the neural basis of maternal behavior in primates, so we really do not know the extent to which the circuits we have mapped out apply to the organization of human maternal behavior and motivation.

In making comparisons between nonprimate mammals and primates, Keverne (2001) has noted that the size of the neocortex and the striatum have increased in primates relative to the rest of the brain and the body, while the size of the MPOA/hypothalamus has decreased. To consider the magnitude of this difference, the volume of the cortex increases roughly 1,000-fold from rats to humans while the hypothalamus increases only nominally in size. Keverne further suggests that the development of a larger neocortex is related to the relative emancipation of primate maternal behavior from endocrine control. In other words, primate maternal behavior is under greater cognitive control by an executive neocortex, and is less influenced by hypothalamic and hormonal mechanisms. He suggests that such an evolutionary development allowed for more strategically correct maternal decisions in the complex social environment of primates. Although Keverne does not explicitly make the following point, his logic implies that the MPOA may be relatively uninvolved in primate maternal behavior.

In support of certain aspects of Keverne's views, as we know from Chapter 5, lesions of the prefrontal neocortex (PFC) have been found to disrupt maternal behavior in primates (Franzen & Myers, 1973). Since there is little support for a role of the PFC in rodent maternal behavior, these findings support the idea of an increased involvement of specific neocortical control mechanisms in the maternal behavior of primates. However, as noted in Chapter 5, the work on PFC involvement in rodent maternal behavior is preliminary and more work should be done. It is also worth pointing out that an early study by Slotnick (1967) found that lesions of the cingulate cortex caused a temporary disruption of maternal behavior in rats. The animals initially showed disorganized retrieval responses (pups were not retrieved to a single location), but this behavior improved over time, presumably as a result of the mother rats' adaptation to the novelty of the testing situation. Since the anterior part of the cingulate cortex receives afferents from the mediodorsal thalamus, as does the PFC, it has been argued that the anterior cingulate cortex should be considered part of the PFC (Numan, 1985; Zilles & Wree, 1995). In this sense, then, there is evidence for PFC involvement in rodent maternal behavior (Interestingly, however, Franzen and Myers [1973] have reported that cingulate lesions do not influence maternal behavior in rhesus monkeys).

The PFC receives input from posterior neocortical association regions (which presumably relay information about memories and current perceptions), and its efferents project to the motor system to influence response selection strategies (Groenewegen, Wright, & Uylings, 1997). On the basis of such connectivity, it could be argued that the PFC allows thought and consciousness to influence voluntary behavior in complex situations where novel adaptive responses may be called for. Importantly, the PFC also receives substantial inputs from subcortical structures involved in motivation and emotion, and Damasio (1995) has been a strong proponent of the view that subcortical projections to the PFC allow motivational and emotional states to influence the planning and deliberative processes that are conducted by the human PFC in its capacity of selecting appropriate voluntary responses in complex social and nonsocial situations (also see Greene, Sommerville, Nystrom, Darley, & Cohen, 2001; LeDoux, 1996).

Within this context, we would like to propose neural models that show that the MPOA could still be involved in human maternal behavior even though hormones may be less important, and even though increased corticalization of maternal response strategies has occurred in primates. The basic idea is that the MPOA may still be the avenue through which the level of maternal motivation is signalled to the PFC so that this information can be used by the PFC in its development of complex (rather than simple) voluntary response strategies. First, in rodents maternally relevant neurons in the MPOA project to both the VTA and to the nucleus accumbens (Numan & Numan, 1996, 1997; Stack et al., 2002). In addition, the VTA projects to both the nucleus accumbens and to the PFC, and the nucleus accumbens, through its projections to the ventral pallidum, is capable of influencing the activity of both the PFC and brainstem motor mechanisms (Groenewegen et al., 1997; Mogenson, 1987; Tzschentke & Schmidt, 2000). As outlined in Figure 9.1, MPOA influences on nucleus accumbens outputs that influence subcortical motor centers may regulate simple

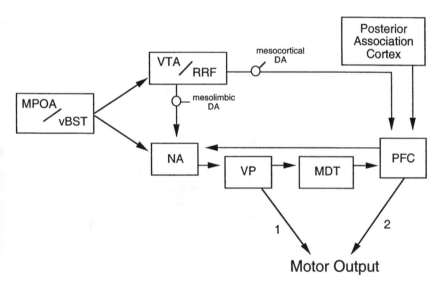

Figure 9.1. Neural connections through which MPOA and vBST might influence the occurrence of simple voluntary maternal responses and more deliberative and strategic responses related to maternal behavior in complex situations. Simple voluntary proactive maternal responses might be regulated though an MPOA/vBST-to-VTA/RRF/NA-to-VP pathway (route 1), while more complex maternal responses might be regulated through neural pathways that include the PFC (route 2). See text for details. Abbreviations: DA = dopamine; MDT = medial dorsal thalamus; MPOA = medial preoptic area; NA = nucleus accumbens; PFC = prefrontal cortex; RRF = retrorubral field; vBST = ventral bed nucleus of stria terminalis; VP = ventral pallidum; VTA = ventral tegmental area.

voluntary maternal responses, such as retrieving behavior, while MPOA projections that ultimately influence prefrontal cortical mechanisms may control more deliberative, strategic responses related to maternal behavior in complex situations. Therefore, even if hormones are not needed to "prepare" the MPOA, it is still certainly possible that the MPOA is needed to signal the level of maternal motivation so that appropriate responses involving infants are made. Of course, the research needed to explore the role of the MPOA and the other neural circuits proposed in Chapter 5 in human and nonhuman primate maternal behavior remains to be done. In humans, lesions from trauma or ischemia may reveal the role of select brain regions. Although there is an abundant literature on brain injury and changes in human social behavior, we know of no evidence that maternal behavior has been selectively lost after brain injury. The location of the MPOA is far from any vascular watershed and far from the cortical surface, so this is not a region that would be selectively damaged by either stroke or trauma. Perhaps the best evidence for the neural circuits involved in human maternal care will come from in vivo neuroimaging. The first study (Loberbaum et al., 2002), focussing on the response of mothers to infant cries, reported high levels of activation not only in the anterior cingulate and prefrontal cortex, regions that show activation in diverse tasks that require emotional monitoring, but also in the hypothalamus (specific nuclei were not resolved), lateral septum, and in dopamine pathways such as the VTA/RRF and NA. Because of its size and location, the MPOA may be difficult to visualize with functional imaging techniques, but the advent of higher-field strength magnets with better resolution may permit a study of MPOA function in the human brain.

The involvement of the PFC in primate maternal behavior will have significant relationships to the other topics covered in this chapter. The PFC projects strongly to the nucleus accumbens (Groenewegen et al., 1997; Tzschentke & Schmidt, 2000), and this is one of the ways in which the PFC influences motor output. More importantly, the norepinephrine neurons of the locus coeruleus project to the entire neocortex, which includes significant inputs to the prefrontal area (Sullivan, Coplan, Kent, & Gorman, 1999), and therefore, as outlined in Figure 8.7, aversive life events that cause a supernormally high and chronic activation of stress-related inputs to the locus coeruleus are likely to cause a dysregulation of PFC activity, with a concomitant negative influence on adaptive maternal responding. As we shall see, although focal lesions from brain injury have not been associated with a specific loss of maternal behavior, women under the more diffuse influences of depression or drugs may show a profound loss of nurturing behavior.

2. Postpartum Depression

The postpartum period is a time of increased vulnerability to mood disorders in women (Devinsky & Bartlik, 1994; Hendrick, Altshuler, & Suri, 1998; Llewellyn, Stowe, & Nemeroff, 1997; Steiner, 1979; Wisner & Stowe, 1997). A mild mood disorder referred to as the "maternity blues" occurs in approximately 50% of women in the first few days postpartum. This event is a short-lasting mood disorder that includes the symptoms of sadness and mild anxiety. Importantly, approximately 20% of women with maternity blues go on to develop major depression, referred to as postpartum depression, within the first 6 weeks postpartum. As Steiner (1979) notes, intense pathological anxiety seems to be a hallmark for the development of this disorder. Significantly, the symptoms of postpartum depression resemble the symptoms of major depression that occur outside the postpartum period, and include sadness, anxiety, and anhedonia (loss of the ability to experience pleasure). Llewellyn et al. (1997) have outlined several risk factors for postpartum depression, which include the following: (a) a family history of depression; (b) a previous episode of postpartum depression or major depression outside the postpartum period; and (c) stressful life events near the time of parturition, such as poor postpartum social support, unwanted pregnancy, and marital instability. An acceptable model for the occurrence of postpartum depression would therefore seem to be that a woman's genotype, previous life experiences, and current life experiences interact with the physiological events that occur around the time of parturition to determine her susceptibility to depression. It is the purpose of most of this chapter to more exactly define what these particular factors are, and their underlying mechanisms.

A large section of this book has emphasized that the postpartum period is a time of decreased fearfulness and increased attraction toward infant-related stimuli, thus allowing for effective maternal responding. The occurrence of postpartum depression in a substantial proportion of women, therefore, should not only be considered an abnormality with respect to general mental function, but should also be considered deleterious to maternal behavior, since the associated anxiety and anhedonia would result in increased fearfulness and decreased maternal pleasure. Indeed, and not surprisingly, we will show that postpartum depression is associated with poor maternal behavior.

We will begin our analysis of postpartum depression by first examining some biological views concerning the etiology of major depression in general, and then we will apply this information to the postpartum condition. The major biological perspectives on depression emphasize the involvement of dysfunctions in central monoamine systems and central CRF systems. The classical monoamine hypothesis of depression argues that depression results from underactivity within central monoamine systems, and important suppport for this hypothesis is derived from the fact that the major clinically effective antidepressant drugs act either by blocking the reuptake of norepinephrine (desipramine, for example), or blocking the reuptake of serotonin (fluoxetine, for example), in this way allowing these monoamines to exert postsynaptic effects for longer periods of time (Ressler & Nemeroff, 1999; Wong & Licinio, 2001). Current evidence is not so clear about how antidepressant drugs eventually produce their therapeutic effects, with some investigators arguing that these drugs do indeed produce their positive effects by upregulating central monoamine activity, while others argue that these drugs ultimately produce their therapeutic effects because they down-regulate monoamine function (Ressler & Nemeroff, 1999; Wong & Licinio, 2001). It is probably safest to say that current evidence indicates that a dysregulation of central monoamine activity underlies depression, and that antidepressant drugs produce some of their therapeutic effects by stabilizing the activity of these systems.

The melancholic type of major depression is characterized by sadness, intense anxiety, anhedonia, and hyperarousal or hypervigilance (Wong et al., 2000). Not surprisingly, it is also associated with hypercortisolemia (increased blood levels of cortisol), which is indicative of a hyperactive hypothalamic-pituitary-adrenal (HPA) system. Another biological hypothesis of depression, therefore, suggests that increased central CRF activity within the hypothalamus and elsewhere in the central nervous system may be involved in inducing depression (Barden, Reul, & Holsboer, 1995; Mitchell, 1998; Nemeroff et al., 1984; Wong et al., 2000). Preliminary evidence in support of this view is that drugs that depress central CRF systems or HPA activity have been found to have antidepressant effects (Kramliger, Peterson, Watson, & Leonard, 1985; Thakore & Dinan, 1995; Zobel et al., 2000).

It is not our purpose to exhaustively review the evidence in favor of each of these hypotheses of depression. Instead, we will present an overview of the evidence that suggests that these two putative causes of melancholic depression may be linked together in that a hyperactive CRF system is probably directly related to the anxiety components of the disorder, and the CRF hyperactivity, in turn, may also cause a dysfunction in central monoamine systems, which then gives rise to mood depression, anhedonia, and hyperarousal. In an important review, Anisman and Zacharko (1982) proposed that stress (aversive life experiences) is a factor that influences the occurrence of depressive illness, and they provided a possible mechanism for this effect. They noted that clinical depression is often preceded by stressful life events, and such antecedent stressful incidents can either be recent or remote. Most importantly, they suggested that early life stresses may sensitize individuals to later stress, in this way exacerbating the likelihood that stress would precipitate depression. In the context of the monoamine hypothesis of depression, they then review the animal literature that suggests that exposure of animals to stress results in a depletion of monoamines from neurons in the brain. On this basis they suggested that stress is a precipitating factor in human depression because it causes underactivity across certain central monoamine synapses. It should be noted that these authors did not propose a mechanism through which stress influences monoamines. Given that genetic factors are critically involved in depression (Wong & Licinio, 2001), we might modify the above views by stating that remote and recent stresses may increase the risk for depression in genetically susceptible individuals.

As others have done (Chrousos, 1998; Geracioti et al., 1990; Ladd et al., 2000), we would like to link stress-induced CRF release to dysregulation of monoamine brain systems, in this way linking together the two neurochemical systems that have been implicated in depression. Our hypothetical model, which builds upon Figure 8.7, is shown in Figure 9.2. The figure shows that stress

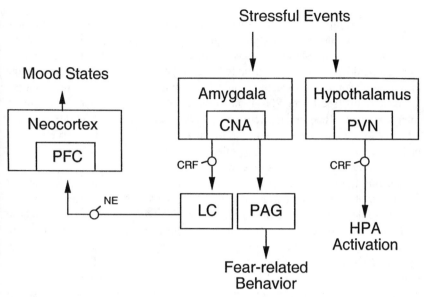

Figure 9.2. A neural model that links stressful life events to hypercortisolemia, anxiety, fear-related behavior, and LC-NE dysregulation and depression. See text for details. Abbreviations: CNA = central nucleus of amygdala; CRF = corticotropin releasing factor; HPA = hypothalamic-pituitary-adrenal system; LC = locus coeruleus; NE = norepinephrine; PAG = periaqueductal gray; PFC = prefrontal cortex; PVN = paraventricular hypothalamic nucleus.

can activate the paraventricular nucleus of the hypothalamus to stimulate CRF release into the anterior pituitary, in this way increasing ACTH and cortisol release. Therefore, remote and recent stressful life events that promote the occurrence of depression should be associated with CRF hyperactivity and hypercortisolemia. Stressful life events are also shown as activating the amygdala, in particular the central nucleus (CNA), which contains CRF neurons. CNA projections include terminations in the periaqueductal gray (PAG) and in the LC, and LC neurons are activated by CRF. We also show LC-NE neurons projecting to the PFC, although this system also has many other projections (Sullivan et al., 1999; also see Chapter 8). Therefore, a hyperactive central CRF system should not only cause hypercortisolemia, but may also cause a dysregulation of the LC-NE system, and such an LC hyperactivity might result in a partial depletion of NE or a functional downregulation of NE action at certain target sites. In accordance with this scheme, it is worth pointing out that clinical depression has been associated with abnormal functional activity in the PFC (Drevets, 2000; Soares & Mann, 1997; Sheline, 2000). This limited hypothetical model, which builds on the data presented in Chapter 8, emphasizes the interaction between CRF systems and NE systems in melancholic depression. Hyperactivation of CNA and PAG may give rise to fear-related behavior, while abnormal functioning of LC-NE input to PFC may cause mood disturbances. Just how serotonin and mesolimbic dopamine systems might be involved remains to be developed. The following points should be considered, however: Many brain regions contain glucocorticoid receptors, which include the serotoninergic raphe nucleus in the brainstem, so hypercortisolemia could have additional effects on brain monoamine systems (see Korte, 2001); the raphe nucleus and the LC are anatomically interconnected (Ressler & Nemeroff, 1999); the prefrontal cortex projects strongly to the nucleus accumbens (Groenewegen et al., 1997; also see the preceding section of this chapter for other relevant anatomical relations between NA, PFC, and the VTA dopamine system).

Can this limited model of depression be applied to postpartum depression? What is it about the postpartum period that appears to make women more vulnerable to depression? We would like to present evidence that suggests that in a *subset of susceptible women,* the precipitous hormonal changes associated with pregnancy termination might cause a persistent hypersensitivity or hyperactivity of central CRF systems, which then precipitates anxiety and depression. Probably the best study showing that the hormonal changes associated with parturition are capable of inducing clinically significant mood disturbances in a subset of susceptible women was performed by Bloch et al. (2000). Women were separated into two groups, one that contained women with a previous history of postpartum depression, and one with individuals who did not have such a history. Importantly, all women in this study were in a normal mood state at the start of the experiment, and the women with the previous history of postpartum depression were at least 1 year removed from their depression. Initially, all women were treated with a gonadotropin-releasing hormone antagonist, which decreased their endogenous steroid hormone levels. During an 8-week addback phase of the experiment, all women were treated with supraphysiological levels of estradiol and progesterone. Finally, during a withdrawal phase, the steroids were withdrawn. This treatment regimen was intended to simulate, over a shorter time scale, the high levels of steroids that occur during pregnancy in women, and the precipitous drop in steroids that occurs at parturition (see Chapter 2). Importantly, only the women with a previous history of depression developed significant depressive symptoms during the withdrawal phase of treatment. Therefore, although the two sets of women were treated with the same hormones at the same dosages, women with a previous history of depression were vulnerable to the mood-destabilizing effects of steroid hormone withdrawal.

Previous reviews of the literature have indicated that it has been difficult to detect differences in endogenous hormone levels over pregnancy and the early postpartum period when postpartum depressed women are compared to their nondepressed counterparts (Hendrick et al., 1998; O'Hara, Schlechte, Lewis, & Varner, 1991; Wisner & Stowe, 1997). The importance of the Bloch et al. (2000) study is that it shows that the absence of such hormone differences does not mean that hormonal factors do not play a precipitating role in the onset of postpartum depression. What appears to vary between women who become depressed and those who do not is their response to the same endocrine changes.

In comparison to their healthy counterparts, how might women who develop postpartum depression differ in their response to the hormonal events of late pregnancy? One possibility is that the depressed women may be less responsive to estradiol. This idea gains significance in light of the preliminary data that show that postpartum estradiol treatment may be clinically effective in ameliorating the symptoms of postpartum depression (Ahokas, Kaukoranta, & Aito, 1999; Epperson, Wisner, & Yamamoto, 1999; Gregoire, Kumar, Everitt, Henderson, & Studd, 1996; also see Galea, Wide, & Barr, 2001, for supportive animal data). Another possibility is that such women overreact to chronic progestereone exposure and to progesterone withdrawal.

Recall from Chapter 2 that in most mammalian species progesterone levels abruptly decline prior to parturition, and this decline is superimposed upon rising estradiol levels. In contrast, for many primates, including humans, although estradiol rises prepartum, progesterone does not show a major decline until after parturition. Also recall from Chapter 4 that in rodents progesterone withdrawal in the absence of an estradiol rise produces an anxiogenic effect, while progesterone withdrawal in conjunction with an estradiol rise causes an anxiolytic effect. Finally, note the finding of Fleming et al. (1997a) that changes in the estradiol-to-progesterone ratio over the course of pregnancy are related to the degree of maternal attachment shown by women toward their infants in the early postpartum period. Putting these pieces of information together, we might suggest that progesterone withdrawal promotes anxiety, which then may precipitate depression in a susceptible group of women, particularly in those who have a subnormal reactivity to estradiol, and that these psychological alterations may also have implications for postpartum maternal behavior (see below). This analysis would suggest that postpartum progesterone treatment might also be a therapy for postpartum depression, but at present there is limited and equivocal evidence for this possibility (Epperson et al., 1999; Lawrie et al., 1998). As we will outline below, however, long-term treatment with progesterone, even in the absence of progesterone withdrawal, may have the potential to exert anxiogenic effects. In this regard, it should be noted that postpartum depression in women is often preceded by increased levels of anxiety and depression during pregnancy (O'Hara et al., 1991).

Although there is no evident difference in progesterone in depressed versus euthymic women postpartum, animal studies suggest that metabolites of progesterone may exert mood-altering effects by modifying the functional activity of the GABAA receptor at synapses within the brain. Such a hypothesis was first proposed by Majewska, Ford-Rice, and Falkay (1989). Recall from Chapter 4 that acute progesterone administration has been found to have anxiolytic effects in animals, and that this effect of progesterone results from its conversion to reduced metabolites such as allopregnanolone (tetrahydroprogesterone). These anxiolytic effects of allopregnanolone are mediated by its action at the GABAA receptor, where it serves as a positive modulator of GABA, in this way increasing Cl$^-$ conductance to a given amount of GABA, resulting in a greater postsynaptic inhibitory effect. Allopregnanolone presumably decreases anxiety by causing increased inhibitory effects on brain regions such as the amygdala, which mediate anxiety (for a review, see Rupprecht, 1997). In other words, allopregnanolone exerts effects similar to the benzodiazepine antianxiety drugs, of which diazepam is an example (Ballenger, 1995). Importantly, recent studies have shown that chronic exposure to allopregnanolone, followed by allopregnanolone withdrawal (in the absence of an estradiol rise), can cause an anxiogenic effect in rats, and this effect has been correlated with a *hypo*sensitivity of neuronal GABAA receptors to GABA and to benzodiazepines (Smith et al., 1998; also see Gulinello, Gong, & Smith, 2001). Significantly, under certain conditions chronic exposure of neurons to allopregnanolone, even in the absence of allopregnanolone withdrawal, can modify the function of GABAA receptors so that they are less responsive to the effects of benzodiazepines (Follesa et al., 2000). These results suggest that chronic exposure to high levels of progesterone and its metabolites, and the termination of such chronic exposure, might result in a neural state where neurons that mediate anxiety and fear are actually under less inhibition. The GABAA receptor is a pentameric receptor, and each of its major subunits (alpha, beta, gamma, delta, and rho) have their own variants (Waxham, 1999). It is the particular subunit composition of the GABAA receptor that determines its electrophysiological and pharmacological properties, and the research presented by Follesa et al. (2000) and Smith et al. (1998) suggests that the exposure of neurons containing GABAA receptors to chronic allopregnanolone alters the subunit composition of the receptors, and that withdrawal from chronic allopregnanolone causes further modifications of these receptors. Smith et al. (1998) have presented evidence that suggests that an increased incorporation of the alpha 4 subunit into the GABAA receptor following withdrawal from long-term allopregnanolone exposure results in a receptor that is hyposensitive to GABA and to benzodiazepines, and that this modification may mediate anxiogenesis.

With respect to the theoretical model we are attempting to build (see Figure 9.2), it is significant to note that *acute* allopregnanolone microinjections into the central nucleus of the amygdala have been found to exert an anxiolytic effect in animals (Akwa, Purdy, Koob, & Britton, 1999). Therefore, we speculate that perhaps chronic exposure of the CNA to allopregnanolone as a result of the high levels of progesterone during pregnancy (see Concas et al., 1998) causes modifications in the subunit composition of the GABAA receptor in this brain region, and that subsequent progesterone (allopregnanolone) withdrawal causes further modifications, the end result being that the output of CNA, and the CRF neurons located in this region, may become hyperactive because of decreased GABAergic inhibition. According to Figures 8.7 and 9.2, such hyperactivity could lead to anxiety, depression, and disruptions in maternal behavior if the effects of allopregnanolone withdrawal are not counteracted by some other process or processes. We can continue to speculate that perhaps the central actions of estradiol, prolactin, and oxytocin serve to counteract the effects of allopregnanolone withdrawal at parturition, so that in the typical puerperal female, normal, adaptive maternal behavior occurs in the absence of a depressed mood state. However, a subset of women may be more vulnerable to this progesterone/ allopregnanolone effect, and a persistent state of anxiety and depression may ensue.

Is there any evidence from humans that allopregnanolone and GABAA receptor function may be related to altered mood states? Some supportive evidence can be derived from women who show the premenstrual syndrome, which is analogous to postpartum depression in certain respects. The premenstrual syndrome (PMS) occurs in a subset of susceptible women during the late luteal phase of the menstrual cycle, and would therefore be preceded by a relatively prolonged period of progesterone exposure followed by progesterone withdrawal. Components of this mood disorder include anxiety, sadness, and sometimes severe depression (Epperson et al., 1999). It has been reported that patients with PMS are less responsive to the effects of pregnanolone (a reduced metabolite of progesterone with properties similar to allopregnanolone) and benzodiazepines during the late luteal phase of the menstrual cycle than are control subjects (Sundstrom et al., 1998; Sundstrom, Ashbrook, & Backstrom, 1997). Interestingly, Sundstrom et al. (1998) found that there were no differences in plasma levels of estradiol, progesterone, and pregnanolone during the menstrual cycle between these two populations. These results are consistent with those of Bloch et al. (2000): Although steroid levels may not be different between PMS and asymptomatic subjects, PMS subjects may be more vulnerable to the effects of chronic steroid exposure and steroid withdrawal. Therefore, these results suggest that late luteal phase PMS subjects may have a hypoactive central GABAergic system.

We hypothesize that decreased GABAergic inhibition may promote hyperactivity in CRF systems, which then triggers an anxiety state that may be followed by depression. Is there any evidence for a relationship between allopregnanolone and CRF systems, and given that melancholic depression has been related to a hyperactive CRF system, is there any evidence that postpartum depression is also associated with increased CRF activity? In animal studies, Patchev, Hassan, Holsboer, and Almeida (1996) have shown that *acute* systemic treatment of rats with allopregnanolone is capable of blocking HPA activation by an emotional stressor, suggesting that the steroid is inhibiting hypothalamic CRF release in response to stress (also see Owens, Ritchie, & Nemeroff, 1992). In support of this view, in vitro studies have shown that allopregnanolone can inhibit the release of CRF from hypothalamic explants (Patchev, Shoaib, Holsboer, & Almeida, 1994). Interestingly, this latter study also showed that intracerebroventricular (ICV) administration of allopregnanolone to rats can decrease the anxiogenic properties of ICV-administered CRF, as measured in the elevated plus maze. These results suggest, therefore, that acute exposure to allopregnanolone can depress stress-induced release of CRF from central neurons, and that the steroid also appears capable of blocking the postsynaptic effects of CRF. Both of these effects are presumed to occur via the GABAA receptor agonist properties of acute allopregnanolone. It is just a short logical step to argue that under certain conditions in a susceptible population, chronic exposure to allopregnanolone, followed by allopregnanolone withdrawal, might result in a hyperactive central CRF system.

The evidence with respect to whether postpartum depression is associated with hyperactive central CRF systems is equivocal. Some studies have reported no differences in postpartum plasma cortisol levels between women showing postpartum mood disorders and healthy control mothers (Magiakou et al., 1996; O'Hara et al., 1991), while other studies have reported increased plasma cortisol levels in postpartum women with dysphoric moods (Okano & Nomura, 1992; Pedersen et al., 1993). However, plasma cortisol levels may not necessarily reflect the activity level of all central CRF systems. Interestingly, Magiakou et al. (1996) have shown that the secretion of ACTH from the anterior pituitary in response to an intravenous injection of ovine CRF was *lower* in postpartum women with mood disorders when compared to healthy control mothers. This finding suggests that a decreased pituitary response to CRF is associated with postpartum depression. Significantly, patients who show a psychiatric disorder referred to as posttraumatic stress disorder (PTSD, a severe form of anxiety that is induced by stimuli related to a previous trauma incident, and that is often associated with depression) also show a reduced ACTH response to CRF. However, such individuals have higher endogenous cerebrospinal fluid levels of CRF than do controls (Bremner et al., 1997). These authors suggest that the blunted ACTH response to CRF might be the result of a downregulation of anterior pituitary CRF receptors caused by exposure of these receptors to chronically elevated levels of endogeous CRF. Perhaps an analogous condition exists in some women with postpartum depression, which would be supportive of the view that hyperactive central CRF systems may be related to postpartum mood disorders. We are not aware of any studies that have measured cerebrospinal fluid levels of CRF in postpartum women with and without mood disorders.

The overall evidence that we have reviewed suggests that a dysregulation of central CRF systems may be involved in the etiology of postpartum depression. In further support of this view, research indicates that the incidence of postpartum depression is positively correlated with the frequency and severity of stressful life events that occur during late pregnancy and the postpartum period (Barnet, Joffe, Duggan, Wilson, & Repke, 1996; Hendrick et al., 1998; O'Hara et al., 1991; Stevens-Simon, Kelly, & Wallis, 2000). Such stressful life events include an unwanted pregnancy, lack of social support, low socioeconomic status, and pregnancy during adolescence. For example, the incidence of postpartum depression can approach 40% in poor adolescent mothers (Barnet et al., 1996). It is expected that increases in stressful life events around the time of parturition will cause a hyperactivation of central CRF systems, which may then interact with steroid withdrawal effects to increase the likelihood of postpartum depression.

3. Child Abuse and Child Neglect

A major theme in our review of the animal literature was that the hormonal events of late pregnancy and parturition appear to produce a reduction in general fear and anxiety levels as well as an attraction to infants, and that such stress hyporesponsiveness and hedonic responses allow for adaptive maternal responses under challenging and demanding environmental conditions. We are arguing here, however, that in a subset of postpartum women the opposite appears to occur, since postpartum depression is associated with increased anxiety and anhedonia. We would suggest that the subset of vulnerable women differ from their normal counterparts both genotypically and experientially, and that such differences might foster the occurrence of postpartum depression for a number of possible reasons, which could include decreased responsiveness to estradiol, prolactin, or oxytocin; a greater degree of GABAA receptor hyporesponsiveness after exposure to chronic progesterone and its withdrawal; a hyperresponsive CRF system; and increased exposure to current or remote stressors. An interesting question in this context is why a mechanism should even exist that would allow progesterone (allopregnanolone) withdrawal to promote anxiety. One possibility is that a regulated and mild increase in vigilance in the immediate postpartum might actually be adaptive for the postpartum female, and that it is only an extreme overactivity in the systems outlined in Figure 9.2 that would result in pathological anxiety and depression (see Majewska et al., 1989). Another possibility that should be considered is that the proposed occurrence of GABAA receptor hyporesponsiveness after exposure to chronic progesterone and its withdrawal may be important for releasing brain regions that play a *positive* role in maternal behavior (such as the MPOA) from GABAergic inhibition (see Numan et al., 1999).

Anxiety and depression, with their associated anhedonia, should obviously cause deficits in maternal behavior, and we will review the evidence that this is indeed the case. However, before we present that evidence we want to continue with our neural model building in Figure 9.3. This model integrates Figures 8.7, 9.1, and 9.2. Without being too redundant, please note that each of these figures is based on hypothetical proposals with only partial supportive evidence. We are actually building hypothetical models on top of hypothetical models, and therefore, one should use caution in evaluating each model. We are trying to integrate animal and human research, and although we believe that the overarching principles upon which these models are built are valid, the particular anatomical and neurochemical details may have to be significantly revised and expanded as future research becomes available. Figure 9.3 shows how an abnormal hyperresponsiveness of the amygdala (the central nucleus, in particular) to neural inputs activated by stressful events could result in anxiety-related behavior (hyperactivation of the PAG), hypervigilance (hyperactivation of LC), depresssion and its associated anhedonia (dysregulation of PFC and its connections to NA), and deficits in maternal motivation and attraction toward infants

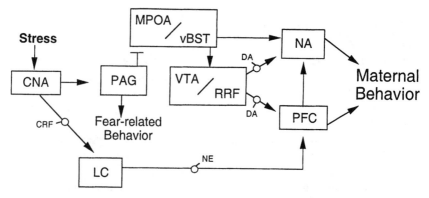

Figure 9.3. A neural model that integrates Figures 8.7, 9.1, and 9.2 to show how hyper-activation of central stress-reactive neural circuits could lead to anxiety, depression, and disruptions in maternal behavior. See text for details. Abbreviations: CNA = central nucleus of amygdala; CRF = corticotropin releasing factor; DA = dopamine; LC = locus coeruleus; MPOA = medial preoptic area; NA = nucleus accumbens; NE = nor-epinephrine; PAG = periaqueductal gray; PFC = prefrontal cortex; RRF = retrorubral field; vBST = ventral bed nucleus of stria terminalis; VTA = ventral tegmental area. (Although the models we have presented have linked CRF systems to anxiety, depression, and maternal dysfunction, for the sake of completeness, it should be noted that stress-induced activation of central tackykinin systems has also been linked to anxiety, depression, and disruptions in maternal behavior [Kramer et al., 1998; Sheehan & Numan, 1997].)

(inhibition of MPOA, abnormal functioning of PFC and its connections to NA, decreased MPOA activation of the mesolimbic and mesocortical DA systems). We should reiterate in this context the evidence that was reviewed in previous chapters, which showed that ICV treatment with CRF depresses maternal be-havior in rats, and that lactation in rats is normally associated with decreases in CRF mRNA in CNA. Also recall the evidence that a putative inhibitory action of oxytocin at the level of CNA may contribute to decreased fearfulness in postpartum female rats, increased maternal aggression, and increased arched-back nursing and licking/grooming of pups. Although we are emphasizing that an abnormal upregulation of CRF systems emanating from CNA may be a pos-sible factor in postpartum disturbances, an upregulation of other CRF systems could also be involved (see Chapter 8). Finally, recall the evidence that central NE systems appear to play a positive role in maternal behavior, and that trans-genic mice with a knockout mutation of the dopamine beta hydroxylase (DBH) gene show severe deficts in maternal behavior (DBH is an enzyme that catalyzes the conversion of dopamine to norepinephrine within NE neurons).

The evidence that postpartum or maternal depression is related to deficits in maternal responding and negative maternal feelings comes from several sources (for good reviews, see Buist [1998a] and Whitbeck et al. [1992]). There have been case studies that have reported that postpartum depression and anxiety are associated with maternal thoughts of harming or killing the infant, and in these instances other individuals (the father or grandparents) had to take over primary infant care responsibilities (Brockington & Brierley, 1984; Smoller & Lewis, 1977). In studies involving larger samples, it has been found that a significantly higher proportion of depressed mothers (41%) *reported* having thoughts of harming their infants than did a control group of nondepressed mothers (7%) (Jennings, Ross, Popper, & Elmore, 1999). Other studies have directly observed mother-infant interactions in middle-class primiparous women (Campbell, Cohn, & Meyers, 1995; Fleming, Ruble, Flett, & Shaul, 1988; also see Teti, Gelfand, Messinger, & Isabella, 1995). Campbell et al. (1995) observed such females with their infants during face-to-face interactions, during toy play, and during feeding, and they report that mothers with persistent (long-lasting) postpartum depression showed less positive interactions (for example, smiling, playful vocalizations) with their infants than did control mothers. Interestingly, negative interactions (expressions of anger, rough handling) were not increased in these depressed mothers. In the Fleming et al. (1988) study, it was found that primiparous middle-class mothers who were mildly depressed expressed less positive feelings about their maternal adequacy, and they also showed fewer affectionate contact behaviors with their infants. Interestingly, prior experience with infants (babysitting, for example) appeared to ameliorate these effects in the depressed mothers. One interpretation of this finding is that the birth of an infant is a stressful life event in itself, and the interaction of a life stress with the occurrence of mild depression may have negative consequences for maternal behavior and feelings. Prior experience with infants may have decreased the stressfulness associated with caring for a newborn. With respect to this last point, and the model we have presented in Figure 9.3, it might be predicted that depressed mothers who are exposed to higher levels of stressful life events above that caused by the birth of a first child might show much more severe disruptions in mood and in maternal behavior, and that such disturbances might lead to physical child abuse (physical violence directed at the infant) or child neglect. There is some evidence to support this view. Note that the Fleming et al. (1988) and Campbell et al. (1995) studies dealt with economically advantaged middle-class women. In contrast, Lahey, Conger, Atkeson, and Treiber (1984) compared the psychological characteristics of a sample of low-socioeconomic-status (SES) mothers who had a documented history (by child welfare agencies) of being physically abusive toward their children with a matched sample of nonabusive low-SES mothers. They found that the abusive mothers had higher levels of depression and anxiety than did the nonabusive mothers, and that the abusive mothers also experienced life as being more stressful. Direct observations of mother-child interactions also showed that the mothers with a prior history of abusive behavior engaged in a higher level of negative physical interactions (for example, pushing, hitting) and a lower level of positive interactions (for example, hugging, expressions of approval) with their children than did the controls. Additional studies have also emphasized the positive correlation between the incidence of stressful life events and the occurrence of child abuse (Straus, 1980; Wolfe, 1985). Therefore, based on the model shown in Figure 9.3, we would suggest that concurrent or recent stressful life events might potentiate maternal anxiety and depression with its associated deficits in maternal behavior and maternal feeling states by causing an overactivation of the depicted stress-related neural circuitry.

In a very interesting review, Maestripieri and Carroll (1998) discuss the relevance of animal research for an understanding of child abuse and neglect in humans. They note that infant abuse (infant dragging, throwing, hitting, etc.) has been observed in several nonhuman primate species, and that infant neglect (infant abandonment, which represents a complete breakdown of maternal responding) has been observed in many vertebrate species. They try to make the case that instances of infant abandonment in animals may be an adaptive response when considered from an evolutionary perspective, while instances of infant abuse may be a pathological (maladaptive) characteristic.

With respect to infant abandonment/neglect, the animal literature suggests that it occurs in a variety of species under environmental conditions that are unfavorable for infant development, such as harsh environmental conditions associated with food shortages (Bronson, 1989; Bronson & Marsteller, 1985; Hrdy, 1999; Maestripieri & Carroll, 1998). Such behavior makes sense from an evolutionary perspective because the mother is conceived of as terminating her investment in an infant that is not likely to survive, in this way conserving her resources so that she can reinitiate a reproductive episode once environmental conditions become favorable again. As outlined by Maestripieri and Carroll (1998; also see Maestripieri, 1998), physical abuse of infants in nonhuman primates is not associated with maternal indifference. Instead, it is associated with maternal anxiety, increased maternal protectiveness, and increased contact time between the mother and her infant (see Chapter 3). Since maternal care continues in these abusive mothers (athough the infants are mishandled, many of them survive to weaning), it is considered a maladaptive or pathological state, particularly since no logic can be constructed to show how such behavior would increase the reproductive success of the abusive parent. In other words, it is considered to be the result of a pathological temperament (increased anxiety and reactivity to stress).

In the human context, infant abuse and infant neglect may map on to the two limbs of the mechanisms for mammalian maternal care: changes in anxiety and changes in reward. In contrast to the high levels of anxiety in the women with infant abuse, the syndrome of maternal neglect (which can be observed immediately after parturition) may result from an absence of hedonic responses to infant stimuli. Women with the syndrome of maternal neglect fail to bond to their infants. In this sense, they resemble rats with central oxytocin blockade or mice with null mutations of the Fos B, DBH, or prolactin receptor genes. They are able to deliver and lactate, but they have no interest in their offspring. In this context, it is interesting to speculate about the fact that the oxytocin receptor belongs to a gene protein receptor family that has been a prototype for human functional mutations. There are multiple mutations in the V2 receptor gene (Merendino, Spiegel, Crawford, Brownstein, & Lolait, 1993; Rosenthal, Siebold, & Antaramian, 1992) as well as in the gene for the prohormone (Rittig et al., 1996). Although there have been no studies of variation in the human oxytocin or oxytocin receptor genes, Hans Zingg at McGill University has identified a pedigree of women with problems in lactation who might be informative for studying how changes in these genes alter the motivation for human maternal behavior (Zingg, personal communication).

A preliminary model with respect to these issues is presented in Figure 9.4. We show that the mother's genotype, current and previous life experiences, and endocrine status can influence her temperament, in particular, her anxiety level, which in turn influences her ability to cope with environmental challenges. High and sustained anxiety could also promote depression with its associated anhedonia, as previously described. We propose that a supernormal level of anxiety may trigger child abuse, and if such activation is further increased in intensity and duration, particularly in primiparous women during the early postpartum period, with the result that severe depression and anhedonia ensue, then it is possible that a complete breakdown in maternal behavior may occur, resulting in infant abandonment or severe child neglect. This model proposes that child abuse and child neglect lie on a continuum with respect to underlying mechanisms. That is, moderately high anxiety may lead to overprotectiveness and abuse, but if anxiety is severe and prolonged, it may precipitate those neural changes associated with depression and anhedonia, and the associated lack of maternal reward may lead to neglect. Alternative models, however, can be presented, and it is certainly possible that child abuse and child neglect have distinct etiologies with distinct underlying mechanisms (see Maestripieri & Carroll, 1998). It is also possible that a variety of etiologies and underlying mechanisms might contribute to potentially different types of abuse and neglect. Based on the animal literature, for example, it is certainly possible to conceive of neurobiological and genetic disturbances that could result in neglect that is independent of generalized anxiety and depression (for example, MPOA dysfunction; also see discussion of the oxytocin receptor gene, above).

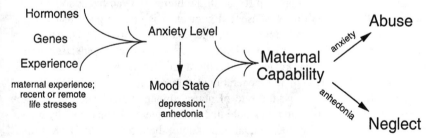

Figure 9.4. A summary diagram that postulates that periparturitional hormones, genetic factors, and previous and current life experiences can all influence the human mother's anxiety level and mood state (depression and anhedonia), which, in turn, can influence her maternal competence. This model proposes that supernormal levels of anxiety may promote abuse, while depression and anhedonia may promote neglect.

Recall the study by Bloch et al. (2000), who showed that women with a previous history of depression reacted with a depressed mood state when they were exposed to a steroid hormone withdrawal regimen, while a group of women without prior depression did not show such a response to the same hormone treatment. In this context, and with respect to the occurrence of infant abuse, the findings of Maestripieri and Megna (2000a, 2000b) are relevant. They compared prepartum and early postpartum estradiol and progesterone levels in groups of abusive and nonabusive rhesus monkeys, and they reported that major differences were *not* evident. These results indicate that a major deficit or abnormality in hormonal stimulation (at least with respect to estradiol and progesterone) was not the cause of the increased maternal anxiety, protectiveness, and abusive behavior, and they further suggest that it was differences in the reactivity of the underlying neural circuitry that probably contributed to the observed behavioral differences.

We will end this section by presenting evidence that abusive human parents show indications of greater stress reactivity in the context of a child rearing situation. In a psychophysiological study, Wolfe and Fairbank (1983) showed a group of mothers with a history of child abuse videotapes of mothers and their children interacting under conditions where the child was misbehaving or behaving appropriately. In comparison to a group of nonabusive mothers, the abusive mothers showed greater sympathetic nervous system arousal when viewing the tapes of the misbehaving child. Given that sympathetic arousal is a correlate of anxiety (see Chapter 3) and anger, one might assume that overreaction to the stress of a misbehaving child leads to abusiveness.

4. Intergenerational Continuity in Abnormal Maternal Behavior

The concept of intergenerational continuity or transmission of maternal responsiveness in humans is a common theme in the literature (Van Ijzendoorn, 1992). Indeed, with respect to child abuse and neglect, it is often argued that there is a cycle of violence (see Widom, 1989), and that those children who have been abused or neglected by their parents tend to become abusive or neglectful parents themselves. In a critical review of the literature, Kaufman and Zigler (1987) suggested that the rate of intergenerational transmission of child abuse is approximately 30%. That is, the rate of abuse among individuals with a prior history of abuse is 30%, and they note that this frequency is substantially higher than the rate observed in the general population (about 5% in 1987). Kaufman and Zigler (1987) have reviewed some factors that may break the cycle of violence, given that many abused children do not become abusive parents, and they suggest that parents who did not repeat the cycle had strong social supports as adults and that they also had a supportive relationship with one of their parents during their childhood.

In addition to the occurrence of child abuse and neglect, there is evidence for intergenerational continuity in other aspects of maternal behavior. For example, Miller, Kramer, Warner, Wickramaratne, and Weissman (1997) have reported that the maternal characteristic referred to as affectionless control is transmitted across generations. Such parents are less attentive to their children and less interested in them, and they also tend to administer harsh discipline. Similar findings have been reported by Whitbeck et al. (1992). Finally, in a review of the literature, Van Ijzendoorn (1992) found that there is good evidence for the intergenerational transmission of parental attachment. That is, adults who recollected having a secure attachment relationship with their parents tended to have children who were securely attached to them, and adults who recollected insecure attachments to their parents tended to have children who were insecurely attached to them.

With respect to the mechanism of a possible intergenerational transmission of maternal responsiveness, a genetic contribution should not be excluded, and in support of such a mechanism we have already indicated that monozygotic twins are more similar in their parental behavior than are dizygotic twins. One should consider the possibility that mutations of certain critical genes could cause disruptions in maternal behavior (see Chapter 6). However, within the context of the evidence presented in this chapter, and the rodent and nonhuman primate data presented in Chapters 3 and 8, an additional ontogenetic mechanism that may influence the intergenerational continuity of child abuse and child neglect can be suggested. It can be proposed that maternal treatment of a child influences the temperamental development of the child so that when the child becomes a parent, her temperamental characteristics influence the way she cares for her offspring (Buist, 1998a; Miller et al., 1997; Whitbeck et al., 1992). In other words, processes similar to those described by Fairbanks, Meaney, Plotsky, and Suomi for nonhuman animals may be operative in humans. In particular, it can be argued that childhood abuse and neglect increase the likelihood that the affected offspring will go on to develop anxiety, anhedonia, and depressive disorders in adulthood, and that such development, along with the parturient and postpartum hormonal milieu, genetics, and current life stresses, may all operate to influence the degree to which a postpartum woman will behave in an abusive or neglectful manner to her own offspring (cf. Buist, 1998a). The evidence in support of this view is not complete, and the best evidence will be that childhood abuse and neglect are risk factors for the development of affective disorders. However, because we have already reviewed the data that postpartum depression is associated with maternal deficits, we can at least see the logic of the cyclic argument. Perhaps the incidence, intensity, and duration of postpartum affective disorders are greater in those women who have been neglected or physically and emotionally abused as children, and the greater degree of stress reactivity, anxiety, depression, and anhedonia in such women may have important ramifications for their maternal responsiveness. Note, however, that humans, as a species, are resilient—neither genes nor developmental experiences are destiny. In considering the effects of experience, we need to remember that individuals differ not only in their perceptions of rearing environments but also in their resilience to adversity.

In partial support for the model that has just been outlined, Whitbeck et al. (1992) have presented correlational evidence that indicates that adults who recollected being rejected by their parents (such parental treatment, of course, is not nearly as severe as abuse or neglect) tended to have a depressed mood that was correlated with deficits in their parental behavior toward their own children. Importantly, their children also tended to have a depressed mood state (also see Miller et al., 1997). In another study, Buist (1998b) reported on the characteristics of women who were admitted to a hospital because of postpartum depression. Some of these women had a history of being abused by their parents, while others did not. The severity of depression and anxiety was greater in the abused subset than in the women without such a history, and the maternal behavior shown by the abused depressed women was inferior to that shown by the nonabused depressed mothers, although the majority of depressed women in both subgroups reported experiencing parental stress. These results suggest that a prior history of child abuse may exacerbate postpartum depression, which then results in poorer maternal behavior. Finally, there is excellent evidence that exposure of a child to adverse experiences increases the incidence of adult depression and anxiety. Such early adverse experiences, which presumably have major influences on the developing brain, include early parental loss (the most extreme case of child neglect or parental deprivation), child neglect, and child abuse (Agid et al., 1999; Breier et al., 1988; Carter, Joyce, Mulder, Luty, & Sullivan, 1999; Heim & Nemeroff, 1999; Hill et al., 2001; Kaufman, Plotsky, Nemeroff, & Charney, 2000; Nemeroff, 1999).

In an interesting study, Teti et al. (1995) examined the maternal behavior and the quality of infant attachment in depressed and nondepressed mothers. As expected, the depressed mothers showed inferior maternal behavior. In addition, the children of the depressed mothers tended to be insecurely attached to their mothers, while the children of the control mothers were securely attached. Child attachment was measured using the Strange Situation procedure. In this procedure, the child is separated from its parent for several minutes while in a novel setting. Afterwards, the child is reunited with the parent and the child's behavior is observed. A securely attached infant usually approaches the mother upon reunion, is briefly reassured, and then uses the mother as a home base for exploration of the new environment. An insecurely attached infant is not comforted by the return of the mother and may even avoid her. In a manner analogous to Suomi's (1999) discussion of attachment theory in relation to emotional development in motherless rhesus monkeys (see Chapter 3), we propose that the results of Teti et al. (1995) suggest that the children of depressed mothers do not use their mothers as a secure home base to facilitate exploration of their environment. This may influence the development of shyness, timidity, and anxiety in these children. Along with other risk factors, such a temperamental outcome might potentiate the development of mood disorders in the affected child, which, in turn, could have an impact on the future parental responsiveness of the child (cf. Van Ijzendoorn, 1992).

If early adverse experiences in children influence the development of anxiety, depression, and related forms of psychopathology, which may then influence the adult parental responsiveness of the affected individuals, is there any evidence pertaining to the possible underlying physiological mechanisms that may be involved? Are such physiological mechanisms similar to those that are induced in animals by maternal deprivation/separation, and that have been related to anxiety and anhedonia in animals? In a pattern somewhat analogous to what appears to occur in adult depression, early adverse experiences appear to cause a dysregulation of HPA activity, which may be indicative of abnormal central CRF activity, and there is also evidence that early trauma may cause a dysregulation in central norepinephrine systems. Kaufman et al. (1997) compared the pituitary ACTH response to a systemic injection of CRF in three groups of children: depressed abused, depressed nonabused, and normal controls (also see Heim & Nemeroff, 1999). The depressed abused group showed the greatest CRF-induced ACTH secretion. The types of child abuse consisted of sexual, physical, and emotional abuse. At the time of the study, the depressed abused children were living in homes that still exposed them to emotional maltreatment, although sexual and physical abuse were no longer occurring. More importantly, De Bellis et al. (1999) have reported that children who have previously been abused but are currently living in stable environments, and who have been diagnosed as currently displaying posttraumatic stress disorder, show higher levels of urinary cortisol and norepinephrine (NE) than do normal control children. These results suggest a hyperactivation of the HPA system and the sympathetic nervous system in the maltreated group. The HPA hyperactivity is presumably reflective of a general increase in central CRF activity, and if peripheral NE activity mirrors central NE activity (see Rogeness & McClure, 1996), these results might also be indicative of hyperactivity in the central NE systems of the abused children. Rogeness and McClure note that *adults* with PTSD sometimes show *lower* levels of endogenous secretion of cortisol than do controls. One possibility, as mentioned previously, is that hyperactivity of CRF release over long durations might cause a downregulation of CRF receptors in the anterior pituitary (Bremner et al., 1997).

Additional work has reported that maltreated children (sexual, physical, or emotional abuse, or neglect) display higher systemic levels of cortisol (Cicchetti & Rogosch, 2001), but other studies show lower levels of cortisol in abused or neglected children when compared to controls (Hart, Gunnar, & Cicchetti, 1995). As these authors indicate, the developmental timing of the abuse and neglect, as well as the intensity and duration of the maltreatment, may influence the measured HPA reactivity. Intense and chronic trauma early in life may cause a downregulation of the pituitary response to CRF even in children. An analogous case may exist with respect to norepinephrine activity. Rogeness and McClure (1996; also see Rogeness, 1991) compared three groups of *emotionally disturbed* children, those with a history of neglect (analogous to maternal deprivation in animals), those with a history of abuse, and those with no history of abuse or neglect. They found that the group with a history of neglect had lower urinary levels of dopamine beta hydroxylase than did the other groups, and this lower level of DBH was associated with a lower systolic blood pressure. The abused group did not differ from the nonabused, nonneglected group on these measures. The authors suggest that neglect may cause a downregulation of sympathetic nervous system activity, and that this peripheral effect may be mirrored by a concomitant downregulation of central NE systems. These results should be compared to those of De Bellis et al. (1999), who found that abused children had higher levels of urinary NE than did *normal* controls. How can one account for the findings that neglected children appear to have lower peripheral NE activity than do abused children? One possibility is that abuse results in a physiological and psychological profile that is distinct from that caused by neglect. Another possibility is that the neglect may have been a more severe trauma than the abuse, or that the developmental time course or the chronicity of the neglect differed from that of the abuse. These differences might have caused a hyperactivity of NE activity in certain abused individuals, but a depletion of NE activity in certain neglected individuals. This NE depletion, however, is conceived of as being induced by an initial period of heightened NE activity caused by the neglect. In other words, as indicated previously when discussing the review by Anisman and Zacharko (1982), in the abused group NE synthesis may have been able to keep up with NE utilization, while in the neglected group NE activation may have been so extreme that synthesis could not maintain normal NE levels. It is interesting to speculate with respect to central NE levels that in cases of both abuse and neglect it is a hyperactive central CRF system that is driving a central NE system, and a likely central NE site would be the locus coeruleus.

In conclusion, therefore, although the evidence is scant, we are making the case that child maltreatment may cause a dysregulation of central CRF and NE systems, and such effects may play a role in the intergenerational transmission of child maltreatment because the anxiety and depression associated with these CRF and NE effects are expected to cause maternal behavior deficits through mechanisms that we have previously described. Finally, be aware of the possibility that child maltreatment may also cause a dysregulation of central dopaminergic systems (see Chapter 8 and Figure 9.3). Indeed, such a dysregulation is likely to contribute to the anhedonia associated with depression, and such an effect would have important implications for adult parental responsiveness. Importantly, such a developmental effect might also influence the propensity for drug abuse in adulthood.

5. Drug Addiction and Human Maternal Behavior

Women who are addicted to opiates or cocaine have been found to have deficits in their maternal behavior when compared to non-drug-abusing controls, even when these controls are matched with the drug users on many variables, including socioeconomic status (Bauman & Dougherty, 1983; Beckwith, Howard, Espinosa, & Tyler, 1999; Black & Mayer, 1980; Burns, Chethik, Burns, & Clark,

1991; Fiks, Johnson, & Rosen, 1985; Hans, Bernstein, & Henson, 1999; Magura & Laudet, 1996; Roland & Volpe, 1989; Suchman & Luthar, 2000). Such maternal deficits include the following: (a) increased occurrence of child abuse or neglect, (b) decreased involvement or interest in the child's activities and decreased sensitivity to the child's needs, (c) increased occurrence of negative behaviors (threats, disapprovals, yelling, ignoring, commanding) directed toward the child, and (d) disruptions in the continuity of maternal care. Over time, a greater proportion of drug-abusing mothers relinquish control of their children when compared to controls. These deficits in the mother-child relationship have been detected through direct observation, structured interviews with the mother (and sometimes with the child), or through inspection of social welfare agency records.

The mechanisms that mediate the association between drug addicition and poor maternal behavior are obviously complex and multifaceted. First, it is possible that cocaine and opiates are acting to directly disrupt the neural mechanisms that we have outlined as contributing to maternal motivation. In rats, direct application of cocaine or opiates to the MPOA is capable of disrupting maternal behavior (Bridges, 1996; Elliott, Lubin, Walker, & Johns, 2001; Rubin & Bridges, 1984; Vernotica, Rosenblatt, & Morrell, 1999; also see Kalin, Shelton, & Lynn, 1995; Misiti, Turillazzi, Zapponi, & Loizzo, 1991). Additionally, most addictive drugs produce their rewarding effects by activating the mesolimbic and mesocortical dopamine systems (Hyman & Malenka, 2001; Koob, 1999). Therefore, chronic opiate or cocaine use may cause dysfunctions in the VTA, nucleus accumbens, or prefrontal cortex, which would have ramifications for maternal responsiveness. Recall from Chapter 5 that too much, as well as too little, telencephalic dopaminergic activity has been associated with deficits in rodent maternal behavior. From a slightly different perspective, note that both infants (for postpartum mothers) and drugs of abuse have reinforcing/attractive qualities, and that these stimuli may actually compete for the mother's attention. If the reward value of a drug, or the compulsion to take the drug, becomes greater than the reward value of caring for the young, then maternal behavior would be expected to be disrupted (see Mattson et al., 2001). Such alterations in motivational inclinations undoubtedly involve modifications in the mesolimbic DA system (see Hyman & Malenka, 2001).

The association between drug abuse and faulty maternal behavior may also entail other mediating mechanisms. For example, stressful living conditions, such as extreme poverty, might promote both drug abuse and faulty maternal behavior. Similarly, the conflict between the craving and compulsion to take a drug and the desire to care for one's young, particularly when this involves obtaining the drug under threatening environmental situations, may cause great stress in the mother, and this increased stress exposure may be the factor that has a disruptive influence on maternal behavior. Finally, cocaine and opiate addiction often co-occur with various forms of psychopathology, which include an increased incidence of depression, anxiety, and personality disorders (paranoia, antisocial personality) (Beckwith et al., 1999; Hans et al., 1999). Since we have already argued that depression and anxiety are associated with deficits in maternal behavior in women, such psychopathology may be important for the observed maternal deficits. A critical issue is whether the psychopathology precedes the onset of the drug abuse or is induced by the drug abuse, and most studies have not been designed to answer this question. If the psychopathology precedes the onset of drug abuse, then drug addicition may simply exacerbate the negative effects of an existing psychopathology on maternal behavior. However, it is also possible that chronic drug abuse precipitates various forms of psychopathology by chronically overactivating, and therefore causing a dysfunction in, the mesolimbic and mesocortical DA systems, which might lead to depressive and/or antisocial symptomology. In this latter case, the psychopathology and the maternal deficits may both be caused by a drug-induced disruption of the same neural system.

At the present time, it is probably safest to conclude that since drugs of abuse, like cocaine and opiates, can cause disruptions in neural systems that are positively involved in controlling maternal behavior, chronic use of such drugs could probably cause some deficits in human maternal behavior independent of other factors, but such deficits would be greatly enhanced by the additional negative effects of remote and recent life stresses, and of any currently existing psychopathologies.

6. Conclusions

Although our knowledge of the biology of human parental behavior is clearly undeveloped, we hope to have shown that an integration of animal and human research findings can provide a strong foundation for future research. One theme running throughout this book is that a decrease in stress reactivity may be critical for appropriate maternal behavior. At least for certain pathological conditions associated with human maternal behavior, a hyperactivity of stress-related neural circuits, caused by a variety of factors, which probably include genetic, hormonal, and experiential influences, may interfere with the normal functioning of central neural mechanisms that are essential for adaptive maternal responding.

Cumulative References

Adamec R (2001) Does long term potentiation in periaqueductal gray (PAG) mediate lasting changes in rodent anxiety-like behavior (ALB) produced by predator stress? Effects of low frequency stimulation (LFS) of PAG on place preference and changes in ALB produced by predator stress. Behav Brain Res 120:111–135.

Adamec RE, McKay D (1993) Amygdala kindling, anxiety, and corticotrophin releasing factor (CRF). Physiol Behav 54:423–431.

Adams F, Schwarting RKW, Boix F, Huston JP (1991) Lateralized changes in behavior and striatal dopamine release following unilateral tactile stimulation of the perioral region: a microdialysis study. Brain Res 553:318–322.

Agid O, Shapira B, Zislin J, Ritsner M, Hanin B, Murad H, Troudart T, Bloch M, Heresco-Levy U, Lerer B (1999) Environment and vulnerability to major psychiatric illness: a case control study of early parental loss in major depression, bipolar disorder and schizophrenia. Mol Psychiatry 4:163–172.

Ahdieh HB, Mayer AD, Rosenblatt JS (1987) Effects of brain antiestrogen implants on maternal behavior and on postpartum estrus in pregnant rats. Neuroendocrinology 46: 522–531.

Ahokas A, Kaukoranta J, Aito M (1999) Effect of estradiol on postpartum depression. Psychopharmacology 146:108–110.

Akwa Y, Purdy RH, Koob GF, Britton KT (1999) The amygdala mediates the anxiolytic-like effect of the neurosteroid allopregnanolone in the rat. Behav Brain Res 106:119–125.

Albert DJ, Chew GL (1980) The septal forebrain and the inhibitory modulation of attack and defense in the rat. A review. Behav Neural Biol 30:357–388.

Albert DJ, Walsh ML, Gorzalka BB, Mendelson S, Zalys C (1986) Intermale social aggression: suppression by medial preoptic area lesions. Physiol Behav 38:169–173.

Alexander G, Stevens D (1981) Recognition of washed lambs by merino ewes. Appl Anim Ethol 7:77–86.

Alheid GF, Heimer L (1988) New perspectives in basal forebrain organization of special relevance for neuropsychiatric disorders: the striatopallidal, amygdaloid, and cortico-petal components of substantia innominata. Neuroscience 27:1–39.

Allin JT, Banks EM (1972) Functional aspects of ultrasound production by infant albino rats (*Rattus norvegicus*). Anim Behav 20:175–185.

Allman J, Rosin A, Kumar R, Hasenstaub A (1998) Parenting and survival in anthropoid primates: caretakers live longer. Proc Natl Acad Sci 95:6866–6869.

Almeida OFX, Yassouridis A, Forgas-Moya I (1994) Reduced availability of milk after central injections of corticotropin-releasing hormone in lactating rats. Neuroendocrinology 59:72–77.

Alston-Mills B, Parker AC, Eisen EJ, Wilson R, Fletcher S (1999) Factors influencing maternal behavior in the hubb/hubb mutant mouse. Physiol Behav 68:3–8.

Altemus M, Deuster PA, Galliven G, Carter, CS, Gold, PW (1995) Suppression of hypothalamic-pituitary-adrenal responses to stress in lactating women. J Clin Endocrinol Metab 80:2954–2959.

Alves SE, Lopez V, McEwen BS, Weiland NG (1998) Differential colocalization of estrogen receptor beta (ERbeta) with oxytocin and vasopressin in the paraventricular and supraoptic nuclei of the female rat brain: an immunocytochemical study. Proc Natl Acad Sci 95:3281–3286.

Amenomori Y, Chen CL, Meites J (1970) Serum prolactin levels in rats during different reproductive states. Endocrinology 86:506–510.

Amico JA, Crowley RS, Insel TR, Thomas A, O'Keefe JA (1995) Effect of gonadal steroids upon hypothalamic oxytocin expression. In: Oxytocin (Ivell R, Russell J, eds), pp 23–35. New York: Plenum Press.

Amico JA, Thomas A, Hollingshead DJ (1997) The duration of estradiol and progesterone exposure prior to progesterone withdrawal regulates oxytocin mRNA levels in the paraventricular nucleus of the rat. Endocrinol Res 23:141–156.

Anderson CH (1982) Changes in dendritic spine density in the preoptic area of the female rat at puberty. Brain Res Bull 8:261–265.

Anderson CO, Zarrow MX, Fuller GB, Denenberg VH (1971) Pituitary involvement in maternal nest building in the rabbit. Horm Behav 2:183–189.

Anisman H, Zacharko RM (1982) Depression: the predisposing influence of stress. Behav Brain Sci 5:89–137.

Anisman H, Zaharia MD, Meaney MJ, Meralis Z (1998) Do early life events permanently alter behavioral and hormonal responses to stressors? Int J Dev Neurosci 16:149–164.

Augert AP, Blaustein JD (1995) Progesterone enhances an estradiol-induced increase in Fos immunoreactivity in localized regions of female rat forebrain. J Neurosci 15: 2272–2279.

Avar Z, Monos E (1969) Biological role of lateral hypothalamic structures participating in the control of maternal behavior in the rat. Acta Physiol Acad Sci Hung 35:285–294.

Bahr NI, Martin RD, Pryce CR (2001) Peripartum sex steroid profiles and the endocrine correlates of postpartum maternal behavior in captive gorillas (Gorilla gorilla gorilla). Horm Behav 40:533–541.

Bakowska JC, Morrell JI (1997) Atlas of the neurons that express mRNA for the long form of the prolactin receptor in the forebrain of the female rat. J Comp Neurol 386: 161–177.

Baldwin BA, Shillito EE (1974) The effects of ablation of the olfactory bulbs on parturition and maternal behavior in soay sheep. Anim Behav 22:220–223.

Bale TL, Contarino A, Smith GW, Chan R, Gold LH, Sawchenko PE, Koob GF, Vale WW, Lee KF (2000) Mice deficient for corticotropin-releasing hormone receptor-2 display anxiety-like behaviour and are hypersensitive to stress. Nature Genet 24:410–414.

Bale TL, Davis AM, Auger AP, Dorsa DM, McCarthy MM (2001) CNS region-specific oxytocin receptor expression: importance in regulation of anxiety and sex behavior. J Neurosci 21:2546–2552.

Bale TL, Picetti R, Contarino A, Koob GF, Vale WW, Lee KF (2002) Mice deficient for both corticotropin-releasing factor receptor 1 (CRFR1) and CRFR2 have an impaired stress response and display sexually dichotomous anxiety-like behavior. J Neurosci 22:193–199.

Ball GF, Balthazart JHJ (2002) Neuroendocrine mechanisms regulating reproductive cycles and reproductive behavior in birds. In: Hormones, brain and behavior, Vol 2 (Pfaff DW, Arnold AP, Etgen AM, Fahrbach SE, Rubin RT, eds), pp 649–798. San Diego: Academic Press.

Ballenger JC (1995) Benzodiazepines. In: Textbook of psychopharmacology (Schatzberg AF, Nemeroff CB, eds), pp 215–230. Washington, DC: American Psychiatric Press.

Balthazart J, Absil P, Gerard M, Appeltants D, Ball GF (1998) Appetitive and consummatory male sexual behavior in Japanese quail are differentially regulated by subregions of the preoptic medial nucleus. J Neurosci 18:6512–6527.

Bamshad M, Novak M, De Vries G (1994) Cohabitation alters vasopressin innervation and paternal behavior in prairie voles (Microtus ochrogaster). Physiol Behav 56:751–758.

Bandler R (1988) Brain mechanisms of aggression as revealed by electrical and chemical stimulation: suggestion of a central role of the midbrain periaqueductal gray region. Prog Psychobiol Physiol Psychol 13:67–153.

Bandler R, Shipley MT (1994) Columnar organization in the midbrain periaqueductal gray: modules for emotional expression? Trends Neurosci 17:379–389.

Banks WA, Kastin AJ (1985) Permeability of the blood-brain barrier to neuropeptides: the case for penetration. Psychoneuroendocrinology 10:385–399.

Barberis C, Tribollet E (1996) Vasopressin and oxytocin receptors in the central nervous system. Crit Rev Neurobiol 10:119–154.

Barden N, Reul JMHM, Holsboer F (1995) Do antidepressants stabilize mood through actions on the hypothalamic-pituitary-adrenocortical system? Trends Neurosci 18:6–11.

Barnet B, Joffe A, Duggan AK, Wilson MD, Repke JT (1996) Depressive symptoms, stress, and social support in pregnant and postpartum adolescents. Arch Pediatr Adolesc Med 150:64–69.

Barofsky AL, Taylor J, Tizabi Y, Kumar R, Jones-Quartey K (1983) Specific neurotoxin lesions of the median raphe serotonergic neurons disrupt maternal behavior in the lactating rat. Endocrinology 113:1884–1889.

Bart J, Tornes A (1989) Importance of monogamous male birds in determining reproductive success. Behav Ecol Sociobiol 24:109–116.

Bauer JH (1983) Effects of maternal state on the responsiveness to nest odors of hooded rats. Physiol Behav 30:229–232.

Baum MJ (1978) Failure of pituitary transplants to facilitate the onset of maternal behavior in ovariectomized virgin rats. Physiol Behav 20:87–89.

Baum MJ, Everitt BJ (1992) Increased expression of c-fos in the medial preoptic area after mating in male rats: role of afferent inputs from the medial amygdala and midbrain central tegmental field. Neuroscience 50:627–646.

Bauman PS, Dougherty FE (1983) Drug-addicted mothers' parenting and their children's development. Int J Addict 18:291–302.

Beach FA (1937) The neural basis of innate behavior: I. The effects of cortical lesions upon the maternal behavior pattern in the rat. J Comp Psychol 24:393–436.

Beach FA, Jaynes J (1956a) Studies on maternal retrieving in rats: II. Effects of practice and previous parturitions. Am Naturalist 90:103–109.

Beach FA, Jaynes J (1956b) Studies of maternal retrieving in rats: I. Recognition of young. J Mammal 37:177–180.

Beach FA, Jaynes J (1956c) Studies of maternal retrieving in rats: III. Sensory cues involved in the lactating female's response to her young. Behaviour 10:104–125.

Beach FA, Wilson JR (1963) Effects of prolactin, progesterone, and estrogen on reactions of nonpregnant rats to foster young. Psychol Rep 13:231–239.

Becker JB, Rudick CN, Jenkins WJ (2001) The role of dopamine in the nucleus accumbens and striatum during sexual behavior in the female rat. J Neurosci 21:3236–3241.

Beckwith L, Howard J, Espinosa M, Tyler R (1999) Psychopathology, mother-child interaction, and infant development: substance-abusing mothers and their offspring. Dev Psychopathol 11:715–725.

Beggs JM, Brown TH, Byrne JH, Crow T, LeDoux JE, LeBar K, Thompson RF (1999) Learning and memory: basic mechanisms. In: Fundamental neuroscience (Zigmond MJ, Bloom FE, Landis SC, Roberts JL, Squire LR, eds), pp 1411–1454. San Diego: Academic Press.

Behbehani MM (1995) Functional characteristics of the midbrain periaqueductal gray. Prog Neurobiol 46:575–605.

Ben-Jonathan N, Arbogast LA, Hyde JF (1989) Neuroendocrine regulation of prolactin release. Prog Neurobiol 33:399–447.

Benuck I, Rowe FA (1975) Centrally and peripherally induced anosmia: influences on maternal behavior in lactating female rats. Physiol Behav 14:439–447.

Berman CM (1990) Intergenerational transmission of maternal rejection rates among free-ranging rhesus monkeys. Anim Behav 39:329–337.

Bermond B (1982) Effects of medial preoptic hypothalamus anterior lesions on three kinds of behavior in the rat: intermale aggressive, male-sexual, and mouse-killing behavior. Aggress Behav 8:335–354.

Bester-Meredith JK, Young LJ, Marler CA (1999) Species differences in paternal behavior and aggression in peromyscus and their associations with vasopressin immunoreactivity and receptors. Horm Behav 36:25–38.

Bitran D, Hilvers RJ, Kellogg CK (1991) Ovarian endocrine status modulates the anxiolytic potency of diazepam and the efficacy of γ-aminobutyric acid-benzodiazepine receptor-mediated chloride ion transport. Behav Neurosci 105:653–662.

Bitran D, Shiekh M, McLeod M (1995) Anxiolytic effect of progesterone is mediated by the neurosteroid allopregnanolone at brain GABAA receptors. J Neuroendocrinol 7:171–177.

Black R, Mayer J (1980) Parents with special problems: alcoholism and opiate addiction. Child Abuse Negl 4:45–54.

Blackburn JR, Pfaus JG, Phillips AG (1992) Dopamine functions in appetitive and defensive behaviours. Prog Neurobiol 39:247–279.

Bloch M, Schmidt PJ, Danaceau M, Murphy J, Nieman L, Rubinow DR (2000) Effects of gonadal steroids in women with a history of postpartum depression. Am J Psychiatry 157:924–930.

Blumer L (1979) Male parental care in the bony fishes. Q Rev Biol 54:149–161.

Blumer L (1982) A bibliography and categorization of bony fishes exhibiting parental care. Zool J Linnean Soc 76:1–22.

Boccia ML, Pedersen CA (1999) Early maternal separation alters postpartum pup-licking and grooming, lactation-associated changes in aggression and anxiety, and central oxytocin binding in female offspring. Neurosci Abstr 25:872.

Boccia ML, Pedersen CA (2001) Brief vs. long maternal separations in infancy: contrasting relationships with adult maternal behavior and lactation levels of aggression and anxiety. Psychoneuroendocrinology 26:657–672.

Bolwerk ELM, Swanson HH (1984) Does oxytocin play a role in the onset of maternal behaviour in the rat? J Endocrinol 101:353–357.

Bonci A, Malenka RC (1999) Properties and plasticity of excitatory synapses on dopaminergic and GABAergic cells in the ventral tegmental area. J Neurosci 19:3723–3730.

Bontempi B, Sharp FR (1997) Systemic morphine-induced Fos protein in the rat striatum and nucleus accumbens is regulated by μ opioid receptors in the substantia nigra and ventral tegmental area. J Neurosci 17:8596–8612.

Bozas E, Tritos N, Phillipidis H, Stylianopoulou F (1997) At least three neurotransmitter systems mediate stress-induced increase in *c-fos* mRNA in different brain areas. Cell Mol Neurobiol 17:157–169.

Braun K, Lange E, Metzger M, Poeggel G (2000) Maternal separation followed by early social deprivation affects the development of monoaminergic fiber systems in the medial prefrontal cortex of *Octodon degus*. Neuroscience 95:309–318.

Breier A, Kelsoe JR, Kirwin PD, Beller SA, Wolkowitz OM, Pickar D (1988) Early parental loss and development of adult psychopathology. Arch Gen Psychiatry 45:987–993.

Bremner JD, Licinio J, Darnell A, Krystal JH, Owens MJ, Southwick SM, Nemeroff CB, Charney DS (1997) Elevated CSF corticotropin-releasing factor concentrations in posttraumatic stress disorder. Am J Psychiatry 154:624–629.

Brennan PA, Kendrick KM, Keverne EB (1995) Neurotransmitter release in the accessory olfactory bulb during and after the formation of an olfactory memory in mice. Neuroscience 69:1075–1086.

Brennan PA, Keverne EB (1997) Neural mechanisms of mammalian olfactory learning. Prog Neurobiol 51:457–481.

Breton C, Pechoux C, Morel G, Zingg HH (1995) Oxytocin receptor messenger ribonucleic acid: characterization, regulation, and cellular localization in the rat pituitary gland. Endocrinology 136:2928–2936.

Breuggeman JA (1973) Parental care in a group of free-ranging rhesus monkeys (*Macaca mulatta*). Folia Primatol 20:178–210.

Bridges RS (1975) Long-term effects of pregnancy and parturition upon maternal responsiveness in the rat. Physiol Behav 14:245–249.

Bridges RS (1977) Parturition: its role in the long term retention of maternal behavior in the rat. Physiol Behav 18:487–490.

Bridges RS (1978) Retention of rapid onset of maternal behavior during pregnancy in primiparous rats. Behav Biol 24:113–117.

Bridges RS (1984) A quantitative analysis of the roles of dosage, sequence, and duration of estradiol and progesterone exposure in the regulation of maternal behavior in the rat. Endocrinology 114:930–940.

Bridges RS (1990) Endocrine regulation of parental behavior in rodents. In: Mammalian parenting (Krasnegor NA, Bridges RS, eds), pp 93–117. New York, Oxford University Press.

Bridges RS (1996) Biochemical basis of parental behavior in the rat. In: Advances in the study of behavior, Vol 25. Parental care: evolution, mechanisms, and adaptive significance (Rosenblatt JS, Snowdon CT, eds), pp 215–242. San Diego: Academic Press.

Bridges RS, Clifton DK, Sawyer CH (1982) Postpartum luteinizing hormone release and maternal behavior in the rat after late-gestational depletion of hypothalamic norepinephrine. Neuroendocrinology 34:286–291.

Bridges RS, DiBiase R, Loundes CD, Doherty PC (1985) Prolactin stimulation of maternal behavor in female rats. Science 227:782–784.

Bridges RS, Freemark M (1995) Human placental lactogen infusions into the medial preoptic area stimulate maternal behavior in steroid-primed, nulliparous female rats. Horm Behav 29:216–226.

Bridges RS, Grimm CT (1982) Reversal of morphine disruption of maternal behavior by concurrent treatment with the opiate antagonist naloxone. Science 218:166–168.

Bridges RS, Mann PE, Coppeta JS (1999) Hypothalamic involvement in the regulation of maternal behavior in the rat: inhibitory roles for the ventromedial hypothalamus and the dorsal/anterior hypothalamic areas. J Neuroendocrinol 11:259–266.

Bridges RS, Numan M, Ronsheim PM, Mann PE, Lupini CE (1990) Central prolactin infusions stimulate maternal behavior in steroid-treated, nulliparous female rats. Proc Natl Acad Sci USA 87:8003–8007.

Bridges RS, Rigero BA, Byrnes EM, Yang L, Walker AM (2001) Central infusions of the recombinant human prolactin receptor antagonist, S179D-PRL, delay the onset of maternal behavior in steroid-primed, nulliparous female rats. Endocrinology 142:730–739.

Bridges RS, Robertson MC, Shiu RPC, Friesen HG, Stuer AM, Mann PE (1996) Endocrine communication between conceptus and mother: placental lactogen stimulation of maternal behavior. Neuroendocrinology 64:57–64.

Bridges RS, Robertson MC, Shiu RPC, Sturgis JD, Henriquez BM, Mann PE (1997) Central lactogenic regulation of maternal behavior in rats: steroid dependence, hormone specificity, and behavioral potencies of rat prolactin and rat placental lactogen I. Endocrinology 138:756–763.

Bridges RS, Ronsheim PM (1990) Prolactin (PRL) regulation of maternal behavior in rats: bromocriptine treatment delays and PRL promotes the rapid onset of behavior. Endocrinology 126:837–848.

Bridges RS, Rosenblatt JS, Feder HH (1978) Stimulation of maternal responsiveness after pregnancy termination in rats: effect of time of onset of behavioral testing. Horm Behav 10:235–245.

Bridges RS, Russell DW (1981) Steroidal interactions in the regulation of maternal behaviour in virgin female rats: effects of testosterone, dihydrotestosterone, oestradiol, progesterone and the aromatase inhibitor 1,4,6-androstatriene-3,17-dione. J Endocrinol 90:31–40.

Bridges R, Zarrow MX, Gandelman R, Denenberg VH (1972) Differences in maternal responsiveness between lactating and sensitized rats. Dev Psychobiol 5:123–127.

Bridges RS, Zarrow MX, Gandelman R, Denenberg VH (1974) A developmental study of maternal responsiveness in the rat. Physiol Behav 12:149–151.

Broad KD, Kendrick KM, Keverne EB, Sirinathsinghji DJS (1993a) Changes in pro-opiomelanocortin and pre-proenkephalin mRNA levels in the ovine brain during pregnancy, parturition and lactation and in response to oestrogen and progesterone. J Neuroendocrinol 5:711–719.

Broad K, Kendrick K, Sirinathsinghji D, Keverne E (1993b) Changes in oxytocin immunoreactivity and mRNA expression in the sheep brain during pregnancy, parturition and lactation and in response to oestrogen and progesterone. J Neuroendocrinol 5: 435–444.

Broad KD, Keverne EB, Kendrick KM (1995) Corticotrophin releasing factor mRNA expression in the sheep brain during pregnancy, parturition and lactation and following exogenous progesterone and oestrogen treatment. Mol Brain Res 29:310–316.

Broad K, Levy F, Evans G, Kimura T, Keverne E, Kendrick K (1999) Previous maternal experience potentiates the effect of parturition on oxytocin receptor mRNA expression in the paraventricular nucleus. Eur J Neurosci 11:3725–3737.

Brockington IF, Brierley E (1984) Rejection of a child by his mother successfully treated after three years. Br J Psychiatry 145:316–318.

Brodzinsky DM, Huffman L (1988) Transition to adoptive parenthood. Marriage Family Rev 12:267–286.

Brog JS, Salyapongse A, Deutch AY, Zahm DS (1993) The pattern of afferent innervation of the core and shell in the "accumbens" part of the rat ventral striatum: immunohistochemical detection of retrogradely transported fluoro-gold. J Comp Neurol 338:255–278.

Broida J, Michael SD, Svare B (1981) Plasma prolactin levels are not related to the initiation, maintenance, and decline of postpartum aggression in mice. Behav Neural Biol 32:121–125.

Broida J, Svare B (1982) Strain-typical patterns of pregnancy-induced nestbuilding in mice: maternal and experiential influences. Physiol Behav 25:153–157.

Bronson FH (1989) Mammalian reproductive biology. Chicago: University of Chicago Press.

Bronson FH, Marstellar FA (1985) Effect of short-term food deprivation on reproduction in female mice. Biol Reprod 33:660–667.

Bronstein PM, Hirsch SM (1976) Ontogeny of defensive reactions in Norway rats. J Comp Physiol Psychol 90:620–629.

Brot MD, Akwa Y, Purdy RH, Koob GF, Britton KT (1997) The anxiolytic-like effects of the neurosteroid allopregnanolone: interactions with GABAa receptors. Eur J Pharmacol 325:1–7.

Brouette-Lahlou I, Godinot F, Vernet-Maury E (1999) The mother rat's vomeronasal organ is involved in detection of dodecyl propionate, the pup's preputial gland pheromone. Physiol Behav 66:427–436.

Brown AE, Mani S, Tobet SA (1999) The preoptic area/anterior hypothalamus of different strains of mice: sex differences and development. Dev Brain Res 115:171–182.

Brown GR (1986) Social and hormonal factors influencing infanticide and its suppression in adult male Long-Evans rats (Rattus norvegicus). J Comp Psychol 100:155–161.

Brown GR, Moger WH (1983) Hormonal correlates of parental behavior in male rats. Horm Behav 17:356–365.

Brown JR, Ye H, Bronson RT, Dikkes P, Greenberg ME (1996) A defect in nurturing in mice lacking the immediate early gene fosB. Cell 86:297–309.

Brown RE (1993) Hormonal and experiential factors influencing parental behaviour in male rodents: an integrative approach. Behav Processes 30:1–28.

Brown RE, Mathieson WB, Stapleton J, Neumann PE (1999) Maternal behavior in female C57BL/6J and DBA/2J inbred mice. Physiol Behav 67:599–605.

Brown RE, Murdoch T, Murphy PR, Moger WH (1995) Hormonal responses of male gerbils to stimuli from their mate and pups. Horm Behav 29:474–491.

Bruce HM (1959) An exteroceptive block to pregnancy in the mouse. Nature 184:105.

Bruce HM (1960) A block to pregnancy in the mouse caused by proximity of strange males. J Reprod Fert 1:96–103.

Brunelli SA, Hofer MA (1990) Parental behavior in juvenile rats: environmental and biological determinants. In: Mammalian parenting (Krasnegor NA, Bridges RS, eds), pp 372–399. New York: Oxford University Press.

Brunelli SA, Shindledecker RD, Hofer MA (1985) Development of maternal behaviors in prepubertal rats at three ages: age-characteristic patterns of responses. Dev Psychobiol 18:309–326.

Brunner D, Buhot MC, Hen R, Hofer M (1999) Anxiety, motor activation, and maternal-infant interactions in 5HT1B knockout mice. Behav Neurosci 113:587–601.

Brussard AB, Herbison AE (2000) Long-term plasticity of postsynaptic GABAA-receptor function in the adult brain: insight from the oxytocin neuron. Trends Neurosci 23: 190–195.

Buck LB (2000) The molecular architecture of odor and pheromone sensing in mammals. Cell 100:611–618.

Buist A (1998a) Childhood abuse, postpartum depression and parenting difficulties: a literature review of associations. Aust NZ J Psychiatry 32:370–378.

Buist A (1998b) Childhood abuse, parenting and postpartum depression. Aust NZ J Psychiatry 32:479–487.

Buntin JD (1996) Neural and hormonal control of parental behavior in birds. In: Advances in the study of behavior, Vol 25. Parental care: evolution, mechanisms, and adaptive significance (Rosenblatt JS, Snowdon CT, eds), pp 161–213. San Diego: Academic Press.

Buntin JD, Becker GM, Ruzycki E (1991) Facilitation of parental behavior in ring doves by systemic or intracranial injections of prolactin. Horm Behav 25:424–444.

Buntin JD, Jaffe S, Lisk RD (1984) Changes in responsiveness to new-born pups in pregnant, nulliparous golden hamsters. Physiol Behav 32:437–439.

Burbach J, van Schaik H, de Bree F, Lopes da Silva S, Adan R (1995) Functional domains in the oxytocin gene for regulation of expression and biosynthesis of gene products. In: Oxytocin: cellular and molecular approaches in medicine and research (Ivell R, Russell J, eds), pp 9–20. New York: Plenum Press.

Burbach JPH, Adan RAH, van Tol HHM, Verbeeck MAE, Axelson JF, van Leeuwen FW, Beekman JM, Ab G (1990) Regulation of rat oxytocin gene by estradiol. J Neuroendocrinol 2:633–639.

Burns K, Chethik L, Burns WJ, Clark R (1991) Dyadic disturbances in cocaine-abusing mothers and their infants. J Clin Psychol 47:316–319.

Butcher RL, Fugo NW, Collins WC (1972) Semicircadian rhythm in plasma levels of prolactin during early pregnancy in the rat. Endocrinology 90:1125–1127.

Byrnes EM, Bridges RS (2000) Endogenous opioid facilitation of maternal memory in rats. Behav Neurosci 114:797–804.

Byrnes EM, Byrnes JJ, Bridges RS (2001) Increased sensitivity of dopamine systems following reproductive experience in rats. Pharmacol Biochem Behav 68:481–489.

Byrnes EM, Rigero BA, Bridges RS (2000) Opioid receptor antagonism during early lactation results in the increased duration of nursing bouts. Physiol Behav 70:211–216.

Caba M, Poindron P, Krehbiel D, Levy F, Romeyer A, Venier G (1995) Naltrexone delays the onset of maternal behavior in primiparous parturient ewes. Pharmacol Biochem Behav 52:743–748.

Calamandrei G, Keverne EB (1994) Differential expression of Fos protein in the brain of female mice dependent upon sensory cues and maternal experience. Behav Neurosci 108:113–120.

Caldji C, Francis D, Sharma S, Plotsky PM, Meaney MJ (2000) The effects of early rearing environment on the development of GABAᴀ and central benzodiazepine receptor levels and novelty-induced fearfulness in the rat. Neuropsychopharmacology 22:219–229.

Caldji C, Tannenbaum B, Sharma S, Francis D, Plotsky PM, Meaney MJ (1998) Maternal care during infancy regulates the development of neural systems mediating the expression of fearfulness in the rat. Proc Natl Acad Sci USA 95:5335–5340.

Caldwell JD, Geer ER, Johnson ME, Prange AJ, Pedersen CA (1987) Oxytocin and vasopressin immunoreactivity in hypothalamic and extrahypothalamic sites in late pregnant and postpartum rats. Neuroendocrinology 46:39–47.

Caldwell JP (1997) Pair bonding in spotted poison frogs. Science 385:211.

Calizo LH, Flanagan-Cato, LM (2000) Estrogen selectively regulates spine density within the dendritic arbor of rat ventromedial hypothalamic neurons. J Neurosci 20:1589–1596.

Cameron AA, Khan IA, Westlund KN, Willis WD (1995) The efferent projections of the periaqueductal gray in the rat: a *Phaseolus vulgaris*-leucoagglutinin study. II. Descending projections. J Comp Neurol 351:585–601.

Campbell SB, Cohn JF, Meyers T (1995) Depression in first-time mothers: mother-infant interaction and depression chronicity. Dev Psychol 31:349–357.

Canteras NS, Chiavegatto S, Ribeiro Do Valle LE, Swanson LW (1997) Severe reduction in rat defensive behavior to a predator by discrete hypothalamic chemical lesions. Brain Res Bull 44:297–305.

Canteras NS, Goto M (1999) Fos-like immunoreactivity in the periaqueductal gray of rats exposed to a natural predator. Neuro Report 10:413–418.

Canteras NS, Simerly RB, Swanson LW (1992) Connections of the posterior nucleus of the amygdala. J Comp Neurol 324:143–179.

Canteras NS, Simerly RB, Swanson LW (1994) Organization of projections from the ventromedial nucleus of the hypothalamus: a *Phaseolus vulgaris*-leucoagglutinin study in the rat. J Comp Neurol 348:41–79.

Canteras NS, Simerly RB, Swanson LW (1995) Organization of projections from the medial nucleus of the amygdala: a PHAL study in the rat. J Comp Neurol 360:213–245.

Canteras NS, Swanson LW (1992) The dorsal premammillary nucleus: an unusual component of the mammillary body. Proc Natl Acad Sci USA 89:10089–10093.

Carlezon WA, Boundy VA, Haile CN, Lane SB, Kalb RG, Neve RL, Nestler EJ (1997) Sensitization to morphine induced by viral-mediated gene transfer. Science 277:812–814.

Carlezon WA, Haile CN, Coopersmith R, Hayashi Y, Malinow R, Neve RL, Nestler EJ (2000) Distinct sites of opiate reward and aversion within the midbrain identified using a herpes simplex virus vector expressing GluR1. J Neurosci 20:RC62 (1–5).

Carlier C, Noirot E (1965) Effects of previous experience on maternal retrieving by rats. Anim Behav 13:423–426.

Carlier M, Roubertoux P, Cohen-Salmon C (1982) Differences in patterns of pup care in *Mus musculus domesticus* 1—comparisons between eleven inbred strains. Behav Neural Biol 35:205–210.

Carter CS, Altemus M (1997) Integrative functions of lactational hormones in social behavior and stress management. Ann NY Acad Sci 807:164–174.

Carter CS, Altemus M, Chrousos GP (2001) Neuroendocrine and emotional changes in the post-partum period. In: Progress in brain research, Vol 133. The maternal brain

(Russell JA, Douglas AJ, Windle RJ, Ingram CD, eds), pp 241–249. Amsterdam: Elsevier.

Carter CS, DeVries A, Getz L (1995) Physiological substrates of mammalian monogamy: the prairie vole model. Neurosci Biobehav Rev 19:303–314.

Carter CS, Keverne B (2002) Affiliation and pair bonding. In: Hormones, brain and behavior, Vol 1 (Pfaff DW, Arnold AP, Etgen AM, Fahrbach SE, Rubin RT, eds), pp 299–338. San Diego: Academic Press.

Carter JD, Joyce PR, Mulder RT, Luty SE, Sullivan PF (1999) Early deficient parenting in depressed outpatients is associated with personality dysfunction and not with depression subtypes. J Affect Disord 54:29–37.

Challis JRG, Lye SJ (1994) Parturition. In: The physiology of reproduction, Vol 2 (Knobil E, Neill JD, eds), pp 985–1031. New York: Raven Press.

Champagne F, Diorio J, Sharma S, Meaney MJ (2001) Naturally occurring variations in maternal behavior in the rat are associated with differences in estrogen-inducible central oxytocin receptors. Proc Natl Acad Sci, USA 98:12736–12741.

Champagne F, Meaney MJ (2000) Prenatal stress effects on maternal behavior. Soc Neurosci Abstr 26:2035.

Champoux M, Coe CL, Schanberg SM, Kuhn CM, Suomi SJ (1989) Hormonal effects of early rearing conditions in the infant rhesus monkey. Am J Primatol 19:111–117.

Chaudhuri A (1997) Neural activity mapping with inducible transcription factors. Neuro Report 8:v–ix.

Chiang CY, Dostrovsky JO, Sessle BJ (1991) Periaqueductal gray matter and nucleus raphe magnus involvement in the anterior pretectal-induced inhibition of jaw-opening reflex in rats. Brain Res 544:71–78.

Chiba T, Murata Y (1985) Afferent and efferent connections of the medial preoptic area in the rat: a WGA-HRP study. Brain Res Bull 14:261–272.

Chiu S, Wise PM (1994) Prolactin receptor mRNA localization in the hypothalamus by in situ hybridization. J Neuroendocrinol 6:191–199.

Chrousos GP (1998) Stressors, stress, and neuroendocrine integration of the adaptive response. Ann NY Acad Sci 851:311–335.

Churchwell G (2000) Expecting: one man's uncensored memoir of pregnancy. New York: HarperCollins.

Cicchetti D, Rogosch FA (2001) Diverse patterns of neuroendocrine activity in maltreated children. Dev Psychopathol 13:677–693.

Clark MM, Galef BG, Jr. (1999) A testosterone-mediated trade-off between parental and sexual effort in male mongolian gerbils (*Meriones unguiculatus*). J Comp Psychol 113:388–395.

Clark MM, Galef BG (2000) Effects of experience on the parental responses of male Mongolian gerbils. Dev Psychobiol 36:177–185.

Clark MM, Galef BG (2001) Age-related changes in paternal responses of gerbils parallel changes in their testosterone concentrations. Dev Psychobiol 39:179–187.

Clark MM, Spencer CA, Galef BG (1986) Responses to novel odors mediate maternal behavior and concaveation in gerbils. Physiol Behav 36:845–851.

Clarke AS (1993) Social rearing effects on HPA axis activity of early development and in response to stress in rhesus monkeys. Dev Psychobiol 26:433–446.

Clutton-Brock TH (1991) The evolution of parental care. Princeton, NJ: Princeton University Press.

Coe CL (1990) Psychobiology of maternal behavior in nonhuman primates. In: Mammalian parenting (Krasnegor NA, Bridges RS, eds), pp 157–183. New York: Oxford University Press.

Cohen J, Bridges RS (1981) Retention of maternal behavior in nulliparous and primiparous rats: effects of duration of previous maternal experience. J Comp Physiol Psychol 95:450–459.

Cohen-Salmon C, Carlier M, Roubertoux P, Jouhaneau J, Semal C, Paillette M (1985) Differences in patterns of pup care in mice-pup ultrasonic emissions and pup care behavior. Physiol Behav 35:167–174.

Cohick CB, Dai G, Xu L, Deb S, Kamei T, Levan G, Szpirer C, Szpirer J, Kwok CM, Soares MJ (1996) Placental lactogen-I variant utilizes the prolactin receptor signaling pathway. Mol Cell Endocrinol 116:49–58.

Concas A, Mostallino MC, Porcu P, Follesa P, Barbaccia ML, Trabucchi M, Purdy RH, Grisenti P, Biggio G (1998) Role of brain allopregnanolone in the plasticity of γ-aminobutyric acid type A receptor in the rat brain during pregnancy and after delivery. Proc Natl Acad Sci USA 95:13284–13289.

Consiglio AR, Lucion AB (1996) Lesion of hypothalamic paraventricular nucleus and maternal aggressive behavior in female rats. Physiol Behav 59:591–596.

Contarino A, Dellu F, Koob GF, Smith GW, Lee KF, Vale W, Gold LH (1999) Reduced anxiety-like and cognitive performance in mice lacking the corticotropin-releasing factor receptor 1. Brain Res 835:1–9.

Cooke BM, Chowanadisai W, Breedlove SM (2000) Post-weaning social isolation of male rats reduces the volume of the medial amygdala and leads to deficits in adult sexual behavior. Behav Brain Res 117:107–113.

Coolen LM, Peters HJPW, Veening JG (1996) Fos immunoreactivity in the rat brain following consummatory elements of sexual behavior: a sex comparison. Brain Res 738:67–82.

Coplan JD, Andrews MW, Rosenblum LA, Owens MJ, Friedman S, Gorman JM, Nemeroff CB (1996) Persistent elevations of cerebrospinal fluid concentrations of corticotropin-releasing factor in adult nonhuman primates exposed to early-life stressors: implications for the pathophysiology of mood and anxiety disorders. Proc Natl Acad Sci USA 93:1619–1623.

Corodimas KP, Rosenblatt JS, Canfield ME, Morrell JI (1993) Neurons in the lateral division of the habenular complex mediate the hormonal onset of maternal behavior in the rat. Behav Neurosci 107:827–843.

Corter CM, Fleming AS (1990) Maternal responsiveness in humans: emotional, cognitive, and biological factors. In: Advances in the study of behavior, Vol 19 (Slater PJB, Rosenblatt JS, Beer C, eds), pp 83–136. San Diego: Academic Press.

Couse JF, Curtis SW, Washburn TF, Lindzey J, Golding TS, Lubahn DB, Smithies O, Korach KS (1995) Analysis of transcription and estrogen insensitivity in the female mouse after targeted disruption of the estrogen receptor gene. Mol Endocrinol 9:1441–1454.

Cowley DS, Roy-Byrne PP (1989) Panic disorder during pregnancy. J Psychosom Obstet Gynaecol 10:193–210.

Crawford SS, Balon EK (1996) Cause and effect of parental care in fishes. In: Parental care: evolution, mechanisms, and adaptive significance (Rosenblatt JS, Snowdon CT, eds), pp 53–107. San Diego: Academic Press.

Crews D (1992) Diversity of hormone-behavior relations in reproductive behavior. In: Behavioral endocrinology (Becker JB, Breedlove SM, Crews D, eds), pp 143–186. Cambridge, MA: MIT Press.

Da Costa APC, Broad KD, Kendrick KM (1997a) Olfactory memory and maternal behaviour-induced c-fos and zif/268 mRNA expression in the sheep brain. Mol Brain Res 46:63–76.

Da Costa APC, De La Riva C, Guevara-Guzman R, Kendrick KM (1999) *C-fos* and *c-jun* in the paraventricular nucleus play a role in regulating peptide gene expression, oxytocin and glutamate release, and maternal behaviour. Eur J Neurosci 11:2199–2210.

Da Costa APC, Guevara-Guzman RG, Ohkura S, Goode JA, Kendrick KM (1996a) The role of oxytocin release in the paraventricular nucleus in the control of maternal behaviour in sheep. J Neuroendocrinol 8:163–177.

Da Costa APC, Kampa RJ, Windle RJ, Ingram CD, Lightman SL (1997b) Region-specific immediate-early gene expression following the administration of corticotropin-releasing hormone in virgin and lactating rats. Brain Res 770:151–162.

Da Costa APC, Ma X, Ingram CD, Lightman SL, Aguilera G (2001) Hypothalamic and amygdaloid corticotropin-releasing hormone (CRH) and CRH receptor-1 mRNA expression in the stress-hyporesponsive late pregnant and early lactating rat. Mol Brain Res 91:119–130.

Da Costa APC, Wood S, Ingram CD, Lightman SL (1996b) Region-specifc reduction in stress-induced *c-fos* mRNA expression during pregnancy and lactation. Brain Res 742: 177–184.

Damasio AR (1995) On some functions of the human prefrontal cortex. Ann NY Acad Sci 769:241–251.

Damsma G, Pfaus JG, Wenkstern D, Phillips AG, Fibiger HC (1992) Sexual behavior increases dopamine transmission in the nucleus accumbens and striatum of male rats: comparison with novelty and locomotion. Behav Neurosci 106:181–191.

Daniel H, Levenes C, Crepel F (1998) Cellular mechanisms of cerebellar LTD. Trends Neurosci 21: 401–407.

Davis CD (1939) The effect of ablations of the neocortex on mating, maternal behavior and the production of pseudopregnancy in the female rat and on copulatory activity in the male. Am J Physiol 127:374–380.

Davis HP, Squire LR (1984) Protein synthesis and memory: a review. Psychol Bull 96: 518–559.

Davis M (1992) The role of the amygdala in fear and anxiety. Annu Rev Neurosci 15: 353–375.

Davis M, Rannie D, Cassell M (1994) Neurotransmission in the rat amygdala related to fear and anxiety. TINS 17:208–214.

Dayas CV, Buller KM, Day TA (1999) Neuroendocrine responses to an emotional stressor: evidence for involvement of the medial but not the central amygdala. Eur J Neurosci 11:2312–2322.

De Almeida RMM, Lucion AB (1994) Effects of intracerebroventricular administration of 5-HT receptor agonists on the maternal aggression of rats. Eur J Pharmacol 264: 445–448.

De Almeida RMM, Lucion AB (1997) 8-OH-DPAT in the median raphe, dorsal periaqueductal gray and corticomedial amygdala nucleus decreases, but in the medial septal area it can increase maternal aggressive behavior in rats. Psychopharmacology 134: 392–400.

De Bellis MD, Baum AS, Birmaher B, Keshaven MS, Eccard CH, Boring AM, Jenkins FJ, Ryan ND (1999) Developmental traumatology part I: biological stress systems. Biol Psychiatry 45:1259–1270.

De Felipe C, Herrero JF, O'Brien JA, Palmer JA, Doyle CA, Smith AJ, Laird JM, Belmonte C, Cervero F, Hunt SP (1998) Altered nociception, analgesia and aggression in mice lacking the receptor for substance P. Nature 392:394–397.

De Lange PJ, De Boer PACM, Ter Maat A, Tensen CP, Van Minnen J (1998) Transmitter identification in neurons involved in male copulation behavior in Lymnaea stagnalis. J Comp Neurol 395:440–449.

Del Cerro MCR (1998) Role of the vomeronasal input in maternal behavior. Psychoneuroendocrinology 23:905–926.

Demarest K, McKay D, Riegle G, Moore K (1983) Biochemical indices of tuberoinfundibular dopaminergic control during lactation: a lack of response to prolactin. Neuroendocrinology 36:130–137.

Denenberg VH (1964) Critical periods, stimulus input, and emotional reactivity: a theory of infantile stimulation. Psychol Rev 71:335–351.

De Olmos JS, Heimer L (1999) The concepts of the ventral striatopallidal system and extended amygdala. Ann NY Acad Sci 877:1–32.

De Olmos JS, Ingram WR (1972) The projection field of the stria terminalis in the rat brain. An experimental study. J Comp Neurol 146:303–333.

Desan PH, Woodmansee WW, Ryan SM, Smock TK, Maier SF (1988) Monoamine neurotransmitters and metabolites during the estrous cycle, pregnancy, and the postpartum period. Pharmacol Biochem Behav 30:563–568.

Deutch AY, Goldstein M, Baldino F, Roth RH (1988) Telencephalic projections of the A8 dopamine cell group. Ann NY Acad Sci 537:27–50.

Devinsky O, Bartlik B (1994) Psychiatric disorders of pregnancy. In: Neurological complications of pregnancy (Devinsky O, Feldmann E, Hainline B, eds), pp 215–230. New York: Raven Press.

DeVito W (1988) Distribution of immunoreactive prolactin in the male and female rat brain: effect of hypophysectomy and intraventricular administration of colchicine. Neuroendocrinology 47:284–289.

De Vries GJ, Buijs RM (1983) The origin of the vasopressinergic and oxytocinergic innervation of the rat brain with special reference to the lateral septum. Brain Res 273:307–317.

Dewsbury DA (1985) Paternal behavior in rodents. Am Zool 25:841–852.

Diano S, Naftolin F, Horvath T (1997) Gonadal steroids target AMPA glutamate receptor-containing neurons in the rat hypothalamus, septum and amygdala: a morphological and biochemical study. Endocrinology 138:778–789.

Dielenberg RA, Hunt GE, McGregor IS (2001) "When a rat smells a cat": The distribution of Fos immunoreactivity in rat brain following exposure to a predatory odor. Neuroscience 104:1085–1097.

Dixson AF, George L (1982) Prolactin and parental behaviour in a male New World primate. Nature 299:551–553.

Dixson AF, Hastings MH (1992) Effects of ibotenic acid-induced neuronal degeneration in the hypothalamus upon proceptivity and sexual receptivity in the female marmoset (Callithrix jacchus). J Neuroendocrinol 4:719–726.

Doerr HK, Siegel HI, Rosenblatt JS (1981) Effects of progesterone withdrawal and estrogen on maternal behavior in nulliparous rats. Behav Neural Biol 32:35–44.

Dohler KD, Wuttke W (1975) Changes with age in serum gonadotropins, prolactin, and gonadal steroids in prepubertal male and female rats. Endocrinology 97:898–907.

Dornan WA, Vink KL, Malen P, Short K, Struthers W, Barrett C (1993) Site-specific effects of intracerebral injections of three neurokinins (neurokinin A, neurokinin K, and neurokinin gamma) on the expression of male rat sexual behavior. Physiol Behav 54:249–258.

Drevets WC (2000) Neuroimaging studies of mood disorders. Biol Psychiatry 48:813–829.

Dudai Y (1989) The neurobiology of memory. Oxford: Oxford University Press.

Duncan GE, Knapp DJ, Breese GR (1996) Neuroanatomical characterization of Fos induction in rat behavioral models of anxiety. Brain Res 713:79–81.

Dunn AJ, Berridge CW (1990) Physiological and behavioral responses to corticotropin-releasing factor administration: is CRF a mediator of anxiety or stress responses? Brain Res Rev 15:71–100.

Dunnett SB, Robbins TW (1992) The functional role of mesotelencephalic dopamine systems. Biol Rev 67:491–518.

Dutt A, Kaplitt M, Kow L, Pfaff DW (1994) Prolactin, central nervous system and behavior: a critical review. Neuroendocrinology 59:413–416.

Dwyer CM, Dingwall WS, Lawrence AB (1999) Physiological correlates of maternal-offspring behaviour in sheep: a factor analysis. Physiol Behav 67:443–454.

Edwards DA, Einhorn LC (1986) Preoptic and midbrain control of sexual motivation. Physiol Behav 37:329–335.

Edwards DA, Nahai FR, Wright P (1993) Pathways linking the olfactory bulbs with medial preoptic anterior hypthalamus are important for intermale aggression in mice. Physiol Behav 53:611–615.

Eghbal-Ahmadi M, Avishai-Eliner S, Hatalski CG, Baram TZ (1999) Differential regulation of the expression of corticotropin-releasing factor receptor type 2 (CRF_2) in hypothalamus and amygdala of the immature rat by sensory input and food intake. J Neurosci 19:3982–3991.

Eghbal-Ahmadi M, Hatalski CG, Avishai-Eliner S, Baram TZ (1997) Corticotropin releasing factor receptor type II (CRF_2) messenger ribonucleic acid levels in the hypothalamic ventromedial nucleus of the infant rat are reduced by maternal deprivation. Endocrinology 138:5041–5048.

Ehret G, Jurgens A, Koch M (1993) Oestrogen receptor occurrence in the male mouse brain: modulation by paternal experience. Neuro Report 4:1247–1250.

Ehret G, Koch M (1989) Ultrasound-induced parental behaviour in house mice is controlled by female sex hormones and parental experience. Ethology 80:81–93.

Eichenbaum H, Otto T, Cohen NJ (1992) The hippocampus—what does it do? Behav Neural Biol 57:2–36.

Elliott JC, Lubin DA, Walker CH, Johns JM (2001) Acute cocaine alters oxytocin levels in the medial preoptic area and amygdala in lactating rat dams: implications for cocaine-induced changes in maternal behavior and maternal aggression. Neuropeptides 35:127–134.

Ellison G (1994) Stimulant-induced psychosis, the dopamine theory of schizophrenia, and the habenula. Brain Res Rev 19:223–239.

Elwood RW (1977) Changes in responsiveness of male and female gerbils (*Meriones unguiculatus*) towards test pups during the pregnancy of the female. Anim Behav 25:46–51.

Elwood RW (1981) Postparturitional reestablishment of pup cannibalism in female gerbils. Dev Psychobiol 14:209–212.

Elwood RW, Kennedy HF (1991) Selectivity in paternal and infanticidal responses by male mice: effects of relatedness, location, and previous sexual partners. Behav Neural Biol 56:129–147.

Elwood RW, Mason C (1994) The couvade and the onset of paternal care: a biological perspective. Ethol Sociobiol 15:145–156.

Emanuele NV, Jurgens JK, Halloran MM, Tentler JJ, Lawrence AM, Kelley MR (1992)

The rat prolactin gene is expressed in brain tissue: detection of normal and alternatively spliced prolactin mRNA. Mol Endocrinol 6:35–42.

Epperson CN, Wisner KL, Yamamoto B (1999) Gonadal steroids in the treatment of mood disorders. Psychosom Med 61:676–697.

Epple G (1978) Reproductive and social behavior of marmosets with special reference to captive breeding. Primatol Med 10:50–62.

Erskine MS (1995) Prolactin release after mating and genitosensory stimulation in females. Endocrinol Rev 16:508–528.

Erskine MS, Barfield RJ, Goldman BD (1978) Intraspecific fighting during late pregnancy and lactation in rats and effects of litter removal. Behav Biol 23:206–213.

Erskine MS, Barfield RJ, Goldman BD (1980a) Postpartum aggression in rats: I. Effects of hypophysectomy. J Comp Physiol Psychol 94:484–494.

Erskine MS, Barfield RJ, Goldman BD (1980b) Postpartum aggression in rats: II. Dependence on maternal sensitivity to young and effects of experience with pregnancy and parturition. J Comp Physiol Psychol 94:495–505.

Everitt BJ (1990) Sexual motivation: a neural and behavioral analysis of the mechanisms underlying appetitive and copulatory responses of male rats. Neurosci Biobehav Rev 14:214–232.

Everitt BJ, Stacey P (1987) Studies of instrumental behavior with sexual reinforcement in male rats (*Rattus norvegicus*): II. Effects of preoptic lesions, castration, and testosterone. J Comp Psychol 101:407–419.

Fabre-Nys C, Meller RE, Keverne EB (1982) Opiate antagonists stimulate affiliative behaviour in monkeys. Pharmacol Biochem Behav 16:653–660.

Factor EM, Mayer AD, Rosenblatt JS (1993) Peripeduncular nucleus lesions in the rat: I. Effects on maternal aggression, lactation, and maternal behavior during pre- and postpartum periods. Behav Neurosci 107:166–185.

Fahrbach SE, Morrell JI, Pfaff DW (1984) Oxytocin induction of short-latency maternal behavior in nulliparous, estrogen-primed female rats. Horm Behav 18:267–286.

Fahrbach SE, Morrell JI, Pfaff DW (1985a) Role of oxytocin in the onset of estrogen-facilitated maternal behavior. In: Oxytocin: clinical and laboratory studies (Amico JA, Robinson AG, eds), pp 372–388. Amsterdam: Elsevier.

Fahrbach SE, Morrell JI, Pfaff DW (1985b) Possible role for endogenous oxytocin in estrogen-facilitated maternal behaviour in rats. Neuroendocrinology 40:526–532.

Fahrbach SE, Morrell JI, Pfaff DW (1986a) Effect of varying the duration of pre-test cage habituation on oxytocin induction of short-latency maternal behavior. Physiol Behav 37:135–139.

Fahrbach SE, Morrell JI, Pfaff DW (1986b) Identification of medial preoptic neurons that concentrate estradiol and project to the midbrain in the rat. J Comp Neurol 247: 364–382.

Fahrbach SE, Pfaff DW (1986) Effect of preoptic region implants of dilute estradiol on the maternal behavior of ovariectomized, nulliparous rats. Horm Behav 20:354–363.

Fairbanks LA (1989) Early experience and cross-generational continuity of mother-infant contact in vervet monkeys. Dev Psychobiol 22:669–681.

Fairbanks LA (1990) Reciprocal benefits of allomothering for female vervet monkeys. Anim Behav 40:553–562.

Fairbanks LA (1996) Individual differences in maternal style. Causes and consequences for mothers and offspring. In: Advances in the study of behavior, Vol 25. Parental care: evolution, mechanisms, and adaptive significance (Rosenblatt JS, Snowdon CT, eds), pp 579–611. San Diego: Academic Press.

Fairbanks LA, McGuire MT (1993) Maternal protectiveness and response to the unfamiliar in vervet monkeys. Am J Primatol 30:119–129.

Fallon JH, Loughlin SE (1995) Substania nigra. In: The rat nervous system, second edition (Paxinos G, ed), pp 215–237. San Diego: Academic Press.

Fanselow MS (1991) The midbrain periaqueductal gray as a coordinator of action in response to fear and anxiety. In: The midbrain periaqueductal gray (Depaulis A, Bandler R, eds), pp 151–173. New York: Plenum Press.

Featherstone RE, Fleming AS, Ivy GO (2000) Plasticity in the maternal circuit: Effects of experience and partum condition on brain astrocyte number in female rats. Behav Neurosci 114:158–172.

Feifel D, Priebe K, Shilling PD (2001) Startle and sensorimotor gating in rats lacking CCK-A receptors. Neuropsychopharmacology 24:663–670.

Felicio LF, Mann PE, Bridges RS (1991) Intracerebroventricular cholecystokinin infusions block beta-endorphin-induced disruption of maternal behavior. Pharmacol Biochem Behav 39:201–204.

Fenelon VS, Herbison AE (1996) Plasticity in GABAA receptor subunit mRNA expression by hypothalamic magnocellular neurons in the adult rat. J Neurosci 16:4872–4880.

Ferguson JN, Aldag JM, Insel TR, Young LJ (2001) Oxytocin in the medial amygdala is essential for social recognition in the mouse. J Neurosci 21:8278–8285.

Ferguson JN, Young LJ, Hearn E, Insel TR, Winslow J (2000) Social amnesia in mice lacking the oxytocin gene. Nature Genet 25:284–288.

Ferguson JN, Young LJ, Insel TR (2002) The neuroendocrine basis of social recognition. Front Neuroendocrinol 23:200–224.

Fernandez-Fewell GD, Meredith M (1998) Olfactory contribution to Fos expression during mating in inexperienced male hamsters. Chem Senses 23:257–267.

Ferreira A, Dahlof L, Hansen S (1987) Olfactory mechanisms in the control of maternal aggression, appetite, and fearfulness: effects of lesions to olfactory receptors, mediodorsal thalamic nucleus, and insular prefrontal cortex. Behav Neurosci 101:709–717.

Ferreira A, Hansen S (1986) Sensory control of maternal aggression in *Rattus norvegicus*. J Comp Psychol 100:173–177.

Ferreira A, Hansen S, Nielsen M, Archer T, Minor BG (1989) Behavior of mother rats in conflict tests sensitive to antianxiety agents. Behav Neurosci 103:193–201.

Ferreira A, Picazo O, Uriarte N, Pereira M, Fernandez-Guasti A (2000) Inhibitory effect of buspirone and diazepam, but not of 8-OH-DPAT, on maternal behavior and aggression. Pharmacol Biochem Behav 66:389–396.

Ferreira G, Meurisse M, Gervais R, Ravel N, Levy F (2001) Extensive immunolesions of basal forebrain cholinergic system impair offspring recognition in sheep. Neuroscience 106:103–115.

Ferreira G, Terrazas A, Poindron P, Nowak R, Orgeur P, Levy F (2000) Learning of olfactory cues is not necessary for early lamb recognition by the mother. Physiol Behav 69:405–412.

Fiks KB, Johnson HL, Rosen TS (1985) Methadone-maintained mothers: 3-year follow-up of parental functioning. Int J Addict 20:651–660.

Findlay ALR, Roth LL (1970) Long-term dissociation of nursing behavior and the condition of the mammary gland in the rabbit. J Comp Physiol Psychol 72:341–344.

Finn PD, Yahr P (1994) Projection of the sexually dimorphic area of the gerbil hypothalamus to the retrorubral field is essential for male sexual behavior: Role of A8 and other cells. Behav Neurosci 108:362–378.

Fisher AE (1956) Maternal and sexual behavior induced by intracranial chemical stimulation. Science 124:228–229.

Fite JE, French JA (2000) Pre- and postpartum sex steroids in female marmosets (*Callithrix kuhlii*): is there a link with infant survivorship and maternal behavior? Horm Behav 38:1–12.

Fitzgerald LW, Ortiz J, Hamedani AG, Nestler EJ (1996) Drugs of abuse and stress increase the expression of GluR1 and NMDAR1 glutamate receptor subunits in the rat ventral tegmental area: common adaptations among cross-sensitizing agents. J Neurosci 16:274–282.

Fleischer S, Slotnick BM (1978) Disruption of maternal behavior in rats with lesions of the septal area. Physiol Behav 21:189–200.

Fleming AS, Cheung US, Barry M (1990) Cycloheximide blocks the retention of maternal experience in postpartum rats. Behav Neural Biol 53:64–73.

Fleming AS, Cheung U, Myhal N, Kessler Z (1989) Effects of maternal hormones on "timidity" and attraction to pup-related odors in female rats. Physiol Behav 46:449–453.

Fleming AS, Corter C, Franks P, Surbey M, Schneider B, Steiner M (1993) Postpartum factors related to the mother's attraction to newborn infant odors. Dev Psychobiol 26:115–132.

Fleming AS, Gavarth K, Sarker J (1992) Effects of transections to the vomeronasal nerves or to the main olfactory bulbs on the initiation and long-term retention of maternal behavior in primiparous rats. Behav Neural Biol 57:177–188.

Fleming AS, Korsmit M (1996) Plasticity in the maternal circuit: effects of maternal experience on Fos-Lir in hypothalamic, limbic, and cortical structures in the postpartum rat. Behav Neurosci 110:567–582.

Fleming AS, Korsmit M, Deller M (1994a) Rat pups are potent reinforcers to the maternal animal: effects of experience, parity, hormones, and dopamine function. Psychobiology 22:44–53.

Fleming AS, Kraemer GW, Gonzalez A, Lovic V, Rees S, Melo A (2002) Mothering begets mothering: the transmission of behavior and its neurobiology across generations. Pharmacol Biochem Behav 73:61–75.

Fleming AS, Luebke C (1981) Timidity prevents the nulliparous female from being a good mother. Physiol Behav 27:863–868.

Fleming AS, Miceli M, Moretto D (1983) Lesions of the medial preoptic area prevent the facilitation of maternal behavior produced by amygdala lesions. Physiol Behav 31:503–510.

Fleming AS, Orpen G (1986) Psychobiology of maternal behavior in rats, selected other species and humans. In: Origns of nurturance (Fogel A, Melson GF, eds), pp 141–207. Hillside, NJ: Lawrence Erlbaum Assoc.

Fleming AS, Rosenblatt JS (1974a) Maternal behavior in the virgin and lactating rat. J Comp Physiol Psychol 86:957–972.

Fleming AS, Rosenblatt JS (1974b) Olfactory regulation of maternal behavior in rats: I. Effects of olfactory bulb removal in experienced and inexperienced lactating and cycling females. J Comp Physiol Psychol 86:221–232.

Fleming AS, Rosenblatt JS (1974c) Olfactory regulation of maternal behavior in rats: II. Effects of peripherally induced anosmia and lesions of the lateral olfactory tract in pup-induced virgins. J Comp Physiol Psychol 86:233–246.

Fleming AS, Ruble DN, Flett GI, Shaul DL (1988) Postpartum adjustment in first-time mothers: relations between mood, maternal attitudes, and mother-infant interactions. Dev Psychol 24:71–81.

Fleming AS, Ruble D, Krieger H, Wong PY (1997a) Hormonal and experiential correlates of maternal responsiveness during pregnancy and the puerperium in human mothers. Horm Behav 31:145–158.

Fleming AS, Sarker J (1990) Experience-hormone interactions and maternal behavior in rats. Physiol Behav 47:1165–1173.

Fleming AS, Steiner M, Corter C (1997b) Cortisol, hedonics, and maternal responsiveness in human mothers. Horm Behav 32:85–98.

Fleming AS, Suh EJ, Korsmit M, Rusak B (1994b) Activation of Fos-like immunoreactivity in MPOA and limbic structures by maternal interactions and social interactions in rats. Behav Neurosci 108:724–734.

Fleming AS, Vaccarino F, Luebke C (1980) Amygdaloid inhibition of maternal behavior in the nulliparous female rat. Physiol Behav 25:731–743.

Fleming AS, Vaccarino F, Tambosso L, Chee P (1979) Vomeronasal and olfactory system modulation of maternal behavior in the rat. Science 203:372–374.

Follesa P, Serra M, Cagetti E, Pisu MG, Porta S, Floris S, Massa F, Sanna E, Biggio G (2000) Allopregnanolone synthesis in cerebellar granule cells: roles in regulation of GABAA receptor expression and function during progesterone treatment and withdrawal. Mol Pharmacol 57:1262–1270.

Francis CM, Anthony ELP, Brunton JA, Kunz TH (1994) Lactation in male fruit bats. Nature 367:691–692.

Francis D, Diorio J, Liu D, Meaney MJ (1999b) Nongenomic transmission across generations of maternal behavior and stress responses in the rat. Science 286:1155–1158.

Francis DD, Brake WG, McEwen B, Allen P, Greengard P, Insel TR (2001) Synaptic rearrangement of the rat MPOA and hippocampus during pregnancy and lactation. Paper presented at 31st Annual Meeting of the Society for Neuroscience, San Diego, CA, November, 2001.

Francis DD, Caldji C, Champagne F, Plotsky PM, Meaney MJ (1999a) The role of corticotropin-releasing factor-norepinephrine systems in mediating the effects of early experience on the development of behavioral and endocrine responses to stress. Biol Psychiatry 46:1153–1166.

Francis DD, Champagne FL, Meaney MJ (2000) Variations in maternal behaviour are associated with differences in oxytocin receptor levels in the rat. J Neuroendocrinol 12:1145–1148.

Francis DD, Szegda K, Campbell G, Martin WD, Insel TR (in press) Epigenetic sources of behavioral differences in mice. Nature Neurosci.

Francis DD, Young LJ, Meaney MJ, Insel TR (2002) Naturally occurring differences in maternal care are associated with the expression of oxytocin and vasopressin (V1a) receptors: gender differences. J Neuroendocrinol 14:349–353.

Franz JJ, Leo RJ, Steuer MA, Kristal MB (1986) Effects of hypothalamic knife cuts and experience on maternal behavior in the rat. Physiol Behav 38:629–640.

Franzen EA, Myers RE (1973) Neural control of social behavior: prefrontal and anterior temporal cortex. Neuropsychologia 11:141–157.

Freeman ME (1994) The neuroendocrine control of the ovarian cycle in the rat. In: The physiology of reproduction, Vol 2 (Knobil E, Neill JD, eds), pp 613–658. New York: Raven Press.

Fride E, Dan Y, Gavish M, Weinstock M (1985) Prenatal stress impairs maternal behavior in a conflict situation and reduces hippocampal benzodiazepine receptors. Life Sci 36: 2103–2109.

Fuchs AR (1983) The role of oxytocin in parturition. In: Current topics in experimental endocrinology, Vol 4. The endocrinology of pregnancy and parturition (Martini L, James VHT, eds), pp 231–265. New York: Academic Press.

Fuchs SAG, Edinger HM, Siegel A (1985) The organization of the hypothalamic pathways mediating affective defense behavior in the cat. Brain Res 330:77–92.

Fujino Y, Nagahama T, Oumi T, Ukena K, Morishita F, Furukawa Y, Matsushima O, Ando M, Takahama H, Satake H, Minakata H, Nomoto K (1999) Possible functions of oxytocin/vasopressin-superfamily peptides in annelids with special reference to reproduction and osmoregulation. J Exp Zool 284:401–406.

Gaffori O, Le Moal M (1979) Disruption of maternal behavior and appearance of cannibalism after ventral mesencephalic tegmentum lesions. Physiol Behav 23:317–323.

Galea LAM, Wide JK, Barr AM (2001) Estradiol alleviates depressive-like symptoms in a novel animal model of post-partum depression. Behav Brain Res 122:1–9.

Galosy SS, Talamantes F (1995) Luteotropic actions of placental lactogens at midpregnancy in the mouse. Endocrinology 136:3993–4003.

Gandelman R (1972) Mice: postpartum aggression elicited by the presence of an intruder. Horm Behav 3:23–28.

Gandelman R (1973a) Maternal behavior in the mouse: effect of estrogen and progesterone. Physiol Behav 10:153–155.

Gandelman R (1973b) Induction of maternal nest building in virgin mice by the presentation of young. Horm Behav 4:191–197.

Gandelman R (1973c) The development of cannibalism in male Rockland-Swiss mice and the influence of olfactory bulb removal. Dev Psychobiol 6:159–164.

Gandelman R, vom Saal FS (1975) Pup-killing in mice: the effects of gonadectomy and testosterone administration. Physiol Behav 15:647–651.

Gandelman R, vom Saal FS (1977) Exposure to early androgen attenuates androgen-induced pup killing in male and female mice. Behav Biol 20:252–260.

Gandelman R, Zarrow MX, Denenberg VH (1970) Maternal behavior: differences between mother and virgin mice as a function of testing procedure. Dev Psychobiol 3:207–214.

Gandelman R, Zarrow MX, Denenberg VH (1971a) Stimulus control of cannibalism and maternal behavior in anosmic mice. Physiol Behav 7:583–586.

Gandelman R, Zarrow MX, Denenberg VH, Myers M (1971b) Olfactory bulb removal eliminates maternal behavior in the mouse. Science 171:210–211.

Gass EK, Leonhardt SA, Nordeen SK, Edwards DP (1998) The antagonists RU 486 and ZK98299 stimulate progesterone receptor binding to deoxyribonucleic acid in vitro and in vivo, but have distinct effects on receptor conformation. Endocrinology 139:1905–1919.

Gavioli EC, Canteras NS, De Lima TCM (1999) Anxiogenic-like effect induced by substance P injected into the lateral septal nucleus. Neuro Report 10:3399–3405.

Gazzaley AH, Weiland NG, McEwen BS, Morrison JH (1996) Differential regulation of NMDAR1 mRNA and protein by estradiol in the rat hippocampus. J Neurosci 16:6830–6838.

Geracioti TD, Kalogeras KT, Pigott TA, Demitrack MA, Altemus M, Chrousos GP, Gold PW (1990) Anxiety and the hypothalamic-pituitary-adrenal axis. In: Handbook of anxiety, Vol 3. The neurobiology of anxiety (Burrows GD, Roth M, Noyes R, eds), pp 355–364. Amsterdam: Elsevier.

Gerlai R (1996) Gene-targeting studies of mammalian behavior: is it the mutation or the background phenotype? TINS 19:177–181.

Getz L, McGuire B, Pizzuto T, Hoffman J, Frase B (1993) Social organization of the prairie vole (*Microtus ochrogaster*). J Mammal 74:44–58.

Getz LL, Carter CS (1980) Social organization in *Microtus ochrogaster* populations. The Biologist 60:134–146.

Gibber JR (1986) Infant-directed behavior of rhesus monkeys during their first pregnancy and parturition. Folia Primatol 46:118–124.

Giesler GJ, Katter JT, Dado RJ (1994) Direct spinal pathways to the limbic system for nociceptive information. Trends Neurosci 17:244–250.

Gilbert AN (1984) Postpartum and lactational estrus: a comparative analysis in rodentia. J Comp Psychol 98:232–245.

Gilbert AN, Burgoon DA, Sullivan KA, Adler NT (1983) Mother-weanling interactions in Norway rats in the presence of a successive litter produced by postpartum mating. Physiol Behav 30:267–271.

Gingrich B, Liu Y, Cascio C, Wang Z, Insel TR (2000) Dopamine D2 receptors in the nucleus accumbens are important for social attachment in female prairie voles. Behav Neurosci 114:173–183.

Gingrich JA, Hen R (2001) Dissecting the role of the serotonin system in neuropsychiatric disorders using knockout mice. Psychopharmacology 155:1–10.

Ginsberg SD, Hof PR, McKinney WT, Morrison JH (1993) Quantitative analysis of tyrosine hydroxylase- and corticotropin-releasing factor-immunoreactive neurons in monkeys raised with differential rearing conditions. Exp Neurol 120:95–105.

Giordano AL, Johnson AE, Rosenblatt JS (1990) Haloperidol-induced disruption of retrieval behavior and reversal with apomorphine in lactating rats. Physiol Behav 48: 211–214.

Giordano AL, Siegel HI, Rosenblatt JS (1984) Effects of mother-litter separation and reunion on maternal aggression and pup mortality in lactating hamsters. Physiol Behav 33:903–906.

Giordano AL, Siegel HI, Rosenblatt JS (1989) Nuclear estrogen receptor binding in the preoptic area and hypothalamus of pregnancy-terminated rats: correlations with the onset of maternal behavior. Neuroendocrinology 50:248–258.

Giordano AL, Siegel HI, Rosenblatt JS (1991) Nuclear estrogen receptor binding in microdissected brain regions of female rats during pregnancy: implications for maternal and sexual behavior. Physiol Behav 50:1263–1267.

Giovenardi M, Padoin MJ, Cadore LP, Lucion AB (1998) Hypothalamic paraventricular nucleus modulates maternal aggression in rats: effects of ibotenic acid lesion and oxytocin antisense. Physiol Behav 63:351–359.

Glaser J, Russell AR, Taljaard JJF (1991) Rat brain hypothalamic and hippocampal monoamine and hippocampal B-adrenergic receptor changes during pregnancy. Brain Res 557:293–299.

Glickstein SB, Schmauss C (2001) Dopamine receptor functions: lessons from knockout mice. Pharmacol Therapeut 91:63–83.

Gonzalez A, Lovic V, Ward GR, Wainwright PE, Fleming AS (2001) Intergenerational effects of complete maternal deprivation and replacement stimulation on maternal behavior and emotionality in female rats. Dev Psychobiol 38:11–32.

González-Mariscal G (2001) Neuroendocrinology of maternal behavior in the rabbit. Horm Behav 40:125–132.

González-Mariscal G, Díaz-Sánchez V, Melo AI, Beyer C, Rosenblatt JS (1994) Maternal behavior in New Zealand white rabbits: quantification of somatic events, motor patterns, and steroid plasma levels. Physiol Behav 55: 1081–1089.

González-Mariscal G, Melo AI, Chirino R, Jiménez P, Beyer C, Rosenblatt JS (1998) Importance of mother/young contact at parturition and across lactation for the expression of maternal behavior in rabbits. Dev Psychobiol 32:101–111.

González-Mariscal G, Melo AI, Jiménez P, Beyer C, Rosenblatt JS (1996) Estradiol, progesterone, and prolactin regulate maternal nest-building in rabbits. J Neuroendocrinol 8:901–907.

González-Mariscal G, Melo AI, Parlow AF, Beyer C, Rosenblatt JS (2000) Pharmacological evidence that prolactin acts from late gestation to promote maternal behaviour in rabbits. J Neuroendocrinol 12:983–992.

González-Mariscal G, Poindron P (2002) Parental care in mammals: immediate internal and sensory factors of control. In: Hormones, brain and behavior, Vol 1 (Pfaff DW, Arnold AP, Etgen AM, Fahrbach SE, Rubin RT, eds), pp 215–298. San Diego: Academic Press.

González-Mariscal G, Rosenblatt JS (1996) Maternal behavior in rabbits. In: Advances in the study of behavior, Vol 25. Parental care: evolution, mechanisms, and adaptive significance (Rosenblatt JS, Snowdon CT, eds), pp 333–360. San Diego: Academic Press.

Goodson JL, Bass AH (2000) Forebrain peptides modulate sexually polymorphic vocal circuitry. Nature 403:769–772.

Goodson JL, Bass AH (2001) Social behavior functions and related anatomical characteristics of vasotocin/vasopressin systems in vertebrates. Brain Res Rev 35:246–265.

Grattan DR (2001) The actions of prolactin in the brain during pregnancy and lactation. In: Progress in brain research, Vol 133. The maternal brain (Russell JA, Douglas AJ, Windle RJ, Ingram CD, eds), pp 151–171. Amsterdam: Elsevier.

Grattan DR, Averill R (1995) Absence of short-loop autoregulation of prolactin during late pregnancy in the rat. Brain Res Bull 36:413–416.

Grattan DR, Xu J, McLachlan M, Kokay IC, Bunn SJ, Hovey RC, Davey HW (2001) Feedback regulation of the prolactin secretion is mediated by the transcription factor, Stat5b. Endocrinology 142:3935–3940.

Gray P, Brooks PJ (1984) Effect of lesion location within the medial preoptic-anterior hypothalamic continuum on maternal and male sexual behaviors in female rats. Behav Neurosci 98:703–711.

Gray P, Chesley S (1984) Development of maternal behavior in nulliparous rats (*Rattus norvegicus*): effects of sex and early maternal experience. J Comp Psychol 98:91–99.

Greene JD, Sommerville RB, Nystrom LE, Darley JM, Cohen JD (2001) An fMRI investigation of emotional engagement in moral judgement. Science 293:2105–2108.

Gregoire AJP, Kumar R, Everitt B, Henderson AF, Sudd JWW (1996) Transdermal oestrogen treatment of severe postnatal depression. Lancet 347:930–933.

Groenewegen HJ, Wright CI, Beijer AVJ, Voorn P (1999) Convergence and segregation of ventral striatal inputs and outputs. Ann NY Acad Sci 877:49–63.

Groenewegen HJ, Wright CI, Uylings HBM (1997) The anatomical relationships of the prefrontal cortex with limbic structures and the basal ganglia. J Psychopharmacol 11:99–106.

Grossman SP (1967) A textbook of physiological psychology. New York: Wiley.

Grundker M, Hrabe de Angelis C, Kirchner C (1993) Placental lactogen-like protein in rabbit placenta. Anat Embryol 188:395–399.

Gubernick DJ (1981) Parent and infant attachment in mammals. In: Parental care in mammals (Gubernick DJ, Klopfer PH, eds), pp 243–305. New York: Plenum Press.

Gubernick DJ (1988) Reproduction in the California mouse, *Peromyscus californicus*. J Mammal 64:857–860.

Gubernick DJ, Alberts JR (1987) The biparental care system of the California mouse, *Peromyscus californicus*. J Comp Psychol 101:169–177.

Gubernick DJ, Alberts JR (1989) Postpartum maintenance of paternal behaviour in the biparental California mouse, *Peromyscus californicus*. Anim Behav 37:656–664.

Gubernick DJ, Nelson RJ (1989) Prolactin and paternal behavior in the biparental California mouse, *Peromyscus californicus*. Horm Behav 23:203–210.

Gubernick DJ, Schneider KA, Jeanotte L (1994) Individual differences in the mechanisms underlying the onset and maintenance of paternal behavior and the inhibition of infanticide in the monogamous biparental mouse, *Peromyscus californicus*. Behav Ecol Sociobiol 34:235–241.

Gubernick DJ, Sengelaub DR, Kurz EM (1993a) A neuroanatomical correlate of paternal and maternal behavior in the biparental California mouse (*Peromyscus californicus*). Behav Neurosci 107:194–201.

Gubernick DJ, Teferi T (2000) Adaptive significance of male parental care in a monogamous mammal. Proceedings of the Royal Society of London—Series B: Biol Sci 267:147–150.

Gubernick DJ, Winslow JT, Jensen P, Jeanotte L, Bowen J (1995) Oxytocin changes in males over the reproductive cycle in the monogamous, biparental California mouse, *Peromyscus californicus*. Horm Behav 29:59–73.

Gubernick DJ, Wright SL, Brown RE (1993b) The significance of father's presence for offspring survival in the monogamous California mouse, *Peromyscus californicus*. Anim Behav 46:539–546.

Gulinello M, Gong QH, Smith SS (2001) Short-term exposure to a neuroactive steroid increases $\alpha 4$ GABA$_A$ receptor subunit levels in association with increased anxiety in the female rat. Brain Res 910:55–66.

Gunnet JW, Freeman ME (1983) The mating-induced release of prolactin: a unique endocrine response. Endocrinol Rev 4:44–61.

Haig D, Graham C (1991) Genomic imprinting and the strange case of the insulin-like growth factor II receptor. Cell 64:1045–1046.

Hall FS (1998) Social deprivation of neonatal, adolescent, and adult rats has distinct neurochemical and behavioral consequences. Crit Rev Neurobiol 12:129–162.

Hall FS, Wilkinson LS, Humby T, Robbins TW (1999) Maternal deprivation of neonatal rats produces enduring changes in dopamine function. Synapse 32:37–43.

Hammer RJ, Bridges RS (1987) Preoptic area opioids opiate receptors increase during pregnancy and decrease during lactation. Brain Res 420:48–56.

Han Y, Shaikh MB, Siegel A (1996) Medial amygdala suppression of predatory attack in the cat: I. Role of substance P pathway from the medial amygdala to the medial hypothalamus. Brain Res 716:59–71.

Hans SL, Bernstein VJ, Henson LG (1999) The role of psychopathology in the parenting of drug-dependent women. Dev Psychopathol 11:957–977.

Hansen S (1989) Medial hypothalamic involvement in maternal aggression of rats. Behav Neurosci 103:1035–1046.

Hansen S (1994) Maternal behavior of female rats with 6-OHDA lesions of the ventral striatum: characterization of the pup retrieval deficit. Physiol Behav 55:615–620.

Hansen S, Bergvall AH, Nyiredi S (1993) Interaction with pups enhances dopamine release in the ventral striatum of maternal rats: a microdialysis study. Pharmacol Biochem Behav 45:673–676.

Hansen S, Ferreira A (1986a) Food intake, aggression, and fear in the mother rat: control by neural systems concerned with milk ejection and maternal behavior. Behav Neurosci 100:64–70.

Hansen S, Ferreira A (1986b) Effects of bicuculline infusions in the ventromedial hypothalamus and amygdaloid complex on food intake and affective behavior in mother rats. Behav Neurosci 100:410–415.

Hansen S, Ferreira A, Selart ME (1985) Behavioural similarities between mother rats and benzodiazepine-treated non-maternal animals. Psychopharmacology 86:344–347.

Hansen S, Harthon C, Wallin E, Lofberg L, Svensson K (1991a) Mesotelencephalic dopamine system and reproductive behavior in the female rat: effects of ventral tegmental 6-hydroxydopamine lesions on maternal and sexual responsiveness. Behav Neurosci 105:588–598.

Hansen S, Harthon C, Wallin E, Lofberg L, Svensson K (1991b) The effects of 6-OHDA-induced dopamine depletions in the ventral or dorsal striatum on maternal and sexual behavior in the female rat. Pharmacol Biochem Behav 39:71–77.

Hansen S, Kohler C (1984) The importance of the peripeduncular nucleus in the neuroendocrine control of sexual behavior and milk ejection in the rat. Neuroendocrinology 39:563–572.

Hard E, Hansen S (1985) Reduced fearfulness in the lactating rat. Physiol Behav 35: 641–643.

Harlow HF, Harlow MK, Hansen EW (1963) The maternal affectional system of rhesus monkeys. In: Maternal behavior in mammals (Rheingold HL, ed), pp 254–281. New York: John Wiley & Sons.

Hart BL, Leedy MG (1985) Neurological basis of male sexual behavior. In: Handbook of behavioral neurobiology, Volume 7 (Adler N, Pfaff D, Goy RW, eds), pp 373–422. New York: Plenum Press.

Hart J, Gunnar M, Cicchetti D (1995) Salivary cortisol in maltreated children: evidence of relations between neuroendocrine activity and social competence. Dev Psychopathol 7:11–26.

Hatton GI, Yang QZ (1990) Activation of excitatory amino acid inputs to supraoptic neurons. I. Induced increases in dye-coupling in lactating, but not virgin or male rats. Brain Res 513:264–269.

Hatton GI, Yang QZ (1994) Incidence of neuronal coupling in supraoptic nuclei of virgin and lactating rats: estimation by neurobiotin and Lucifer yellow. Brain Res 650:63–69.

Hatton JD, Ellisman MH (1982) A restructuring of hypothalamic synapses is associated with motherhood. J Neurosci 2:704–707.

Hauser H, Gandelman R (1985) Lever pressing for pups: evidence for hormonal influence upon maternal behavior in mice. Horm Behav 19:454–468.

Hausfater G, Hrdy SB (1984) Infanticide: comparative and evolutionary perspectives. New York: Aldine.

Heeb MM, Yahr P (1996) C-Fos immunoreactivity in the sexually dimorphic area of the hypothalamus and related brain regions of male gerbils after exposure to sex-related stimuli or performance of specific sexual behaviors. Neuroscience 72:1049–1071.

Heim C, Nemeroff CB (1999) The impact of early adverse experiences on brain systems involved in the pathophysiology of anxiety and affective disorders. Biol Psychiatry 46:1509–1522.

Heimer L, Zahm DS, Churchill L, Kalivas PW, Wohltmann C (1991) Specificity in the projection patterns of accumbal core and shell in the rat. Neuroscience 41:89–125.

Hendrick V, Altshuler LL, Suri R (1998) Hormonal changes in the postpartum and implications for postpartum depression. Psychosomatics 39:93–101.

Hennessy MB, Harney KS, Smotherman WP, Coyle S, Levine S (1977) Adrenalectomy-induced deficits in maternal retrieval in the rat. Horm Behav 9:222–227.

Herbison AE (1997) Estrogen regulation of GABAA transmission in rat preoptic area. Brain Res Bull 44:321–326.

Herbison AE, Fenelon VS (1995) Estrogen regulation of GABAA receptor subunit mRNA expression in preoptic area and bed nucleus of the stria terminalis of female rat brain. J Neurosci 15:2328–2337.

Herrada G, Dulac C (1997) A novel family of putative pheromone receptors in mammals with a topographically organized and sexually dimorphic distribution. Cell 90:763–773.

Herrenkohl LR, Rosenberg PA (1972) Exteroceptive stimulation of maternal behavior in the naive rat. Physiol Behav 8:595–598.

Herrera DG, Robertson HA (1996) Activation of c-fos in the brain. Prog Neurobiol 50:83–107.

Higley JD, Suomi SJ, Linnoila M (1992) A longitudinal assessment of CSF monoamine metabolites and plasma cortisol concentrations in young rhesus monkeys. Biol Psychiatry 32:127–145.

Higuchi T, Honda K, Fukuoka T, Negoro H, Wakabayashi K (1985) Release of oxytocin during suckling and parturition in the rat. J Endocrinol 105:339–346.

Hill J, Pickles A, Burnside E, Byatt M, Rollinson L, Davis R, Harvey K (2001) Child sexual abuse, poor parental care and adult depression: evidence for different mechanisms. Br J Psychiatry 179:104–109.

Hinde RA (1970) Animal behaviour. New York: McGraw-Hill.

Hofer MA (1984) Relationships as regulators: a psychobiologic perspective on bereavement. Psychosom Med 46:183–197.

Hofer MA, Shair HN, Masmela JR, Brunelli SA (2001) Developmental effects of selective breeding for an infantile trait: the rat pup ultrasonic isolation call. Dev Psychobiol 39:239–246.

Holman SD, Goy RW (1980) Behavioral and mammary responses of adult female rhesus to strange infants. Horm Behav 14:348–357.

Holman SD, Goy RW (1995) Experiential and hormonal correlates of care-giving in rhesus macaques. In: Motherhood in human and nonhuman primates (Pryce CR, Martin RD, Skuse D, eds), pp 87–92. Basel: Karger.

Holmes WG, Sherman PW (1982) The ontogeny of kin recognition in two species of ground squirrels. Am Zool 22:491–517.

Horseman ND, Zhao W, Montecino-Rodriguez E, Tanaka M, Nakashima K, Engle SJ, Smith F, Markoff E, Dorshkind K (1997) Defective mammopoiesis, but normal hematopoiesis, in mice with a targeted disruption of the prolactin gene. EMBO J 16:6926–6935.

Hoshina Y, Takeo T, Nakano K, Sato T, Sakuma Y (1994) Axon-sparing lesion of the preoptic area enhances receptivity and diminishes proceptivity among components of female rat sexual behavior. Behav Brain Res 61:197–204.

Hrabovszky E, Kallo I, Hajszan T, Shughrue P, Merchenthaler I (1998) Expression of estrogen receptor-beta messenger ribonucleic acid in oxytocin and vasopressin neurons of the rat supraoptic and paraventricular nuclei. Endocrinology 139:2600–2604.

Hrdy SB (1977) The langurs of Abu. Cambridge, MA: Harvard University Press.

Hrdy SB (1999) Mother nature. New York: Pantheon.

Hu Z, Zhuang L, Dufau M (1996) Multiple and tissue-specific promoter control of gonadal and non-gonadal prolactin receptor gene expression. J Biol Chem 271:10242–10246.

Hull EM, Lorrain DS, Du J, Matuszewich L, Lumley LA, Putnam SK, Moses J (1999) Hormone-neurotransmitter interactions in the control of sexual behavior. Behav Brain Res 105:105–116.

Hyman SE, Malenka RC (2001) Addiction and the brain: the neurobiology of compulsion and its persistence. Nature Rev Neurosci 2:695–702.

Ichikawa M, Matsuoka M, Mori Y (1993) Effect of differential rearing on synapses and soma size in rat medial amygdaloid nucleus. Synapse 13:50–56.

Ihnat R, White NR, Barfield RJ (1995) Pup's broadband vocalizations and maternal behavior in the rat. Behav Proc 33:257–272.

Ikemoto S, Panksepp J (1999) The role of nucleus accumbens dopamine in motivated behavior: a unifying interpretation with special reference to reward seeking. Brain Res Rev 31:6–41.

Ingram CD, Moos F (1992) Oxytocin-containing pathway to the bed nuclei of the stria terminalis of the lactating rat brain: immunocytochemical and in vitro electrophysiological evidence. Neuroscience 47:439–452.

Insel TR (1986) Postpartum increases in brain oxytocin binding. Neuroendocrinology 44:515–518.

Insel TR (1990a) Oxytocin and maternal behavior. In: Mammalian parenting (Krasnegor NA, Bridges RS, eds), pp 260–280. New York: Oxford University Press.

Insel TR (1990b) Regional induction of c-fos-like protein in rat brain after estradiol administration. Endocrinology 126:1849–1853.

Insel TR (1990c) Regional changes in brain oxytocin receptor post-partum: time-course and relationship to maternal behavior. J Neuroendocrinol 2:539–545.

Insel TR (1992) Oxytocin—a neuropeptide for affiliation: evidence from behavioral, receptor autoradiographic, and comparative studies. Psychoneuroendocrinology 17:3–35.

Insel TR, Gelhard RE, Shapiro LE (1991) The comparative distribution of neurohypophyseal peptide receptors in monogamous and polygamous mice. Neuroscience 43:623–630.

Insel TR, Gingrich B, Young LJ (2001) Oxytocin: who needs it? In: The maternal brain: neurobiological and neuroendocrine adaptation and disorders in pregnancy and postpartum (Russell JA, Douglas A, Windle R, Ingram C, eds), pp 59–67. Amsterdam: Elsevier.

Insel TR, Harbaugh CR (1989) Lesions of the hypothalamic paraventricular nucleus disrupt the initiation of maternal behavior. Physiol Behav 45:1033–1041.

Insel TR, Hulihan TJ (1995) A gender specific mechanism for pair bonding: oxytocin and partner preference formation in monogamous voles. Behav Neurosci 109:782–789.

Insel TR, Shapiro LE (1992) Oxytocin receptor distribution reflects social organization in monogamous and polygamous voles. Proc Natl Acad Sci USA 89:5981–5985.

Insel TR, Wang Z, Ferris CF (1994) Patterns of brain vasopressin receptor distribution associated with social organization in microtine rodents. J Neurosci 14:5381–5392.

Insel TR, Winslow JT (1998) Serotonin and neuropeptides in affiliative behaviors. Biol Psychiatry 44:207–219.

Insel TR, Young LJ (2000) Neuropeptides and the evolution of social behavior. Curr Opin Neurobiol 10:784–789.

Insel TR, Young LJ (2001) The neurobiology of attachment. Nature Rev Neurosci 2: 129–136.

Insel TR, Young LJ, Witt D, Crews D (1993) Gonadal steroids have paradoxical effects on brain oxytocin receptors. J Neuroendocrinol 5:619–628.

Jacobson CD, Terkel J, Gorski RA, Sawyer CH (1980) Effects of small medial preoptic area lesions on maternal behavior: retrieving and nestbuilding in the rat. Brain Res 194:471–478.

Jacobson M (1991) Developmental neurobiology. New York: Plenum Press.

Jakubowski M, Terkel J (1982) Infanticide and caretaking in nonlactating *Mus musculus*: influence of genotype, family group and sex. Anim Behav 30:1029–1035.

Jakubowski M, Terkel J (1985) Incidence of pup killing and parental behavior in virgin female and male rats (*Rattus norvegicus*): differences between Wistar and Sprague-Dawley stocks. J Comp Psychol 99:93–97.

Jakubowski M, Terkel J (1986) Establishment and maintenance of maternal responsiveness in postpartum Wistar rats. Anim Behav 34:256–262.

Jay P (1963) Mother-infant relations in langurs. In: Maternal behavior in mammals (Rheingold HL, ed), pp 282–304. New York: John Wiley & Sons.

Jenkins WJ, Becker JB (2001) Role of the striatum and nucleus accumbens in paced copulatory behavior in the female rat. Behav Brain Res 121:119–128.

Jennings KD, Ross S, Popper S, Elmore M (1999) Thoughts of harming infants in depressed and nondepressed mothers. J Affect Disord 54:21–28.

Jirik-Babb P, Manaker S, Tucker AM, Hofer M (1984) The role of the accessory and main olfactory systems in maternal behavior of the primiparous rat. Behav Neural Biol 40:170–178.

Johns JM, Noonan LR, Zimmerman LI, Li L, Pedersen CA (1994) Effects of chronic and acute cocaine treatment on the onset of maternal behavior and aggression in Sprague-Dawley rats. Behav Neurosci 108:107–112.

Johns JM, Noonan LR, Zimmerman LI, Li L, Pedersen CA (1997) Effects of short- and long-term withdrawal from gestational cocaine treatment on maternal behavior and aggression in Sprague-Dawley rats. Dev Neurosci 19:368–374.

Johnson AE, Coirini A, Ball GF, McEwen BS (1989) Anatomical localization of the effects of 17b-estradiol on oxytocin receptor binding in the ventromedial hypothalamic nucleus. Endocrinology 124:207–211.

Johnson LD, Petto AJ, Sehgal PK (1991) Factors in the rejection and survival of captive cotton top tamarins. Am J Primatol 25:91–102.

Johnston RE (1998) Pheromones, the vomeronasal system, and communication. Ann NY Acad Sci 855:333–348.

Jones JS, Wynne-Edwards KE (2000) Paternal hamsters mechanically assist the delivery, consume amniotic fluid and placenta, remove fetal membranes, and provide parental care during the birth process. Horm Behav 37:116–125.

Jones PM, Robinson ICAF, Harris MC (1983) Release of oxytocin into blood and cerebrospinal fluid by electrical stimulation of the hypothalamus or neural lobe in the rat. Neuroendocrinology 37:454–458.

Joppa MA, Meisel RL, Garber MA (1995) c-Fos expression in the female hamster brain following sexual and aggressive behaviors. Neuroscience 68:783–792.

Kaba H, Nakanishi S (1995) Synaptic mechanisms of olfactory recognition memory. Rev Neurosci 6:125–141.

Kalin NH, Shelton SE, Lynn DE (1995) Opiate systems in mother and infant primates coordinate intimate contact during reunion. Psychoneuroendocrinology 20:735–742.

Kalinichev M, Rosenblatt JS, Morrell JI (2000a) The medial preoptic area, necessary for adult maternal behavior in rats, is only partially established as a component of the neural circuit that supports maternal behavior in juvenile rats. Behav Neurosci 114: 196–210.

Kalinichev M, Rosenblatt JS, Nakabeppu Y, Morrell JI (2000b) Induction of c-Fos-like and FosB-like immunoreactivity reveals forebrain neuronal populations involved differentially in pup-mediated maternal behavior in juvenile and adult rats. J Comp Neurol 416:45–78.

Kalivas PW (1985) Interactions between neuropeptides and dopamine neurons in the ventromedial mesencephalon. Neurosci Biobehav Rev 9:573–587.

Kalivas PW, Churchill L, Romanides A (1999) Involvement of the pallidal-thalamocortical circuit in adaptive behavior. Ann NY Acad Sci 877:64–70.

Kalivas PW, Stewart J (1991) Dopamine transmission in the initiation and expression of drug- and stress-induced sensitization of motor activity. Brain Res Rev 16:223–244.

Kandel ER, Schwartz JH, Jessell TM (2000) Principles of neural science. New York: McGraw-Hill.

Karlsson U, Sundgren AK, Nasstrom J, Johansson S (1997) Glutamate-evoked currents in acutely dissociated neurons from the rat medial preoptic nucleus. Brain Res 759: 270–276.

Kaufman J, Birmaher B, Perel J, Dahl RE, Moreci P, Nelson B, Wells W, Ryan ND (1997) The corticotropin-releasing hormone challenge in depressed abused, depressed nonabused, and normal control children. Biol Psychiatry 42:669–679.

Kaufman J, Plotsky PM, Nemeroff CB, Charney DS (2000) Effects of early adverse experiences on brain structure and function: clinical implications. Biol Psychiatry 48: 778–790.

Kaufman J, Zigler E (1987) Do abused children become abusive parents? Am J Orthopsychiatry 57:186–192.

Keer SE, Stern JM (1999) Dopamine receptor blockade in the nucleus accumbens inhibits maternal retrieval and licking, but enhances nursing behavior in lactating rats. Physiol Behav 67:659–669.

Kelley AE (1999) Functional specificity of ventral striatal compartments in appetitive behaviors. Ann NY Acad Sci 877:71–90.

Kellogg CK, Barrett KA (1999) Reduced progesterone metabolites are not critical for plus-maze performance of lactating female rats. Pharmacol Biochem Behav 63:441–448.

Kelz MB, Chen J, Carlezon WA, Whisler K, Gilden L, Beckmann AM, Steffen C, Zhang Y, Marotti L, Self DW, Tkatch T, Baranauskas G, Surmeier DJ, Neve RL, Duman RS, Picciotto MR, Nestler EJ (1999) Expression of ΔFosB in the brain controls sensitivity to cocaine. Nature 401:272–276.

Kemble ED, Blanchard DC, Blanchard RJ (1990) Effects of regional amygdaloid lesions on flight and defensive behaviors of wild black rats (*Rattus rattus*). Physiol Behav 48:1–5.

Kemps A, Timmermans P, Vossen J (1989) Effects of mother's rearing condition and multiple motherhood on early development of mother-infant interactions in java macaques (*Macaca fascicularis*). Behaviour 111:61–76.

Kendler KS (1996) Parenting: a genetic-epidemiologic perspective. Am J Psychiatry 153: 11–20.

Kendrick KM (2000) Oxytocin, motherhood, and bonding. Exp Physiol 85 (Spec Issue): 111S–124S.

Kendrick KM, Da Costa APC, Broad KD, Ohkura S, Guevara R, Levy F, Keverne EB (1997a) Neural control of maternal behaviour and olfactory recognition of offspring. Brain Res Bull 44:383–395.

Kendrick KM, Dixson AF (1986) Anteromedial hypothalamic lesions block proceptivity but not receptivity in the female common marmoset (*Callithrix jacchus*). Brain Res 375:221–229.

Kendrick KM, Guevara-Guzman R, Zorrilla J, Hinton MR, Broad KD, Mimmack M, Ohkura S (1997b) Formation of olfactory memories mediated by nitric oxide. Nature 388:670–674.

Kendrick KM, Keverne EB (1989) Effects of intracerebroventricular infusions of naltrexone and phentolamine on central and peripheral oxytocin release and on maternal behaviour induced by vaginocervical stimulation in the ewe. Brain Res 505:329–332.

Kendrick KM, Keverne EB (1991) Importance of progesterone and estrogen priming for the induction of maternal behavior by vagino-cervical stimulation in sheep: effects of maternal experience. Physiol Behav 49:745–750.

Kendrick KM, Keverne EB, Baldwin BA (1987) Intracerebroventricular oxytocin stimulates maternal behaviour in the sheep. Neuroendocrinology 46:56–61.

Kendrick KM, Keverne EB, Baldwin BA, Sharman DF (1986) Cerebrospinal fluid levels of acetylcholinesterase, monoamines and oxytocin during labor, parturition, vaginocervical stimulation, lamb separation and suckling in sheep. Neuroendocrinology 44: 149–156.

Kendrick MK, Keverne EB, Hinton MR, Goode JA (1992a) Oxytocin, amino acid and monoamine release on the region of the medial preoptic area and bed nucleus of the stria terminalis of the sheep during parturition and suckling. Brain Res 569:199–209.

Kendrick MK, Levy F, Keverne EB (1991) Importance of vaginocervical stimulation for the formation of maternal bonding in primiparous and multiparous ewes. Physiol Behav 50:595–600.

Kendrick MK, Levy F, Keverne EB (1992b) Changes in sensory processing of olfactory signals induced by birth in sheep. Science 256:833–836.

Kenyon P, Cronin P, Keeble S (1981) Disruption of maternal retrieving by perioral anesthesia. Physiol Behav 27:313–321.

Kenyon P, Cronin P, Keeble S (1983) Role of the infraorbital nerve in retrieving behavior in lactating rats. Behav Neurosci 97:255–269.

Ketterson E, Nolan V (1994) Male parental care in birds. Annu Rev Ecol Systems 25: 601–628.

Keverne EB (1995) Olfactory learning. Curr Opin Neurobiol 5:482–488.

Keverene EB (1996) Psychopharmacology of maternal behaviour. J Psychopharmacol 10:16–22.

Keverne EB (1999) The vomeronasal organ. Science 286:716–720.

Keverne EB (2001) Genomic imprinting, maternal care, and brain evolution. Horm Behav 40:146–155.

Keverne EB, Kendrick KM (1991) Morphine and corticotrophin-releasing factor potentiate maternal acceptance in multiparous ewes after vaginocervical stimulation. Brain Res 540:55–62.

Keverne EB, Kendrick KM (1992) Oxytocin facilitation of maternal behavior in sheep. Ann NY Acad Sci 652:83–101.

Keverne EB, Levy F, Poindron P, Lindsay DR (1983) Vaginal stimulation: an important component of maternal bonding in sheep. Science 219:81–83.

Keverne EB, Martensz N, Tuite B (1989) B-endorphin concentrations in CSF of monkeys are influenced by grooming relationships. Psychoneuroendocrinology 14:155–161.

Kevetter G, Winans SS (1981a) Connections of the corticomedial amygdala in the golden hamster. I. Efferents of the "vomeronasal amygdala". J Comp Neurol 197:81–98.

Kevetter G, Winans SS (1981b) Connections of the corticomedial amygdala in the golden hamster. II. Efferents of the "olfactory amygdala". J Comp Neurol 197:99–111.

Keyser-Marcus L, Stafisso-Sandoz G, Gerecke K, Jasnow A, Nightingale L, Lambert KG, Gatewood J, Kinsley CH (2001) Alterations of medial preoptic area neurons following pregnancy and pregnancy-like steroidal treatment in the rat. Brain Res Bull 55:737–745.

Khan M, McNabb FMA, Walters JR, Sharp PJ (2001) Patterns of testosterone and prolactin concentrations and reproductive behavior of helpers and breeders in the cooperatively breeding red-cockaded woodpecker (Picoides borealis). Horm Behav 40:1–13.

Khatib S, Insel TR, Young LJ (2001) The effects of pup exposure on parental responsiveness, serum prolactin, and prolactin receptor gene expression in the biparental prairie vole. Soc Neurosci Abstr 27:746.4.

Kimble DP, Rodgers L, Hendrickson CW (1967). Hippocampal lesions disrupt maternal, not sexual, behavior in the albino rat. J Comp Physiol Psychol 63:401–407.

Kinsley CH, Bridges RS (1986) Opiate involvement in postpartum aggression in rats. Pharmacol Biochem Behav 25:1007–1011.

Kinsley CH, Bridges RS (1988a) Prolactin modulation of the maternal-like behavior displayed by juvenile rats. Horm Behav 22:49–65.

Kinsley CH, Bridges RS (1988b) Prenatal stress and maternal behavior in intact virgin rats: response latencies are decreased in males and increased in females. Horm Behav 22:76–89.

Kinsley CH, Bridges RS (1990) Morphine treatment and reproductive condition alter olfactory preferences for pup and adult male odors in female rats. Dev Psychobiol 23:331–347.

Kinsley CH, Turco D, Bauer A, Beverly M, Wellman J, Graham AL (1994) Cocaine alters the onset and maintenance of maternal behavior in lactating rats. Pharmacol Biochem Behav 47:857–864.

Kinsley CH, Wellman JC, Carr DB, Graham A (1993) Opioid regulation of parental behavior in juvenile rats. Pharmacol Biochem Behav 44:763–768.

Kirkpatrick B, Carter C, Newman S, Insel T (1994a) Axon sparing lesions of the medial nucleus of the amygdala decrease affiliative behaviors in the prairie vole (Microtus ochrogaster): behavioral and anatomic specificity. Behav Neurosci 108:501–513.

Kirkpatrick B, Kim JW, Insel TR (1994b) Limbic system fos expression associated with paternal behavior. Brain Res 658:112–118.

Kleiman DG, Malcolm J (1981) The evolution of male parental investment in mammals. In: Parental care in mammals (Gubernick DJ, Klopfer PH, eds), pp 347–387. New York: Plenum Press.

Klein DF, Skrobala AM, Garfinkel RS (1994/1995) Preliminary look at the effects of pregnancy on the course of panic disorder. Anxiety 1:227–232.

Klindt J, Robertson MC, Friesen HG (1981) Secretion of placental lactogen, growth hormone, and prolactin in late pregnant rats. Endocrinology 109:1492–1495.

Klopfer PH, Gamble J (1966) Maternal "imprinting" in goats: the role of chemical senses. Z Tierpsychol 23:588–592.

Koch M, Ehret G (1989a) Estradiol and parental experience, but not prolactin are necessary for ultrasound recognition and pup-retrieving in the mouse. Physiol Behav 45: 771–776.

Koch M, Ehret G (1989b) Immunocytochemical localization and quantitation of estrogen-binding cells in the male and female (virgin, pregnant, lactating) mouse brain. Brain Res 489:101–112.

Kohlert JG, Meisel RL (1999) Sexual experience sensitizes mating-related nucleus accumbens dopamine responses of female Syrian hamsters. Behav Brain Res 99:45–52.

Kollack-Walker S, Watson SJ, Akil H (1997) Social stress in hamsters: defeat activates specific neurocircuits within the brain. J Neurosci 17:8842–8855.

Kolunie JM, Stern JM (1995) Maternal aggression in rats: effects of olfactory bulbectomy, ZnSO4-induced anosmia, and vomeronasal organ removal. Horm Behav 29: 492–518.

Kolunie JM, Stern JM, Barfield RJ (1994) Maternal aggression in rats: effects of visual or auditory deprivation of the mother and dyadic pattern of ultrasonic vocalizations. Behav Neural Biol 62:41–49.

Komisaruk BR, Rosenblatt JS, Barona ML, Chinapen S, Nissanov J, O'Bannon RT, Johnson BM, Del Cerro MCR (2000) Combined c-fos and ^{14}C-2-deoxyglucose method to differentiate site-specific excitation from disinhibition: analysis of maternal behavior in the rat. Brain Res 859:262–272.

Koob GF (1999) Drug reward and addiction. In: Fundamental neuroscience (Zigmond MJ, Bloom FE, Landis SC, Roberts JL, Squire LR, eds), pp 1261–1279. San Diego: Academic Press.

Korte SM (2001) Corticosteroids in relation to fear, anxiety and psychopathology. Neurosci Biobehav Rev 25:117–142.

Kovaks GL, Sarnyai Z, Szabo G (1998) Oxytocin and addiction: a review. Psychoneuroendocrinology 23:945–962.

Kramer MS, Cutler N, Feighner J, Shrivastava R, Carman J, Sramek JJ, Reines SA, Liu G, Snavely D, Wyatt-Knowles E, Hale JJ, Mills SG, MacGross M, Swain CJ, Harrison T, Hill RG, Hefti F, Scolnick EM, Cascieri MA, Chicchi GG, Sadowski S, Williams AR, Hewson L, Smith D, Carlson EJ, Hargraves RJ, Rupniak NM (1998) Distinct mechanism for antidepressant activity by blockade of central substance P receptors. Science 281:1640–1645.

Kramlinger KG, Peterson GC, Watson PK, Leonard LL (1985) Metyrapone for depression and delirium secondary to Cushing's syndrome. Psychosomatics 26:67–71.

Krehbiel D, Poindron P, Levy F, Prud'Homme MJ (1987) Peridural anesthesia disrupts maternal behavior in primiparous and multiparous parturient ewes. Physiol Behav 40: 463–472.

Kremarik P, Freund-Mercier MJ, Stoeckel ME (1991) Autoradiographic detection of oxytocin- and vasopressin-binding sites in various subnuclei of the bed nucleus of the stria terminalis in the rat. Effects of functional and experimental sexual steroid variations. J Neuroendocrinol 3:689–698.

Kremarik P, Freund-Mercier MJ, Stoeckel ME (1995) Oxytocin and vasopressin binding sites in the hypothalamus of the rat: histautoradiographic detection. Brain Res Bull 36:195–203.

Krettek JE, Price JL (1978) Amygdaloid projections to subcortical structures with the basal forebrain and brainstem of the rat and cat. J Comp Neurol 178:225–254.

Krieger J, Schmitt A, Lobel D, Gudermann T, Schulz G, Breer H, Boekhoff I (1999)

Selective activation of G protein subtypes in the vomeronasal organ upon stimulation with urine-derived compounds. J Biol Chem 274:4655–4662.

Kuiper GGJM, Carlsson B, Grandien K, Enmark E, Haggblad J, Nilsson S, Gustafsson J (1997) Comparison of the ligand binding specificity and transcript tissue distribution of estrogen receptors α and β. Endocrinology 138:863–870.

Kumar A, Dudley CA, Moss RL (1999) Functional dichotomy within the vomeronasal system: distinct zones of neuronal activity in the accessory olfactory bulb correlate with sex-specific behaviors. J Neurosci 19:RC32.

Ladd CO, Huot RL, Thrivikraman KV, Nemeroff CB, Meaney MJ, Plotsky PM (2000) Long-term behavioral and neuroendocrine adaptations to adverse early experience. In: Progress in brain research, Vol 122. The biological basis of mind body interactions (Mayer EA, Saper CB, eds), pp 81–103. Amsterdam: Elsevier.

Ladd CO, Owens MJ, Nemeroff CB (1996) Persistent changes in corticotropin-releasing factor neuronal systems induced by maternal deprivation. Endocrinology 137:1212–1218.

Lahey BB, Conger RD, Atkeson BM, Treiber FA (1984) Parenting behavior and emotional status of physically abusive mothers. J Consult Clin Psychol 52:1062–1071.

Lancaster JB (1971) Play mothering: the relations between juvenile females and young infants among free-ranging vervet monkeys. Folia Primatol 15:161–182.

Landgraf R, Neumann I, Pittman QJ (1991) Septal and hippocampal release of vasopressin and oxytocin during late pregnancy and parturition in the rat. Neuroendocrinology 54:378–383.

Lang RE, Heil J, Ganten D, Hermann K, Rascher W, Unger T (1983) Effects of lesions in the paraventricular nucleus of the hypothalamus on vasopressin and oxytocin contents in brainstem and spinal cord of rat. Brain Res 260:326–329.

Lawrie T, Hofmeyer GJ, De Jager M, Berk M, Paiker J, Viljoen E (1998) A double-blind randomised placebo controlled trial of postnatal norethisterone enanthate: the effect on postnatal depression and serum hormones. Br J Obstet Gynaecol 105:1082–1090.

Leblond CP (1938) Extra-hormonal factors in maternal behavior. Proc Soc Exp Biol Med 38:66–70.

Leblond CP (1940) Nervous and hormonal factors in the maternal behavior of the mouse. J Genet Psychol 57:327–344.

Leblond CP, Nelson WO (1937) Maternal behavior in hypophysectomized male and female mice. Am J Physiol 120:167–172.

Leckman JF, Herman AE (2001) Maternal behavior and developmental psychopathology. Biol Psychiatry 51:27–43.

LeDoux J (1996) The emotional brain. New York: Simon and Schuster.

LeDoux JE (1993) Emotional memory systems in the brain. Behav Brain Res 58:69–79.

LeDoux JE (2000) Emotion circuits in the brain. Annu Rev Neurosci 23:155–184.

Lee A, Clancy S, Fleming AS (2000) Mother rats bar-press for pups: effects of lesions of the MPOA and limbic sites on maternal behavior and operant responding for pup-reinforcement. Behav Brain Res 108:215–231.

Lee A, Li M, Watchus J, Fleming AS (1999) Neuroanatomical basis of maternal memory in postpartum rats: selective role for the nucleus accumbens. Behav Neurosci 113:523–538.

Lee MHS, Williams DI (1974) Changes in licking behaviour of rat mother following handling of young. Anim Behav 22:679–681.

Lefebvre L, Viville S, Barton SC, Ishino F, Keverne EB, Surani MA (1998) Abnormal maternal behaviour and growth retardation associated with loss of the imprinted gene Mest. Nature Genet 20:163–169.

Lehmann J, Stohr T, Feldon J (2000) Long-term effects of prenatal stress and postnatal maternal separation on emotionality and attentional processes. Behav Brain Res 107: 133–144.

Lehrman DS (1961) Hormonal regulation of parental behavior in birds and infrahuman mammals. In: Sex and internal secretions (Young WC, ed), pp 1268–1382. Baltimore: Williams and Wilkins.

Le Neindre P, Poindron P, Delouis C (1979) Hormonal induction of maternal behavior in non-pregnant ewes. Physiol Behav 22:731–734.

Leon M (1992) The neurobiology of filial learning. Annu Rev Psychol 43:377–398.

Leon M, Numan M, Chan A (1975) Adrenal inhibition of maternal behavior in virgin female rats. Horm Behav 6:165–171.

Leong DA, Frawley LS, Neill JD (1983) Neuroendocrine control of prolactin secretion. Annu Rev Physiol 45:109–127.

Lepri JJ, Wysocki CJ, Vandenbergh JG (1985) Mouse vomeronasal organ: effects on chemosignal production and maternal behavior. Physiol Behav 35:809–814.

LeRoy LM, Krehbiel DA (1978) Variations in maternal behavior in the rat as a function of sex and gonadal state. Horm Behav 11:232–247.

Levine S (1967) Maternal and environmental influences on the adrenocortical response to stress in weanling rats. Science 156:258–260.

Levine S (1975) Psychosocial factors in growth and development. In: Society, stress, and disease (Levy L, ed), pp 43–50. London: Oxford University Press.

Levine S (2002) Enduring effects of early experience on adult behavior. In: Hormones, brain and behavior, Vol 4 (Pfaff DW, Arnold AP, Etgen AM, Fahrbach SE, Rubin RT, eds), pp 535–542. San Diego: Academic Press.

Levy F, Gervais R, Kindermann U, Orgeur P, Piketty V (1990) Importance of β-noradrenergic receptors in the olfactory bulb of sheep for recognition of lambs. Behav Neurosci 104:464–469.

Levy F, Guevara-Guzman R, Hinton MR, Kendrick KM, Keverne EB (1993) Effects of parturition and maternal experience on noradrenaline and acetylcholine release in the olfactory bulb of sheep. Behav Neurosci 107:662–668.

Levy F, Kendrick KM, Goode JA, Guevara-Guzman R, Keverne EB (1995a) Oxytocin and vasopressin release in the olfactory bulb of parturient ewes: changes with maternal experience and effects on acetylcholine, γ-aminobutyric acid, glutamate, and norad-renaline release. Brain Res 669:197–206.

Levy F, Kendrick KM, Keverne EB, Piketty V, Poindron P (1992) Intracerebral oxytocin is important for the onset of maternal behavior in inexperienced ewes delivered under peridural anesthesia. Behav Neurosci 106:427–432.

Levy F, Locatelli A, Piketty V, Tillet Y, Poindron P (1995b) Involvement of the main but not the accessory olfactory system in maternal behavior of primiparous and multiparous ewes. Physiol Behav 57:97–104.

Levy F, Meurisse M, Ferreira G, Thibault J, Tillet Y (1999) Afferents to the rostral olfactory bulb in sheep with special emphasis on the cholinergic, noradrenergic and serotonergic connections. J Chem Neuroanat 16:245–263.

Levy F, Poindron P (1987) Importance of amniotic fluids for the establishment of maternal behaviour in relation with maternal experience in sheep. Anim Behav 35:1188–1192.

Levy F, Poindron P, Le Neindre P (1983) Attraction and repulsion by amniotic fluids and their olfactory control in the ewe around parturition. Physiol Behav 31:687–692.

Levy F, Porter RH, Kendrick KM, Keverne EB, Romeyer A (1996) Physiological, sensory, and experiential factors of parental care in sheep. In: Advances in the study of behavior, Vol 25. Parental care: evolution, mechanisms, and adaptive significance (Rosenblatt JS, Snowdon CT, eds), pp 385–422. San Diego: Academic Press.

Lewis MH, Gluck JP, Beauchamp AJ, Keresztury MF, Mailman RB (1990) Long-term effects of early social isolation in *Macaca mulatta*: changes in dopamine receptor function following apomorphine challenge. Brain Res 513:67–73.

Lewis SE, Pusey AE (1997) Factors influencing the occurrence of communal care in plural breeding mammals. In: Cooperative breeding in mammals (Solomon NG, French JA, eds), pp 335–363. Cambridge: Cambridge University Press.

Li H, Weiss SRB, Chuang D, Post RM, Rogawski MA (1998) Bidirectional synaptic plasticity in the rat basolateral amygdala: characterization of an activity-dependent switch sensitive to the presynaptic metabotropic glutamate receptor antagonist 2S-α-ethylglutamic acid. J Neurosci 18:1662–1670.

Li L, Keverne EB, Aparicio S, Ishino F, Barton S, Surani M (1999) Regulation of maternal behavior and offspring growth by paternally expressed Peg 3. Science 284:330–333.

Li Y, Takada M, Shinonaga Y, Mizuno N (1993) Direct projections from the midbrain periaqueductal gray and the dorsal raphe nucleus to the trigeminal sensory complex in the rat. Neuroscience 54:431–443.

Liberzon I, Chalmers DT, Mansour A, Lopez JF, Watson SJ, Young EA (1994) Glucocorticoid regulation of hippocampal oxytocin receptor binding. Brain Res 650:317–322.

Liberzon I, Trujillo KA, Akil H, Young EA (1997) Motivational properties of oxytocin in the conditioned place preference paradigm. Neuropsychopharmacology 17:353–359.

Licht G, Meredith M (1987) Convergence of main and accessory olfactory pathways onto single neurons in the hamster amygdala. Exp Brain Res 69:7–18.

Lightman SL, Windle RJ, Wood SA, Kershaw YM, Shanks N, Ingram CD (2001) Peripartum plasticity within the hypothalamo-pituitary-adrenal axis. In: The maternal brain. Progress in brain research, Vol 133 (Russell JA, Douglas AJ, Windle RJ, Ingram CD, eds), pp 111–129. Amsterdam: Elsevier.

Lightman SL, Young WSI (1987) Vasopressin, oxytocin, enkephalin, dynorphin, corticotrophin releasing factor mRNA stimulation in the rat. J Physiol 394:23–39.

Lightman SL, Young WS (1989) Lactation inhibits stress-mediated secretion of corticosterone and oxytocin and hypothalamic accumulation of corticotropin-releasing factor and enkephalin messenger ribonucleic acids. Endocrinology 124:2358–2364.

Lin S, Miyata S, Weng W, Matsunaga W, Ichikawa J, Furuya K, Nakashima T, Kiyohara T (1998) Comparison of the expression of two immediate early gene proteins, FosB and Fos in the rat preoptic area, hypothalamus and brainstem during pregnancy, parturition and lactation. Neurosci Res 32:333–341.

Lincoln DW (1974) Suckling: a time constant in the nursing behavior of the rabbit. Physiol Behav 13:711–714.

Linden A, Uvnas-Moberg K, Eneroth P, Sodersten P (1989) Stimulation of maternal behaviour in rats with cholecystokinin octapeptide. J Neuroendocrinol 1:389–392.

Lindzey J, Korach KS (1997) Developmental and physiological effects of estrogen receptor gene disruption in mice. Trends Endocr Metab 8:137–145.

Lisk RD (1971) Oestrogen and progesterone synergism and elicitation of maternal nest-building in the mouse (*Mus musculus*). Anim Behav 19:606–610.

Liu D, Caldji C, Sharma S, Plotsky PM, Meaney MJ (2000) Influence of neonatal rearing conditions on stress-induced adrenocortical responses and norepinephrine release in the hypothalamic paraventricular nucleus. J Neuroendocrinol 12:5–12.

Liu D, Diorio J, Tannenbaum B, Caldji C, Francis D, Freedman A, Sharma S, Pearson D, Plotsky PM, Meaney MJ (1997) Maternal care, hippocampal glucocorticoid receptors, and hypothalamic-pituitary-adrenal responses to stress. Science 277:1659–1662.

Liu Y, Curtis JT, Wang Z (2001) Vasopressin in the lateral septum regulates pair bond formation in male prairie voles (*Microtus ochrogaster*). Behav Neurosci 115:910–919.

Llewellyn AM, Stowe ZN, Nemeroff CB (1997) Depression during pregnancy and the puerperium. J Clin Psychiatry 58 (suppl 15):26–32.

Loberman JP, Newman JD, Horwitz AR, Dubno JR, Lydiard RB, Hamner MB, Bohning DE, George MS (2002) A potential role for thalamocingulate circuitry in human maternal behavior. Biol Psychiatry 51:431–445.

Lonstein JS, De Vries GJ (1999a) Sex differences in the parental behaviour of adult virgin prairie voles: independence from gonadal hormones and vasopressin. J Neuroendocrinol 11:441–449.

Lonstein JS, De Vries GJ (1999b) Comparison of the parental behavior of pair-bonded female and male prairie voles (*Microtus ochrogaster*). Physiol Behav 66:33–40.

Lonstein JS, De Vries GJ (2000a) Sex differences in the parental behavior of rodents. Neurosci Biobehav Rev 24:669–686.

Lonstein JS, De Vries GJ (2000b) Influence of gonadal hormones on the development of parental behavior in adult virgin prairie voles (*Microtus ochrogaster*). Behav Brain Res 114:79–87.

Lonstein JS, De Vries GJ (2000c) Maternal behaviour in lactating rats stimulates *c-fos* in glutamate decarboxylase-synthesizing neurons of the medial preoptic area, ventral bed nucleus of the stria terminalis, and ventrocaudal periaqueductal gray. Neuroscience 100:557–568.

Lonstein JS, De Vries GJ (2001) Social influences on parental and nonparental responses toward pups in virgin female prairie voles (*Microtus ochrogaster*). J Comp Psychol 115:53–61.

Lonstein JS, Greco B, De Vries GJ, Stern JM, Blaustein JD (2000) Maternal behavior stimulates c-fos activity within estrogen receptor alpha-containing neurons in lactating rats. Neuroendocrinology 72:91–101.

Lonstein JS, Simmons DA, Stern JM (1998a) Functions of the caudal periaqueductal gray in lactating rats: kyphosis, lordosis, maternal aggression, and fearfulness. Behav Neurosci 112:1502–1518.

Lonstein JS, Simmons DA, Swann JM, Stern JM (1998b) Forebrain expression of *c-fos* due to active maternal behaviour in lactating rats. Neuroscience 82:267–281.

Lonstein JS, Stern JM (1997a) Role of midbrain periaqueductal gray in maternal nurturance and aggression: *c-fos* and electrolytic lesion studies in lactating rats. J Neurosci 17:3364–3378.

Lonstein JS, Stern JM (1997b) Somatosensory contributions to c-fos activation within the caudal periaqueductal gray of lactating rats: effects of perioral, rooting, and suckling stimuli from pups. Horm Behav 32:155–166.

Lonstein JS, Stern JM (1998) Site and behavioral specificity of periaqueductal gray lesions on postpartum sexual, maternal, and aggressive behaviors in rats. Brain Res 804:21–35.

Lonstein JS, Wagner CK, De Vries GJ (1999) Comparison of the "nursing" and other parental behaviors of nulliparous and lactating female rats. Horm Behav 36:242–251.

Lott DF, Fuchs SS (1962) Failure to induce retrieving by sensitization or the injection of prolactin. J Comp Physiol Psychol 55:1111–1113.

Loundes DD, Bridges RS (1986) Length of prolactin priming differentially affects maternal behavior in female rats. Biol Reprod 34:495–501.

Lovic V, Gonzalez A, Fleming AS (2001) Maternally separated rats show deficits in maternal care in adulthood. Dev Psychobiol 39:19–33.

Lowry CA, Rodda JE, Lightman SL, Ingram CD (2000) Corticotropin-releasing factor increases in vitro firing rates of serotonergic neurons in the rat dorsal raphe nucleus: evidence for activation of a topographically organized mesolimbocortical serotonergic system. J Neurosci 20:7728–7736.

Lubahn DB, Motyer JS, Golding TS, Couse JF, Korach KS, Smithies O (1993) Alteration of reproductive function but not prenatal sexual development after insertional disruption of the mouse estrogen receptor gene. Proc Natl Acad Sci 90:11162–11166.

Lubin M, Leon M, Moltz H, Numan M (1972) Hormones and maternal behavior in the male rat. Horm Behav 3:369–374.

Lucas BK, Ormandy CJ, Binart N, Bridges RS, Kelly PA (1998) Null mutation of the prolactin receptor gene produces a defect in maternal behavior. Endocrinology 139: 4102–4107.

Luiten PGM, Koolhaas JM, de Boer S, Koopmans SJ (1985) The cortico-medial amygdala in the central nervous system organization of agonistic behavior. Brain Res 332: 283–297.

Luscher C, Nicoll RA, Malenka RC, Muller D (2000) Synaptic plasticity and dynamic modulation of the postsynaptic membrane. Nature Neurosci 3:545–550.

Lydon JP, DeMayo FJ, Conneely OM, O'Malley BW (1997) Reproductive phenotypes of the progesterone receptor null mutant mouse. J Steroid Biochem Mol Biol 56:67–77.

Mackinnon DA, Stern JM (1977) Pregnancy duration and fetal number: effects on maternal behavior in rats. Physiol Behav 18:793–797.

Madison DM (1978) Movement indications of reproductive events among female meadow voles as revealed by radiotelemetry. J Mammal 59:835–843.

Maestripieri D (1993) Maternal anxiety in rhesus macaques (Macaca mulatta) II. Emotional bases of individual differences in mothering style. Ethology 95:32–42.

Maestripieri D (1994a) Social structure, infant handling, and mothering styles in group living old world monkeys. Int J Primatol 15:531–553.

Maestripieri D (1994b) Influence of infants on female social relationships in monkeys. Folia Primatol 63:192–202.

Maestripieri D (1998) Parenting styles of abusive mothers in group-living rhesus macaques. Anim Behav 55:1–11.

Maestripieri D (1999) The biology of human parenting: insights from nonhuman primates. Neurosci Biobehav Rev 23:411–422.

Maestripieri D, Alleva E (1991) Do male mice use parental care as a buffering strategy against maternal aggression? Anim Behav 41:904–906.

Maestripieri D, Badiani A, Puglisi-Allegra S (1991) Prepartal chronic stress increases anxiety and decreases aggression in lactating female mice. Behav Neurosci 105:663–668.

Maestripieri D, Carroll KA (1998) Child abuse and neglect: usefulness of the animal data. Psychol Bull 123:211–223.

Maestripieri D, D'Amato FR (1991) Anxiety and maternal aggression in house mice (*mus musculus*): a look at interindividual variability. J Comp Psychol 105:295–301.

Maestripieri D, Megna NL (2000a) Hormones and behavior in abusive and nonabusive rhesus macaque mothers: 1. Social interactions during late pregnancy and lactation. Physiol Behav 71:35–42.

Maestripieri D, Megna NL (2000b) Hormones and behavior in abusive and nonabusive rhesus macaque mothers: 2. Mother-infant interactions. Physiol Behav 71:43–49.

Maestripieri D, Wallen K, Carroll KA (1997) Infant abuse runs in families of group-living pigtail macaques. Child Abuse Negl 21:465–471.

Maestripieri D, Zehr JL (1998) Maternal responsiveness increases during pregnancy and after estrogen treatment in macaques. Horm Behav 34:223–230.

Magiakou MA, Mastorakos G, Rabin D, Dubbert B, Gold PW, Chrousos GP (1996) Hypothalamic corticotropin-releasing hormone suppression during the postpartum period: implications for the increase in psychiatric manifestations at this time. J Clin Endocrinol Metab 81:1912–1917.

Magnusson JE, Fleming AS (1995) Rat pups are reinforcing to the maternal rat: role of sensory cues. Psychobiology 23:69–75.

Magura S, Laudet AB (1996) Parental substance abuse and child maltreatment: review and implications for intervention. Child Youth Serv Rev 18:193–220.

Maillard-Gutekunst C, Edwards DA (1994) Preoptic and subthalamic connections with the caudal brainstem are important for copulation in the male rat. Behav Neurosci 108:758–766.

Majewska MD, Ford-Rice F, Falkay G (1989) Pregnancy-induced alterations in GABAA receptor sensitivity in maternal brain: an antecedent of post-partum "blues"? Brain Res 482:397–401.

Malenfant SA, Barry M, Fleming AS (1991a) Effects of cycloheximide on the retention of olfactory learning and maternal experience effects in postpartum rats. Physiol Behav 49:289–294.

Malenfant SA, O'Hearn S, Fleming AS (1991b) MK801, an NMDA antagonist, blocks acquisition of a spatial task but does not block maternal experience effects. Physiol Behav 49:1129–1137.

Malick A, Burstein R (1998) Cells of origin of the trigeminohypothalamic tract in the rat. J Comp Neurol 400:125–144.

Mangurian LP, Walsh RJ, Posner BI (1992) Prolactin enhancement of its own uptake at the choroid plexus. Endocrinology 131:698–702.

Mani S (2001) Ligand-independent activation of progestin receptors in sexual receptivity. Horm Behav 40:183–190.

Mani SK, Allen JMC, Clark JH, Blaustein JD, O'Malley BW (1994) Convergent pathways for steroid hormone- and neurotransmitter-induced rat sexual behavior. Science 265:1246–1249.

Mani SK, Fienberg AF, O'Callaghan JP, Snyder GL, Allen PB, Dash PK, Moore AN, Mitchell AJ, Bibb J, Greengard P, O'Malley BW (2000) Requirement of DARPP-32 in progesterone-facilitated sexual receptivity in female mice. Science 287: 1053–1056.

Mann M, Michael SD, Svare B (1980) Ergot drugs suppress plasma prolactin and lactation but not aggression in parturient mice. Horm Behav 14:319–328.

Mann MA, Kinsley C, Broida J, Svare B (1983) Infanticide exhibited by female mice: genetic, developmental and hormonal influences. Physiol Behav 30:697–702.

Mann MA, Konen C, Svare B (1984) The role of progesterone in pregnancy-induced aggression in mice. Horm Behav 18:140–160.

Mann PE, Bridges RS (1992) Neural and endocrine sensitivities to opioids decline as a function of multiparity in the rat. Brain Res 580:241–248.

Mann PE, Felicio LF, Bridges RS (1995) Investigation into the role of cholecystokinin in the induction and maintenance of maternal behavior in rats. Horm Behav 29:392–406.

Mann PE, Kinsley CH, Bridges RS (1991) Opioid receptor subtype involvement in maternal behavior in lactating rats. Neuroendocrinology 53:487–492.

Mann PE, Rubin BS, Bridges RS (1997) Differential proopiomelanocortin gene expression in the medial basal hypothalamus of rats during pregnancy and lactation. Mol Brain Res 46:9–16.

Manning CJ, Dewsbury DA, Wakeland EK, Potts WK (1995) Communal nesting and communal nursing in house mice, *Mus musculus domesticus*. Anim Behav 50:741–751.

Maren S, Fanselow MS (1996) The amygdala and fear conditioning: has the nut been cracked? Neuron 16:237–240.

Marques DM (1979) Roles of main olfactory and vomeronasal systems in the response of the female hamster to young. Behav Neural Biol 26:311–329.

Marques DM, Valenstein ES (1976) Another hamster paradox: more males carry pups and fewer kill and cannibalize young than do females. J Comp Physiol Psychol 90:653–657.

Martel FL, Nevison CM, Rayment FD, Simpson MJ, Keverne EB (1993) Opioid receptor blockade reduces maternal affect and social grooming in rhesus monkeys. Psychoneuroendocrinology 18:307–321.

Martin LJ, Spicer DM, Lewis MH, Gluck JP, Cork LC (1991) Social deprivation of infant rhesus monkeys alters the chemoarchitecture of the brain: I. Subcortical regions. J Neurosci 11:3344–3358.

Maselli MA, Piepoli AL, Pezzolla F, Guerra V, Caruso ML, Mennuni L, Lorusso D, Makovec F (2001) Effect of three nonpeptide cholecystokinin antagonists on human isolated gallbladder. Dig Dis Sci 46:2773–2778.

Mathieson WB, Taylor SW, Marshall M, Neumann PE (2000) Strain and sex differences in the morphology of the medial preoptic nucleus of mice. J Comp Neurol 428:254–265.

Matsuoka M, Kaba H, Mori Y, Ichikawa M (1997) Synaptic plasticity in olfactory memory formation in female mice. Neuro Report 8:2501–2504.

Matthews K, Dalley JW, Matthews C, Tsai TH, Robbins TW (2001) Periodic maternal separation of neonatal rats produces region- and gender-specific effects on biogenic amine content in postmortem adult brain. Synapse 40:1–10.

Matthews K, Hall FS, Wilkinson LS, Robbins TW (1996a) Retarded acquisition and reduced expression of conditioned locomotor activity in adult rats following repeated early maternal separation: effects of prefeeding, d-amphetamine, dopamine antagonists and clonidine. Psychopharmacology 126:75–84.

Matthews K, Wilkinson LS, Robbins TW (1996b) Repeated maternal separation of pre-weanling rats attenuates behavioral responses to primary and conditioned incentives in adulthood. Physiol Behav 59:99–107.

Matthews Felton T, Corodimas KP, Rosenblatt JS, Morrell JI (1995) Lateral habenula neurons are necessary for the hormonal onset of maternal behavior and for the display of postpartum estrus in naturally parturient female rats. Behav Neurosci 109:1172–1188.

Matthews Felton T, Linton LN, Rosenblatt JS, Morrell JI (1998a) Estrogen implants in the lateral habenular nucleus do not stimulate the onset of maternal behavior in female rats. Horm Behav 35:71–80.

Matthews Felton T, Linton LN, Rosenblatt JS, Morrell JI (1998b) First order and second order maternal behavior related afferents of the lateral habenula. Neuro Report 10: 883–887.

Mattson BJ, Williams S, Rosenblatt JS, Morrell JI (2001) Comparison of two positive reinforcing stimuli: pups and cocaine throughout the postpartum period. Behav Neurosci 115:683–694.

Mayer AD, Ahdieh HB, Rosenblatt JS (1990a) Effects of prolonged estrogen-progesterone treatment and hypophysectomy on the stimulation of short-latency maternal behavior and aggression in female rats. Horm Behav 24:152–173.

Mayer AD, Carter L, Jorge WA, Mota MJ, Tannu S, Rosenblatt JS (1987a) Mammary stimulation and maternal aggression in rodents: thelectomy fails to reduce pre- or postpartum aggression in rats. Horm Behav 21:501–510.

Mayer AD, Faris PL, Komisaruk BR, Rosenblatt JS (1985) Opiate antagonism reduces placentophagia and pup cleaning by parturient rats. Pharmacol Biochem Behav 22: 1035–1044.

Mayer AD, Monroy MA, Rosenblatt JS (1990b) Prolonged estrogen-progesterone treatment of nonpregnant ovariectomized rats: factors stimulating home-cage and maternal aggression and short-latency maternal behavior. Horm Behav 24:342–364.

Mayer AD, Reisbick S, Siegel HI, Rosenblatt JS (1987b) Maternal aggression in rats: changes over pregnancy and lactation in a Sprague-Dawley strain. Aggress Behav 13: 29–43.

Mayer AD, Rosenblatt JS (1975) Olfactory basis for the delayed onset of maternal behavior in virgin female rats: experiential effects. J Comp Physiol Psychol 89:701–710.

Mayer AD, Rosenblatt JS (1979) Ontogeny of maternal behavior in the laboratory rat: early origins in 18-to 27-day-old young. Dev Psychobiol 12:407–424.

Mayer AD, Rosenblatt JS (1984) Prepartum changes in maternal responsiveness and nest defense in *Rattus norvegicus*. J Comp Psychol 98:177–188.

Mayer AD, Rosenblatt JS (1987) Hormonal factors influence the onset of maternal aggression in laboratory rats. Horm Behav 21:253–267.

Mayer AD, Rosenblatt JS (1993) Contributions of olfaction to maternal aggression in laboratory rats (*Rattus norvegicus*): effects of peripheral deafferentation of the primary olfactory system. J Comp Psychol 107:12–24.

McCarthy MM (1990) Oxytocin inhibits infanticide in female house mice (*Mus domesticus*). Horm Behav 24:365–375.

McCarthy MM, Curran GF, Siegel HI (1994a) Evidence for the involvement of prolactin in the maternal behavior of the hamster. Physiol Behav 55:181–184.

McCarthy MM, Kleopoulos SP, Mobbs CV, Pfaff DW (1994b) Infusion of antisense oligodeoxynucleotides to the oxytocin receptor in the ventromedial hypothalamus reduces estrogen-induced sexual receptivity and oxytocin receptor binding in the female rat. Neuroendocrinology 59:432–440.

McCarthy MM, McDonald CH, Brooks PJ, Goldman D (1996) An anxiolytic action of oxytocin is enhanced by estrogen in the mouse. Physiol Behav 60:1209–1215.

McCormick JA, Lyons V, Jacobson MD, Noble J, Diorio J, Nyirenda M, Weaver S, Ester W, Yau JL, Meaney MJ, Seckl JR, Chapman KE. (2000) 5-heterogeneity of glucocorticoid receptor messenger RNA is tissue specific: differential regulation of variant transcripts by early-life events. Mol Endocrinol 14:506–517.

McEwen BS, Alves SE (1999) Estrogen action in the central nervous system. Endocrinol Rev 20:279–307.

McGuire B (1988) Effects of cross-fostering on parental behavior of meadow voles (*Microtus pennsylvanicus*). J Mammal 69:332–341.

McGuire B, Getz LL, Hoffmann JE, Pizzuto T, Frase B (1993) Natal dispersal and philopatry in prairie voles (*Microtus ochrogaster*) in relation to population density, season, and natal social environment. Behav Ecol Sociobiol 32:293–302.

McGuire B, Novak M (1984) A comparison of maternal behaviour in the meadow vole (*Microtus pennsylvanicus*), prairie vole (*M. ochrogaster*) and pine vole (*M. pinetorum*). Anim Behav 32:1132–1141.

Meaney MJ, Brake W, Gratton A (2002) Environmental regulation of the development of mesolimbic dopamine systems: a neurobiological mechanism for vulnerability to drug abuse. Psychoneuroendocrinology 27:127–138.

Meaney MJ, Champagne FA (2000) Latency to maternal behavior in high and low licking and grooming mothers/offspring. Soc Neurosci Abstr 26:2035.

Meaney MJ, Diorio J, Francis D, Widdowson J, LaPlante P, Caldji C, Sharma S, Seckl J, Plotsky PM (1996) Early environmental regulation of forebrain glucocorticoid receptor gene expression: implications for adrenocortical responses to stress. Dev Neurosci 18:49–72.

Meisel RL, Camp DM, Robinson TE (1993) A microdialysis study of ventral striatal dopamine release during sexual behavior in female Syrian hamsters. Behav Brain Res 55:151–157.

Meisel RL, Sachs BD (1994) The physiology of male sexual behavior. In: The physiology of reproduction, Vol 2, second edition (Knobil E, Neill, JD, eds), pp 3–105. New York: Raven Press.

Melia KR, Sananes CB, Davis M (1992) Lesions of the central nucleus of the amygdala block the excitatory effects of septal ablation on the acoustic startle reflex. Physiol Behav 51:175–180.

Mennella JA, Moltz H (1989) Pheromonal emission by pregnant rats protects against infanticide by nulliparous conspecifics. Physiol Behav 46:591–595.

Mens WBJ, Witter A, Greidanus TBVM (1983) Penetration of neurohypophyseal hormones from plasma into cerebrospinal fluid (CSF): half-times of disappearance of these neuropeptides from CSF. Brain Res 262:143–149.

Merendino JJ, Spiegel AM, Crawford JD, Brownstein MJ, Lolait SJ (1993) A mutation in the V2 vasopressin receptor gene in a kindred with X-linked nephrogenic diabetes insipidus. N Engl J Med 328:1538–1541.

Miceli MO, Fleming AS, Malsbury CW (1983) Disruption of maternal behaviour in virgin and postparturient rats following sagittal plane knife cuts in the preoptic area-hypothalamus. Behav Brain Res 9:337–360.

Miceli MO, Malbury CW (1982) Sagittal knife cuts in the near and far lateral preoptic area-hypothalamus disrupt maternal behaviour in female hamsters. Physiol Behav 28: 857–867.

Miller L, Kramer R, Warner V, Wickramaratne P, Weissman M (1997) Intergenerational transmission of parental bonding among women. J Am Acad Child Adolesc Psychiatry 36:1134–1139.

Misiti A, Turillazzi PG, Zapponi GA, Loizzo A (1991) Heroin induces changes in mother-infant monkey communication and subsequent disruption of their dyadic interaction. Pharmacol Res 24:93–104.

Mitani JC, Watts D (1997) The evolution of non-maternal caretaking among anthropoid primates: do helpers help? Behav Ecol Sociobiol 40:213–220.

Mitchell AJ (1998) The role of corticotropin releasing factor in depressive illness: a critical review. Neurosci Biobehav Rev 22:635–651.

Modney BK, Hatton GI (1994) Maternal behaviors: evidence that they feed back to alter brain morphology and function. Acta Paediatr Suppl 397:29–32.

Modney BK, Yang QZ, Hatton GI (1990) Activation of excitatory amino acid inputs to supraoptic neurons. II. Increased dye-coupling in maternally behaving virgin rats. Brain Res 513:270–273.

Moffat SD, Suh EJ, Fleming AS (1993) Noradrenergic involvement in the consolidation of maternal experience in postpartum rats. Physiol Behav 53:805–811.

Mogenson GJ (1987) Limbic-motor integration. Prog Psychobiol Physiol Psychol 12: 117–170.

Molewijk HE, van der Poel AM, Olivier B (1995) The ambivalent behaviour "stretched approach posture" in the rat as a paradigm to characterize anxiolytic drugs. Psychopharmacology 121:81–89.

Moltz H, Geller D, Levin R (1967) Maternal behavior in the totally mammectomized rat. J Comp Physiol Psychol 64:225–229.

Moltz H, Levin R, Leon M (1969) Differential effects of progesterone on the maternal behavior of primiparous and multiparous rats. J Comp Physiol Psychol 67:36–40.

Moltz H, Lubin M, Leon M, Numan M (1970) Hormonal induction of maternal behavior in the ovariectomized nulliparous rat. Physiol Behav 5:1373–1377.

Moltz H, Robbins D (1965) Maternal behavior of primiparous and multiparous rats. J Comp Physiol Psychol 60:417–421.

Moltz H, Robbins D, Parks M (1966) Caesarean delivery and maternal behavior of primiparous and multiparous rats. J Comp Physiol Psychol 61:445–460.

Moltz H, Weiner E (1966) Effects of ovariectomy on maternal behavior of primiparous and multiparous rats. J Comp Physiol Psychol 62:382–387.

Monassi CR, Leite-Panissi CRA, Menescal-de-Oliveira L (1999) Ventrolateral periaqueductal gray matter and the control of tonic immobility. Brain Res Bull 50:201–208.

Montagnese CM, Poulain DA, Vincent J, Theodosis DT (1987) Structural plasticity in the rat supraoptic nucleus during gestation, post-partum lactation and suckling-induced pseudogestation and lactation. J Endocrinol 115:97–105.

Moore F, Lowry C (1998) Comparative neuroanatomy of vasotocin and vasopressin in amphibians and other vertebrates. Comp Biochem Physiol, Part C Pharmacol Toxicol Endocrinol 119:251–260.

Moore F, Richardson C, Lowry C (2000) Sexual dimorphism in numbers of vasotocin-immunoreactive neurons in brain areas associated with reproductive behaviors in the roughskin newt. Gen Comp Endocrinol 117:281–298.

Moos F, Ingram C, Wakerley J, Guerne Y, Freund-Mercier M, Richard P (1991) Oxytocin in the bed nucleus of the stria terminalis and lateral septum facilitates bursting of hypothalamic oxytocin neurons in suckled rats. J Neuroendocrinol 3:163–171.

Moretto D, Paclik L, Fleming A (1986) The effects of early rearing environments on maternal behavior in adult female rats. Dev Psychobiol 19:581–591.

Morgan HD, Fleming AS, Stern JM (1992) Somatosensory control of the onset and retention of maternal responsiveness in primiparous Sprague-Dawley rats. Physiol Behav 51:541–555.

Morgan HD, Wachtus JA, Milgram NW, Fleming AS (1999) The long lasting effects of electrical stimulation of the medial preoptic area and medial amygdala on maternal behavior in female rats. Behav Brain Res 99:61–73.

Morgan JI, Curran T (1991) Stimulus-transcription coupling in the nervous system: involvement of the inducible proto-oncogenes *fos* and *jun*. Annu Rev Neurosci 14:421–451.

Mori K, Nagao H, Yoshihara Y (1999) The olfactory bulb: coding and processing of odor molecule information. Science 286:711–715.

Morin SM, Ling N, Liu X, Kahl SD, Gehlert DR (1999) Differential distribution of urocortin- and corticotropin-releasing factor-like immunoreactivities in the rat brain. Neuroscience 92:281–291.

Morishige WK, Pepe GJ, Rothchild I (1973) Serum luteinizing hormone, prolactin and progesterone levels during pregnancy in the rat. Endocrinology 92:1527–1530.

Morris M, Barnard RR Jr, Sain LE (1984) Osmotic mechanisms regulating cerebrospinal fluid vasopressin and oxytocin in the conscious rat. Neuroendocrinology 39:377–383.

Mos J, Olivier B (1986) RO 15-1788 does not influence postpartum aggression in lactating female rats. Psychopharmacology 90:278–280.

Murphy AZ, Rizvi TA, Ennis M, Shipley MT (1999) The organization of preoptic-medullary circuits in the male rat: evidence for interconnectivity of neural structures involved in reproductive behavior, antinociception and cardiovascular regulation. Neuroscience 91:1103–1116.

Murphy DD, Cole NB, Greenberger V, Segal M (1998) Estradiol increases dendritic spine density by reducing GABA neurotransmission in hippocampal neurons. J Neurosci 18:2550–2559.

Murphy DD, Segal M (1996) Regulation of dendritic spine density in cultured rat hippocampal neurons by steroid hormones. J Neurosci 16:4059–4068.

Murphy DD, Segal M (1997) Morphological plasticity of dendritic spines in central neurons is mediated by activation of cAMP response element binding protein. Proc Natl Acad Sci USA 94:1482–1487.

Murphy MR, Seckl JR, Burton S, Checkley SA, Lightman SL (1987) Changes in oxytocin and vasopressin secretion during sexual activity in men. J Clin Endocrinol Metab 65:738–741.

Murtra P, Sheasby AM, Hunt SP, De Felipe C (2000) Rewarding effects of opiates are absent in mice lacking the receptor for substance P. Nature 405:180–183.

Myers MM, Denenberg VH, Thoman E, Holloway WR, Bowerman DR (1975) The effects of litter size on plasma corticosterone and prolactin response to ether stress in the lactating rat. Neuroendocrinology 19:54–58.

Myers RE, Swett C, Miller M (1973) Loss of social group affinity following prefrontal lesions in free-ranging macaques. Brain Res 64:257–269.

Nagano M, Kelly PA (1994) Tissue distribution and regulation of rat prolactin receptor gene expression. Quantitative analysis by polymerase chain reaction. J Biol Chem 269:13337–13345.

Nagata A, Ito M, Iwata N, Kuno J, Takano H, Minowa O, Chihara K, Matsui T, Noda T (1996) G protein-coupled cholecystokinin-B/gastrin receptors are responsible for physiological cell growth of the stomach mucosa in vivo. Proc Natl Acad Sci 93:11825–11830.

Nelson E, Panksepp J (1996) Oxytocin and infant-mother bonding in rats. Behav Neurosci 110:583–592.

Nemeroff CB (1999) The preeminent role of early untoward experience on vulnerability to major psychiatric disorders: the nature-nurture controversy revisited and soon to be resolved. Mol Psychiatry 4:106–108.

Nemeroff CB, Widerlov E, Bissette G, Walleus H, Karlsson I, Eklund K, Kilts CD, Loosen PT, Vale W (1984) Elevated concentrations of CSF corticotropin-releasing factor-like immunoreactivity in depressed patients. Science 226:1342–1343.

Nestler EJ, Kelz MB, Chen J (1999) ΔFosB: a molecular mediator of long-term neural and behavioral plasticity. Brain Res 835:10–17.

Neumann ID (2001) Alterations in behavioral and neuroendocrine stress coping strategies in pregnant, parturient and lactating rats. In: Progress in brain research, Vol 133. The maternal brain (Russell JA, Douglas AJ, Windle RJ, Ingram CD, eds), pp 143–152. Amsterdam: Elsevier.

Neumann ID, Johnstone HA, Hatzinger M, Liebsch G, Shipston M, Russell JA, Landgraf R, Douglas AF (1998a) Attenuated neuroendocrine responses to emotional and physical stressors in pregnant rats involve adenohypophyseal changes. J Physiol 508: 289–300.

Neumann ID, Russell JA, Landgraf R (1993) Oxytocin and vasopressin release within the supraoptic and paraventricular nuclei of pregnant, parturient, and lactating rats: a microdialysis study. Neuroscience 53:65–75.

Neumann ID, Torner L, Wigger A (2000) Brain oxytocin: differential inhibition of neuroendocrine stress responses and anxiety-related behaviour in virgin, pregnant and lactating rats. Neuroscience 95:567–575.

Neumann ID, Toschin N, Ohl F, Torner L, Kromer SA (2001) Maternal defence as an emotional stressor in female rats: correlation of neuroendocrine and behavioural parameters and involvement of brain oxytocin. Eur J Neurosci 13:1016–1024.

Neumann ID, Wigger A, Liebsch G, Holsboer F, Landgraf R (1998b) Increased basal activity of the hypothalamo-pituitary-adrenal axis during pregnancy in rats bred for high anxiety-related behaviour. Psychoneuroendocrinology 23:449–463.

Newman SW (1999) The medial extended amygdala in male reproductive behavior. Ann NY Acad Sci 877:242–257.

Newman SW (2002) Pheromonal signals access the medial extended amygdala: one node in a proposed social behavior network. In: Hormones, brain and behavior, Vol 2 (Pfaff DW, Arnold AP, Etgen AM, Fahrbach SE, Rubin RT, eds), pp 17–32. San Diego: Academic Press.

Nicola SM, Surmeier DJ, Malenka RC (2000) Dopaminergic modulation of neuronal excitability in the striatum and nucleus accumbens. Annu Rev Neurosci 23:185–215.

Nicoll CS, Tarpey JF, Mayer GL, Russell SM (1986) Similarities and differences among prolactins and growth hormones and their receptors. Am Zool 26:965–983.

Nishimori K, Young LJ, Guo Q, Wang Z, Insel TR, Matzuk MM (1996) Oxytocin is required for nursing but is not essential for parturition or reproductive behavior. Proc Natl Acad Sci (USA) 93:777–783.

Nissen E, Gustavsson P, Widstrom AM, Uvnas-Moberg K (1998) Oxytocin, prolactin, milk production and their relationship with personality traits in women after vaginal delivery or caesarean section. J Psychosom Obstet Gynaecol 19:49–58.

Noirot E (1972) The onset of maternal behavior in rats, hamsters, and mice: a selective review. In: Advances in the study of behavior, Vol 4 (Lehrman DS, Hinde RA, Shaw E, eds), pp 107–145. New York: Academic Press.

Noirot E, Goyens J, Buhot M (1975) Aggressive behavior of pregnant mice toward males. Horm Behav 6:9–17.

Numan M (1974) Medial preoptic area and maternal behavior in the female rat. J Comp Physiol Psychol 87:746–759.

Numan M (1978) Progesterone inhibition of maternal behavior in the rat. Horm Behav 11:209–231.

Numan M (1985) Brain mechanisms and parental behavior. In: Handbook of behavioral neurobiology, Vol 7 (Adler N, Pfaff D, Goy RW, eds), pp 537–605. New York: Plenum Press.

Numan M (1988) Maternal behavior. In: The physiology of reproduction (Knobil E, Neill JD, eds), pp 1569–1645. New York: Raven Press.

Numan M (1990) Long-term effects of preoptic area knife cuts on the maternal behavior of rats. Behav Neural Biol 53:284–290.

Numan M (1994) Maternal behavior. In: The physiology of reproduction, Vol 2 (Knobil E, Neill JD, eds), pp 221–302. New York: Raven Press.

Numan M (1999) Parental behavior, mammals. In: Encyclopedia of reproduction, Vol 3 (Knobil E, Neill JD, eds), pp 684–694. San Diego: Academic Press.

Numan M, Callahan EC (1980) The connections of the medial preoptic region and maternal behavior in the rat. Physiol Behav 25:653–665.

Numan M, Corodimas KP (1985) The effects of paraventricular hypothalamic lesions on maternal behavior in rats. Physiol Behav 35:417–425.

Numan M, Corodimas KP, Numan MJ, Factor EM, Piers WD (1988) Axon-sparing lesions of the preoptic region and substantia innominata disrupt maternal behavior in rats. Behav Neurosci 102:381–396.

Numan M, McSparren J, Numan MJ (1990) Dorsolateral connections of the medial preoptic area and maternal behavior in rats. Behav Neurosci 104:964–979.

Numan M, Morrell JI, Pfaff DW (1985) Anatomical identification of neurons in selected brain regions associated with maternal behavior deficits induced by knife cuts of the lateral hypothalamus in rats. J Comp Neurol 237:552–564.

Numan M, Numan MJ (1991) Preoptic-brainstem connections and maternal behavior in rats. Behav Neurosci 105:1013–1029.

Numan M, Numan MJ (1994) Expression of Fos-like immunoreactivity in the preoptic area of maternally behaving virgin and postpartum rats. Behav Neurosci 108:379–394.

Numan M, Numan MJ (1995) Importance of pup-related sensory inputs and maternal performance for the expression of Fos-like immunoreactivity in the preoptic area and ventral bed nucleus of the stria terminalis of postpartum rats. Behav Neurosci 109:135–149.

Numan M, Numan MJ (1996) A lesion and neuroanatomical tract-tracing analysis of the role of the bed nucleus of the stria terminalis in retrieval behavior and other aspects of maternal responsiveness in rats. Dev Psychobiol 29:23–51.

Numan M, Numan MJ (1997) Projection sites of medial preoptic area and ventral bed nucleus of the stria terminalis neurons that express Fos during maternal behavior in female rats. J Neuroendocrinol 9:369–384.

Numan M, Numan MJ, English JB (1993) Excitotoxic amino acid injections into the medial amygdala facilitate maternal behavior in virgin female rats. Horm Behav 27:56–81.

Numan M, Roach JK, del Cerro MCR, Guillamon A, Segovia S, Sheehan TP, Numan MJ (1999) Expression of intracellular progesterone receptors in rat brain during different reproductive states, and involvement in maternal behavior. Brain Res 830:358–371.

Numan M, Rosenblatt JS, Komisaruk BR (1977) Medial preoptic area and onset of maternal behavior in the rat. J Comp Physiol Psychol 91:146–164.

Numan M, Sheehan TP (1997) Neuroanatomical circuitry for mammalian maternal behavior. Ann NY Acad Sci 807:101–125.

Numan M, Smith HG (1984) Maternal behavior in rats: evidence for the involvement of preoptic projections to the ventral tegmental area. Behav Neurosci 98:712–727.

Nunes S, Fite JE, Patera KJ, French JA (2001) Interactions among paternal behavior, steroid hormones, and parental experience in male marmosets (*Callithrix kuhlii*). Horm Behav 39:70–82.

O'Day DH, Payne L, Drmic I, Fleming AS (2001) Loss of calcineurin from the medial preoptic area of primiparous rats. Biochem Biophys Res Commun 281:1037–1040.

Ogawa S, Chan J, Chester AE, Gustafsson J-A, Korach KS, Pfaff DW (1999) Survival of reproductive behaviors in estrogen receptor-beta gene-deficient male and female mice. Proc Natl Acad Sci 96:12887–12892.

Ogawa S, Eng V, Taylor JA, Lubahn DB, Korach KS, Pfaff DW (1998) Roles of estrogen receptor-α gene expression in reproduction-related behaviors in female mice. Endocrinology 139:5070–5081.

Ogawa S, Taylor JA, Lubahn DB, Korach KS, Pfaff DW (1996) Reversal of sex roles in genetic female mice by disruption of estrogen receptor gene. Neuroendocrinology 64:467–470.

O'Hara MW, Schlechte JA, Lewis DA, Varner MW (1991) Controlled prospective study of postpartum mood disorders: psychological, environmental, and hormonal variables. J Abnorm Psychol 100:63–73.

Okano T, Nomura J (1992) Endocrine study of the maternity blues. Prog Neuropsychopharmacol Biol Psychiatry 16:921–932.

Okere CO, Kaba H, Higuchi T (1996) Formation of an olfactory recognition memory in mice: reassessment of the role of nitric oxide. Neuroscience 71:349–354.

Oliveras D, Novak M (1986) A comparison of paternal behavior in the meadow vole, the pine vole, and the prairie vole. Anim Behav 34:519–526.

Olivier B, Mos J, van Oorschot R (1985) Maternal aggression in rats: effects of chlordiazepoxide and fluprazine. Psychopharmacology 86:68–76.

Olivier B, Mos J, van Oorschot R (1986) Maternal aggression in rats: lack of interaction between chlordiazepoxide and fluprazine. Psychopharmacology 88:40–43.

Olton DS, Walker JA, Wolf WA (1982) A disconnection analysis of hippocampal function. Brain Res 233:241–253.

Ormandy CJ, Camus A, Barra J, Damotte D, Lucas BK, Buteau H, Edery M, Brousse N, Binart N, Kelly PA (1997) Null mutation of the prolactin receptor gene produces multiple reproductive defects in the mouse. Genes Devel 11:167–178.

Orpen BG, Fleming AS (1987) Experience with pups sustains maternal responding in postpartum rats. Physiol Behav 40:47–54.

Ostermeyer MC (1983) Maternal aggression. In: Parental behaviour of rodents (Elwood RW, ed), pp 151–179. Chichester: John Wiley & Sons.

Ostermeyer MC, Elwood RW (1983) Pup recognition in *Mus musculus*: parental discrimination between their own and alien young. Dev Psychobiol 16:75–82.

Owens MJ, Ritchie JC, Nemeroff CB (1992) 5α-pregnane-3α,21-diol-20-one (THOC) attenuates mild stress-induced increases in plasma corticosterone via a nonglucocorticoid mechanism: comparison with alprazolam. Brain Res 573:353–355.

Oxley G, Fleming AS (2000) The effects of medial preoptic area lesions on maternal behavior in the juvenile rat. Dev Psychobiol 37:253–265.

Packer C, Lewis S, Pusey A (1992) A comparative analysis of non-offspring nursing. Anim Behav 43:265–281.

Paech K, Webb P, Kuiper GGJM, Nilsson S, Gustafsson J, Kushner PJ, Scanlan TS (1997) Differential ligand activation of estrogen receptors ERα and ERβ at AP1 sites. Science 277:1508–1510.

Panksepp J (1981) Hypothalamic integration of behavior. In: Handbook of the hypothalamus, Vol 3, Part B, behavioral studies of the hypothalamus (Morgane PJ, Panksepp J, eds), pp 289–431. New York: Marcel Dekker.

Panksepp JB, Herman B, Conner R, Bishop P, Scott JP (1978) The biology of social attachments: opiates alleviate separation distress. Biol Psychiatry 13:607–613.

Pardon M, Gerardin P, Joubert C, Perez-Diaz F, Cohen-Salmon C (2000) Influence of prepartum chronic ultramild stress on maternal pup care behavior in mice. Biol Psychiatry 47:858–863.

Parducz A, Perez J, Garcia-Segura LM (1993) Estradiol induces plasticity of GABAergic synapses in the hypothalamus. Neuroscience 53:395–401.

Paredes R, Highland L, Karam P (1993) Socio-sexual behavior in male rats after lesions of the medial preoptic area: evidence for reduced sexual motivation. Brain Res 618: 271–276.

Paredes RG, Tzschentke T, Nakach N (1998) Lesions of the medial preoptic/anterior hypothalamus (MPOA/AH) modify partner preference in male rats. Brain Res 813:1–8.

Parker KJ, Kinney LF, Phillips KM, Lee TM (2001) Paternal behavior is associated with central neurohormone receptor binding patterns in meadow voles (*Microtus pennsylvanicus*). Behav Neurosci 115:1341–1348.

Parker KJ, Lee TM (2001a) Central vasopressin administration regulates the onset of facultative paternal behavior in *Microtus pennsylvanicus* (Meadow voles). Horm Behav 39:285–294.

Parker KJ, Lee TM (2001b) Social and environmental factors influence the suppression of pup-directed aggression and development of paternal behavior in captive *Microtus pennsylvanicus* (meadow voles). J Comp Psychol 115:331–336.

Parmigiani S, Ferrari PF, Palanza P (1998) An evolutionary approach to behavioral pharmacology: using drugs to understand proximate and ultimate mechanisms of different forms of aggression in mice. Neurosci Biobehav Rev 23:143–153.

Parmigiani S, Palanza P, Mainardi D, Brain PF (1994) Infanticide and protection of young in house mice (*Mus domesticus*): female and male strategies. In: Infanticide and parental care (Parmigiani S, vom Saal FS, eds), pp 341–363. Chur, Switzerland: Harwood Academic Publishers.

Parmigiani S, Palanza P, Rodgers J, Ferrari PF (1999) Selection, evolution and animal models in behavioral neuroscience. Neurosci Biobehav Rev 23:957–970.

Parr LA, Winslow JT, Davis M (2002) Rearing experience differentially affects somatic and cardiac startle responses in rhesus monkeys (*Macaca mulatta*). Behav Neurosci 116:378–386.

Patchev VK, Hassan AHS, Holsboer F, Almeida OFX (1996) The neurosteroid tetrahydroprogesterone attenuates the endocrine response to stress and exerts glucocorticoid-like effects on vasopressin gene transcription in the rat hypothalamus. Neuropsychopharmacology 15:533–540.

Patchev VK, Shoaib M, Holsboer F, Almeida OFX (1994) The neurosteroid tetrahydroprogesterone counteracts corticotropin-releasing hormone-induced anxiety and alters the release and gene expression of corticotropin-releasing hormone in the rat hypothalamus. Neuroscience 62:265–271.

Paut-Pagano L, Roky R, Valatx JL, Kitahama K, Jouvet M (1993) Anatomical distribution of prolactin-like immunoreactivity in the rat brain. Neuroendocrinology 58:682–695.

Paylor R, Johnson RS, Papaioannou V, Spiegelman BM, Wehner JM (1994) Behavioral assessment of c-fos mutant mice. Brain Res 651:275–282.

Pedersen CA (1997) Oxytocin control of maternal behavior. Ann NY Acad Sci 807: 126–145.

Pedersen CA, Ascher JA, Monroe YL, Prange AJ Jr (1982) Oxytocin induces maternal behaviour in virgin female rats. Science 216:648–649.

Pedersen CA, Caldwell JD, Fort SA, Prange AJ (1985) Oxytocin antiserum delays onset of ovarian steroid-induced maternal behaviour. Neuropeptides 6:175–182.

Pedersen CA, Caldwell JD, McGuire M, Evans D (1991) Corticotropin-releasing hormone inhibits maternal behavior and induces pup killing. Life Sci 48:1537–1546.

Pedersen CA. Caldwell JD, Walker C, Ayers G, Mason GA (1994) Oxytocin activates the postpartum onset of rat maternal behavior in the ventral tegmental area and medial preoptic areas. Behav Neurosci 108:1163–1171.

Pedersen CA, Johns JM, Musiol I, Perez-Delgado M, Ayers G, Faggin B, Caldwell JD (1995) Interfering with somatosensory stimulation from pups sensitizes experienced, postpartum rat mothers to oxytocin antagonist inhibition of maternal behavior. Behav Neurosci 109:980–990.

Pedersen CA, Prange AJ Jr (1979) Induction of maternal behavior in virgin rats after intracerebroventricular administration of oxytocin. Proc Natl Acad Sci USA 76:6661–6665.

Pedersen CA, Stern RA, Pate J, Senger MA, Bowes WA, Mason G (1993) Thyroid and adrenal measures during late pregnancy and the puerperium in women who have been major depressed or who become dysphoric postpartum. J Affect Disord 29:201–211.

Perrigo G, Belvin L, Quindry P, Kadir T, Becker J, van Look C, Niewoehner J, vom Saal FS (1993) Genetic mediation of infanticide and parental behavior in male and female domestic and wild stock house mice. Behav Genet 23:525–531.

Perrigo G, Belvin L, vom Saal FS (1992) Time and sex in the male mouse: temporal regulation of infanticide and parental behavior. Chronobiol Interntl 9:421–433.

Perrigo G, Bryant WC, vom Saal FS (1989) Fetal, hormonal, and experiential factors influencing the mating-induced regulation of infanticide in male house mice. Physiol Behav 46:121–128.

Perusse D, Neale MC, Heath AC, Eaves LJ (1994) Human parental behavior: evidence for genetic influence and potential implication for gene-culture transmission. Behav Genet 24:327–335.

Peterson G, Mason GA, Barakat AS, Pedersen CA (1991) Oxytocin selectively increases holding and licking of neonates in preweanling but not postweanling juvenile rats. Behav Neurosci 105:470–477.

Petrulis A, Johnston RE (1999) Lesions centered on the medial amygdala impair scent-marking and sex-odor recognition but spare discrimination of individual odors in female golden hamsters. Behav Neurosci 113:345–357.

Pettibone DJ, Clineschmidt BV, Kishel MT, Lis EV, Reiss DR, Woyden CJ, Evans BE, Freidinger RM, Veber DF, Cook MJ, Haluska GJ, Novy MJ, Lowensohn RI (1993) Identification of an orally active, nonpeptidyl oxytocin antagonist. J PET 264:308–314.

Pezzone MA, Lee W, Hoffman GE, Rabin BS (1992) Induction of c-Fos immunoreactivity in the rat forebrain by conditioned and unconditioned aversive stimuli. Brain Res 597:41–50.

Pfaff D, Keiner M (1973) Atlas of estradiol-concentrating cells in the central nervous system of the female rat. J Comp Neurol 151:121–158.

Pfaus JG, Damsma G, Wenkstern D, Fibiger HC (1995) Sexual activity increases dopamine transmission in the nucleus accumbens and striatum of female rats. Brain Res 693:21–30.

Pfaus JG, Kleopoulos SP, Mobbs CV, Gibbs RB, Pfaff DW (1993) Sexual stimulation activates c-fos within estrogen-concentrating regions of the female rat forebrain. Brain Res 624:253–267.

Pfaus JG, Phillips AG (1991) Role of dopamine in anticipatory and consummatory aspects of sexual behavior in the male rat. Behav Neurosci 105:727–743.

Pi XJ, Grattan DR (1998) Distribution of prolactin receptor immunoreactivity in the brain of estrogen-treated, ovariectomized rats. J Comp Neurol 394:462–474.

Pi XJ, Grattan DR (1999) Increased prolactin receptor immunoreactivity in the hypothalamus of lactating rats. J Neuroendocrinol 11:693–705.

Picazo O, Fernandez-Guasti A (1993) Changes in experimental anxiety during pregnancy and lactation. Physiol Behav 54:295–299.

Picazo O, Fernandez-Guasti A (1995) Anti-anxiety effects of progesterone and some of its reduced metabolites: an evaluation using the burying behavior test. Brain Res 680: 135–141.

Pihoker C, Robertson MC, Freemark M (1993) Rat placental lactogen-I binds to the choroid plexus and hypothalamus of the pregnant rat. J Endocrinol 139:235–242.

Pissonnier D, Thierry JC, Fabre-Nys C, Poindron P, Keverne EB (1985) The importance of olfactory bulb noradrenalin for maternal recognition in sheep. Physiol Behav 35: 361–363.

Pitkanen A (2000) Connectivity of the rat amygdaloid complex. In: The amygdala (Aggleton JP, ed), pp 31–115. New York: Oxford University Press.

Pitkanen A, Savander V, LeDoux JE (1997) Organization of intra-amygdaloid circuitries in the rat: an emerging framework for understanding functions of the amygdala. Trends Neurosci 20:517–523.

Pitkow L, Sharer C, Ren X, Insel T, Terwilliger E, Young L (2001) Facilitation of affiliation and pair-bond formation by vasopressin receptor gene transfer into the ventral forebrain of a monogamous vole. J Neurosci 21:7392–7396.

Plotsky PM, Meaney MJ (1993) Early, postnatal experience alters hypothalamic corticotropin-releasing factor (CRF) mRNA, median eminence CRF content and stress-induced release in adult rats. Mol Brain Res 18:195–200.

Poeggel G, Lange E, Hase C, Metzger M, Gulyaeva N, Braun K (1999) Maternal separation and early social deprivation in Octodon degus: quantitative changes of nicotinamide adenine dinucleotide phosphate-diaphorase-reactive neurons in the prefrontal cortex and nucleus accumbens. Neuroscience 94:497–504.

Poindron P (1976) Mother-young relationships in intact or anosmic ewes at the time of suckling. Biol Behav 2:161–177.

Poindron P, Le Neindre P (1980) Endocrine and sensory regulation of maternal behavior in the ewe. In: Advances in the study of behavior, Vol 2 (Rosenblatt JS, Hinde RA, Beer C, Bunsel MC, eds), pp 75–119. New York: Academic Press.

Poindron P, Orgeur P, Le Neindre P, Kann G, Raksanyi I (1980) Influences of blood concentration of prolactin on the length of the sensitive period for establishing maternal behavior in sheep at parturition. Horm Behav 14:173–177.

Polston EK, Erskine MS (2001) Excitotoxic lesions of the medial amygdala differentially disrupt prolactin secretory responses in cycling and mated female rats. J Neuroendocrinol 13:13–21.

Pournajafi Nazarloo H, Takao T, Nanamiya W, Asaba K, De Souza EB, Hashimoto K (2001) Effect of non-peptide corticotropin-releasing factor receptor type 1 antagonist on adrenocorticotropic hormone release and interleukin-1 receptors followed by stress. Brain Res 902:119–126.

Powell RA, Fried JJ (1992) Helping by juvenile pine voles (*Microtus pinetorum*), growth and survival of younger siblings, and the evolution of pine vole sociality. Behav Ecol 3:325–333.

Powers JB, Valenstein ES (1972) Sexual receptivity: facilitation by medial preoptic lesions in female rats. Science 175:1003–1005.

Price JL, Slotnick BM, Revial MF (1991) Olfactory projections to the hypothalamus. J Comp Neurol 306:447–461.

Pryce CR (1992) A comparative systems model of the regulation of maternal motivation in mammals. Anim Behav 43:417–441.

Pryce CR (1993) The regulation of maternal behaviour in marmosets and tamarins. Behav Proc 30:201–224.

Pryce CR (1995) Determinants of motherhood in human and nonhuman primates. In: Motherhood in human and nonhuman primates (Pryce CR, Martin RD, Skuse D, eds), pp 1–15. Basel: Karger.

Pryce CR (1996) Socialization, hormones, and the regulation of maternal behavior in nonhuman simian primates. In: Advances in the study of behavior, Vol 25. Parental care: evolution, mechanisms, and adaptive significance (Rosenblatt JS, Snowdon CT, eds), pp 423–473. San Diego: Academic Press.

Pryce CR, Abbott DH, Hodges JK, Martin RD (1988) Maternal behavior is related to prepartum urinary estradiol levels in red-bellied tamarin monkeys. Physiol Behav 44: 717–726.

Pryce CR, Bettschen D, Feldon J (2001) Comparison of the effects of early handling and early deprivation on maternal care in the rat. Dev Psychobiol 38:239–251.

Pryce CR, Dobeli M, Martin RD (1993) Effects of sex steroids on maternal motivation in the common marmoset (*Callithrix jacchus*): development and application of an operant system with maternal reinforcement. J Comp Psychol 107:99–115.

Pryce CR, Mutschler T, Dobeli M, Nievergelt C, Martin RD (1995) Prepartum sex steroid hormones and infant-directed behavior in primiparous marmoset mothers (*Callithrix jacchus*). In: Motherhood in human and nonhuman primates (Pryce CR, Martin RD, Skuse D, eds), pp 78–86. Basel: Karger.

Qureshi GA, Hansen S, Sodersten P (1987) Offspring control cerebrospinal fluid GABA concentrations in lactating rats. Neurosci Lett 75:85–88.

Radulovic Ruhmann A, Liepold T, Spiess J (1999) Modulation of learning and anxiety by corticotropin-releasing factor (CRF) and stress: differential roles of CRF receptors 1 and 2. J Neurosci 19:5016–5025.

Ramirez VD, Zheng J (1996) Membrane sex-steroid receptors in the brain. Front Neuroendocrinol 17:402–439.

Reburn CJ, Wynne-Edwards KE (1999) Hormonal changes in males of a naturally biparental and a uniparental mammal. Horm Behav 35:163–176.

Rees SL, Fleming AS (2001) Early maternal separation and juvenile experience with pups affects maternal behavior in adult rats. Paper presented at the 31st Annual Meeting of the Society for Neuroscience, San Diego, CA.

Reisbick S, Rosenblatt JS, Mayer AD (1975) Decline of maternal behavior in the virgin and lactating rat. J Comp Physiol Psychol 89:722–732.

Ressler KJ, Nemeroff CB (1999) Role of norepinephrine in the pathophysiology and treatment of mood disorders. Biol Psychiatry 46:1219–1233.

Rheingold HL (1963) Maternal behavior in mammals. New York: Wiley.

Richards MPM (1966) Maternal behaviour in the golden hamster: responsiveness to young in virgin, pregnant, and lactating females. Anim Behav 14:310–313.

Riedman, ML (1982) The evolution of alloparental care and adoption in mammals and birds. Q Rev Biol 57:405–435.

Risold PY, Canteras NS, Swanson LW (1994) Organization of projections from the anterior hypothalamic nucleus: a *Phaseolus vulgaris*-leucoagglutinin study in the rat. J Comp Neurol 348:1–40.

Risold PY Swanson LW (1997) Connections of the lateral septal complex. Brain Res Rev 24:115–195.

Risold PY, Thompson RH, Swanson LW (1997) The structural organization of connections between hypothalamus and cerebral cortex. Brain Res Rev 24:197–254.

Rittig S, Robertson GL, Siggaard C, Kovács L, Gregersen N, Nyborg J, Pedersen EB (1996) Identification of 13 new mutations in the vasopressin-neurophysin II gene in 17 kindreds with familial autosomal dominant neurohypophyseal diabetes insipidus. Am J Hum Genet 58:107–117.

Rizvi TA, Ennis M, Shipley MT (1992) Reciprocal connections between the medial preoptic area and the midbrain periaqueductal gray in the rat: a WGA-HRP and PHA-L study. J Comp Neurol 315:1–15.

Rizvi TA, Murphy AZ, Ennis M, Bebehani MM, Shipley MT (1996) Medial preoptic area afferents to periaqueductal gray medullo-output neurons: a combined Fos and tract tracing study. J Comp Neurol 16:333–344.

Roberts RL, Jenkins KT, Lawler T, Wegner FH, Newman JD (2001a) Bromocriptine administration lowers serum prolactin and disrupts parental responsiveness in common marmosets (*Callithrix jacchus*). Horm Behav 39:106–112.

Roberts RL, Jenkins KT, Lawler T, Wegner FH, Norcross JL, Bernhards DE, Newman JD (2001b) Prolactin levels are elevated after infant carrying in parentally inexperienced common marmosets. Physiol Behav 72:713–720.

Roberts RL, Miller AK, Taymans SE, Carter CS (1998a) Role of social and endocrine factors in alloparental behavior of prairie voles (*Microtus ochrogaster*). Can J Zool 76:1862–1868.

Roberts RL, Williams JR, Wang AK, Carter CS (1998b) Cooperative breeding and monogamy in prairie voles: influence of the sire and geographical variation. Anim Behav 55:1131–1140.

Roberts RL, Zullo A, Gustafson EA, Carter CS (1996) Perinatal steroid treatments alter alloparental and affiliative behavior in prairie voles. Horm Behav 30:576–582.

Robertson LM, Kerppola TK, Vendrell M, Luk D, Smeyne RJ, Bocchiaro C, Morgan JI, Curran T (1995) Regulation of c-fos expression in transgenic mice requires multiple independent transcription control elements. Neuron 14:241–252.

Robinson TE, Kolb B (1997) Persistent structural modifications in nucleus accumbens and prefrontal cortex neurons produced by previous experience and amphetamine. J Neurosci 17:8491–8497.

Rodgers RJ, Haller J, Holmes A, Halasz J, Walton TJ, Brain PF (1999) Corticosterone response to plus-maze: high correlation with risk assessment in rats and mice. Physiol Behav 68:47–53.

Rogeness GA (1991) Psychosocial factors and amine systems. Psychiatry Res 37:215–217.

Rogeness GA, McClure EB (1996) Development and neurotransmitter-environment interactions. Dev Psychopathol 8:183–199.

Roky R, Paut-Pagano L, Goffin V, Kitahama K, Valatx JL, Kelly PA, Jouvet M (1996) Distribution of prolactin receptors in the rat forebrain. Neuroendocrinology 63:422–429.

Roland EH, Volpe JJ (1989) Effect of maternal cocaine use on the fetus and newborn: review of the literature. Pediatr Neurosci 15:88–94.

Romeyer A, Poindron P, Orgeur P (1994) Olfaction mediates the establishment of selective bonding in goats. Physiol Behav 56:693–700.

Rosenberg KM (1974) Effect of pre- and post-pubertal castration and testosterone on pup killing behavior in the male rat. Physiol Behav 13:159–161.

Rosenberg KM, Denenberg VH, Zarrow MX, Frank BL (1971) Effects of neonatal castration and testosterone on the rat's pup-killing behavior and activity. Physiol Behav 7:363–368.

Rosenberg KM, Herrenkohl LR (1976) Maternal behavior in male rats: critical times for the suppressive effects of androgens. Physiol Behav 16:293–297.

Rosenberg KM, Sherman GF (1975) Influence of testosterone on pup killing in the rat is modified by prior experience. Physiol Behav 15:669–672.

Rosenberg P, Halaris A, Moltz H (1977) Effects of central norepinephrine depletion on the initiation and maintenance of maternal behavior in the rat. Pharmacol Biochem Behav 6:21–24.

Rosenblatt JS (1967) Nonhormonal basis of maternal behavior in the rat. Science 156:1512–1514.

Rosenblatt JS, Ceus K (1998) Estrogen implants in the medial preoptic area stimulate maternal behavior in male rats. Horm Behav 33:23–30.

Rosenblatt JS, Factor EM, Mayer AD (1994) Relationship between maternal aggression and maternal care in the rat. Aggress Behav 20:243–255.

Rosenblatt JS, Hazelwood S, Poole J (1996) Maternal behavior in male rats: effects of medial preoptic area lesions and presence of maternal aggression. Horm Behav 30:201–215.

Rosenblatt JS, Lehrman DS (1963) Maternal behavior in the laboratory rat. In: Maternal behavior in mammals (Rheingold H, ed), pp 8–57. New York: Wiley.

Rosenblatt JS, Mayer AD (1995) An analysis of approach/withdrawal processes in the initiation of maternal behavior in the laboratory rat. In: Behavioral development (Hood KE, Greenberg G, Tobach E, eds), pp 177–230. New York: Garland Press.

Rosenblatt JS, Mayer AD, Siegel HS (1985) Maternal behavior among nonprimate mammals. In: Handbook of behavioral neurobiology, Vol 7. Reproduction (Adler N, Pfaff D, Goy RW, eds), pp 229–298. New York: Plenum Press.

Rosenblatt JS, Siegel HI (1975) Hysterectomy-induced maternal behavior during pregnancy in the rat. J Comp Physiol Psychol 89:685–700.

Rosenblatt JS, Snowdon CT (1996) Advances in the study of behavior, Vol 25. Parental care: evolution, mechanisms, and adaptive significance. San Diego: Academic Press.

Rosenblum LA (1972) Sex and age differences in response to infant squirrel monkeys. Brain Behav Evol 5:30–40.

Rosenblum LA, Andrews MW (1994) Influences of environmental demand on maternal behavior and infant development. Acta Paediatr 83 (Suppl 397):57–63.

Rosenthal W, Siebold A, Antaramian A (1992) Molecular identification of the gene responsible for nephrogenic congenital diabetes insipidus. Nature 359:233–235.

Ross S, Denenberg VH, Frommer GP, Sawin PB (1959) Genetic, physiological, and behavioral background of reproduction in the rabbit. V. Nonretrieving of neonates. J Mammal 40:91–96.

Ross S, Sawin PB, Zarrow MX, Denenberg VH (1963) Maternal behavior in the rabbit. In: Maternal behavior in mammals (Rheingold H, ed), pp 94–121. New York: Wiley.

Roth H, Darms K (1993) The social organization of marmosets: a critical evaluation of recent concepts. In: Marmosets and tamarins: systematics, behavior and ecology (Rylands AB, ed), pp 176–199. Oxford: Oxford University Press.

Rowell TE (1960) On the retrieving of young and other behaviour in lactating golden hamsters. Proc Zool Soc Lond 135:265–282.

Rowell TE (1961) Maternal behaviour in non-maternal golden hamsters. Anim Behav 9:11–15.

Rozen F, Russo C, Banville D, Zingg HH (1995) Structure, characterization, and expression of the rat oxytocin receptor gene. Proc Natl Acad Sci 92:200–204.

Rubin BS, Bridges RS (1984) Disruption of ongoing maternal responsiveness by central administration of morphine sulfate. Brain Res 307:91–97.

Rubin BS, Bridges RS (1989) Immunoreactive prolactin in the cerebrospinal fluid of estrogen-treated and lactating rats as determined by push-pull perfusion of the lateral ventricles. J Neuroendocrinol 1:345–349.

Rubin BS, Menniti FS, Bridges RS (1983) Intracerebral administration of oxytocin and maternal behavior in rats after prolonged and acute steroid pretreatment. Horm Behav 17:45–53.

Rupniak NM, Carlson EC, Harrison T, Oates B, Seward E, Owen S, de Felipe C, Hunt S, Wheeldon A (2000) Pharmacological blockade or genetic deletion of substance P (NK(1)) receptors attenuates neonatal vocalisation in guinea-pigs and mice. Neuropharmacology 39:1413–1421.

Rupniak NM, Kramer MS (1999) Discovery of the antidepressant and anti-emetic efficacy of substance P receptor (NK1) antagonists. TIPS 20:485–490.

Ruppenthal GC, Arling GL, Harlow HF, Sackett GP, Suomi SJ (1976) A 10-year perspective of motherless-mother behavior. J Abnorm Psychol 85:341–349.

Rupprecht R (1997) The neuropsychopharmacological potential of neuroactive steroids. J Psychiatry Res 31:297–314.

Russell JA, Leng G (1998) Sex, parturition and motherhood without oxytocin? J Endocrinol 157:343–359.

Sakanaka M, Magari S, Shibasaki T, Lederis K (1988) Corticotropin releasing factor-containing afferents to the lateral septum of the rat brain. J Comp Neurol 270:404–415.

Sakanaka M, Shibasaki T, Lederis K (1986) Distribution and efferent projections of corticotropin-releasing factor-like immunoreactivity in the rat amygdaloid complex. Brain Res 382:213–238.

Sakanaka M, Shibasaki T, Lederis K (1987) Corticotropin releasing factor-like immunoreactivity in the rat brain as revealed by a modified cobalt-glucose oxidase-diaminobenzidine method. J Comp Neurol 260:256–298.

Salamone JD (1996) The behavioral neurochemistry of motivation: methodological and conceptual issues in studies of the dynamic activity of nucleus accumbens dopamine. J Neurosci Meth 64:137–149.

Samuels M, Bridges RS (1983) Plasma prolactin concentrations in parental male and female: effects of exposure to rat young. Endocrinology 113:1647–1654.

Sanchez-Toscano F, Sanchez M, Garzon J (1991) Changes in the number of dendritic spines in the medial preoptic area during premature long-term social isolation in rats. Neurosci Lett 122:1–3.

Sapolsky RM (1992) Stress, the aging brain, and the mechanisms of neuron death. Cambridge, MA: MIT Press.

Satake H, Takuwa K, Minakata H, Matsushima O (1999) Evidence for conservation of the vasopressin/oxytocin superfamily in Annelida. J Biol Chem 274:5605–5611.

Sauvage M, Steckler T (2001) Detection of corticotropin-releasing hormone receptor 1 immunoreactivity in cholinergic, dopaminergic and noradrenergic neurons in the murine basal forebrain and brainstem nuclei—potential implication for arousal and attention. Neuroscience 104:643–652.

Scalia F, Winans SS (1975) The differential projections of the olfactory bulb and accessory olfactory bulb in mammals. J Comp Neurol 161:31–55.

Schanberg S, Field T (1987) Sensory deprivation stress and supplemental stimulation in the rat pup and preterm human neonate. Child Dev 58:1431–1447.

Schino G, D'Amato FR, Troisi A (1995) Mother-infant relationships in Japanese macaques: sources of inter-individual variation. Anim Behav 49:151–158.

Schneider JS, Stone MK, Wynne-Edwards KE, Horton TH, Lydon J, O'Malley B, Levine JE (2003) Progesterone receptors mediate male aggression toward infants. Proc Natl Acad Sci USA 100:2951–2956.

Schneirla TC (1959) An evolutionary and developmental theory of biphasic processes underlying approach and withdrawal. In: Nebraska symposium on motivation, Vol VII (Jones MR, ed), pp 1–42. Lincoln: University of Nebraska Press.

Schoech SJ (1998) Physiology of helping in Florida scrub-jays. Am Sci 86:70–77.

Schoech SJ, Ketterson ED, Nolan V, Sharp PJ, Buntin JD (1998) The effect of exogenous testosterone on parental behavior, plasma prolactin, and prolactin binding sites in dark-eyed juncos. Horm Behav 34:1–10.

Schuchard M, Landers JP, Sandhu NP, Spelsberg TC (1993) Steroid hormone regulation of nuclear proto-oncogenes. Endocrinol Rev 14:659–669.

Schulkin J (1998) Fear and its neuroendocrine basis. Prog Psychobiol Physiol Psychol 17:35–66.

Schumacher M, Coirini H, Robert F, Guennoun R, El-Etr M (1999) Genomic and membrane actions of progesterone: implications for reproductive physiology and behavior. Behav Brain Res 105:37–52.

Seegal RF, Denenberg VH (1974) Maternal experience prevents pup-killing in mice induced by peripheral anosmia. Physiol Behav 13:339–341.

Shapiro LE, Insel TR (1989) Ontogeny of oxytocin receptors in rat forebrain: a quantitative study. Synapse 4:259–266.

Shapiro LE, Leonard CM, Sessions CE, Dewsbury DA, Insel TR (1991) Comparative neuroanatomy of the sexually dimorphic hypothalamus in monogamous and polygamous voles. Brain Res 541:232–240.

Sharp FR, Gonzalez MF, Sharp JW, Sagar SM (1989) c-fos expression and (^{14}C) 2-deoxyglucose uptake in the caudal cerebellum of the rat during motor/sensory cortex stimulation. J Comp Neurol 284:621–636.

Sheehan TP (2000) An investigation into the neural and hormonal inhibition of maternal behavior in rats. Unpublished doctoral dissertation, Boston College.

Sheehan TP, Cirrito J, Numan MJ, Numan M (2000) Using c-Fos immunocytochemistry to identify forebrain regions that may inhibit maternal behavior in rats. Behav Neurosci 114:337–352.

Sheehan TP, Numan M (1997) Microinjection of the tachykinin neuropeptide K into the ventromedial hypothalamus disrupts the hormonal onset of maternal behavior in female rats. J Neuroendocrinol 9:677–687.

Sheehan T, Numan M (2000) The septal region and social behavior. In: The behavioral neuroscience of the septal region (Numan R, ed), pp 175–207. New York: Springer-Verlag.

Sheehan T, Numan M (2002) Estrogen, progesterone, and pregnancy termination alter neural activity in brain regions that control maternal behavior in rats. Neuroendocrinology 75:12–23.

Sheehan T, Paul M, Amaral E, Numan MJ, Numan M (2001) Evidence that the medial amygdala projects to the anterior/ventromedial hypothalamic nuclei to inhibit maternal behavior in rats. Neuroscience 106:341–356.

Sheline YI (2000) 3D MRI studies of neuroanatomic changes in unipolar major depression: the role of stress and medical comorbidity. Biol Psychiatry 48:791–800.

Sheng M, Greenberg ME (1990) The regulation and function of c-fos and other immediate early genes in the nervous system. Neuron 4:477–485.

Shimura T, Yamamoto T, Shimokochi M (1994) The medial preoptic area is involved in both sexual arousal and performance in male rats: re-evaluation of neuron activity in freely moving animals. Brain Res 640:215–222.

Shipley MT, Ennis M (1996) Functional organization of the olfactory system. J Neurobiol 30:123–176.

Shirota M, Banville D, Ali S, Jolicoeur C, Boutin JM, Edery M, Djiane J, Kelly PA (1990) Expression of two forms of prolactin receptor in rat ovary and liver. Mol Endocrinol 4:1136–1143.

Shivers B, Harlan R, Pfaff D (1989) A subset of neurons containing immunoreactive prolactin is a target for estrogen regulation of gene expression in rat hypothalamus. Neuroendocrinology 49:23–27.

Shughrue PJ, Lane MV, Merchenthaler I (1997) Comparative distribution of estrogen receptor-α and-β mRNA in the rat central nervous system. J Comp Neurol 388:507–525.

Siegel HI, Clark MC, Rosenblatt JS (1983a) Maternal responsiveness during pregnancy in the hamster (Mesocricetus auratus). Anim Behav 31:497–502.

Siegel HI, Doerr HK, Rosenblatt JS (1978) Further studies on estrogen-induced maternal behavior in hysterectomized-ovariectomized virgin rats. Physiol Behav 21:99–103.

Siegel HI, Giordano AL, Mallafre CM, Rosenblatt JS (1983b) Maternal aggression in hamsters: effects of stage of lactation, presence of pups, and repeated testing. Horm Behav 17:86–93.

Siegel HI, Greenwald GS (1975) Prepartum onset of maternal behavior in hamsters and the effects of estrogen and progesterone. Horm Behav 6:237–245.

Siegel HI, Rosenblatt JS (1975a) Hormonal basis of hysterectomy-induced maternal behavior during pregnancy in the rat. Horm Behav 6:211–222.

Siegel HI, Rosenblatt JS (1975b) Estrogen-induced maternal behavior in hysterectomized-ovariectomized virgin rats. Physiol Behav 14:465–471.

Siegel HI, Rosenblatt JS (1975c) Progesterone inhibition of estrogen-induced maternal behavior in hysterectomized-ovariectomized virgin rats. Horm Behav 6:223–230.

Siegel HI, Rosenblatt JS (1978a) Duration of estrogen stimulation and progesterone inhibition of maternal behavior in pregnancy-terminated rats. Horm Behav 11:12–19.

Siegel HI, Rosenblatt JS (1978b) Effects of adrenalectomy on maternal behavior in pregnancy-terminated rats. Physiol Behav 21:831–833.

Siegel HI, Rosenblatt JS (1980) Hormonal and behavioral aspects of maternal care in the hamster. Neurosci Biobehav Rev 4:17–26.

Silk JB (1999) Why are infants so attractive to others? The form and function of infant handling in bonnet macaques. Anim Behav 57:1021–1032.

Silva MRP, Bernardi MM, Felicio LF (2001) Effects of dopamine receptor antagonists on ongoing maternal behavior in rats. Pharmacol Biochem Behav 68:461–468.

Silveira MCL, Graeff FG (1992) Defense reaction elicited by microinjection of kainic acid into the medial hypothalamus of the rat: antagonism by a GABAA receptor agonist. Behav Neural Biol 57:226–232.

Silveira MCL, Sandner G, Graeff FG (1993) Induction of Fos immunoreactivity in the brain by exposure to the elevated plus-maze. Behav Brain Res 56:115–118.

Silverman AJ, Livne I, Witkin JW (1994) The gonadotropin-releasing hormone (GnRH) neuronal systems: immunocytochemistry and in situ hybridization. In: The physiology of reproduction, Vol 1 (Knobil E, Neill JD, eds), pp 1683–1709. New York: Raven Press.

Simerly RB, Chang C, Muramatsu M, Swanson LW (1990) Distribution of androgen and estrogen receptor mRNA-containing cells in the rat brain: an in situ hybridization study. J Comp Neurol 294:76–95.

Simerly RB, Gorski RA, Swanson LH (1986) Neurotransmitter specificity of cells and fibers in the medial preoptic nucleus: an immunohistochemical study in the rat. J Comp Neurol 246:343–363.

Simerly RB, Swanson LW (1986) The organization of neural inputs to the medial preoptic nucleus of the rat. J Comp Neurol 246:312–342.

Simerly RB, Swanson LW (1988) Projections of the medial preoptic nucleus: a *Phaseolus vulgaris* leucoagglutinin anterograde tract-tracing study in the rat. J Comp Neurol 270:209–242.

Simerly RB, Zee MC, Pendleton JW, Lubahn DB, Korach KS (1997) Estrogen receptor-dependent sexual differentiation of dopaminergic neurons in the preoptic region of the mouse. Proc Natl Acad Sci 94:14077–14082.

Simon H, Le Moal M, Calas A (1979) Efferents and afferents of the ventral tegmental-A10 region studied after local injection of [^{3}H]leucine and horseradish peroxidase. Brain Res 178:17–40.

Singer LM, Brodzinsky DM, Ramsay D, Steir M, Waters E (1985) Mother-infant attachment in adoptive families. Child Dev 56:1543–1551.

Sjogren B, Widstrom AM, Edman G, Uvnas-Moberg K (2000) Changes in personality pattern during the first pregnancy and lactation. J Psychosom Obstet Gynaecol 21:31–38.

Slawski BA, Buntin JD (1995) Preoptic area lesions disrupt prolactin-induced parental and feeding behavior in ring doves. Horm Behav 29:248–266.

Slotnick BM (1967) Disturbances of maternal behavior in the rat following lesions of the cingulate cortex. Behaviour 29:204–236.

Slotnick BM, Carpenter ML, Fusco R (1973) Initiation of maternal behavior in pregnant nulliparous rats. Horm Behav 4:53–59.

Smith MO, Holland RC (1975) Effects of lesions of the nucleus accumbens on lactation and postpartum behavior. Physiol Psychol 3:331–336.

Smith MS (1993) Lactation alters neuropeptide-Y and pro-opiomelanocortin gene expression in the arcuate nucleus of the rat. Endocrinology 133:1258–1265.

Smith MS, Neill JD (1976) Termination at midpregnancy of two daily surges of plasma prolactin initiated by mating in the rat. Endocrinology 98:696–701.

Smith R (1999) The timing of birth. Sci Am March:68–75.

Smith SS (2002) Novel effects of neuroactive steroids in the CNS. In: Hormones, brain and behavior, Vol 3 (Pfaff DW, Arnold AP, Etgen AM, Fahrbach SE, Rubin RT, eds), pp 747–778. San Diego: Academic Press.

Smith SS, Gong QH, Li X, Moran MH, Bitran D, Frye CA, Hsu F (1998) Withdrawal from 3α-OH-5α-pregnan-20-one using a pseudopregnancy model alters the kinetics of hippocampal GABAA-gated current and increases the GABAA receptor α4 subunit in association with increased anxiety. J Neurosci 18:5275–5284.

Smithson KG, Weiss ML, Hatton GI (1989) Supraoptic nucleus afferents from the main olfactory bulb. I. Anatomical evidence from anterograde and retrograde tracers in rat. Neuroscience 31:277–287.

Smoller B, Lewis AB (1977) A psychological theory of child abuse. Psychiatr Q 49: 38–44.

Smotherman WP, Bell RW, Starzec J, Elias J, Zachman TA (1974) Maternal responses to infant vocalizations and olfactory cues in rats and mice. Behav Biol 12:55–66.

Snowdon CT (1996) Infant care in cooperatively breeding species. In: Advances in the study of behavior, Vol 25. Parental care: evolution, mechanisms, and adaptive significance (Rosenblatt JS, Snowdon CT, eds), pp 643–689. San Diego: Academic Press.

Soares JC, Mann JJ (1997) The functional neuroanatomy of mood disorders. J Psychiatr Res 31:393–432.

Soloff MS, Alexandrova M, Fernstrom MJ (1979) Oxytocin receptors: triggers for parturition and lactation? Science 204:1313–1314.

Solomon NG, French JA (1997) Cooperative breeding in mammals. Cambridge: Cambridge University Press.

Soroker V, Terkel J (1988) Changes in incidence of infanticidal and parental responses during the reproductive cycle in male and female wild mice Mus musculus. Anim Behav 36:1275–1281.

Southard JN, Talamantes F (1991) Placental prolactin-like proteins in rodents: variations on a structural theme. Mol Cell Endocrinol 79:C133–C140.

Sparks PD, LeDoux JE (2000) The septal complex as seen through the context of fear. In: The behavioral neuroscience of the septal region (Numan R, ed), pp 234–269. New York: Springer-Verlag.

Spielewoy C, Roubert C, Hamon M, Nosten-Bertrand M, Betancur C, Giros B (2000) Behavioural disturbances associated with hyperdopaminergia in dopamine-transporter knockout mice. Behav Pharmacol 11:279–290.

Spitz LM, Croxatto HB, Robbins A (1996) Antiprogestins: mechanisms of action and contraceptive potential. Annu Rev Pharmacol Toxicol 36:47–81.

Squire LR (1992) Memory and the hippocampus: a synthesis from findings with rats, monkeys, and humans. Psychol Rev 99:195–231.

Stack EC, Balakrishnan R, Numan MJ, Numan M (2002) A functional neuroanatomical investigation of the role of the medial preoptic area in neural circuits regulating maternal behavior. Behav Brain Res 131:17–36.

Stack EC, Numan M (2000) The temporal course of expression of c-Fos and Fos B within the medial preoptic area and other brain regions of postpartum female rats during prolonged mother-young interactions. Behav Neurosci 114:609–622.

Stafisso-Sandoz G, Holt E, Polley D, Lambert KG, Kinsley CH (1998) Opiate disruption of maternal behavior: morphine reduces and naloxone restores c-fos activity in the medial preoptic area of lactating rats. Brain Res Bull 45:307–313.

Stamm JS (1955) The function of the median cerebral cortex in maternal behavior in rats. J Comp Physiol Psychol 48:347–356.

Steckler T, Holsboer F (1999) Corticotropin-releasing hormone receptor subtypes and emotion. Biol Psychiatry 46:1480–1508.

Steele MK, Rowland D, Moltz H (1979) Initiation of maternal behavior in the rat: possible involvement of limbic norepinephrine. Pharmacol Biochem Behav 11:123–130.

Steimer T, Driscoll P, Schulz PE (1997) Brain metabolism of progesterone, coping behaviour and emotional reactivity in male rats from two psychogenetically selected lines. J Neuroendocrinol 9:169–175.

Steiner M (1979) Psychobiology of mental disorders associated with childbearing. Acta Psychiatr Scand 60:449–464.

Stern JM (1977) Effects of ergocryptine on postpartum maternal behavior, ovarian cyclicity, and food intake in rats. Behav Biol 21:134–140.

Stern JM (1983) Maternal behavior priming in virgin and caesarean-delivered Long-Evans rats: effects of brief contact or continuous exteroceptive pup stimulation. Physiol Behav 31:757–763.

Stern JM (1987) Pubertal decline in maternal responsiveness in Long-Evans rats: maturational influences. Physiol Behav 41:93–98.

Stern JM (1989) Maternal behavior: sensory, hormonal, and neural determinants. In: Psychoendocrinology (Bush SR, Levine S, eds), pp 105–226. New York: Academic Press.

Stern JM (1990) Multisensory regulation of maternal behavior and masculine sexual behavior: a revised view. Neurosci Biobehav Rev 14:183–200.

Stern JM (1991) Nursing posture is elicited rapidly in maternally naive, haloperidol-treated female and male rats in response to ventral trunk stimulation from active pups. Horm Behav 25:504–517.

Stern JM (1996a) Somatosensation and maternal care in Norway rats. In: Advances in the study of behavior, Vol 25. Parental care: evolution, mechanisms, and adaptive significance (Rosenblatt JS, Snowdon CT, eds), pp 243–294. San Diego: Academic Press.

Stern JM (1996b) Trigeminal lesions and maternal behavior in Norway rats: II. Disruption of parturition. Physiol Behav 60:187–190.

Stern JM (1997a) Offspring-induced nurturance: animal-human parallels. Dev Psychobiol 31:19–37.

Stern JM (1997b) Trigeminal lesions and maternal behavior in Norway rats: III. Experience with pups facilitates recovery. Dev Psychobiol 30:115–126.

Stern JM, Dix L, Bellomo C, Thramann C (1992) Ventral trunk somatosensory determinants of nursing behavior in Norway rats. II. Role of nipple and surrounding sensations. Psychobiology 20:71–80.

Stern JM, Goldman L, Levine S (1973) Pituitary-adrenal responsiveness during lactation in rats. Neuroendocrinology 12:179–191.

Stern JM, Johnson SK (1989) Perioral somatosensory determinants of nursing behavior in Norway rats (Rattus norvegicus). J Comp Psychol 103:269–280.

Stern JM, Johnson SK (1990) Ventral somatosensory determinants of nursing behavior in Norway rats. I. Effects of variations in the quality and quantity of pup stimuli. Physiol Behav 47:993–1011.

Stern JM, Kolunie JM (1989) Perioral anesthesia disrupts maternal behavior during early lactation in Long-Evans rats. Behav Neural Biol 52:20–38.

Stern JM, Kolunie JM (1991) Trigeminal lesions and maternal behavior in Norway rats: I. Effects of cutaneous rostral snout denervation on maintenance of nurturance and maternal aggression. Behav Neurosci 105:984–997.

Stern JM, Kolunie JM (1993) Maternal aggression of rats is impaired by cutaneous anesthesia of the ventral trunk, but not by nipple removal. Physiol Behav 54:861–868.

Stern JM, Levine S (1972) Pituitary-adrenal activity in the postpartum rat in the absence of suckling stimulation. Horm Behav 3:237–246.

Stern JM, Lonstein JS (2001) Neural mediation of nursing and related maternal behaviors. Prog Brain Res 133:263–278.

Stern JM, MacKinnon DA (1976) Postpartum, hormonal, and nonhormonal induction of maternal behavior in rats: effects on T-maze retrieval of pups. Horm Behav 7:305–316.

Stern JM, McDonald C (1989) Ovarian hormone-induced short-latency maternal behavior in ovariectomized virgin Long-Evans rats. Horm Behav 23:157–172.

Stern JM, Protomastro M (2000) Effects of low dosages of apomorphine on maternal responsiveness in lactating rats. Pharmacol Biochem Behav 66:353–359.

Stern JM, Rogers L (1988) Experience with younger siblings facilitates maternal responsiveness in pubertal Norway rats. Dev Psychobiol 21:575–589.

Stern JM, Taylor LA (1991) Haloperidol inhibits maternal retrieval and licking, but enhances nursing behavior and litter weight gains in lactating rats. J Neuroendocrinol 3:591–596.

Stevens-Simon C, Kelly L, Wallis J (2000) The timing of norplant insertion and postpartum depression in teenagers. J Adolesc Health 26:408–413.

Stone CP (1925) Preliminary note on the maternal behavior of rats living in parabiosis. Endocrinology 9:505–512.

Stone CP (1938) Effects of cortical destruction on reproductive behavior and maze learning in albino rats. J Comp Psychol 26:217–236.

Storey AE, Joyce TL (1995) Pup contact promotes paternal responsiveness in male meadow voles. Anim Behav 49:1–10.

Storey AE, Walsh CJ, Quinton R, Wynne-Edwards KE (2000) Hormonal correlates of paternal responsiveness in new and expectant fathers. Evol Hum Behav 21:79–95.

Straus MA (1980) Stress and physical child abuse. Child Abuse Negl 4:75–88.

Sturgis JD, Bridges RS (1997) N-methyl-DL-aspartic acid lesions of the medial preoptic area disrupt ongoing parental behavior in male rats. Physiol Behav 62:305–310.

Suchman NE, Luthar SS (2000) Maternal addiction, child maladjustment and sociodemographic risks: implications for parenting behaviors. Addiction 95:1417–1428.

Sugiyama T, Minoura H, Kawabe N, Tanaka M, Nakashima K (1994) Preferential expression of long form prolactin receptor mRNA in the rat brain during the oestrous cycle, pregnancy and lactation: hormones involved in its gene expression. J Endocrinol 141:325–333.

Sugiyama T, Minoura H, Toyoda N, Sakaguchi K, Tanaka M, Sudo S, Nakashima K (1996) Pup contact induces the expression of long form prolactin receptor mRNA in the brain of female rats: effects of ovariectomy and hypophysectomy on receptor gene expression. J Endocrinol 149:335–340.

Sullivan GM, Coplan JD, Kent JM, Gorman JM (1999) The noradrenergic system in pathological anxiety: a focus on panic and relevance to generalized anxiety and phobias. Biol Psychiatry 46:1205–1218.

Sumner BEH, Douglas AJ, Russell AR (1992) Pregnancy alters the density of opioid binding sites in the supraoptic nucleus and posterior pituitary glands of rats. Neurosci Newslett 137:216–220.

Sundstrom I, Andersson A, Nyberg S, Ashbrook D, Purdy RH, Backstrom T (1998) Patients with premenstrual syndrome have a different sensitivity to a neuroactive steroid during the menstrual cycle compared to control subjects. Neuroendocrinology 67:126–138.

Sundstrom I, Ashbrook D, Backstrom T (1997) Reduced benzodiazepine sensitivity in patients with premenstrual syndrome, a pilot study. Psychoneuroendocrinology 22: 25–38.

Suomi SJ (1978) Maternal behavior by socially incompetent monkeys. J Pediatr Psychol 3:28–34.

Suomi SJ (1991) Primate separation models of affective disorders. In: Neurobiology of learning, emotion and affect (Madden J, ed), pp 195–214. New York: Raven Press.

Suomi SJ (1997) Early determinants of behaviour: evidence from primate studies. Br Med Bull 53:170–184.

Suomi SJ (1999) Attachment in rhesus monkeys. In: Handbook of attachment. Theory, research, and clinical applications (Cassidy J, Shaver PR, eds), pp 181–197. New York: Guilford Press.

Suomi SJ, Ripp C (1983) A history of motherless mother monkey mothering at the University of Wisconsin Primate Laboratory. In: Child abuse: the nonhuman primate data (Reite M, Caine NG, eds), pp 49–78. New York: Alan R. Liss.

Svare B (1981) Maternal aggression in mammals. In: Parental care in mammals (Gubernick DJ, Klopfer PH, eds), pp 179–210. New York: Plenum Press.

Svare B (1990) Maternal aggression: hormonal, genetic, and developmental determinants. In: Mammalian parenting (Krasnegor NA, Bridges RS, eds), pp 118–132. New York: Oxford University Press.

Svare B, Gandelman R (1973) Postpartum aggression in mice: experiential and environmental factors. Horm Behav 4:323–334.

Svare B, Gandelman R (1976a) A longitudinal analysis of maternal aggression in Rockland-Swiss albino mice. Dev Psychobiol 9:437–446.

Svare B, Gandelman R (1976b) Postpartum aggression in mice: the influence of suckling stimulation. Horm Behav 7:407–416.

Svare B, Kinsley CH, Mann MA, Broida J (1984) Infanticide: accounting for genetic variation in mice. Physiol Behav 33:137–152.

Svare B, Mann M, Broida J, Michael SD (1982) Maternal aggression exhibited by hypophysectomized parturient mice. Horm Behav 16:455–461.

Swanson LJ, Campbell CS (1979a) Induction of maternal behavior in nulliparous golden hamsters (*Mesocricetus auratus*). Behav Neural Biol 26:364–371.

Swanson LJ, Campbell CS (1979b) Maternal behavior in the primiparous and multiparous golden hamster. Z Tierpsychol 50:96–104.

Swanson LJ, Campbell CS (1981) The role of the young in the control of the hormonal events during lactation and behavioral weaning in the golden hamster. Horm Behav 15:1–15.

Swanson LW (1977) Immunohistochemical evidence for a neurophysin-containing autonomic pathway arising in the paraventricular nucleus of the hypothalamus. Brain Res 128:346–353.

Swanson LW (1982) The projections of the ventral tegmental area and adjacent regions: a combined retrograde fluorescent and immunofluorescence study in the rat. Brain Res Bull 9:321–353.

Swanson LW (1992) Brain maps: structure of the rat brain. Amsterdam: Elsevier.

Swanson LW (2000) Cerebral hemisphere regulation of motivated behavior. Brain Res 886:113–164.

Swanson LW, Cowan WM (1979) The connections of the septal region. J Comp Neurol 186:621–656.

Swanson LW, Mogenson GJ, Gerfen GJ, Robinson P (1984) Evidence for a projection from the lateral preoptic area and substantia innominata to the "mesencephalic locomotor region" in the rat. Brain Res 295:161–178.

Swanson LW, Petrovich GD (1998) What is the amygdala? Trends Neurosci 21:323–331.

Swanson LW, Sawchenko PE (1983) Hypothalamic integration: organization of paraventricular and supraoptic nuclei. Annu Rev Neurosci 6:269–324.

Tardif SD, Richter CB, Carson RL (1984) Effects of sibling-rearing experience on future reproductive success in two species of callitrichidae. Am J Primatol 6:377–380.

Terkel J, Bridges RS, Sawyer CH (1979) Effects of transecting the lateral neural connections of the medial preoptic area on maternal behavior in the rat: nest building, pup retrieval, and prolactin secretion. Brain Res 169:369–380.

Terkel J, Rosenblatt JS (1968) Maternal behavior induced by maternal blood plasma injected into virgin rats. J Comp Physiol Psychol 65:479–482.

Terkel J, Rosenblatt JS (1971) Aspects of nonhormonal maternal behavior in the rat. Horm Behav 2:161–171.

Terkel J, Rosenblatt JS (1972) Humoral factors underlying maternal behavior at parturition. J Comp Physiol Psychol 80:365–371.

Terlecki LJ, Sainsbury RS (1978) Effects of fimbria lesions on maternal behavior in the rat. Physiol Behav 21:89–97.

Teti DM, Gelfand DM, Messinger DS, Isabella R (1995) Maternal depression and the quality of early attachment: an examination of infants, preschoolers, and their mothers. Dev Psychol 31:364–376.

Thakore JH, Dinan TG (1995) Cortisol synthesis inhibition: a new treatment strategy for clinical and endocrine manifestations of depression. Biol Psychiatry 37:364–368.

Theodosis DT, Poulain DA (1984) Evidence for structural plasticity in the supraoptic nucleus of the rat hypothalamus in relation to gestation and lactation. Neuroscience 11:183–193.

Theodosis DT, Poulain DA (1987) Oxytocin-secreting neurones: a physiological model for structural plasticity in the adult mammalian brain. TINS 10(10):426–430.

Theodosis DT, Poulain DA (1989) Neuronal-glial and synaptic plasticity in the adult rat paraventricular nucleus. Brain Res 484:361–366.

Thoman EB, Arnold WJ (1968) Effects of incubator rearing with social deprivation on maternal behavior in rats. J Comp Physiol Psychol 65:441–446.

Thomas SA, Palmiter RD (1997) Impaired maternal behavior in mice lacking norepinephrine and epinephrine. Cell 91:583–592.

Thompson AC, Kristal MB (1996) Opioid stimulation in the ventral tegmental area facilitates the onset of maternal behavior in rats. Brain Res 743:184–201.

Thompson WR (1953) The inheritance of behaviour: behavioural differences in fifteen strains. Can J Psychol 7:145–155.

Timberlake W, Silva KM (1995) Appetitive behavior in ethology, psychology, and behavior systems. In: Perspectives in ethology, Vol 11. Behavioral design (Thompson NS, ed), pp 211–253. New York: Plenum Press.

Torner L, Toschi N, Nava G, Clapp C, Neumann ID (2002) Increased hypothalamic expression of prolactin in lactation: involvement in behavioral and neuroendocrine stress responses. Eur J Neurosci 15:1381–1389.

Torner L, Toschi N, Pohlinger A, Landgraf R, Neumann ID (2001) Anxiolytic and anti-stress effects of brain prolactin: improved efficacy of antisense targeting of the prolactin receptor by molecular modeling. J Neurosci 21:3207–3214.

Toufexis DJ, Rochford J, Walker C (1999) Lactation-induced reduction in rats' acoustic startle is associated with changes in noradrenergic transmission. Behav Neurosci 113: 176–184.

Toufexis DJ, Thrivikraman KV, Plotsky PM, Morilak DA, Huang N, Walker C (1998) Reduced noradrenergic tone to the hypothalamic paraventricular nucleus contributes to the stress hyporesponsiveness of lactation. J Neuroendocrinol 10:417–427.

Toufexis DJ, Walker C (1996) Noradrenergic facilitation of the adrenocorticotropin response to stress is absent during lactation in the rat. Brain Res 737:71–77.

Trainor BC, Marler CA (2001) Testosterone, paternal behavior, and aggression in the monogamous California mouse (Peromyscus californicus). Horm Behav 40:32–42.

Treit D, Menard J (2000) The septum and anxiety. In: The behavioral neuroscience of the septal region (Numan R, ed), pp 210–233. New York: Springer-Verlag.

Tribollet E, Charpak S, Schmidt A, Dubois-Dauphin M, Dreifuss JJ (1989) Appearance and transient expression of oxytocin receptors in fetal, infant, and peripubertal rat brain studied by autoradiography and electrophysiology. J Neurosci 9:1764–1773.

Trivers R (1972) Parental investment and sexual selection. In: Sexual selection and the descent of man, 1871–1971 (Campbell BG, ed), pp 136–179. Chicago: Aldine.

Troisi A, D'Amato FR (1991) Anxiety in the pathogenesis of primate infant abuse: a pharmacological study. Psychopharmacology 103:571–572.

Troisi A, D'Amato FR (1994) Mechanisms of primate infant abuse: the maternal anxiety hypothesis. In: Infanticide and parental care (Parmigiani S, vom Saal FS, eds), pp 199–210. Chur, Switzerland: Harwood Academic Publishers.

Troisi A, D'Amato FR, Carnera A, Trinca L (1988) Maternal aggression by lactating group-living Japanese macaque females. Horm Behav 22:444–452.

Tucker HA (1994) Lactation and its hormonal control. In: The physiology of reproduction, Vol 2 (Knobil E, Neill JD, eds), pp 1065–1098. New York: Raven Press.

Tzschentke TM, Schmidt WJ (2000) Functional relationship among medial prefrontal cortex, nucleus accumbens, and ventral tegmental area in locomotion and reward. Crit Rev Neurobiol 14:131–142.

Uvnas-Moberg K (1997) Physiological and endocrine effects of social contact. Ann NY Acad Sci 807:146–163.

Uvnas-Moberg K, Eklund M, Hillergaart V, Ahlenius S (2000) Improved conditioned avoidance learning by oxytocin administration in high-emotional male Sprague-Dawley rats. Regul Pept 88:27–32.

Uvnas-Moberg K, Widstrom A, Nissen E, Bjorvell H (1990) Personality traits in women 4 days postpartum and their correlation with plasma levels of oxytocin and prolactin. J Psychosom Obstet Gynaecol 11:261–273.

Valentino RJ, Curtis AL, Page ME, Pavovich LA, Lechner SM, Van Bockstaele E (1998) The locus coeruleus-noradrenergic system as an integrator of stress responses. Prog Psychobiol Physiol Psychol 17:91–126.

Vallee M, Mayo W, Dellu F, Le Moal M, Simon H, Maccari S (1997) Prenatal stress induces high anxiety and postnatal handling induces low anxiety in adult offspring: correlation with stress-induced corticosterone secretion. J Neurosci 17:2626–2636.

Van de Kar LD, Blair ML (1999) Forebrain pathways mediating stress-induced hormone secretion. Front Neuroendocrinol 20:1–48.

Vandenbergh JG (1973) Effects of central and peripheral anosmia on reproduction of female mice. Physiol Behav 10:257–261.

Van Ijzendoorn MH (1992) Intergenerational transmission of parenting: a review of studies in nonclinical populations. Dev Rev 12:76–99.

van Kesteren RE, Smit AB, Dirks RW, de With ND, Geraerts WPM, Joosse J (1992) Evolution of the vasopressin/oxytocin superfamily: characterization of a cDNA encoding a vasopressin-related precursor, preproconopressin, from the mollusc *Lymnaea stagnalis*. Proc Natl Acad Sci 89:4593–4597.

van Leengoed E, Kerker E, Swanson HH (1987) Inhibition of postpartum maternal behaviour in the rat by injecting an oxytocin antagonist into the cerebral ventricles. J Endocrinol 112:275–282.

Vernotica EM, Lisciotto CA, Rosenblatt JS, Morrell JI (1996) Cocaine transiently impairs maternal behavior in the rat. Behav Neurosci 110:315–323.

Vernotica EM, Rosenblatt JS, Morrell JI (1999) Microinfusion of cocaine into the medial preoptic area or nucleus accumbens transiently impairs maternal behavior in the rat. Behav Neurosci 113:377–390.

Voci VE, Carlson NR (1973) Enhancement of maternal behavior and nest building following systemic and diencephalic administration of prolactin and progesterone in the mouse. J Comp Physiol Psychol 83:388–393.

vom Saal FS (1985) Time-contingent change in infanticide and parental behavior induced by ejaculation in male mice. Physiol Behav 34:7–15.

vom Saal FS, Grant W, McMullen C, Laves K (1983) High fetal estrogen titers correlate with enhanced sexual performance and decreased aggression in male mice. Science 220:1306–1309.

vom Saal FS, Howard LS (1982) The regulation of infanticide and parental behavior: implications for reproductive success in male mice. Science 215:1270–1272.

Von Krosigk M, Smith AD (1991) Descending projections from the substantia nigra and retrorubral field to the medullary and pontomedullary reticular formation. Eur J Neurosci 3:260–273.

Wagner CK, Morrell JI (1995) In situ analysis of estrogen receptor mRNA expression in the brain of female rats during pregnancy. Mol Brain Res 33:127–135.

Wagner CK, Morrell JI (1996) Levels of estrogen receptor immunoreactivity are altered in behaviorally-relevant brain regions in female rats during pregnancy. Mol Brain Res 42:328–336.

Wakerley JB, Clarke G, Summerlee AJS (1994) Milk ejection and its control. In: The physiology of reproduction, Vol 2 (Knobil E, Neill JD, eds), pp 1131–1177. New York: Raven Press.

Wakerley JB, Jiang QB, Housham SJ, Terenzi MG, Ingram CD (1995) Influence of reproductive state and ovarian steroids on facilitation of the milk-ejection reflex by central oxytocin. In: Oxytocin: cellular and molecular approaches in medicine and research (Ivell R, Russell JA, eds), pp 117–132. New York: Plenum Press.

Walker C, Lightman SL, Steele MK, Dallman MF (1992) Suckling is a persistent stimulus to the adrenocortical system of the rat. Endocrinology 130:115–125.

Walker CD, Toufexis DJ, Burlet A (2001) Hypothalamic and limbic expression of CRF and vasopressin during lactation: implications for the control of ACTH secretion and stress hyporesponsiveness. In: The maternal brain. Progress in brain research, Vol 133 (Russell JA, Douglas AJ, Windle RJ, Ingram CD, eds), pp 99–110. Amsterdam: Elsevier.

Walker CD, Welberg LAM, Plotsky PM (2002) Glucocorticoids, stress, and development. In: Hormones, brain and behavior, Vol 4 (Pfaff DW, Arnold AP, Etgen AM, Fahrbach SE, Rubin RT, eds), pp 487–534. San Diego: Academic Press.

Wallen K (1990) Desire and ability: hormones and the regulation of female sexual behavior. Neurosci Biobehav Rev 14:233–241.

Walsh CJ, Fleming AS, Lee A, Magnusson JE (1996) The effects of olfactory and somatosensory desensitization on Fos-like immunoreactivity in the brains of pup-exposed postpartum rats. Behav Neurosci 110:134–153.

Walsh R, Slaby F, Posner B (1987) A receptor-mediated mechanism for the transport of prolactin from blood to cerebrospinal fluid. Endocrinology 120:1846–1850.

Walton JM, Wynne-Edwards KE (1997) Paternal care reduces maternal hyperthermia in Djungarian hamsters (*Phodopus campbelli*). Physiol Behav 63:41–47.

Wamboldt MZ, Gelhard RE, Insel TR (1988) Gender differences in caring for infant *Cebuella pygmaea*: the role of infant age and relatedness. Dev Psychobiol 21:187–202.

Wamboldt MZ, Insel TR (1987) The ability of oxytocin to induce short latency maternal behavior is dependent on peripheral anosmia. Behav Neurosci 101:439–441.

Wan XST, Liang F, Moret V, Wiesendanger M, Rouiller EM (1992) Mapping of motor pathways in rats: c-Fos induction by intracortical microstimulation of the motor cortex correlated with efferent connectivity of the site of cortical stimulation. Neuroscience 49:749–761.

Wang MW, Crombie DL, Hayes JS, Heap RB (1995) Aberrant maternal behaviour in mice treated with a progesterone receptor antagonist during pregnancy. J Endocrinol 145:371–377.

Wang Z, De Vries GJ (1993) Testosterone effects on paternal behavior and vasopressin immunoreactive projections in prairie voles (*Microtus ochrogaster*). Brain Res 631:156–160.

Wang Z, Ferris CF, De Vries GJ (1994) Role of septal vasopressin innervation in paternal behavior in prairie voles (*Microtus ochrogaster*). Proc Natl Acad Sci USA 91:400–404.

Wang Z, Insel TR (1996) Parental behavior in voles. In: Advances in the study of behavior, Vol 25. Parental care: evolution, mechanisms, and adaptive significance (Rosenblatt JS, Snowdon CT, eds), pp 361–384. San Diego: Academic Press.

Wang Z, Novak MA (1992) Influence of the social environment on parental behavior and pup development of meadow voles (*Microtus pennsylvanicus*) and prairie voles (*M. ochrogaster*). J Comp Psychol 106:163–171.

Wang Z, Novak MA (1994) Parental care and litter development in primiparous and multiparous prairie voles (*Microtus ochrogaster*). J Mammal 75:18–23.

Wang ZX, Liu Y, Young LJ, Insel TR (2000) Hypothalamic vasopressin gene expression increases in both males and females postpartum in a biparental rodent. J Neuroendocrinol 12:111–120.

Wang ZX, Young LJ, Yu GZ, Smith MT, DeVries GJ, Insel TR (1997) Vasopressin gene expression increases associated with paternal behavior in prairie voles. Soc Neurosci Abs 23:1073.

Wardlaw SL, Frantz AG (1983) Brain β-endorphin during pregnancy, parturition, and the post-partum period. Endocrinology 113:1664–1668.

Warren SG, Humphreys AG, Juraska JM, Greenough WT (1995) LTP varies across the estrous cycle: enhanced synaptic plasticity in proestrus rats. Brain Res 703:26–40.

Watanabe Y, Ikegaya Y, Saito H, Abe K (1996) Opposite regulation by β-adrenoceptor-cyclic AMP system of synaptic plasticity in the medial and lateral amygdala in vitro. Neuroscience 71:1031–1035.

Waxham NM (1999) Neurotransmitter receptors. In: Fundamental neuroscience (Zigmond MJ, Bloom FE, Landis SC, Roberts JL, Squire LR, eds), pp 235–267. San Diego: Academic Press.

Weiland N (1992) Estradiol selectively regulates agonist binding sites on N-methyl-D-aspartate receptor complex in the CA1 region of the hippocampus. Endocrinology 131:662–668.

Weisbard C, Goy RW (1976) Effect of parturition and group composition on competitive drinking order in stumptail macaques (Macaca arctoides). Folia Primatol 25:95–121.

Weisner BP, Sheard NM (1933) Maternal behaviour in the rat. Edinburgh: Oliver and Boyd.

Weiss IC, Domeney AM, Heidbreder CA, Moreau J, Feldon J (2001) Early social isolation, but not maternal separation, affects behavioral sensitization to amphetamine in male and female rats. Pharmacol Biochem Behav 70:397–409.

Wellman J, Carr DB, Graham A, Jones H, Humm JL, Ruscio M, Billack B, Kinsley CH (1997) Preoptic area infusions of morphine disrupt—and naloxone restores—parental-like behavior in juvenile rats. Brain Res Bull 44:183–191.

Wenkstern D, Pfaus JG, Fibiger HC (1993) Dopamine transmission increases in the nucleus accumbens of male rats during their first exposure to sexually receptive female rats. Brain Res 618:41–46.

Wersinger SR, Baum MJ, Erskine MS (1993) Mating-induced FOS-like immunoreactivity in the rat forebrain: A sex comparison and a dimorphic effect of pelvic nerve transection. J Neuroendocrinol 5:557–568.

Whitbeck LB, Hoyt DR, Simons RL, Conger RD, Elder GH, Lorenz FO, Huck S (1992) Intergenerational continuity of parental rejection and depressed affect. J Person Soc Psychol 63:1036–1045.

White N, Carr G (1985) The conditioned place preference is affected by two independent reinforcement processes. Pharmacol Biochem Behav 23:37–42.

White NR, Adox R, Reddy A, Barfield RJ (1992) Regulation of rat maternal behavior by broadband pup vocalizations. Behav Neural Biol 58:131–137.

Whitney J (1986) Effect of medial preoptic lesions on sexual behavior of female rats is determined by test situation. Behav Neurosci 100:230–235.

Widom CS (1989) The cycle of violence. Science 244:160–166.

Wiesenfeld AR, Malatesta CZ, Whitman PB, Granrose C, Uli R (1985) Psychophysiological response of breast- and bottle-feeding mothers to their infants' signals. Psychophysiology 22:79–86.

Wigger A, Neumann ID (1999) Periodic maternal deprivation induces gender-dependent alterations in behavioral and neuroendocrine responses to emotional stress in adult rats. Physiol Behav 66:293–302.

Williams J, Insel T, Harbaugh C, Carter C (1994) Oxytocin administered centrally facilitates formation of a partner preference in female prairie voles (Microtus ochrogaster). J Neuroendocrinol 6:247–250.

Wilson DA, Sullivan RM (1994) Neurobiology of associative learning in the neonate: Early olfactory learning. Behav Neural Biol 61:1–18.

Windle RJ, Brady MM, Kunanandam T, Da Costa APC, Wilson BC, Harbuz M, Lightman SL, Ingram CD (1997a) Reduced response of the hypothalamo-pituitary-adrenal axis to α1-agonist stimulation during lactation. Endocrinology 138:3741–3748.

Windle RJ, Shanks N, Lightman SL, Ingram CD (1997b) Central oxytocin administration reduces stress-induced corticosterone release and anxiety behavior in rats. Endocrinology 138:2829–2834.

Windle RJ, Wood S, Shanks N, Perks P, Conde GL, Da Costa APC, Ingram CD, Lightman SL (1997c) Endocrine and behavioural responses to noise stress: comparison of virgin and lactating female rats during non-disrupted maternal activity. J Neuroendocrinol 9:407–414.

Wingfield JC, Hegner R, Dufty A, Ball GF (1990) The "challenge hypothesis": Theoretical implications for patterns of testosterone secretion, mating systems, and breeding strategies. Am Natural 136:835–846.

Winn P, Brown VJ, Inglis WL (1997) On the relationships between the striatum and the pedunculopontine tegmental nucleus. Crit Rev Neurobiol 11:241–261.

Winslow JT, Hastings N, Carter CS, Harbaugh CR, Insel TR (1993) A role for central vasopressin in pair bonding in monogamous prairie voles. Nature 365:545–548.

Wise DA (1974) Aggression in the female golden hamster: Effects of reproductive state and social isolation. Horm Behav 5:235–250.

Wisner KL, Stowe ZN (1997) Psychobiology of postpartum mood disorders. Sem Reprod Endocrinol 15:77–89.

Witt DM (1995) Oxytocin and rodent sociosexual responses: from behavior to gene expression. Neurosci Biobehav Rev 19:315–324.

Witt DM, Winslow JT, Insel TR (1992) Enhanced social interactions in rats following chronic, centrally infused oxytocin. Pharmacol Biochem Behav 43:855–861.

Wolfe DA (1985) Child-abusive parents: an empirical review and analysis. Psychol Bull 97:462–482.

Wolfe DA, Fairbank JA (1983) Child abusive parents' physiological responses to stressful and non-stressful behavior in children. Behav Assess 5:363–371.

Wolff JO (1985) Maternal aggression as a deterrent to infanticide in *Peromyscus leucopus* and *P. maniculatus*. Anim Behav 33:117–123.

Wong M, Kling MA, Munson PJ, Listwak S, Licinio J, Prolo P, Karp B, McCutcheon IE, Geracioti TD, DeBellis MD, Rice KC, Goldstein DS, Veldhuis J, Chrousos GP, Oldfield EH, McCann SM, Gold PW (2000) Pronounced and sustained central hypernoradrenergic function in major depression with melancholic features: relation to hypercortisolism and corticotropin-releasing hormone. Proc Natl Acad Sci, USA 97:325–330.

Wong M, Licinio J (2001) Research and treatment approaches to depression. Nat Rev Neurosci 2:343–351.

Woolley CS, McEwen BS (1992) Estradiol mediates fluctuation in hippocampal synapse density during the estrous cycle in the adult rat. J Neurosci 12:2549–2554.

Woolley CS, Weiland NG, McEwen BS, Schwartzkroin PA (1997) Estradiol increases the sensitivity of hippocampal CA1 pyramidal cells to NMDA receptor-mediated synaptic input: correlation with dendritic spine density. J Neurosci 17:1848–1859.

Wynne-Edwards KE (2001) Hormonal changes in mammalian fathers. Horm Behav 40:139–145.

Wynne-Edwards KE, Lisk RD (1989) Differential effects of paternal presence on pup survival in two species of dwarf hamster (*Phodopus sungorus* and *Phodopus camp-belli*). Physiol Behav 45:465–469.

Wysocki CJ (1979) Neurobehavioral evidence for the involvement of the vomeronasal system in mammalian reproduction. Neurosci Biobehav Rev 3:301–341.

Xerri C, Stern JM, Merzenich MM (1994) Alterations of the cortical representation of the rat ventrum induced by nursing behavior. J Neurosci 14:1710–1721.

Xia Z, Dudek H, Miranti CK, Greenberg ME (1996) Calcium influx via the NMDA receptor induces immediate early gene transcription by a MAP kinase/ERK-dependent mechanism. J Neurosci 16:5425–5436.

Yang L, Clemens LG (2000) MPOA lesions affect female pacing of copulation in rats. Behav Neurosci 114:1191–1202.

Yang QZ, Hatton GI (1988) Direct evidence for electrical coupling among rat supraoptic nucleus neurons. Brain Res 463:47–56.

Yang QZ, Smithson KG, Hatton GI (1995) NMDA and non-NMDA receptors on the rat supraoptic nucleus neurons activated monosynaptically by olfactory afferents. Brain Res 680:207–216.

Yeo JAG, Keverne EB (1986) The importance of vaginal-cervical stimulation for maternal behavior in the rat. Physiol Behav 37:23–26.

Yezierski RP (1991) Somatosensory input to the periaqueductal gray: a spinal relay to a descending control center. In: The midbrain periaqueductal gray (Depaulis A, Bandler R, eds), pp 365–386. New York: Plenum Press.

Yoshimura H, Ogawa N (1991) Ethopharmacology of maternal aggression in mice: effects of diazepam and SM-3997. Eur J Pharmacol 200:147–153.

Young L, Nilsen R, Waymire K, MacGregor G, Insel T (1999) Increased affiliative response to vasopressin in mice expressing the V1a receptor from a monogamous vole. Nature 400:766–768.

Young LJ, Huot B, Nilsen R, Wang Z, Insel TR (1996) Species differences in central oxytocin receptor gene expression: comparative analysis of promoter sequences. J Neuroendocrinol 8:777–783.

Young LJ, Lim MM, Gingrich B, Insel TR (2001) Cellular mechanisms of attachment. Horm Behav 40:133–138.

Young LJ, Muns S, Wang Z, Insel TR (1997a) Changes in oxytocin receptor mRNA in the rat brain during pregnancy and the effects of estrogen and interleukin-6. J Neuroendocrinol 9:859–865.

Young LJ, Winslow JT, Wang Z, Gingrich B, Guo B, Matzuk MM, Insel TR (1997b) Gene targeting approaches to neuroendocrinology: oxytocin, maternal behavior, and affiliation. Horm Behav 31:221–231.

Young WS, Shepard E, Amico J, Hennighausen L, Wagner K-U, LaMarca ME, McKinney C, Ginns EI (1996) Deficiency in mouse oxytocin prevents milk ejection, but not fertility or parturition. J Neuroendocrinol 8:847–853.

Yu GZ, Kaba H, Okutani F, Takahashi S, Higuchi T (1996a) The olfactory bulb: a critical site of action for oxytocin on the induction of maternal behaviour. Neuroscience 72:1083–1088.

Yu GZ, Kaba H, Okutani F, Takahashi S, Higuchi T, Seto K (1996b) The action of oxytocin originating in the hypothalamic paraventricular nucleus on mitral and granule cells in the rat main olfactory bulb. Neuroscience 72:1073–1082.

Zahm DS (1999) Functional-anatomical implications of the nucleus accumbens core and shell subterritories. Ann NY Acad Sci 877:113–128.

Zaias J, Okimoto L, Trivedi A, Mann PE, Bridges RS (1996) Inhibitory effects of naltrexone on the induction of parental behavior in juvenile rats. Pharmacol Biochem Behav 53:987–993.

Zarrow MX, Denenberg VH, Anderson CO (1965) Rabbit: frequency of suckling in the pup. Science 150:1835–1836.

Zarrow MX, Farooq A, Denenberg VH, Sawin PB, Ross S (1963) Maternal behaviour in the rabbit: endocrine control of maternal nest building. J Reprod Fertil 6:375–383.

Zarrow MX, Gandelman R, Denenberg VH (1971) Prolactin: is it an essential hormone for maternal behavior in the mammal? Horm Behav 2:343–354.

Zeigler HP, Jacquin MF, Miller MG (1985) Trigeminal orosensation and ingestive behavior in the rat. Prog Psychobiol Physiol Psychol 11:63–196.

Ziegler TE, Snowdon CT (1997) Role of prolactin in paternal care in a monogamous New World primate, *Saguinus oedipus*. Ann NY Acad Sci 807:599–601.

Ziegler TE, Snowdon CT (2000) Preparental hormone levels and parenting experience in male cotton-top tamarins, *Saguinus oedipus*. Horm Behav 38:159–167.

Zilles K, Wree A (1995) Cortex: areal and laminar structure. In: The rat nervous system, second edition (Paxinos G, ed), pp 649–685. San Diego: Academic Press.

Zobel AW, Nickel T, Kunzel HE, Ackl N, Sonntag A, Ising M, Holsboer F (2000) Effects of the high-affinity corticotropin-releasing hormone receptor 1 antagonist R121919 in major depression: the first 20 patients treated. J Psychiatr Res 34:171–181.

Index